D1653678

Advanced Micro & Nanosystems
Volume 5

Micro Process Engineering

Related Titles

Other AMN Volumes

Baltes, H., Brand, O., Fedder, G.K., Hierold, C., Korvink, J.G., Tabata, O., Löhe, D., Haußelt, J. (eds.)

Microengineering of Metals and Ceramics
Part I
Design, Tooling, and Injection Molding

2005
ISBN 3-527-31208-0

Baltes, H., Brand, O., Fedder, G.K., Hierold, C., Korvink, J.G., Tabata, O., Löhe, D., Haußelt, J. (eds.)

Microengineering of Metals and Ceramics
Part II
Special Replication Techniques, Automation, and Properties

2005
ISBN 3-527-31493-8

Baltes, H., Brand, O., Fedder, G.K., Hierold, C., Korvink, J.G., Tabata, O. (eds.)

CMOS-MEMS

2005
ISBN 3-527-31080-0

Baltes, H., Brand, O., Fedder, G.K., Hierold, C., Korvink, J.G., Tabata, O. (eds.)

Enabling Technologies for MEMS and Nanodevices
Advanced Micro and Nanosystems

2004
ISBN 3-527-30746-X

Kumar, C.S.S.R., Hormes, J., Leuschner, C. (eds.)

Nanofabrication Towards Biomedical Applications
Techniques, Tools, Applications, and Impact

2005
ISBN 3-527-31115-7

Advincula, R.C., Brittain, W.J., Caster, K.C., Rühe, J. (eds.)

Polymer Brushes
Synthesis, Characterization, Applications

2004
ISBN 3-527-31033-9

Köhler, M., Fritzsche, W.

Nanotechnology
An Introduction to Nanostructuring Techniques

2004
ISBN 3-527-30750-8

Korvink, J.G., Greiner, A.

Semiconductors for Micro- and Nanotechnology
An Introduction for Engineers

2002
ISBN 3-527-30257-3

Advanced Micro & Nanosystems
Volume 5

Micro Process Engineering

Fundamentals, Devices, Fabrication, and Applications

Volume Editor
Norbert Kockmann

WILEY-VCH Verlag GmbH & Co. KGaA

Volume Editor

Dr.-Ing. Norbert Kockmann
Laboratory for Design of Microsystems
University of Freiburg – IMTEK
Georges-Köhler-Allee 102
79110 Freiburg
Germany

Series Editors

Oliver Brand
School of Electrical and Computer Engineering
Georgia Institute of Technology
777 Atlantic Drive
Atlanta, GA 30332-0250
USA

Prof. Dr. Gary K. Fedder
ECE Department & Robotics Institute
Carnegie Mellon University
Pittsburgh, PA 15213-3890
USA

Prof. Dr. Christopher Hierold
Chair of Micro- and Nanosystems
ETH Zürich
ETH Zentrum, CLA H9
Tannenstrasse 3
8092 Zürich
Switzerland

Prof. Dr. Jan G. Korvink
Laboratory for Design of Microsystems
University of Freiburg – IMTEK
Georges-Köhler-Allee 102
79110 Freiburg
Germany

Prof. Dr. Osamu Tabata
Department of Mechanical Engineering
Kyoto University
Yoshida Honmachi Sakyo-ku
606-8501 Kyoto
Japan

Cover picture
Arrangement of microstructured devices, starting bottom left: mixing of aqueous solutions with Bromothymol Blue pH-indicator (Dr. Kockmann, Germany, Chapter 3); generation of monodisperse emulsion droplets in microchannels (Dr. Kobayashi, Chapter 5, courtesy of Nat. Food Res. Inst., Japan); assembly of microstructured platelets with housing for a heat pump (Dr. Stenkamp, Chapter 13, courtesy of PNNL, USA); mounting a chemical reactor with microstructured elements (Dr. Schirrmeister, Chapter 7; courtesy of Uhde GmbH and Degussa AG, Germany)

Library of Congress Card No.: applied for

British Library Cataloguing-in-Publication Data:
A catalogue record for this book is available from the British Library

Bibliographic information published by Die Deutsche Bibliothek
Die Deutsche Bibliothek lists this publication in the Deutsche Nationalbibliografie; detailed bibliographic data is available in the Internet at http://dnb.ddb.de

Typesetting K+V Fotosatz GmbH, Beerfelden
Printing Strauss GmbH, Mörlenbach
Binding Litges & Dopf Buchbinderei GmbH, Heppenheim
Cover Grafik-Design Schulz, Fußgönheim

Printed on acid-free paper
Printed in the Federal Republic of Germany

ISBN-13: 978-3-527-31246-7
ISBN-10: 3-527-31246-3

Preface

Besides the development of new devices the main goal of engineering activities is to achieve a high performance in technical systems with low effort for optimized processes and products. An incredible performance increase was achieved in communication and information technology by the miniaturization of electronic equipment down to the nanometer scale during the last decades. Moore's law of doubling the number of circuits in electronic devices in 18 months by miniaturization still holds since decades and is expected to last.

Process technology is a wide field where small processes down to the molecular scale happen in devices having a length of several meters. The scale-up of chemical production or power plants has led to high energy efficiencies and affordable consumer products. Around 1920, cryogenic air separation units produced an amount of about 1.3 t/h oxygen with 98–99% purity. 30 years later, the largest air separation units delivered about 5.2 t/h oxygen with 99% purity. Nowadays, the largest air separation units are supplying large customers with about 65 t/h oxygen with 99.5% purity and higher. As the throughput increases, the specific energy consumption decreases from about 1.5 kW/kg oxygen to about 0.4 kW/kg oxygen. Besides the development of large units, the consumer specific supply was also addressed by small and adjusted plants for flexible production satisfying the costumer's demand. Additionally, some branches of the chemical industry are not subjected to the economy of scale like the pharmaceutical industry or fine chemicals; flexibility as well as the product price and quality are the important factors.

The combination of process engineering and micro system engineering with the design, fabrication, and integration of functional microstructures is one of the most promising research and development areas of the last two decades. This is reflected in the publishing of scientific journals like "Sensors and Actuators" (since 1981) as well as in the growing field of international conferences like µTAS (Micro Total Analysis Systems, since 1994), the IMRET (International Conference on Micro Reaction Technology by AIChE and DECHEMA, since 1997), or the ICMM (International Conference on Micro and Mini Channels by ASME, since 2003). This can also be seen in the growing industrial activities using microstructured equipment in process development and production of chemicals. Some activities can be summarized under the concept of process intensification, such as compact heat exchangers or structured packing in separation columns for intensified heat and mass transfer. With characteristic lengths of the devices in the size of boundary layers, the transfer processes can be en-

Advanced Micro and Nanosystems Vol. 5. Micro Process Engineering. Edited by N. Kockmann

ISBN: 3-527-31246-3

hanced and controlled in the desired way. Other activities include modular platforms and entire chemical plants consisting of several microstructure elements and devices, mainly for laboratory and process development.

This book on micro process engineering is divided into four sections: fundamentals (Chapter 1 to 6), the design and system integration (Chapter 7 to 9), fabrication technologies and materials (Chapter 10 to 12), and, finally, the applications of microstructured devices and systems (Chapter 13 to 16). Each chapter has review character and stands on its own, but is also integrated into the whole book. A common nomenclature and index will help the orientation of the reader. In Chapter 1 to 6 the fundamentals and tools of process engineering are presented with single-phase and multiphase fluid flow, heat and mass transfer as well as the treatment of chemical reactions following the concept of unit operations. The equipment and process design is organized by project management methods and assisted by modeling and simulation as well as the integration of sensors and analytical equipment, described in Chapter 7, 8, and 9. The broad fabrication variety of microstructured devices for micro process engineering is illustrated in Chapter 10, 11, and 12 grouped according the materials metal, polymers, silicon, glass, and ceramics. Some typical examples of microstructured devices illustrate the various fabrication methods. Even more examples are given in Chapter 13 to 15 with industrial applications in Europe, Japan and the US. Last but not least Chapter 16 emphasizes the application of microstructured devices in education and laboratory research work. This gives students a deeper insight into the complex behavior of chemical plants and will lead to a more sophisticated view of continuous flow processing in education, laboratory experiments, and chemical synthesis.

The aim of this book is the comprehensive description of actual knowledge and competence for microfluidic and chemical process fundamentals, design rules, related fabrication technology, as well as an overview of actual and future applications. This work is located at the boundary of at least two different disciplines, trying to collect and unify some of the special knowledge from different areas, driven by the hope that innovation happens at the interfaces between the disciplines. From this, the team of authors of various engineers, physicists and chemists, from universities, research institutes, and industry in different countries contributes an embracing part of detailed know-how about processes in and applications of microstructures. I hope that this knowledge will help to look out of the box to other related areas of chemical engineering, micro system engineering and to other engineering, physical, chemical, or biological areas.

Finally, I want to thank all the contributors for their enduring work, besides their actual work and activities. I hope that this enthusiasm can be read throughout the book, will spread further on to the readers and will help to enlarge the knowledge and activities on this new and gap-filling area of micro process engineering.

Norbert Kockmann
Volume Editor
November 2005

Foreword

We hereby present the fifth volume of *Advanced Micro & Nanosystems* (*AMN*), entitled *Micro Process Engineering*.

Usually, when engineering devices get smaller, we expect higher speeds, more accuracy, or less power consumption, but typically we do not associate smaller devices to successfully compete with larger ones when it comes to material throughput. Not so in micro process engineering. This research area has quietly grown in the flanks, and promises to become one of the most profitable areas in microtechnology. Why is this so? It turns out that micro process engineering targets the more efficient manufacture of chemical substances, no less than miniaturized chemical factories that match the throughput of their macroscopic counterparts.

The volume editor, Dr. Norbert Kockmann, has assembled a notable international authors hip to bring to us the state of the art in this very exciting application area. At the microscale, many physical and chemical effects have to be reevaluated as they apply to chemical engineering manufacturing processes, and in this volume six chapters guide us through the most important fundamental concepts. The revised theory implies the need for new design methods, and so three chapters consider simulation, modelling, and system design. Device fabrication sets specific challenges, for all resulting production surfaces must be chemically and thermally resistant, and must target high throughput of liquids and gases. Finally, because micro process engineering is driven by its exciting applications, four chapters cover the most important topics from a completely international perspective.

We are happy to report here that the decision to produce topical volumes such as *CMOS-MEMS* or *Microengineering of Metals and Ceramics* is finding tremendous acceptance with our readers and hence we will continue to plan further relevant topics from either an application area or a specific manufacturing technology.

Looking ahead, we hope to welcome you back, dear reader, to the upcoming sixth member of the *AMN* series, in which we take a close look at the fascinating field of LIGA and its application.

Oliver Brand, Gary K. Fedder, Christofer Hierold, Jan G. Korvink, and Osamu Tabata
Series Editors
October 2005
Atlanta, Pittsburgh, Zurich, Freiburg and Kyoto

Advanced Micro and Nanosystems Vol. 5. Micro Process Engineering. Edited by N. Kockmann

ISBN: 3-527-31246-3

Contents

Advanced Micro and Nanosystems Vol. 5. Micro Process Engineering. Edited by N. Kockmann

ISBN: 3-527-31246-3

List of Contributors

Thomas Bayer
Siemens AG
A&D SP Solutions Process Industries
Business Development
Industriepark Höchst, Bldg. G 811
D-65926 Frankfurt am Main
Germany

Theodorian Borca-Tasciuc
Department of Mechanical, Aerospace and Nuclear Engineering
Rensselaer Polytechnic Institute
110 8th Street
Troy, NY 12180-3590
USA

Jürgen J. Brandner
Forschungszentrum Karlsruhe
Institute for Micro Process Engineering (IMVT)
Hermann-von-Helmholtz-Platz 1
D-76344 Eggenstein-Leopoldshafen
Germany

Robert A. Dagle
Pacific Northwest National Laboratory
902 Battelle Blvd., K8-93
Richland, WA 99354
USA

Thomas R. Dietrich
Mikroglas chemtech GmbH
Galileo-Galilei-Str. 28
D-55129 Mainz
Germany

Alexander Doll
Laboratory for Design of Microsystems
Department of Microsystems Engineering
University of Freiburg – IMTEK
Georges-Köhler-Allee 102
D-79110 Freiburg
Germany

Michael Engler
Laboratory for Design of Microsystems
Department of Microsystems Engineering
University of Freiburg – IMTEK
Georges-Köhler-Allee 102
D-79110 Freiburg
Germany

Thomas Gietzelt
Forschungszentrum Karlsruhe
Institute for Micro Process Engineering (IMVT)
Hermann-von-Helmholtz-Platz 1
D-76344 Eggenstein-Leopoldshafen
Germany

Advanced Micro and Nanosystems Vol. 5. Micro Process Engineering. Edited by N. Kockmann

ISBN: 3-527-31246-3

Frank Goldschmidtböing
Laboratory for Design
of Microsystems
Department of Microsystems
Engineering
University of Freiburg – IMTEK
Georges-Köhler-Allee 102
D-79110 Freiburg
Germany

Mike Günther
Institute of Physics
Technical University Ilmenau
Weimarer Straße 32
D-98684 Ilmenau
Germany

Torsten Henning
Forschungszentrum Karlsruhe
Institute for Micro Process
Engineering (IMVT)
Hermann-von-Helmholtz-Platz 1
D-76344 Eggenstein-Leopoldshafen
Germany

Heinz Herwig
Technical Thermodynamics
Technical University
Hamburg-Harburg
Denickestr. 17
D-21073 Hamburg
Germany

Michael K. Jensen
Department of Mechanical, Aerospace
and Nuclear Engineering
Rensselaer Polytechnic Institute
110 8th Street
Troy, NY 12180-3590
USA

Goran N. Jovanovic
Department of Chemical Engineering
Oregon State University
102 Gleeson Hall
Corvallis, OR 97331
Microproducts Breakthrough Institute
Corvallis, OR 97330
USA

Satish G. Kandlikar
Department of Mechanical
Engineering
Rochester Institute of Technology
Rochester, NY 14623-5604
USA

Markus Kinzl
Siemens AG
A&D SP Solutions Process Industries
Business Development
Industriepark Höchst, Bldg. G 811
D-65926 Frankfurt am Main
Germany

Lioubov Kiwi-Minsker
Institute of Chemical Sciences
and Engineering
Swiss Federal Institute of Technology
(EPFL)
CH – LGRC
CH-1015 Lausanne
Switzerland

Walther Klemm
Institute of Technical Chemistry
and Environmental Chemistry
Friedrich Schiller University of Jena
Lessingstr. 12
D-07743 Jena
Germany

Regina Knitter
Forschungszentrum Karlsruhe
Institute for Material
Science III (IMF III)
Hermann-von-Helmholtz-Platz 1
D-76344 Eggenstein-Leopoldshafen
Germany

Isao Kobayashi
Food Engineering Devision
National Food Research Institute
Kannondai 2-1-12, Tsukuba
Ibaraki 305-8642
Japan

Norbert Kockmann
Laboratory for Design
of Microsystems
Department of Microsystems
Engineering
University of Freiburg – IMTEK
Georges-Köhler-Allee 102
D-79110 Freiburg
Germany

Michael Köhler
Institute of Physics
Technical University Ilmenau
Weimarer Straße 32
D-98684 Ilmenau
Germany

Stefan Löbbecke
Fraunhofer Institute for Chemical
Technology (ICT)
Joseph-von-Fraunhofer-Straße 7
D-76327 Pfinztal
Germany

Manfred Kraut
Forschungszentrum Karlsruhe
Institute for Micro Process
Engineering (IMVT)
Hermann-von-Helmholtz-Platz 1
D-76344 Eggenstein-Leopoldshafen
Germany

Holger Moritz
Forschungszentrum Karlsruhe
Institute for Microstructure
Technology (IMT)
Hermann-von-Helmholtz-Platz 1
D-76344 Eggenstein-Leopoldshafen
Germany

Mitsutoshi Nakajima
Food Engineering Devision
National Food Research Institute
Kannondai 2-1-12, Tsukuba
Ibaraki 305-8642
Japan

Hideho Okamoto
Department of Synthetic Chemistry
and Biological Engineering
Graduate School of Engineering
Kyoto University
Kyoto 615-8510
Japan

Bernd Ondruschka
Institute of Technical Chemistry
and Environmental Chemistry
Friedrich Schiller University of Jena
Lessingstr. 12
D-07743 Jena
Germany

Daniel R. Palo
Pacific Northwest National Laboratory
902 Battelle Blvd., K8-93
Richland, WA 99354
Microproducts Breakthrough Institute
Corvallis, OR 97330
USA

Yoav Peles
Department of Mechanical, Aerospace and Nuclear Engineering
Rensselaer Polytechnic Institute
110 8th Street
Troy, NY 12180-3590
USA

Wilhelm Pfleging
Forschungszentrum Karlsruhe
Institute for Material Science I (IMF I)
Hermann-von-Helmholtz-Platz 1
D-76344 Eggenstein-Leopoldshafen
Germany

Albert Renken
Institute of Chemical Sciences and Engineering
Swiss Federal Institute of Technology (EPFL)
CH – LGRC
CH-1015 Lausanne
Switzerland

Steffen Schirrmeister
Uhde GmbH
Friedrich-Uhde-Str. 15
D-44141 Dortmund
Germany

Victoria S. Stenkamp
Pacific Northwest National Laboratory
902 Battelle Blvd., K6-28
Richland, WA 99354
USA

Osamu Tonomura
Department of Chemical Engineering
Kyoto University
Katsura Campus, Nishikyo-ku
Kyoto 615-8510
Japan

Jun-ichi Yoshida
Department of Synthetic Chemistry and Biological Engineering
Graduate School of Engineering
Kyoto University
Kyoto 615-8510
Japan

Nomenclature

List of main parameters

If not indicated in the text.

Name	Unit	Description
A	m^2	area or cross section
a	$\frac{m^2}{s}$	temperature conductivity $a = \frac{\lambda}{\rho c_p}$
b	m	geometrical factor, channel width
C	–	constant
C_i	–	ratio of heat capacity fluxes
c	$\frac{m}{s}$	absolute velocity
c_i	–	concentration of component i
c_p	$\frac{kJ}{kg\,K}$	isobaric specific heat capacity
c_v	$\frac{kJ}{kg\,K}$	isochoric specific heat capacity
D	$\frac{m^2}{s}$	diffusion coefficient
D	m	diameter
d_h	m	hydraulic diameter $= \frac{4A}{U}$
E	J	energy
e	$\frac{J}{kg}$	specific energy
F	N	force
f	–	probability distribution function
G	$\frac{kg}{m^2 s}$	mass velocity (US literature)
g	m/s^2	gravity constant

Advanced Micro and Nanosystems Vol. 5. Micro Process Engineering. Edited by N. Kockmann

ISBN: 3-527-31246-3

g	m	temperature jump coefficient
H	J	enthalpy
H_i	Pa	Henry coefficient
h	$\frac{\text{J}}{\text{kg}}$	specific enthalpy
h_{LV}	$\frac{\text{J}}{\text{kg}}$	latent heat of vaporization
h	$\frac{\text{W}}{\text{m}^2\text{K}}$	heat transfer coefficient (US literature)
h	m	geometrical factor, height
I	A	electrical current
J	J/s	general energy current
K	–	general coefficient
k	$\frac{\text{W}}{\text{m}^2\text{K}}$	overall heat transfer coefficient
k	$\frac{\text{W}}{\text{mK}}$	thermal conductivity (US literature)
k	*)	reaction rate constant, *) unit depends on reaction order
k_M	$\frac{\text{kg}}{\text{sPa}}$	membrane conductivity
k	$\frac{\text{J}}{\text{K}}$	Boltzmann constant $\left(1.380662 \times 10^{-23} \frac{\text{J}}{\text{K}}\right)$
L	m	length
L_{ij}	–	general transport coefficient
L_p	mol/s	rate of production
l	m	length, characteristic
M	$\frac{\text{kg}}{\text{kmol}}$	molar mass
M^*	kg	mass of an atom or molecule
m	kg	mass
m	–	reaction order
$\dot{m}$	$\frac{\text{kg}}{\text{s}}$	mass flow
N	–	number
N_i	–	ratio of transferred heat to heat capacity, number of transfer units
n	1/s	rotation speed
n	mol	amount of substance
$\dot{n}$	mol/s	molar flow rate
P	W	power
p	bar, $\text{Pa} = \frac{\text{N}}{\text{m}^2}$	pressure

p_i	bar, $\mathrm{Pa} = \frac{\mathrm{N}}{\mathrm{m}^2}$	partial pressure
Q	$\mathrm{J} = \mathrm{Nm} = \frac{\mathrm{kg\,m}^2}{\mathrm{s}^2}$	heat
$\dot{Q}$	$\mathrm{W} = \frac{\mathrm{J}}{\mathrm{s}}$	heat flux
q	$\frac{\mathrm{J}}{\mathrm{kg}}$	specific heat
$\dot{q}$	$\frac{\mathrm{W}}{\mathrm{m}^2}$	specific heat flux
R	m	radius
R	$\frac{\mathrm{J}}{\mathrm{kg\,K}}$	individual gas constant $= \frac{R_m}{M}$
R_i	$\frac{\mathrm{mol}}{\mathrm{m}^3\mathrm{s}}$	transformation rate
R_m	$\frac{\mathrm{J}}{\mathrm{kmol\,K}}$	universal gas constant $\left(8.314\,\frac{\mathrm{kJ}}{\mathrm{kmol\,K}}\right)$
R_{el}	$\frac{\mathrm{V}}{\mathrm{A}}$	electrical resistance
r	m	location coordinate, radius
r	1/s	reaction rate
S	$\frac{\mathrm{J}}{\mathrm{K}}$	entropy
S'	–	heat production potential
s	$\frac{\mathrm{J}}{\mathrm{kg\,K}}$	specific entropy
T	K	temperature (Kelvin)
t	s	time
U	J	inner energy
U	m	circumference, perimeter
U	V	electric voltage
u	$\frac{\mathrm{J}}{\mathrm{kg}}$	specific inner energy
u	$\frac{\mathrm{m}}{\mathrm{s}}$	velocity in x-direction
V	m^3	volume
$\dot{V}$	m^3/s	volume flow rate
v	$\frac{\mathrm{m}}{\mathrm{s}}$	velocity in y-direction
$\vec{v}$	$\frac{\mathrm{m}}{\mathrm{s}}$	vector of velocity
W	J	work
W_{diss}	J	dissipation work

w	$\frac{m}{s}$	velocity in z-direction
w	$\frac{J}{kg}$	specific work
X	–	general transport variable
X	–	conversion
$\vec{x}$	m	position vector
x	m	cartesian coordinate
x	–	vapor quality or void fraction
x_i	–	liquid concentration of component i
y	m	cartesian coordinate
y_i	–	vapor concentration of component i
z	m	cartesian coordinate, main flow direction in a channel

Greek letters

α	$\frac{W}{m^2K}$	heat transfer coefficient
α	–	mixing quality
β	$\frac{m}{s}$	mass transfer coefficient
γ	–	Arrhenius number
γ	$\frac{kg}{m^2s}$	momentum transfer coefficient
γ	$\frac{N}{m}$	interfacial tension
γ_i	–	relative amount of substance, mole fraction
δ	m	boundary layer thickness
ε	$\frac{m^2}{s^3}$	specific energy dissipation
ε	–	porosity
ζ	–	friction factor of channel fitting or installation
ζ	m	slip length of rarefied gas flow
η	$\frac{kg}{m\,s} = \frac{Ns}{m^2}$	dynamic viscosity
Θ	–	dimensionless temperature
θ	deg	contact angle
θ	–	heat exchanger efficiency
κ	–	isentropic exponent
Λ	m	mean free path length
λ	$\frac{W}{mK}$	heat conductivity

λ_R	–	channel friction factor
μ	Pa s	dynamic viscosity (US literature)
ν	$\frac{m^2}{s}$	kinematic viscosity $= \frac{\eta}{\rho}$
ν_i	–	stoichiometric coefficient
π_i	Pa	mole fraction acc. Raoult's law
ρ	$\frac{kg}{m^3}$	density $= \frac{m}{V}$
σ	m	diameter of an atom or molecule
σ	$\frac{N}{m}$	surface tension
$\dot{\sigma}$	$\frac{J}{Ks}$	dissipation function or local entropy production
τ	s	space time
τ	$\frac{N}{m^2}$	shear stress
φ	–	probability
χ	–	dispersion
ω	1/s	angular velocity

Sub- and superscripts

$\vec{x}$	vector description of x
$\bar{x}$	mean value of x
x^*	dimensionless form of x

Dimensionless numbers of fluid mechanics, heat and mass transfer

Name	Description
$\mathrm{Bi} = \frac{al}{\lambda}$	Biot number ($\lambda = \lambda_{\mathrm{solid}}$)
$\mathrm{Bo} = \frac{wl}{D_{\mathrm{ax}}}$	Bodenstein number
$\mathrm{Bo} = \frac{q''}{Gh_{fg}}$	boiling number (US literature)
$\mathrm{Bo} = \frac{\mathrm{We}}{\mathrm{Fr}} = \frac{d^2 g\rho}{\sigma}$	Bond number (US literature)
$\mathrm{Ca} = \frac{\eta w}{\sigma}$	capillary number

$\mathrm{Co} = \left(\frac{1-x}{x}\right)^{0.8}\left(\frac{\rho_V}{\rho_L}\right)^{0.5}$	convection number
$\mathrm{DaI} = \frac{t_r}{\tau_{res}}$	1st Damköhler number
$\mathrm{DaII} = \frac{k_s}{k_D} = \frac{t_r}{t_D}$	2nd Damköhler number
$\mathrm{Dn} = \mathrm{Re}\left(\frac{D}{R_c}\right)^{1/2}$	Dean number
$\mathrm{Ec} = \frac{w^2}{c_p \Delta T}$	Eckert number
$\mathrm{Eu} = \frac{\Delta p}{\rho w^2}$	Euler number
$\mathrm{Fo} = \frac{at}{l^2}$	Fourier number
$\mathrm{Fr} = \frac{w^2}{dg}$	Froude number
$\mathrm{Gr} = \frac{g_z \beta_p s^3 \Delta T}{\nu^2}$	Grashoff number
$\mathrm{Kn} = \frac{\Lambda}{L}$	Knudsen number
$\mathrm{Le} = \frac{a}{D}$	Lewis number
$\mathrm{Ma} = \frac{w}{c}$	Mach number
$\mathrm{Ne} = \frac{P}{\rho n^3 d^5}$	Newton number
$\mathrm{Nu} = \frac{al}{\lambda}$	Nußelt number ($\lambda = \lambda_{Fluid}$)
$\mathrm{Pe} = \mathrm{RePr} = \frac{wl}{a}$	Péclet number
$\mathrm{Pr} = \frac{\nu}{a}$	Prandtl number
$\mathrm{Ra} = \mathrm{GrPr}$	Rayleigh number
$\mathrm{Re} = \frac{wl}{\nu}$	Reynolds number
$\mathrm{Sc} = \mathrm{LePr} = \frac{\nu}{D}$	Schmidt number

$\mathrm{Sh} = \frac{\beta l}{D}$ Sherwood number

$\mathrm{We} = \frac{w^2 d\rho}{\sigma}$ Weber number

$= \frac{n^2 d^3 \rho}{\sigma}$

Mathematical operators

D	substantial differential
d	general differential
∂	partial differential
Δ	delta, divergence, difference
$\nabla = \vec{i}\,\frac{\partial}{\partial x} + \vec{j}\,\frac{\partial}{\partial y} + \vec{k}\,\frac{\partial}{\partial z}$	Nabla operator
grad $\varphi = \nabla\varphi$	gradient of φ
div $\vec{v} = \nabla\,\vec{v}$	divergence of $\vec{v}$
rot $\vec{v} = \nabla \times \vec{v}$	rotation of $\vec{v}$
$\Delta\varphi = \frac{\partial^2\varphi}{\partial x^2} + \frac{\partial^2\varphi}{\partial y^2} + \frac{\partial^2\varphi}{\partial z^2}$	Laplace operator

Main conferences and their abbreviation

IMRET 1: W. Ehrfeld (Ed.), *Microreaction Technology, Proc. of the 1st Int. Conf. on Microreaction Technology 1997*, Springer, Berlin, **1998**.

IMRET 2: *Proc. of the 2nd Int. Conf. on Microreaction Technology*, AIChE National Spring Meeting, New Orleans, **1998**.

IMRET 3: W. Ehrfeld (Ed.), *Microreaction Technology: Industrial Prospects, Proc. of the 3rd Int. Conf. on Microreaction Technology 1999, Frankfurt*, Springer, Berlin, **2000**.

IMRET 4: *Microreaction Technology: Proc. of the 4th Int. Conf. on Microreaction Technology*, AIChE National Spring Meeting, Atlanta, **2000**.

IMRET 5: M. Matlosz, W. Ehrfeld, P. Baselt (Eds.), *Microreaction Technology, Proc. of the 5th Int. Conf. on Microreaction Technology 2001, Straßbourg*, Springer, Berlin, **2002**.

IMRET 6: P. Baselt, U. Eul, R.S. Wegeng, I. Rinard, B. Horch (Eds.), *Proc. of the 6th Int. Conf. on Microreaction Technology*, AIChE National Spring Meeting, New Orleans, **2002**.

IMRET 7: A. Renken, M. Matlosz (Eds.), *Proc. of the 7th Int. Conf. on Microreaction Technology 2003*, Lausanne, **2003**.

IMRET 8: *Proc. of the 8th Int. Conf. on Microreaction Technology*, AIChE National Spring Meeting, Atlanta, **2005**.

ICMM2003: S.G. Kandlikar, *ASME 1st Int. Conf. on Microchannels and Minichannels*, Rochester NY, **2003**.

ICMM2004: S.G. Kandlikar, *ASME 2nd Int. Conf. on Microchannels and Minichannels*, Rochester NY, **2004**.

ICMM2005: S.G. Kandlikar, M. Kawaji, *ASME 3rd Int. Conf. on Microchannels and Minichannels*, Toronto, Canada, **2005**.

1
Process Engineering Methods and Microsystem Technology

Norbert Kockmann, Laboratory for Design of Microsystems, Department of Microsystem Engineering (IMTEK), University of Freiburg, Germany

Abstract
The fundamentals of chemical engineering are presented with the aim of applications in microsystem technology, microfluidics, and transport processes in microstructures. After a general overview about both disciplines and common areas the concept of unit operations is briefly introduced. The balance equations are derived from statistical mechanics and applied to other relevant systems of process engineering together with the kinetic description of main transfer processes. Engineering tools like dimensional analysis, order of magnitude estimations, or lumped element modeling are explained, which are very helpful for dealing with complex nonlinear systems. Concluding this chapter, the benefits and limits of miniaturization of various unit operations and typical issues are explained that might serve as a plentiful source for the future development.

Keywords
Unit operations, balance equations, transport equations, engineering modeling, scaling process

Advanced Micro and Nanosystems Vol. 5. Micro Process Engineering. Edited by N. Kockmann

ISBN: 3-527-31246-3

1.1 Introduction

Process technology and microsystem technology are both interdisciplinary engineering and natural science branches connecting physics, chemistry, biology, engineering arts, and management techniques to an enabling toolbox for various applications. Process engineering embraces orientating calculations for process and equipment design under general orientation, and system-orientated, cross-linked thinking. Process engineers are working in various areas ranging from the food industry through biotechnology to pharmaceutical products, from analytical and laboratory equipment through energy conversion to industrial chemistry for the production of millions of tons of chemicals [1, Chapter 1]. Chemical process engineering covers not only the design and implementation of chemical production and analytical processes but also deals with the equipment design, the appropriate materials, the fabrication, and operation of various chemical production processes. The aims of process technology are the economical and safe production of the desired products with the intended form and composition.

Microsystem technology, coming from information technology and miniaturization of data-processing devices, has now entered many fields in our daily life. Silicon chips and sensors can be found in cars, washing machines or smart cards with various functions. Besides the data-processing function, microsystems have taken over other tasks like sensing and analyzing, actuating or controlling larger systems. Microsystem engineering comprises besides engineering skills like design, simulation, or material knowledge also a deep physical and chemical knowledge for the fabrication and functional design issues. Also medical and biological skills are useful for the growing application fields for analysis, diagnostics, and therapeutics. A good overview about the state-of-the-art in microsystem technology is given in [2]. For the control and manipulation of still smaller systems, microsystem technology is a major link to nanotechnology [3, 4].

Figure 1.1 gives an impression of the wide field and complexity of both disciplines, but also illustrates the multiple interfaces and common fields. The fruitful ideas from both sides may inspire the further development in both disciplines and result in an enlargement of possibilities and applications for the innovation across the borderlines.

Chemistry in *miniaturized equipment* is an emerging discipline coming together from microsystem technology and from chemical engineering, but also an established discipline of chemical analytics. Starting at the end of the nineteenth century a group of researchers at the University of Delft around Behrens [4a] and at the Technical University of Graz around Prof. Emich and Prof. Pregl developed the chemical analysis of very small amounts of reagents. In 1900 Prof. Behrens wrote his book "Mikrochemische Technik" [4a] about micro chemical techniques. In 1911 Prof. Friedrich Emich published the textbook "Lehrbuch der Mikrochemie" [5] and Prof. Fritz Pregl was rewarded in 1923 by the Nobel price for his fundamental work in microchemical analysis. In the middle of the last century in nuclear science small structures were developed for the separation of isotopes, see [6]. From this work, among others, the LIGA technology emerges at German research institutes.

Dealing with very small geometrical structures is also a well-known area in process engineering. The adsorption technology and chemical reactions at catalytic surfaces are based on the flow and adhesion processes in nanoscale pores [7, Chapter 4]. Transformations and transfer processes on the molecular scale are called "micro processes" in contrast to a "macro process" where convection

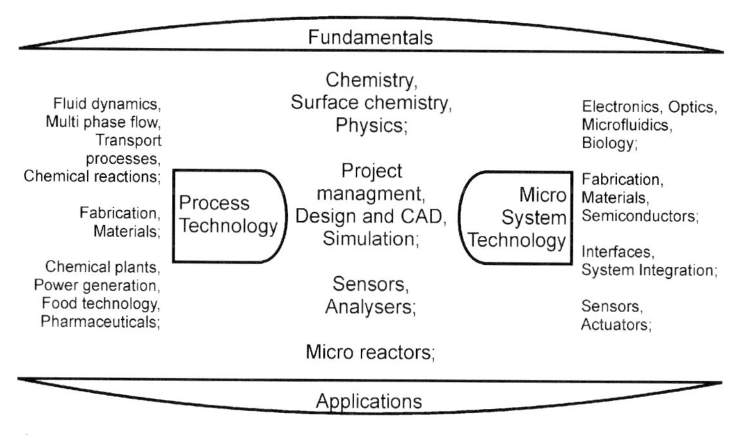

Fig. 1.1 Disciplines of process engineering and microsystem technology, differing and common overlapping areas (middle column). The lists are not complete and the future will certainly bring new applications and new common fields and applications.

plays the major role. Some typical length scales for process technology, chemistry and microtechnology are given in Fig. 1.2.

Figure 1.2 illustrates the different wording in process engineering, microsystem technology, and nanotechnology, especially the different meaning of "micro". The micropores in adsorption media are one characteristic example on the nanometer scale. Microstructured equipment has internal characteristic dimensions like channel diameter or gap height within the micrometer range. A clear definition of "micro" does not exist, but it is not necessarily required for all applications and areas.

In the process industry, there are several applications of structures with typical dimensions below 1 mm, like compact plate and fin heat exchangers or structured packings in separation columns for enhanced heat and mass transfer. This is often summarized under the key word of process intensification. However, the miniaturization of conventional technology is limited by two major restrictions: the fabrication possibilities for the small structures at reasonable costs and the increased fouling probability, the high danger of blocking, and total failure of these structures. The first restriction has been widened with the enhanced fabrication possibilities, but the risk of fouling and blocking is still there and should not be underestimated.

The elementary setup of microstructured and conventional equipment is similar and displayed in Fig. 1.3. Process plants consist of process units, which themselves are made of equipment like heat exchangers or vessels with internal structures. The basic geometrical elements of the internal structures in conven-

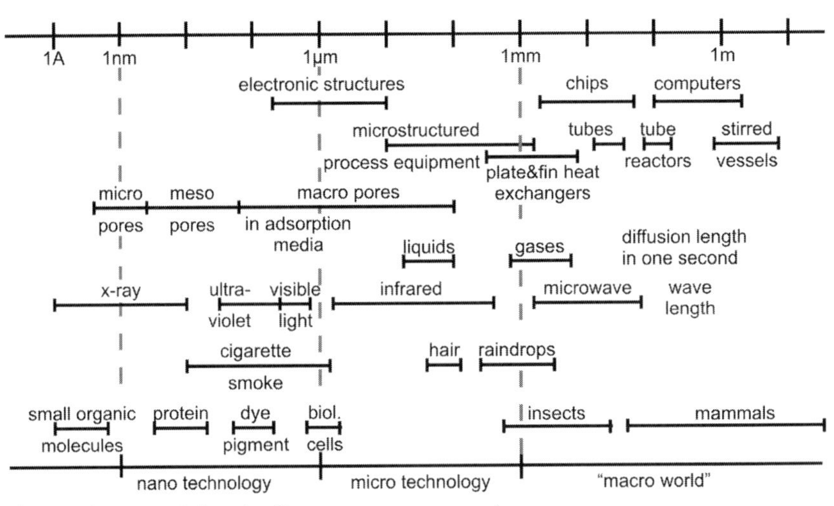

Fig. 1.2 Characteristic length of important processes and equipment in chemical engineering and microsystem technology. The top and bottom line indicate also the different wording of micro processes in the two disciplines, adopted from [8].

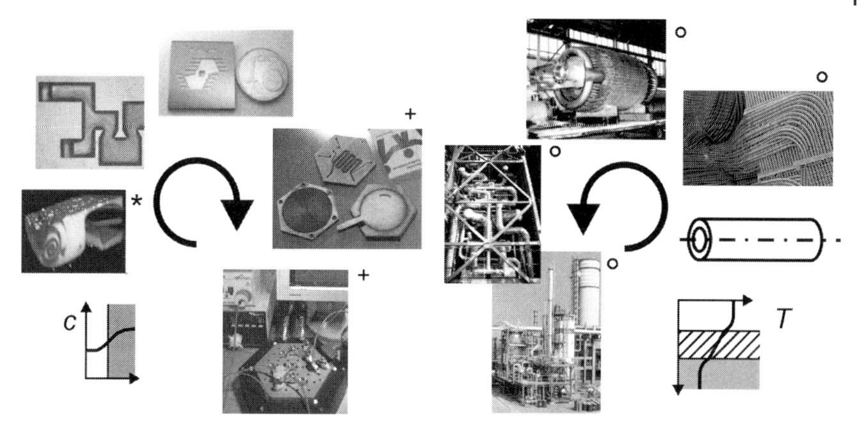

Fig. 1.3 Setup of microstructured and conventional equipment from single microstructures through combined elements to an entire plant. The principle of the active area can be applied in both cases (Sources: * from [9]; + courtesy of FhG-ICT, Pfinztal, Germany; o courtesy of Linde AG, Wiesbaden, Germany).

tional technology are the tube, the plate, and the film, on or in which the transport processes and transformations happen. The layout of process equipment and process steps follow this scheme from small elementary active areas ("micro process") over the process space of the device ("macro process") to the balancing of the complete process.

The parallel arrangement of microstructured channels or elements is called internal numbering-up, which is the most frequent way to increase the throughput of an apparatus. The parallelization of microstructured devices is called external numbering-up, applied to bypass the flow distribution problems within the equipment. A relatively new concept is the equal-up concept, the parallelization of similar effects [8]. The numbering-up and equal-up concepts facilitate the scale-up process from laboratory equipment to production equipment, but still have their own problems of flow distribution in manifolds, see Chapter 8.

1.2
Unit Operations and Beyond

The consecutive groups or steps in a process plant can frankly be named for many cases as

- pretreatment or conditioning of the incoming substances,
- transformation of the reagents in chemical, physical, or biological processes,
- separation of the received components, and
- purification and conditioning of the products, see Fig. 1.4.

The physical and chemical processes in the various steps may be the same or similar, like heat transfer or extraction. They are called *unit operations* that are playing a major role in the research and development of process engineering. The unit operations can be combined and connected in different forms. The concept of unit operation combines a macro process with the apparatus to a process unit. It allows us to treat all micro processes within the process space in the same manner and to derive scientifically based design rules and calculation instructions. For an entire process plant the unit operations are combined and switched in a proper way and integrated for efficient material and energy use. Besides the energy and mass flow integration the appropriate process control and automation determines the economical performance and safety of the plant. This gives a very complex picture of a chemical or process technology plant, which is illustrated in Fig. 1.4. For a proper design and operation of a plant, many disciplines have to work closely together.

The unit operations can be categorized into three major groups according the employed physical effects and major driving forces for combination or separation of substances: the mechanical, the electromagnetic, and the thermal unit operations (molecular driving forces) see Table 1.1. This list does not claim to be complete, especially the separation processes from analytics are only shown schematically. Probably in the next years further operations will be developed enabled by enhanced fabrication and integration possibilities. In adsorption of species or membrane separation, chemical processes may also be involved for mass-transfer processes in microstructures, see [11, Chapter 3]. The consequent treatment of unit operations allows the methodological design with help of the following principles. The *principle of continuity* of substances, phases, energy and momentum includes the preference of continuous processes opposite to batch processes. The *principle of balancing* of the relevant transport processes gives the energy, momentum, and mass fluxes in differential or black-box form. The *principle of scaling* and *similarity* of processes gives a calculation tool for transferring experimental, analytical, and numerical results to processes on different scales with the help of dimensionless numbers and groups.

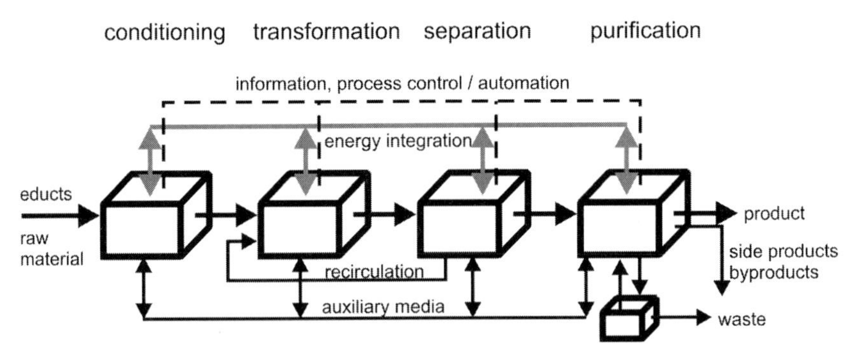

Fig. 1.4 Main process steps in a chemical production plant with pretreatment, conversion, separation, and purification of the products, adapted from [10]. The system integration includes the energy management, auxiliary media as well as information for the process control and automation.

Table 1.1 Main mixing and separation unit operations.

Unit operations	Molecular/thermal	Mechanical/ext. force	Electromagnetic
Mixing and aggregation Combination Control of segregation	diffusion[1)] dissolving[2)] extracting[2c)] desorption[2c)]	spraying[2)] aeration[2)] stirring[2)] mixing[1, 2)] dosing[1, 2)]	electrophoretic mixing[1)] mixing with magnetic beads[2)]
Separation *Employed phases:* 1) single phase 2) multiple phase a) with own cophase b) own + additional cophase c) additional cophase	thermodiffusion[1)] countercurrent diffusion[1)] condensation[2a)] evaporation[2a)] crystallization[2a)] distillation/ rectification[2a)] drying[2b)] absorption[2c)] adsorption[2c)] extraction[2c)] ion exchange[2c)] membrane processes	sedimentation[2)] cycloning[2)] centrifucation[2)] pressure diffusion[1, 2)] (ultracentrifuge) filtration osmosis gas permeation classification sorting	eledcodeposition[2)] magnetodeposition[2)] electrofiltration electrodialysis electroosmosis electrophoresis magnetostriction

More detailed operation description and further reading in [10, 13–15]; additional thermal unit operations, which are closely related to the listed:
Condensation: partial condensation[2a)]
Evaporation: flash evaporation[2a)], vacuum evaporation[2a)]
Drying: freeze-drying[2a)], radiation drying[2a)], superheated steam drying[2a)]
Distillation: outside/secondary steam distillation[2b)], molecular distillation[2a)], reactive distillation
Rectification: extractive rectification[2b)], azeotropic rectification[2b)]
Absorption: chromatography, desorption
Membrane processes: permeation, pervaporation, dialysis, osmosis and reverse osmosis, micro- and ultrafiltration.

The *principle of an active area* indicates the platform of the driving forces in molecular and thermal processes. It provides a description for the transfer processes with linear correlations between the flux and the driving force, also called the kinetic approach [12, Chapter 1]. The processes act in basic geometrical elements like the vessel, the tube, a channel, pipe, pores, or plates, which are combined to form the process space in the chemical equipment. Within these elements the fluid itself forms geometrical elements like beads, drops, bubbles, films or thin layers, which determine the transfer processes and which are confined by the geometry, see Fig. 1.5. The three phases of a pure substance allow the following combinations for phase mixtures of a carrier fluid and a dissolved

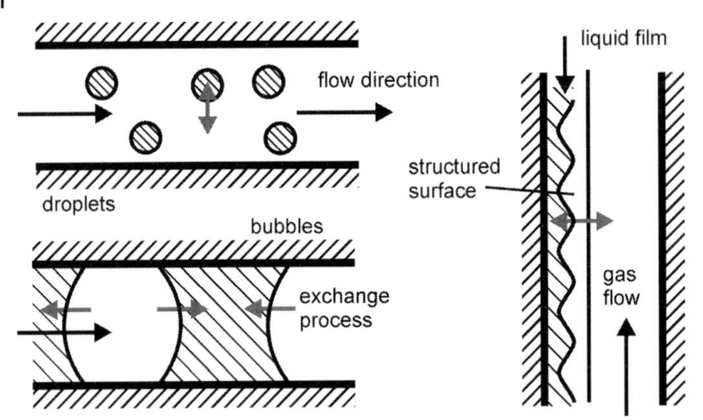

Fig. 1.5 Geometrical elements for transport processes between two phases: droplets, bubbles and a falling film on a structured surface.

phase: gas – gas, gas – liquid (droplets and aerosols), gas – solid (aerosols) and liquid – gas (bubbles and foam), liquid – liquid (miscible or immiscible, emulsion), and liquid – solid (suspension).

The *principle of technical enhancement* and *process intensification* compared to the natural driving forces gives the opportunity to control the transfer rates and state conditions in a way that is optimal for the desired results. The *principle of a selective phase* enormously enlarges the process space by adding a new component, which enforces a new equilibrium within a process (drying, extraction, stripping etc.). The *principle of flow guidance* in the equipment and process space (cocurrent, countercurrent, crosscurrent, mixed arrangement or recirculated flow) in addition with various switching possibilities (series, parallel, cascading) gives the basis for the effective exploitation of the existing driving force. The heuristic application of these principles can facilitate the system design and encourage the process engineering and microsystem engineering research and development [10].

Mixing can be treated as a major unit operation, which is fundamental for many other processes. Mixing can adopt many forms like homogenization, dispersing, or suspending; mixing can occur with or without chemical reactions, or as a precursor for chemical reactions, in combustion, or polymerization. A modern, general definition of mixing is the control of segregation, which describes the general role of mixing in process technology. The potential of microstructures in mixing processes, like the short diffusion length, a fast mixing or controlled flows, will be shown in various parts of this book.

A further major process step, the transport of fluids, is not listed in Table 1.1, but is partially included in the unit operation description. Active devices for fluid transport are pumps and compressors, which possess a wide variety of possibilities depending of the fluid, the viscosity, the required pressure increase, and the volume flow. Inside conventional equipment, field forces are employed for the fluid transport: a density or pressure difference, centrifugal or inertial forces. Additionally, capillary forces can be used for fluid transport due to the channel geometry and the surface characteristics of the material, see [16].

Chemical reactions are more heterogeneous than the above presented unit operations. There exist some segmentation proposals similar to the unit operations that follow physical or physicochemical aspects like heat release (exo-, endothermic), rate constants (fast, slow), kind of initiation (photo, electro, ...), or the component phases. Vauck and Müller [1] count up to 27 chemical reaction types, which can hardly be classified. The reactors, the equipment with chemical reactions inside, can be categorized with the help of the operation: batchwise within a vessel, stopped flow for many analytical applications, or continuous flow in a pipe, in a fixed or fluidized bed. The continuous-flow operation is the predominant mode for microstructured equipment due to the small hold up of media inside the reactor. A comprehensive overview of chemical reactions in microstructures is given in [17, Chapters 3 to 5], which describe 21 different reactor types with 23 gas-phase reactions, 36 different reactor types with 95 liquid-phase reactions, and 12 different reactor types with 28 gas-liquid reactions.

A large field of chemical reactions deals with catalytic transformations. In homogeneous catalysis, the catalyst acts in the same phase as the reagents. Examples are enzymatic reactions where the liquid catalyst enforces biological transformations. Hessel et al. [17, Chapter 3] count 24 reactor types for catalyst screening, which is a popular application for microstructured devices. The majority of catalytic reactions (>80%) are dealing with more than one phase, especially with solid or immobilized catalysts. In these reactions the mass transfer is the major limiting process. Fuel-cell reactors are summarized by Hessel et al. in [18, Chapter 2] with about 12 reactor types and 63 reactions.

At the end of this chapter, the scaling behavior of processes as well as the benefits and the potential for miniaturization is sketched for selected unit operations together with a more detailed description. The combination of the various units leads to a complex, hierarchic unit of processes and equipment that can show emerging abilities not possible within a single unit. The whole system is more than the sum of all elements. Hence, the concept of unit operations has its limits and should be complemented by a holistic integrated process design.

1.3 Balances and Transport Equations

The starting point of process engineering calculations and the design of process equipment are the conservation and balance equations of mass, species, momentum, and energy as well as the definition of the entropy. The conservation laws of mass (continuity equations) and energy (First Law of thermodynamics) hold in the scope of chemical processes dealt with in this book. They can be described by words in the following scheme:

$$\begin{bmatrix}\text{System change}\\ \text{with time}\end{bmatrix} = \begin{bmatrix}\text{Incoming}\\ \text{Flow}\end{bmatrix} - \begin{bmatrix}\text{Outgoing}\\ \text{Flow}\end{bmatrix} + \begin{bmatrix}\text{Source or}\\ \text{Sink}\end{bmatrix} \tag{1.1}$$

The source or sink of the flow property depends on the system and the parameter itself and is described later together with other possible simplifications.

1.3.1 Statistical Mechanics and Boltzmann Equation

Before introducing the balance equations of the various parameters, a short excursion to the molecular origin of these equations starts with the derivation of the Boltzmann transport equation for the thermodynamic equilibrium. In an ideal gas the molecules are regarded as hard spheres interacting only by very short hits with other molecules or with the boundary (wall, surface, or other limiting elements). It can be assumed that the probability of a molecule moving with the velocity w in a certain direction is equal for all three space coordinates, see also [19, p. 148]. This can be expressed by the constant ratio of the derivation of the probability distribution function (PDF) to the function $f(w)$ itself and the velocity component w,

$$\frac{f'(w)}{wf(w)} = \frac{\mathrm{d}\ln f(w)}{w\mathrm{d}w} = -2\gamma, \quad \Rightarrow \ln f(w) = c_i - \gamma \cdot w^2 \tag{1.2}$$

The integration constant is set to -2γ and determined with the kinetic energy of the molecules,

$$\gamma = \frac{M^*}{2kT} = \frac{M}{2R_\mathrm{m}T} \tag{1.3}$$

The integration constant c_i is determined by normalizing the sum of the probability to unity. The integration gives the probability distribution for one velocity component w, which stands for the other components as well.

$$f(w) = \left(\frac{M}{2\pi R_\mathrm{m}T}\right)^{\frac{1}{2}} \mathrm{e}^{-\frac{M}{2R_\mathrm{m}T}w^2} \tag{1.4}$$

The integration over a sphere of the three space coordinates gives the probability of the absolute velocity c, independent of the direction,

$$\varphi(c)\mathrm{d}c = 4\pi c^2 F(c)\mathrm{d}c \Rightarrow \varphi(c) = 4\pi c^2 \left(\frac{1}{2\pi RT}\right)^{\frac{3}{2}} \mathrm{e}^{-\frac{1}{2RT}c^2} \tag{1.5}$$

The most probable velocity of a molecule is determined by $c_\mathrm{mp} = \sqrt{2RT}$. The mean velocity from the kinetic energy is given by $\bar{w} = \sqrt{3RT}$. With the number of molecules in a unit volume N_A and the collision cross section of the spherical molecule $\pi\sigma^2$, the number of hits between the molecules and the mean time between these hits can be determined. Multiplied with the mean velocity, an estimation of the mean free path of a molecule, the average length between two collisions, can be derived:

$$\Lambda = \frac{kT}{\sqrt{2}\pi p \sigma^2} \; . \tag{1.6}$$

The mean free path divided by a characteristic length gives the dimensionless Knudsen number Kn, which is used later to estimate the influence of the molecular mobility on the fluid behavior inside microstructures, see Section 3.3. A closer look at the probability distribution f for the location $\vec{x} = \vec{x}\,(x, y, z)$ and the velocity space of a particle $\vec{w} = \vec{w}\,(u, v, w)$ varying with time gives a better image of the forces and energy distributions in an ideal gas. The integration of the PDF over the velocity space results in the number of particles in the control volume [20, p. 4],

$$\int f(t, \vec{x}, \vec{w})\mathrm{d}\vec{w} = N(t, \vec{x}) \tag{1.7}$$

The integration of the PDF over the velocity space, divided by the mass, results in the fluid density,

$$\rho = \frac{1}{m}\int f(t, \vec{x}, \vec{w})\mathrm{d}\vec{w} \tag{1.8}$$

The integration of a state variable multiplied by the PDF over the velocity space gives the mean value of this variable. The total derivative of the PDF in an external field (for example a gravitation field) is determined by the collisions of the molecules in a control volume, in detail, the current of gain and loss due to the molecule collisions,

$$\frac{\mathrm{d}f}{\mathrm{d}t} = \frac{\partial f}{\partial t} + \vec{w}\cdot\frac{\partial f}{\partial \vec{x}} + \frac{F}{m}\frac{\partial f}{\partial \vec{w}} = J_{\text{gain}} - J_{\text{loss}} = \Delta J_{\text{coll}} \tag{1.9}$$

This equation is called the Maxwell–Boltzmann transport equation, an integro-differential equation. The determination of the loss and gain current, the collision integral, and the construction of the PDF are the main problems in solving Eq. (1.9), see also [21, 22]. Regarding the collision of two molecules with the relative velocity w_{rel}, with the probability f and f_1 before and the probability f' and f_1 after the collision, the integration over the volume element and the velocity space of both molecules after the collision determines the left side of Eq. (1.9), the collision integral, see [19, p. 263].

$$\Delta J_{\text{coll}} = \int \frac{\sigma^2}{2}\,\vec{w}_{\text{rel}}(f' f_1' - f f_1)\mathrm{d}\vec{w}_1\,\mathrm{d}\vec{w}'\,\mathrm{d}\vec{w}_1' \tag{1.10}$$

This collision term is also used and implemented in the numerical lattice methods, see [23]. The first moments of the PDF f_J, for which the collision term will vanish in the case of local equilibrium, are also called the collision invariants and will be used to yield the first solutions of Eq. (1.9). With these variables f_J the Boltzmann equation can be simplified to:

$$\frac{\partial}{\partial t}(\rho f_J) + \frac{\partial}{\partial \vec{x}}(\rho f_J \,\vec{w}) + \frac{\rho \vec{F}}{m}\frac{\partial f_J}{\partial \vec{w}} = 0 \tag{1.11}$$

For $f_J = 1$ the mass conservation equation is derived,

$$\frac{\partial \rho}{\partial t} + \operatorname{div}(\rho \vec{w}) = 0 \tag{1.12}$$

For the velocity components $f_J = u, v, w$, the momentum conservation equation for all three space coordinates is derived,

$$\rho\left(\frac{\partial \vec{w}}{\partial t} + \vec{w}\frac{\partial \vec{w}}{\partial \vec{x}}\right) + \frac{\partial p_J}{\partial \vec{x}} + \operatorname{div}\left(\frac{\partial \vec{w}}{\partial \vec{x}}\right) - \frac{\rho \vec{F}}{m} = 0 \tag{1.13}$$

This equation is the basis for the Navier–Stokes equations of fluid dynamics. The collision invariant $f_J = 1/2\ w^2$, which describes the continuity of the kinetic energy during the collision process, leads to the following equation,

$$\frac{\partial}{\partial t}\left(\frac{\rho}{2}\overline{\vec{w}^2}\right) + \frac{\partial}{\partial \vec{x}}\left(\frac{\rho}{2}\overline{\vec{w}^2\,\vec{w}}\right) + \frac{\rho \vec{F}}{m}\overline{\vec{w}} = 0 \tag{1.14}$$

With the following simplifications for the velocity mean values [19, p. 266]

$$\overline{\vec{w}^2} = \frac{3p}{\rho} + w^2 \quad \text{and} \quad \overline{\vec{w}^2\,\vec{w}} = \frac{2}{\rho}(-\lambda \operatorname{grad} T) \tag{1.15}$$

the energy conservation equation can be written as

$$\rho\frac{\partial \dot{q}}{\partial t} + \vec{w}\frac{\partial \dot{q}}{\partial \vec{x}} + \operatorname{div}(-\lambda \operatorname{grad} T) + p \operatorname{div} \vec{w} + \dot{e}_q = 0 \tag{1.16}$$

where the term $\dot{e}_q$ describes the shear and deformation tensors in the energy equation, which can be summarized with the viscous dissipation. A more detailed discussion on the energy equation is given in [19, p. 266]. Here, the illustration of the linkage between the microstate Boltzmann equation (1.9) and the macrostate balance Eqs. (1.12) to (1.16) is sufficient and gives a good insight for molecular processes. The macroscopic equations are valid, if enough molecules act together to receive a smoothed signal from the collisions ($N > 10^6$). A further and detailed discussion of special microeffects can be found in [24]. In Chapter 2 of this book the limits of continuity are treated in more detail.

1.3.2
Macroscopic Balance Equations

The macroscopic balance equations for process equipment can be formulated on various levels as shown in Fig. 1.6. The equipment level is the basic calculation niveau for plant design and layout. Elements of the equipment like tubes or trays can be balanced on the level of a differential length, which refers to the "micro process" description. The third and base level is represented by the differential element (here for all three dimensions with flow in the z-direction), which allows the continuum approach and the integration over the balance space. These three levels may take different forms regarding the relevant problem.

In Fig. 1.6 the balanced variables in a differential element are given by the parameter X, which stands for the mass, species, momentum, or energy. The general balance equation for a system and a differential element is written as:

$$\frac{\partial}{\partial t}X = X\mathrm{d}y\,\mathrm{d}x - \left(X\mathrm{d}y\mathrm{d}x + \frac{\partial}{\partial z}X\,\mathrm{d}z\,\mathrm{d}y\,\mathrm{d}x\right) \tag{1.17}$$

with X as the general balanced value.

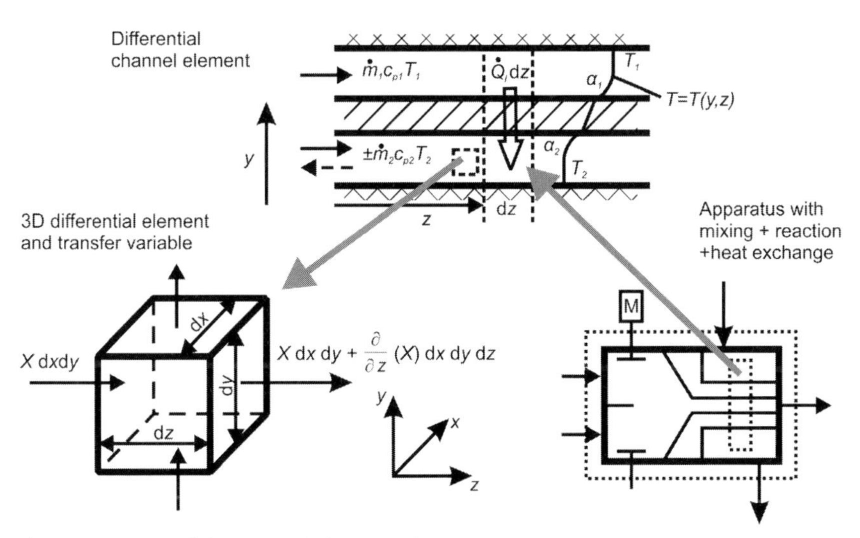

Fig. 1.6 Overview of the various balancing volumes in process engineering: 3D differential element for general calculations, 1D differential element depending on the geometry of the active area (channel, tube, plate, film, interface, see Fig. 1.5), and entire equipment for process balances (right).

1.3.3
The Mass Balance

Within the systems treated in this section no mass is generated or destroyed, hence no sink or source appears in the balance equation (1.1). The mass balance for a process element is written as

$$\sum m_{\mathrm{in}} = \sum m_{\mathrm{out}} - \sum m_{\mathrm{conv}} = \mathrm{const.} \tag{1.18}$$

For time-dependent mass flow rates, Eq. (1.8) can be written as

$$\frac{\partial m}{\partial t} = \dot{m}_{\mathrm{in}} - \dot{m}_{\mathrm{out}} \tag{1.19}$$

In a one-dimensional differential element, the balanced value X is set to the density multiplied by the velocity. The balance equation (1.17) can accordingly be written as

$$\frac{\partial(\rho A)}{\partial t} = -\frac{\partial(\rho w A)}{\partial z} \tag{1.20}$$

For a constant cross section or for a cubic differential element the mass balance can be written as

$$\frac{\partial \rho}{\partial t} + \mathrm{div}\,(\rho \vec{w}) = 0 \tag{1.21}$$

This equation is also known as the continuity equation [25, p. 2] and is one of the fundamental equations of fluid dynamics. For an incompressible fluid the continuity equation can be simplified to

$$\frac{\partial(\rho A)}{\partial t} = 0 = -\frac{\partial(\rho w A)}{\partial z} \tag{1.22}$$

This equation is also called the equation of continuity for the mass in a system. It will help to simplify other balance equations.

1.3.4
The Species Equation

If the fluid density is replaced by the concentration of the relevant species, the species balance of a system can be derived from the mass balance. For an arbitrary process volume this equation can be written as

$$\sum \dot{n}_{\mathrm{in}} = \sum \dot{n}_{\mathrm{out}} - \sum \dot{n}_{\mathrm{conv}} = \mathrm{const.} \tag{1.23}$$

The balance value X for a differential element is $X = \rho c_i w_i$ for the species formulation, which results in

$$\frac{\partial(\rho c_i A)}{\partial t} = -\frac{\partial(\rho c_i w A)}{\partial z} . \tag{1.24}$$

Considering chemical reactions [7, p. 297] in the balance space, a species source or sink must be added according the individual reaction rate r_i. For constant density and cross section of the channel the above equation can be rewritten

$$\frac{\partial c_i}{\partial t} = -\frac{\partial(c_i w)}{\partial z} + \sum_s v_{is} r_s . \tag{1.25}$$

The species transport over the boundary of the balance element is obtained by convection with the velocity w and additionally by diffusion. Hence, for a complete picture of the species transport, the diffusion term must be added to Eq. (1.25) and rewritten as

$$\frac{\partial c_i}{\partial t} = -\frac{\partial(c_i w)}{\partial z} + \frac{\partial(D_i \Delta c_i)}{\partial z} + \sum_s v_{is} r_s . \tag{1.26}$$

A closer look at the transport processes will be given in the next section as well as in Chapters 3 and 6 of this book.

1.3.5
The Momentum Equation and Force Balance

The momentum of a moving fluid can be expressed as the product of the mass and the flow velocity. The integral of the momentum over the volume results in a net force of the fluid on the volume boundary or on the equipment. In general, the momentum balance of a device can be written as

$$\sum J_{\text{in}} = \sum J_{\text{out}} - \sum J_{\text{loss}} = \text{const.} \tag{1.27}$$

The momentum loss can be interpreted as the viscous momentum loss, which is expressed as pressure loss along the channel or device flow. For a nonviscous fluid flow through an arbitrary channel with an external force, the momentum change can be expressed by the momentum, the pressure, and the gravity part:

$$\frac{\partial(mw)}{\partial t} = (\dot{m}w)_{\text{in}} - (\dot{m}w)_{\text{out}} + (pA_i)_{\text{in}} - (pA_i)_{\text{out}} + mg + F_z \tag{1.28}$$

For a one-dimensional differential channel element with viscous flow, the momentum balance can be written as

$$\frac{\partial(\rho A w)}{\partial t} = -\frac{\partial(\rho w A w)}{\partial z} - \frac{\partial(p A)}{\partial z} + p\frac{\partial A}{\partial z} - \tau L_c - \rho A g \,, \tag{1.29}$$

where L_c means the perimeter of the differential channel element. With the continuity equation (1.22), the above equation can be simplified to

$$\rho A \frac{\partial w}{\partial t} = -\rho w A \frac{\partial w}{\partial z} - A\frac{\partial p}{\partial z} - \tau L_c - \rho A g \,. \tag{1.30}$$

For constant density, nonviscous flow, and a cubic differential element for all three directions, the momentum balance can be written as

$$\frac{\partial \vec{w}}{\partial t} = -(\vec{w}\nabla)\vec{w} - \frac{\nabla p}{\rho} + \vec{g} \,. \tag{1.31}$$

This equation is also called the Euler equation for a nonviscous flow and is one of the basic equations of hydrodynamics [25, p. 4]. Additional forces in microflows may occur with surface effects in multiphase flow like bubbles and droplets, see also Chapters 4 and 5.

1.3.6 The Energy Balance

Energy can take various forms in process technology. But like the mass, the energy itself is conserved according the First Law of thermodynamics.

$$\sum \dot{E}_{\text{in}} = \sum \dot{E}_{\text{out}} - \sum \dot{E}_{\text{loss}} = \text{const.} \tag{1.32}$$

The energy loss takes into account that the energy conversion from one form into another is accompanied by natural losses. These losses are characterized by the entropy change during a process according the Second Law of thermodynamics. For a differential element, which describes the situation for heat transfer, the balanced value X can be set for the heat convection to $X = \rho w_i e_i$ in the z-direction and for the heat conduction to $X=q$ in the y-direction. This leads to the energy equation

$$\frac{\partial(\rho w_i)}{\partial t} e = -\frac{\partial(qA)}{\partial z} \,. \tag{1.33}$$

For a process device with mass flow, heat transfer over the boundary, and with technical work (for example mechanical work from an active mixer), and a chemical reaction, the energy equation is written as

$$\frac{\partial(me)}{\partial t} = (\dot{m}e)_{\text{in}} - (\dot{m}e)_{\text{out}} + (p\dot{V})_{\text{in}} - (p\dot{V})_{\text{out}} - p\frac{\partial V}{\partial t} + \dot{Q} + \dot{E}_{\text{q}} + P \,. \tag{1.34}$$

In words: the energy change in a system with time consists of the energy flowing in and out, the volume flow in and out, the volume change inside the system, the heat flux over the boundary by convection, conduction, or radiation, the energy produced inside the system (for example from chemical reactions or flow dissipation), and on the very right-hand side the technical work brought into or produced by the system. This relatively complex equation can be simplified with the help of assumptions and simplifications regarding the actual system.

Considering a channel with a constant cross section, without chemical reactions, and technical work consumed or produced, the energy of the fluid can be expressed according the First Law with kinetic and potential energy

$$me = \rho A(u + w^2/2 + gz) \,. \tag{1.35}$$

The dissipated energy originates from the shear stress τ at the wall. With the First Fourier law $\dot{q} = -\lambda \frac{\partial T}{\partial z}$ for the conductive heat transfer, the energy equation can be written in the following form

$$\frac{\partial}{\partial t}\left(\rho\left(u + \frac{w^2}{2}\right)\right) = -\frac{\partial}{\partial z}\left(\rho w\left(u + \frac{w^2}{2}\right)\right) + \rho w g - \frac{\partial}{\partial z}(p w) - \frac{\partial}{\partial z}\left(\lambda \frac{\partial T}{\partial z}\right) - \frac{\partial}{\partial z}(\tau w) + \dot{w} \,, \tag{1.36}$$

the so-called Bernoulli equation for the energy balance in a channel flow. This complex equation (not taking into account other energy-transfer processes like radiation, technical work, or the chemical reactions) can be simplified by adjusting to the actual process.

For open systems and flow processes, the inner energy is replaced by the enthalpy $h = u + p/\rho$. For flow processes the total enthalpy $h_\mathrm{t} = u + p/\rho + \frac{w^2}{2} - gz$ is used with the kinetic and potential energy part [26, p. 66]. With the energy dissipation coming from the shear stress and the velocity gradient $\varepsilon = \frac{\partial}{\partial z}(\tau w)$ (here only the z-direction) the enthalpy form of the energy equation can be written as

$$\rho \frac{\mathrm{d}h}{\mathrm{d}t} = \frac{\mathrm{d}p}{\mathrm{d}t} + \varepsilon - \nabla \dot{q} \,. \tag{1.37}$$

With the caloric equation of state, the correlation between the inner energy or enthalpy and the temperature

$$\mathrm{d}u = c_\mathrm{v} \mathrm{d}T \,; \; \mathrm{d}h = c_\mathrm{p} \mathrm{d}T \,, \tag{1.38}$$

the energy equation can be rewritten as

$$\rho c_\mathrm{p} \frac{\mathrm{d}T}{\mathrm{d}t} = \frac{\mathrm{d}p}{\mathrm{d}t} + \varepsilon - \mathrm{div}\,(\lambda \,\mathrm{grad}\, T)\ . \tag{1.39}$$

The solution of this equation gives the temperature distribution for the actual process. Besides the conduction and convection, heat may also be transported by radiation, which is proportional to T^4 and the emission properties of a surface. Radiation heat transfer is not treated here; interested readers are referred to [27, Chapter 5]. A chemical reaction within the system touches not only the species equation but the reaction enthalpy has also to be considered in the energy balance due to the apparent heat consumption or release.

$$+\Delta H_\mathrm{R} = H_\mathrm{r} - H_\mathrm{p} \tag{1.40}$$

The net heat balance is calculated from the enthalpy of the reactants and the products, see also [28, p. 371].

1.3.7 The Entropy Equation and the Efficiency of a System

On dividing the reversible heat flux q by the temperature T a new state variable, the entropy s, is derived for further characterization of states and processes. With the thermodynamic correlations

$$\begin{aligned} \mathrm{d}s &= \frac{\mathrm{d}q}{T} = \frac{\mathrm{d}u + p\mathrm{d}v}{T}\ , \\ T\mathrm{d}s &= \mathrm{d}u + p\mathrm{d}v \\ &= \mathrm{d}h - v\mathrm{d}p\ , \end{aligned} \tag{1.41}$$

the entropy correlation can be derived from Eq. (1.37)

$$\frac{\mathrm{d}s}{\mathrm{d}t} = \frac{1}{T}\left\{\frac{\varepsilon}{\rho} - \frac{1}{\rho}\nabla\dot{q}\right\}. \tag{1.42}$$

The dissipation function ε is the friction loss per volume and time unit. The entropy production is a major indicator of the efficiency or effectiveness of the process. Generally, the efficiency of a process is described by the ratio of the profit to the effort put into the process. The effort for a process may come from the input like energy, pressure loss, or heat loss. Furthermore, the effort may be defined from the fabrication effort for the device, the used materials, the effort to maintain a correct operation, or reliability efforts. The profit from a process depends on the task. It may be the throughput, the energy transfer, the mixing intensity, or the conversion rate, selectivity or time-space yield of a chemical reactor. Due to the large variety of devices in micro process engineering, the application task is also very important. A mixing device for bioanalytical purposes possesses different specifications from a mixer for chemical reactions in the production. A micromixer for a fast chemical reaction may be inadequate for a slow reaction and lead to insufficient results. Therefore, the application of the device must be exactly specified and, hence, the effectiveness as well. These various applications may come from analysis systems (μTAS), from the lab-on-a-chip field, from process or material screening, from energy conversion, or from chemical production.

1.3.8
Elementary Transport Processes and their Description

The balance equations are derived from equilibrium states, which do not reflect the transport processes between the single states. If the gradients are not too high (valid in our case), the transport processes can be described with linear correlations and phenomenological coefficients. The entire change of a state variable is described by the transport processes of the conduction in the immobile phase (solids or resting fluids), of the convection in the fluid phase (gases and liquids), and by the generation or depletion in the control volume.

$$\begin{bmatrix}\text{Total flow}\\ \text{density}\end{bmatrix} = \begin{bmatrix}\text{Conduction}\\ \text{over the}\\ \text{boundary}\end{bmatrix} + \begin{bmatrix}\text{Convection}\\ \text{over the}\\ \text{boundary}\end{bmatrix} + \begin{bmatrix}\text{Source}\\ \text{or}\\ \text{Sink}\end{bmatrix} \tag{1.43}$$

The transport correlations can be expressed in the fundamental form

$$J = L \cdot A \cdot X \tag{1.44}$$

In words: the flux (intensity) equals the intensity (transport coefficient) multiplied by the active area (cross section) and the driving force (parameter difference). The main transport phenomena in process engineering are given in Table 1.2 supplemented by the electrical transport. The geometrical situation of the flow, the temperature and concentration distribution at the channel wall is displayed in Fig. 1.7, left side.

Both single-phase transport as well as transport processes at interfaces often occur in process engineering. For the fluid/fluid exchange of mass, species and energy, there are several models to describe the transport process. The *two-film model* is one of the most common models, which gives good analytical results and shows a physical complete picture, see Fig. 1.7, right side.

For a liquid A and a fluid B, which may be gaseous or liquid with a low miscibility in A, the species transport equations may be written in the following form

$$\dot{n}_i^A = \beta_i^A A(c_{i,\text{bulk}}^A - c_{i,\text{interf}}^A) \tag{1.45}$$

$$\dot{n}_i^B = \beta_i^B A(c_{i,\text{interf}}^B - c_{i,\text{bulk}}^B) \tag{1.46}$$

Table 1.2 Transport mechanisms and transport properties.

Flow process	Mass flow	Species	Momentum	Heat	Electrons
Conduction	through porous membrane $\dot{m} = k\Delta p$	$\dot{n}_i = -DA\nabla c_i$	$\dot{J} = F = -\eta A \nabla w$	$\dot{Q} = -\lambda A \nabla T = -a\rho c_p A \nabla T$	$I = \dot{E} = \frac{1}{R_{el}} \nabla U$
Convection	$\dot{m} = \rho A w$	$\dot{n}_i = c_i A w$	$\dot{J} = F = \frac{\rho}{2} A w^2$	$\dot{Q} = \rho c_p T A w$	$P = U \cdot I$
Transfer flow	$\dot{m} = \rho A w$	$\dot{n}_i = \beta A \Delta c_i$	$\dot{J} = F = \gamma A \Delta w$	$\dot{Q} = a A \Delta T$	Flow of electrolytes [29]

Transfer coefficient $\beta = \frac{D}{\delta}$, m/s $a = \frac{\lambda}{\delta}$, W/m²K $\gamma = \frac{\eta}{\delta}$, kg/m² s, with the boundary layer thickness δ

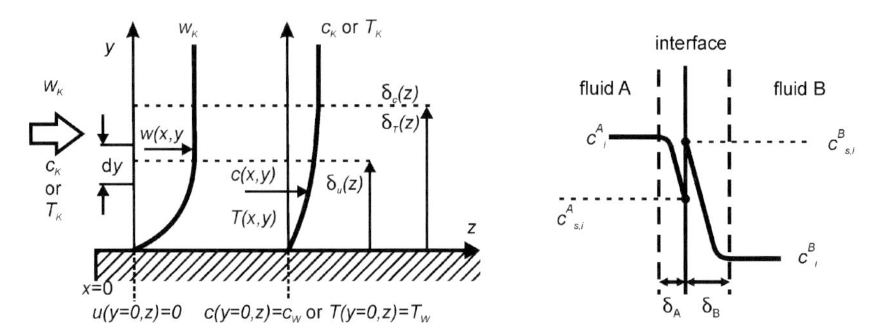

Fig. 1.7 Left: Transport process in a differential element in channel flow with velocity w, temperature T, and concentration c profiles at the wall; right: Two-film model with two different interface conditions [7, p. 93].

When fluid B is an ideal gas, the concentration can be replaced by the partial pressure and the ideal gas law can substitute the transfer coefficient. For the absorption of a gas into a liquid, Henry's law describes the solubility of the gas [30, p. 535]

$$p_{i,g} = H_i c_{i,i} \tag{1.47}$$

with the Henry coefficient H_i. For the evaporation or condensation of a binary mixture (distillation, rectification) the relation between the concentrations at the interface is described by the Raoult's law [30, p. 278]

$$p_i = p_t y_i = \pi_i x_i \,, \tag{1.48}$$

with the partial pressure p_i, the vapor concentration y_i, and the liquid concentration x_i. The concentration difference at the interface of emulsions can be described with the Nernst distribution for extraction applications [30, p. 634].

1.3.9 Additional Remarks to Balance Equations

In chemical engineering the values of the process parameters from the balance equations are often displayed in tables combined with process flow diagrams. This combined display gives a comprehensive overview and is the output from basic engineering design. The design and development of processes is described in Chapter 7 in more detail, which includes the usage of balance sheets, process flow diagrams, and pipe and instrumentation diagrams (PI diagram).

The rather complex balance equations can be simplified considering the actual process. In a steady process the time derivative vanishes $\partial/\partial t = 0$. In systems with high velocities, the convection transport is dominant compared to conductive and diffusive fluxes. For systems with dominant chemical reactions, only the change of substance has to be considered. Incompressible flow does not count for density changes. Also, simple pipe and channel flows without chemical reactions or technical work are often found in technical systems.

The limit of the continuity assumption is reached for the diffusivity in porous media [7, p. 74] and rarefied gas flow for Knudsen numbers Kn larger than 0.001 [2]. The linear transport coefficients are limited to moderate process conditions and low transport rates. The relaxation time of processes in liquids and gases under normal conditions is about 10^{-9} s, which gives an order of magnitude where the processes can be treated with linear correlations and transport coefficients. Faster processes are briefly mentioned in Chapter 3 and can be treated, for example, with the model of extended irreversible thermodynamics [31].

1.4 Calculation Methods and Simulation

Various steps of knowledge for technical processes exist: first the basics as a rough, but generally accurate interpretation of the observed phenomena, second, a detailed analysis, elaborated in the framework of measurement rules or empirical correlations that deliver a reliable solution for certain phenomena in the praxis, and third, as most exact knowledge, a theory that shows the physical background of the investigated process [1, Chapter 1]. Calculation for process engineering can be classified according to the knowledge and the needs of praxis to

- exact mathematical calculations with complete knowledge of the correlations and solutions of the differential equations (ideal case),
- calculation with simplified models, where the simplification grade determines the exactness (numerical simulations are always simplifications and approximations),
- inter- and extrapolation of experimental results,
- application of dimensional analysis and similarity theory for general laws and for scale up/down.

Besides the knowledge and application of mathematical and engineering tools, the accurate, critical judgment and interpretation of the results are essential for the reasonable and responsible work of an engineer. The main sources of errors in that process are:

- incomplete or erroneous information of states and behavior of technical systems. This plays an important role in complex systems, where emergent behavior cannot be predicted in an exact way;
- numerical procedures are always general approximation methods that only get close to the exact solution and possess an inherent error. These errors have to be estimated by the numerical code;
- the treatment of numbers by the programs and routines can lead to errors, for example by truncations that may influence the calculation of complex, iterative systems.

Hence, an approach from different sides, analytical and numerical calculations as well as experimental validation, will help in establishing a comprehensive treatment of microstructured devices.

1.4.1 Physical Variables and Dimensional Analysis

A numerical value and a related dimension define physical state variables. The dimension is the generalization of a physical parameter. For example, the mechanical theory can be explained with the help of three main dimensions: length (meter, L), mass (kilogram, M), and time (second, T). From these values derived units are the volume (M^3), the density (ML^{-3}), the velocity (LT^{-1}), or the acceleration (LT^{-2}). Other basic units are the temperature (Kelvin, Θ), the species amount (mol, N), or the electrical current (ampere, I). The following calculation rules are valid for the units:

$$L + L = L;\ LT^{-1} - LT^{-1} = LT^{-1};\ L \times L = L^2;\ L/T = LT^{-1};\ L/L = 0 \tag{1.49}$$

The important physical parameters of a process can be combined to dimensionless numbers. There are four ways to derive the dimensionless numbers from the problem description [32]:

1. dimensionless groups from the dimensionless differential equations, for example the Reynolds number Re from the dimensionless Navier–Stokes equations;
2. the fractional analysis of the important forces, of the energy currents and of the characteristic times, for example the ratio of the inertia force to the viscous force leads to the Re number;
3. the reduction of the variables by transformation to dimensionless variables, like the dimensionless time of unsteady heat conduction, which is also the Fourier number;
4. and with the Π-theorem of Buckingham applied on the set of important parameters describing the physical process.

A systematic procedure to derive dimensionless numbers is given by Wetzler [33], who lists about 400 numbers to describe various physical processes.

The Π-theorem of Buckingham says that every equation with correct dimensions can be written as a correlation of a complete set of dimensionless parameters.

$$f(x_1, x_2, \ldots x_n) = 0 \Rightarrow F(\Pi_1, \Pi_2, \ldots \Pi_s) \tag{1.50}$$

The dimensionless numbers are the potential product of the physical variables

$$\Pi = x_1^{a_1} x_2^{a_2} \ldots x_n^{a_n} \tag{1.51}$$

The general procedure to set up the dimensionless numbers is to:

1. collect a list of the important physical variables x_i,
2. prepare the matrix of basic dimensions for the variables x_i,
3. draw the conditional equations for the exponents a_i,
4. select reasonably of one or more independent variables,
5. use known dimensionless numbers for first simplifications [32],
6. write down the potential equations,
7. determine the exponents with the help of experimental data, analytical, or numerical models.

The above-described procedure gives many possibilities of the dimensionless characterization. A few hints for simplification of dimensionless groups:

- let the interesting variable only appear in one number,
- use simple numbers (simplex) of similar variables from the beginning,
- systematic testing of various sets of numbers.

The dimensionless numbers give the following opportunities for the treatment of physical systems:

- processes are completely similar, if all of their corresponding dimensionless numbers are coinciding,
- if a process is described by several dimensionless numbers, each number is a function of the other numbers,
- to scale a parameter of a process from a test model to the real system, all other parameters should be varied and adjusted to keep the corresponding dimensionless numbers constant. If not all numbers can be kept constant, a partial similarity may be sufficient and an engineering approach with an error estimation has to be conducted to evaluate the scaling.

Based on the dimensionless numbers, the similarity of processes plays an important role in process engineering. The benefits of dimensionless numbers and groups are the possibility of scaling, the reduction of variables, the independence from the unit system (SI or British), and the similarity of processes with the same values for dimensionless numbers. One major drawback is their limited application range. The numbers and their combination are only valid for their model system they were defined for. For example, the Re number for pipe flow is somewhat different from the Re number of a flow over a plate, and should not be intermingled. Also the correlations between the dimensionless numbers are only valid for a certain range. The friction factor for laminar channel flow is inverse proportional to the Re number, which is not valid for turbulent channel flow ($Re > Re_{crit}$). The choice of the influence parameters has to be very cautious and adjusted to the existing problem. The dimensionless numbers do not give any new physical insight, but give a more structured description of the process.

1.4.2 Similarity Laws and Scaling Laws

The basic idea of the scaling is the insight that all physical rules are independent of the choice of the system of dimensions. The application of the scaling laws employs the following relations:

- geometric similarity of the systems;
- dynamic similarity of the systems, the relative values of temperature, pressure, velocity, and others in a system should be the same on both scales;
- boundary conditions should be the same, like constant temperature or zero velocity at the wall.

Miniaturization is assisted very much by the usage of dimensional analysis and the transfer of knowledge into the smaller dimensions. In microstructured equipment the so-called numbering-up process partly replaces the scale-up process, which describes the data transfer from model systems to real plant dimensions. Increasing the number of microstructured elements like channels, mixing elements, or heat-transfer plates in a device means an internal numbering-up. An external numbering-up is the parallel setup of some microstructured devices to a whole group. Both internal and external numbering-up are mainly limited by the appropriate fluid distribution to each element for a proper operation. This drawback can be overcome by a good design of the manifolds or by the equal-up method, where the microstructured elements are not fabricated on small devices like chips or platelets, but integrated on plates or tubes with appropriate fabrication technology, which are themselves also integrated into a whole apparatus [8]. The aim is to reach similar process conditions on different geometrical scales.

1.4.3 Order of Magnitude

To show scaling relations, the Trimmer brackets [2, Chapter 4] were introduced to give an overview about the consequences of a length variation. However, for more complex or coupled processes, this method needs a high effort for displaying the relations between length and other parameters. An alternative way to yield information about the order of magnitude for the relevant effects is the scale analysis described by Bejan [34]. Scale analysis goes beyond dimensional analysis and is a good supplement and a relatively facile method to gain information about the system behavior.

To estimate the order of magnitude for the quantities of interest, the governing equations like balances and the geometrical situation must be known. This has also to be done for the "exact" analysis and is the first step for an engineering analysis. The basic equations, often in the form of partial differential equations, are transformed into equations with geometrical and state parameters, for example a gradient is expressed as difference ratio:

$$\frac{\partial T}{\partial t} \rightarrow \frac{\Delta T}{t_c} \quad \text{or} \quad \frac{\partial^2 T}{\partial x^2} \rightarrow \frac{\Delta T}{L_c^2} \tag{1.52}$$

where t_c and L_c define the characteristic time and length, respectively, for the actual problem. With these simplifications, the governing equations are resolved for the interesting parameters like characteristic time or penetration depth. The relation between two dominant parameters can be found from one equation. If we have more than two relevant parameters, the following rules for estimation of the order of magnitude can be applied [34]:

- Summation: If $O(a) > O(b)$ and $c = a + b$, the order of c is $O(c) = O(a)$. If $O(a) = O(b)$ and $c = a + b$, the order of c is $O(c) \sim O(a) \sim O(b)$. The same holds for the difference and the negation.
- Product: For $c = a \cdot b$, the order of c is $O(c) = O(a) \cdot O(b)$.
- Ratio: For $c = a/b$, the order of c is $O(c) = O(a)/O(b)$.

According to Bejan [34], scale analysis is widely employed in heat transfer and one of the first steps in engineering analysis. It is easy to get a first impression and "house number" results without the effort of the "exact" analysis, which also depends on the model prerequisites.

1.4.4 Lumped Element Modeling

For complex systems and network arrangements, the analytical treatment is still possible, but the amount of elements cannot be handled in the traditional way. The numerical treatment of complex systems has advanced during the last decades, the treatment of complex and coupled systems is established, but nevertheless, the computational effort is extremely high to treat transport processes in a fluidic network, if even possible. From electronic engineering, systems with millions of individual, but analytically describable, elements are known that are treated with lumped element programs like P-Spice [35, 36]. The analogy of transport processes and the corresponding equations, see Table 1.2, helps to translate complex flow networks or heat-management problems into a mathematical description, which can be handled by electronic simulation programs. The individual processes of mass, species, heat and energy transfer have their equivalent process in the electrical current transport, like the resistance, the capacity (storage), and the inductivity (kinetic energy). The method is described in Chapter 8 of this volume. Other applications in microsystem technology can be found in [35, Chapter 6; 36, Chapter 12; 37, Chapter 4].

1.4.5
Numerical Simulation and Analytical Modeling

Basically, one can distinguish between continuum modeling of transport processes in microstructures [38, Chapter 4; 39, Chapter 2] and statistical methods like the Monte Carlo Methods DSMC [40, Chapter 10; 41, Chapters 7 and 8] or the lattice Boltzmann methods LBM [42, 43]. The tasks in process engineering demand the solution of the conservation and balance equation of convection, diffusion, and/or reaction problems in various geometries under certain boundary conditions, which means solving the partial differential equations on discrete domains. For the majority of problems this can be done with continuum models, described in the following.

The finite-element method (FEM) was initially derived from solving structure-mechanical problems using simple piecewise functions, but is also applied in fluid mechanics [44]. The finite-difference method (FD) generates approximations for the derivatives of the unknowns at each grid point with the help of truncated Taylor series expansions [45]. Originally developed as a special FD formulation, the finite-volume method (VOF) solves by an iterative method the algebraic equations, which have been derived from the integral equations governing the transport processes. Additionally, spectral methods approximate the unknowns with the help of truncated Fourier series or Chebychev polynomials on the entire domain [44].

In the context of process engineering, continuum modeling and analytical modeling are preferred to bridge the gap between micro processes, equipment transfer processes and complete process design. The analytical modeling of process engineering phenomena is based on physically partial founded models with adjustment parameters. A complete calculation comprises four working levels:
- physic level for the process principles of unit operations,
- equipment level of the process technical and mechanical,
- process structure for controlling and regulation programs of automation, and
- the validation level for economical calculations.

The risk and flexibility of the assumptions and simplifications of the model determine the system modeling. There has to be a compromise between the mathematical and physical exactness and the real technical complexity. The mathematical models have to be appropriate for the experimental test units or the real plant, from which the data are gathered.

1.5 Miniaturization and its Application to Transport Processes

The miniaturization of chemical equipment emphasizes mainly the length reduction of the main dimensions and simultaneously constant process conditions like pressure, temperature, or concentrations. In thermodynamics, these parameters are called intensive state variables, which are not changed by reducing the size of the system. The length reduction influences many aspects in process engineering that are treated in the following. This listing sheds light on the actual state of the art, but is hopefully just the beginning of a larger process of gaining knowledge and technical skills in an emerging field of new technology and application. The picture is certainly not complete, but shows the complexity of the interwoven fields and gives many hints and much advice for miniaturization aspects.

1.5.1 Length

Typical length scales in microelectronics, process technology, and other areas are shown in Fig. 1.2. With the decreasing length, the operation and efficiency of unit operations is affected in various ways, which is shown in the following.

Cyclone The density difference between dispersed particles and the surrounding fluid leads to a separation within a force field like centrifugal forces. They are generated in a rotating system like a centrifuge or a cyclone [46, Chapter 6].

In a mechanical centrifuge the acceleration scales proportional to the radius. Hence, a miniaturization will reduce the radial acceleration and therefore the separation factor. The cyclone is a mechanical separation process of dispersions of solid–gas or solid–liquid mixtures without moving parts. The separation of immiscible liquids with different densities is also possible, but other separation processes like settling, filtering, or coalescing are more effective, see [46, Chapter 6]. The separation factor z is defined as the ratio between the acceleration induced by the force field and the gravity g.

The vortex in a cyclone exhibits a high tangential velocity near the vortex center, which decreases with increasing radius r. The tangential velocity of a real viscous vortex, like an Hamel–Oseen vortex [47], scales with r^{-n}, where n ranges from 0.5 to 0.8 [46, Chapter 6]. The centrifugal acceleration is proportional to the square of the tangential velocity w_t divided by the radius r, which gives for the separation factor z.

$$z = \frac{w_t^2}{rg} = \frac{\text{const.}}{r^2 g} \tag{1.53}$$

The separation factor increases with a smaller radius of the cyclone. Hence, small particles are more likely to be separated by smaller cyclones. In a feasibil-

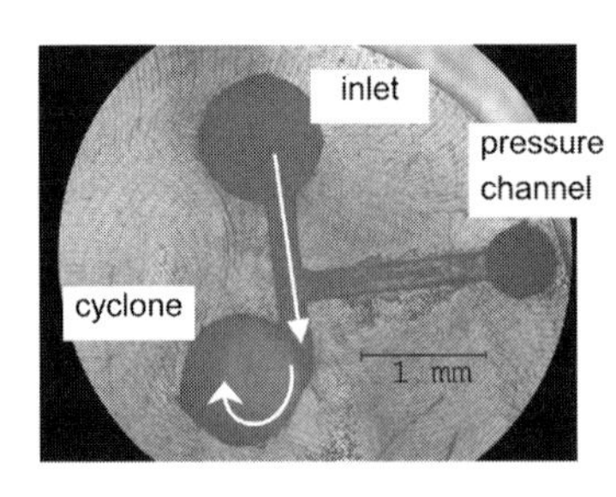

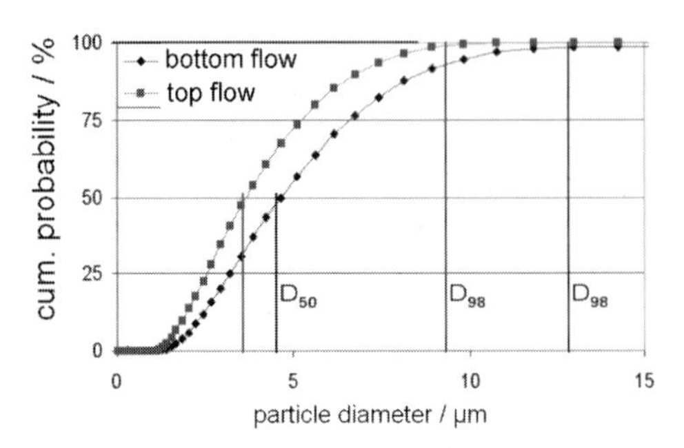

Fig. 1.8 Left: Cross section of the microcyclone with inlet and channel for pressure measurement; right: Cumulative probabilities of the particle diameter for the top and bottom flow of the cyclone, small particles are collected in the top flow.

ity study, Kockmann et al. [48] fabricated a microcyclone with an inner cyclone diameter of 1 mm from PMMA and separated an aqueous SiO_2 suspension (mean diameter ~4.5 µm, particle density ~2500 kg/m^3).

The suspension enters the cyclone hole tangentially. Driven by the centrifugal force the large particles flow to the outer radius; smaller particles are collected in the center and leave the cyclone through the top flow. The separation efficiency of one cyclone can be seen in Fig. 1.8 at the D_{98} value (9 µm and 13 µm for the top and bottom flow, respectively). A cascading of many microcyclones will give a higher separation factor but requires a higher pressure difference. The cyclone process can be extended to pressure diffusion, a coupled process described by the thermodynamics of irreversible processes, see Chapter 3, Section 4.

In two papers, Ookawara et al. [49, 50] presented the centrifugal separation of an aqueous suspension flow in curved rectangular channels (200×150 µm cross section, 30° and 180° curves, 20 mm radius). With a pressure loss of about 200 Pa, the typical Dean numbers were between 30 and 45, which means Re numbers varied from 1400 to 2200. The experimental results and numerical evaluation by CFD simulation show a good separation of particles larger than 15 µm from the fluid. Contrary to centrifugal force, the Dean vortices push the particles inwards by friction forces and disturb the separation flow. Hence, a shorter bend may be more effective.

As part of the product catalogue of the German microchemtec project, two companies (Little Things Factory, Ilmenau [51], and mikroglas, Mainz) offer a microcyclone, which fits into their series of process equipment.

Molecular Distillation In molecular distillation and vacuum rectification, the gap width between the evaporating and condensing falling film plays a mayor role, where the mean free path of the molecules is larger than the distance between evaporating and condensing surfaces due to the low pressure. These processes usually separate liquid mixtures, which are thermally sensitive like vegetable oils, vitamins or fatty acids, see [52] for the historical development and [53, Chapter 4; 54, Chapter 10]. The Knudsen regime serves for a gentle thermal separation and low product degradation. A lower distance requires a lower process vacuum, less energy consumption, and will probably result in a higher product purity. The design length is limited by the film thickness of the often high viscous liquids (about 1 to 5 mm). Even though an application has not yet been reported, a falling film micro device under vacuum would be an interesting test object.

Flame Distance and Explosion Limits The typical dimension of microstructured equipment is just below the extinction length or quench distance of many fast reactions, which is about 1 mm. Veser [55] reports on a calm hydrogen oxidation with flame suppression in channels smaller than 500 μm. The author determined three different explosion limits for a stoichiometric H_2/O_2 mixture dependent on the pressure, temperature, and the channel diameter. The main mechanisms are thermal quenching by wall heat conduction and "radical" quenching by kinetic effects where the mean free path of the molecules is in the range of the channel dimensions.

Miesse and coworkers [56] observed small flame cells in flat microchannels. They regard the coupled heat and mass transfer and the reaction kinetics as a self-organized system that exhibits characteristic features like flame length or periodicity. For cold walls, the thermal quenching of the reaction is the dominant stopping mechanism for the combustion. They give three design rules for successful microburner layout and complete combustion. The wall material should not quench the radicals to keep the gas-phase reaction undisturbed. The thermal insulation must be sufficient to hold a high temperature for the combustion. The flow and temperature management must ensure a sufficient high bulk temperature and prevent a melting of the wall material.

For fuel-cell applications Hessel et al. [18, Chapter 2] show in a review that reactions with strong exo- or endothermic behavior and high-energy transfer, like explosive reactions, take profit from the high transfer rates in microchannels for precise temperature control.

Gradients dX/dx For constant differences of the process parameters like temperature, pressure, or concentration, a reduction of the transfer distance increases the gradient, the driving force for the transport processes. Therefore, the corresponding transfer flux is increased and the equilibrium state is reached much faster.

Not only are the main transport processes of mass, momentum, and energy increased by the high gradients, also many minor coupled processes become more important [57, Chapters 4 and 5]. The thermoelectric effect is exploited for sensing (thermocouple, no electrical current, NiCr-Ni 0.04 mV/K), for cooling (Peltier element) or for the generation of electrical energy (Seebeck element, about 0.5 mV/K). The generation of a species current with the help of a temperature gradient is called the Soret effect, the thermodiffusion. The Soret coefficient D' has an order of 10^{-12} to 10^{-14} m^2/s K for liquids and of 10^{-8} to 10^{-10} m^2/s K in gases [58, Chapter 13]. Mainly used for the separation of isotopes, the Clusius–Dickel column employs the thermodiffusion assisted by natural convection in a heated gap, and enhanced by cascading many of the columns. The species concentration gradient due to a pressure difference is called pressure diffusion, which was originally investigated also for isotope separation [6]. The fabrication of the separation nozzles for pressure diffusion combined with centrifugal forces is one of the birthplaces of microfluidic technology [59]. The exploitation of these processes is described more detailed in Chapter 3.

Thermal Insulation Length In small devices, the thermal insulation of devices with different temperatures is very difficult. One major problem of some modules in the "backbone system" (microchemtec, Germany [60]) was the thermal insulation of the different reactors with operating temperature from below 260 K to over 470 K. Technical polymers like PEEK ($\lambda = 0.25$ W/m K) can be used as the construction material for low temperature differences up to 100 K. For higher temperature differences or very short distances, multilayer superinsulation from cryogenic technology can help to overcome the difficulties [61, Chapter 7]. Although the multilayer insulations are very expensive and difficult to apply to complicated shapes, they offer the best performance of all technical insulations ($\lambda \sim 10^{-8}$ W/m K). The multilayer insulation consists of evacuated alternating layers of highly reflecting material (aluminum foil or copper foil) with low-conductivity spacers, like glass or polymers. Their extremely low thermal conductivity depends on the reduction of all three modes of heat transfer (radiation, gaseous and solid conduction) to a bare minimum by evacuation of the gaps. The low insulation of small elements is the main reason why Peterson et al. [62] give the size limits of small regenerative heat engines as a few millimeters. Small heat engines will suffer from an insufficient heat recovery in the regenerator and from the difficult buildup of the necessarily high temperature difference.

Characteristic Times The length scaling often goes along with a time scaling of the relevant processes. In general, the shorter the length, the shorter the characteristic time for transport processes will be, and the higher the frequencies of changes are. The diffusion of a species in a surrounding fluid is displaying this process. The mean path of a molecule or small particle in a surrounding fluid is given by the Einstein equation [63],

$$x^2 = 2Dt\,. \tag{1.54}$$

The typical diffusion length within one second is about 7 mm in gases (air) and about 70 μm in liquids like water. A similar conduction length can be derived from the basic balance equation for the momentum ($x_p = \sqrt{2\nu t}$) and the heat transfer ($x_q = \sqrt{2at}$). The characteristic time is proportional to the square of the length variation and the transport coefficient.

From unsteady heat transfer, the characteristic time for heating or cooling of a body is proportional to the temperature difference and the ratio of the heat capacity to the heat transfer with the environment,

$$t = \frac{mc_{\mathrm{p}}}{\alpha A} = \frac{\rho c_{\mathrm{p}} V}{\alpha A} . \tag{1.55}$$

In a similar manner, relaxation times of other unsteady processes can be estimated.

In process engineering, other characteristic times determine the process conditions. The channel length and the mean flow velocity determine the mean residence time $t_{\mathrm{P}}=l/w$ of a fluid element in the channel. With shorter length, also the residence time decreases. The residence time must be designed appropriately for the actual process. A very important characteristic time in process engineering belongs to the chemical reaction: Besides the concentration c and the reaction order m the reaction time scale depends mainly on the reaction rate constant $k\,(t_{\mathrm{R}} \propto 1/k)$. The reaction rate constant mainly depends on the temperature. Inside devices with short characteristic length and corresponding times like mixing or residence time, a slow reaction may be incomplete at the channel outlet. Fast reactions, on the other hand, take profit from the fast mixing and fast heat exchange. In particular, combined reactions with slower side-reactions will show a higher selectivity and higher yield in microstructured devices. Also, reactions with high energy demand or release are suitable for micro devices. The concentrations can be increased to intensify the reaction or new reaction paths can be addressed where fast mixing plays an important role. Within a chemical reactor, the three characteristic times of mixing t_{M}, of residence in the device t_{P} and of the chemical reaction t_{R} have to be adjusted to the entire process to yield an optimum result, see Chapter 6.

Free Convection The heat transfer due to free or natural convection is induced by the thermal expansion and density differences of a fluid under temperature gradient. The heat transfer between two plates with different temperatures and a gap width is characterized by the Grashoff number Gr, the ratio of the volume force to the viscous force,

$$\mathrm{Gr} = \frac{g\beta_{\mathrm{p}} s^3 (T_{w1} - T_{w2})}{\nu^2} \tag{1.56}$$

with the thermal expansion coefficient

$$\beta_{\mathrm{p}} = -\frac{1}{\rho}\left(\frac{\partial \rho}{\partial T}\right)_{p=\mathrm{const.}} \tag{1.57}$$

Depending on the temperature difference between the walls and on the distance s between the walls, the fluid will circulate in the gap. A prominent example of natural convection is the Bénard convection in a horizontal layer, which is heated from below [58, Chapter 14]. If the layer or the gap is too thin, the viscous forces damp the convection and the heat is transferred solely by conduction. If the Rayleigh number Ra (=Gr Pr) is smaller than 3×10^8 the free convection flow is laminar and ruled by the viscous forces [27, Chapter 3; 34, Chapter 5]. For Ra numbers lower than 175, the heat-transfer augmentation by natural convection is smaller than 1%. In a gap filled with air or water, the influence of natural convection on the heat transfer is smaller than 1% for a gap width smaller than 8 cm or 4 cm, respectively. Therefore, natural convection may be neglected for most cases in micro process technology, but is the major cause of heat losses of equipment to the environment. For the measurement of an incline or acceleration, a sensing device developed by Billat et al. [64a] employs the temperature and flow field due to free convection around a heated wire.

1.5.2
Area

When miniaturizing the length, the area decreases proportional to the square of the length, L^2. The area plays a mayor role as a cross section for transport fluxes like mass or energy flow or as an active area, where processes take place. For a constant flow rate like fluid velocity, species transport, or heat exchange, the total amount of the quantity decreases proportional to L^2 with a smaller length scale. For a constant driving force, the area specific flux remains constant. The pressure loss in channel flow grows with smaller cross section, see Chapter 2, Section 5. The order of magnitude of the pressure loss increase depends on the variables, which remain constant.

With a constant absolute flux like energy transfer (in Watt), the specific flux dramatically increases [in W/m^2]. For example, the energy load in modern microelectronic devices increases with their performance improvement, which leads to heat loads of already above 100 W/cm^2 [64]. Finally, a small cross section of immersed bodies shows a small flow resistance, which appears in low settlement velocities and nearly Lagrangian particle behavior (massless particles).

High Specific Area for Transfer Processes Many transport processes take place at phase boundaries, which have the form of films, bubbles, or droplets in emulsions, dispersions, or foams. With a higher specific area and high gradients, the transport rates are increased and the equilibrium state is reached much faster. Temperature and concentration differences are homogenized much earlier, which plays a prominent role in mixing and heat transfer.

Considering the simple geometrical figure of a sphere, the specific surface-to-volume ratio a_v is inverse proportional to the characteristic length, $a_v \propto 6/L$. The smaller the channel structures, the larger grows the specific interface and therefore the active area for transport processes. This effect will be shown in the following for major separation unit operations, together with other specific miniaturization effects.

Filter and Membrane Processes The variety of membrane processes covers a broad area ranging from food technology through potable-water generation to blood and serum cleaning. A good overview of membrane types and specific application is given by Melin and Rautenbach [11]. Conventional membrane processes already work among others with hollow fibers with a diameter of 40 to 400 μm and a high specific surface up to 20 000 m^2/m^3 [10]. These hollow fibers reach a specific permeate volume flow (fluid passing through the membrane) of about 300 m^3/h per cubic meter equipment volume. Microstructured membranes from tantalum, palladium and its alloys are used for carbon monoxide cleanup in fuel-cell applications [18, Chapter 2].

A new field in membrane technology is covered by liquid membranes, where a liquid inside a porous structure absorbs components from one side and transfers them to the other side. They exhibit a good product selectivity, low pressure difference, and high transfer rates [11, Chapter 2]. Besides the appropriate material properties one crucial point is the protection of the enclosed liquid against the high pressure difference across the two membrane sides. The fabrication of the structures for liquid membranes can be assisted by microfabrication technology, but has not yet been reported.

Distillation and Rectification The separation of a liquid mixture with subsequent evaporation and condensation enriches the low boiling component in the vapor phase. This is the fundamental mechanism of distillation and rectification, the most frequently used separation technique in process technology [10, Chapter 7]. The coupled heat and mass transfer is performed in small laboratory equipment like the microdistillation apparatus, as well as in huge rectification towers like the 60-m high cryogenic rectification columns of air-separation units. A detailed treatment of the underlying transfer processes can be found in textbooks [30, 65] or concise handbooks [12].

Sowata and Kusakabe [66] demonstrated the vapor–liquid separation of a water-ethanol mixture on a chip (20×40 mm^2) with mixer-settler arrangement. Both streams come into contact in a Y-shaped mixer and the bubbly flow streams through a meandering channel. At the outlet, the two phases are separated by gravity in a separation chamber (8×8 mm^2). The experimental data indicate the equilibrium state for the gaseous phase and a liquid concentration of 0.65 mole fraction methanol, which means a stage efficiency of about 65% for the liquid side. A better mixing in the meandering channel may enhance the stage efficiency and lead to a better separation. Hence, a good mixing process enhances a separation process, too, due to the fast achievement of the thermodynamic equilibrium.

Fink and Hampe [67] presented a design study of a distillation column to illustrate the setup of an entire micro device. The experimental investigations show that a flow rate of about 20 ml/h has been realized with 2.5 theoretical separation stages. Even small temperature changes lead to operational disturbances and uncontrolled concentration changes [8, 60].

In our own design study [48], a distillation chip with various geometries was fabricated and tested by evaporation and condensation of a water-ethanol mix-

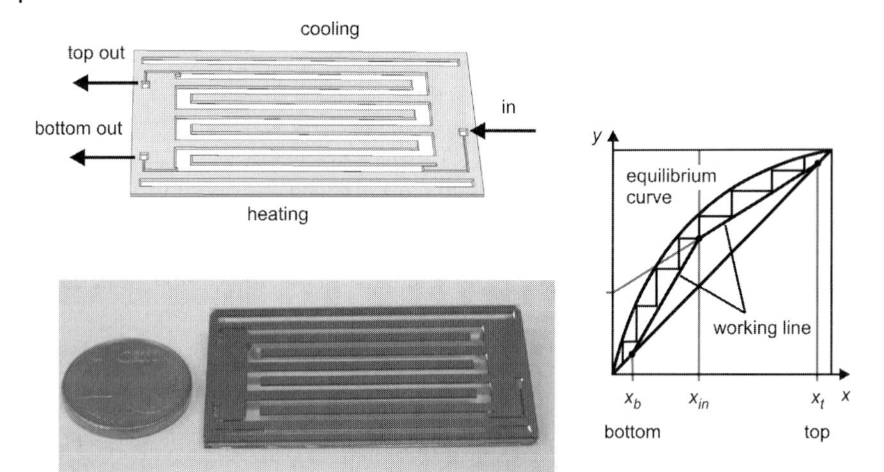

Fig. 1.9 Left top: Geometrical setup of a distillation/rectification chip ($20\times40\times1.6\ mm^3$) made of a silicon wafer, anisotropic KOH etched and covered with two glass lids; bottom: image of a rectification chip with meandering separation channel; Right: McCobe-Thiele diagram stagewise operation of rectification process, see [69].

ture. The chip, with a meandering separation channel is made of an anisotropic KOH-etched silicon wafer covered with two Pyrex glass lids, see Fig. 1.9. From the heated side (hot water, 93°C) the mixture partly evaporates and condenses on the opposite side of the channel. The McCabe–Thiele diagram on the right side of Fig. 1.9 explains the stagewise operation of a rectification column. The operating line in the upper part of the diagram describes the enriching section of the separation column, where the low-boiling substance enriches in the vapor phase. A theoretical distillation stage or tray is described by the horizontal line for the condensation of the rising vapor and by the vertical line for the evaporation of liquid from the tray. In the ideal case of reaching an equilibrium state, the lines touch the equilibrium curve and give the number of theoretical trays. The graphical treatment of these operations is almost faster than analytical or numerical calculations and gives a better physical insight into the process.

The first experimental results show a low mass flow rate and a poor thermal insulation to the ambient. From the design study the following main issues have been identified:

- the fluid transport can not rely on gravity, other principles must be employed, i.e. capillary transport or external force fields like rotation [68];
- the contact between liquids of different concentration has to be avoided, which also limits the minimal size of the gaps for evaporation, condensation, and vapor flow;
- the fabrication of suitable geometries to guide the liquids and vapor currents is a crucial point. Besides the silicon techniques, alternative fabrication techniques like metal foil structuring should be investigated, see [8];

- an internal numbering-up may help to increase the volume flow, which is a considerable drawback of the actual devices.

These points are generally valid for multiphase flow transport problems and coupled heat and mass transfer. However, the potential of a high specific area provides a high transfer area and high exchange rate between the phases. A short distance between the phases and concentration differences allows a fast approach to the thermodynamic equilibrium and leads to smaller equipment, but low throughput and complex fluid management have to be considered.

Absorption and Desorption Similar to the rectification and the distillation process, absorption is a coupled heat and mass-transfer process, where a gaseous component is dissolved in the liquid phase [69]. Desorption or stripping is the opposite process, where a dissolved component is released from the liquid phase into the gaseous phase. The state changes and the stagewise operation is displayed in the operation diagram with the equilibrium line, given by Henry's law, and the operating lines for desorption and absorption, see Fig. 1.10.

The orientation of the operating line depends on the process conditions and is explained in [26, Chapter 1; 69]. In an absorption process a component of a gaseous mixture is step-wise received into a liquid, partially chemically bound in the liquid by additives. For successful operation, a high mass transfer between the phases leads to a fast establishment of the thermodynamic equilibrium. Prominent examples are the synthesis gas (syngas) cleaning in the chemical industry (CO_2 removal) or the CO_2 enrichment in water for sparkling water.

The opposite process of absorption, the stripping of a dissolved component from a liquid (toluene in water) was investigated by Cypes and Engstrom [70].

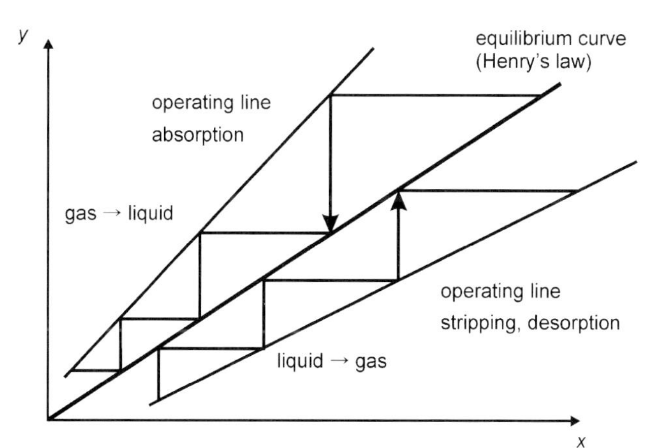

Fig. 1.10 Operation diagram for absorption and desorption/stripping, concentration of the key component in the gas phase y over the concentration in the liquid phase x. The display of the stagewise operation is similar to the McCabe-Thiele diagram for rectification, see Fig. 1.9.

The liquid mixture flowing in a silicon channel was treated with superheated steam entering from a second channel through a perforated wall. After leaving the channel the bubbly flow was separated in a chamber and analyzed. The comparison between a microfabricated stripping column and a conventional packed tower shows a transfer rate nearly an order of magnitude higher than conventional packed towers, indicating a high specific interface. This large interface enhances the convective mass-transfer process by reducing the diffusive length.

Both absorption and desorption integrated to a process cycle result in the absorption heat pump process [26, Chapter 6], or in the closed absorption and regeneration cycle to reuse the washing fluid. An absorption heat pump process with microstructured elements was presented by Ameel et al. [71] and Stenkamp and teGrotenhuis [72].

Extraction One can distinguish between liquid–liquid and liquid–solid extraction (leaching) [13, Chapter 15; 14, Chapter 1; 69, Chapter 3]. Extraction employs the different solubility of a component C within other components A and B. Often, the low solubility of component A in component C leads to a two-phase flow with mass transfer at the interface between the phases. The separation of the component B from A with the extracting agent C writes as follows:

$$(A + B) + C \Rightarrow A + (B + C)\,. \tag{1.58}$$

Many applications of extraction processes can be found in the pharmaceutical industry, for various foods and detergents, vitamins, or in the petrochemical industry. The Nernst distribution describes the equilibrium concentration between the three phases, see Fig. 1.11. Material properties can be found in the literature [14, Section 1.9; 69].

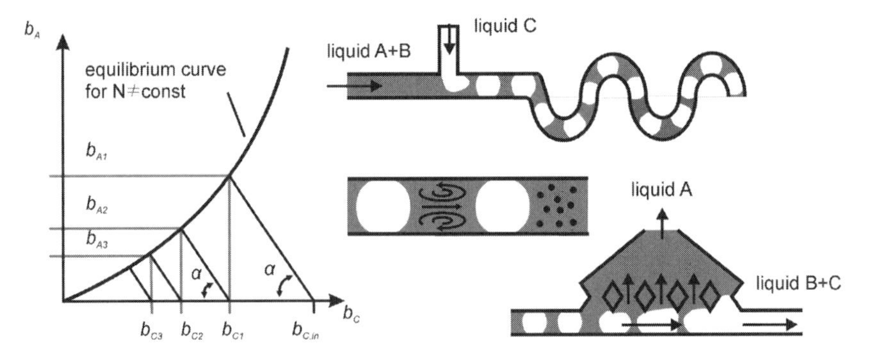

Fig. 1.11 Left: Loading of the component B b_A in A over the loading in C b_C; stepwise cross-flow extraction with four stages, equilibrium line is given by the Nernst distribution N; right:) geometrical examples of microcontacting equipment (adopted from [73, 74]), contacting and mixing in a serpentine channel, flow characteristics in the bubbly flow, and separation by geometry and surface properties using capillary forces.

Ehrfeld et al. [75] and Klemm et al. [8] describe the extraction process in microstructured equipment. Recently, Okubo et al. [76] presented the rapid extraction process on a chip. TeGrotenhuis et al. [77] developed micro devices for transport processes in solvent extraction. Kusakabe et al. [78] investigated the flow details and transport processes within a droplet in a micro contactor for solvent extraction. Their comparison with experimental data shows a good agreement for the extraction of ethyl acetate from n-hexane to water. Specific problems for solvent extraction are the fluid transport and phase separation after the mixing. In conventional equipment the driving forces for the fluid transport are the density difference and gravity, external pumps as well as capillary effects. In microstructures, the pressure-driven flow and capillary effects are the most predominant pumping mechanisms. A design study for microstructured extraction equipment is shown in Fig. 1.11 and 1.12.

The combination of microstructures with macro devices for fluid transport, for example, the mixer–settler chip with radial arrangement and the stepwise transport by a pump is displayed in Fig. 1.12. The general setup was adopted

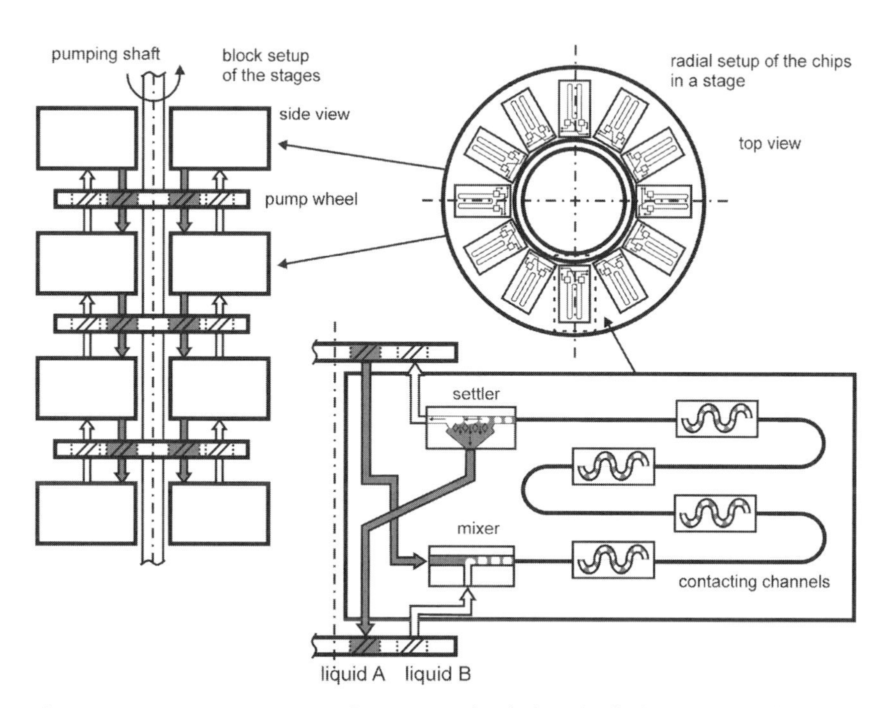

Fig. 1.12 Micro/macro integration of a mixer–settler extraction apparatus; bottom right: chip layout for contacting, mixing in a meandering channel and separation (settler) of a liquid–gas flow. The connections to further processing like the pump wheel are sketched, in the further layout, the fluids are guided perpendicular to the chip plane; top right: radial arrangement of different mixer-settler chips; left: side view of the chip blocks and intermittent pump wheels for fluid transport.

from conventional extraction equipment with rotating internals for mixing and fluid transport, like the Scheibel or the Kühni column, from Schlünder and Thuner [69, Chapter 3]. For the design and fabrication, the main effort has to be spent in the packing of the chips in the radial mount and in the sealing between the block and the pump wheel.

The integration of microstructures into macro equipment has to be carried out to use the benefits and opportunities of both fields and additional emergent effects. Hence, as guidelines for the next steps: Not only as small as useful for the designated application, but also as many microstructured devices as necessary for the actual process.

1.5.3 Volume

The volume scales with the cubic length L^3 and is directly connected over the material density with the mass. A low mass means low capacity of the elements (see Section 1.5.1, Characteristic Times) and corresponding high frequencies. A low volume in process equipment is useful for the handling of toxic and hazardous products, for expensive and valuable goods. With smaller plants a distributed onsite production is possibly a production at the point of use or point of care (in medicine or diagnostics). The production-on-demand of small amounts of chemicals is much easier and represents a flexible production [17, Chapter 1].

The high throughput screening is a valuable tool for catalysis research [18, Chapter 3] and chemical synthesis [79], but the extremely low volumes in microstructured equipment also affect other processes. With small devices with process conditions close to larger equipment, pilot plants can assist process development and process routes discovery with fast response on parameter variations and optimizing the equipment.

Adsorption The adsorption is the arrangement and attachment of volatile molecules to a solid, but porous surface from a gaseous or liquid phase. In conventional equipment porous pellets or beads provide the porous surface. The pellets are ordered in packed beds in larger vessels. Therefore, a miniaturization of the entire plant results in a low volume, which is quickly loaded and therefore negative for the amount of adsorbed species. Smaller pellets give a higher specific area per volume, but also increase the pressure loss of the flow through the bed.

The calculation and design procedure of the adsorption isotherm and the Knudsen regime for transport processes in pores is not treated here. Interested readers are referred to [13, Chapter 16; 30, Chapter 10; 46, Chapter 13; 65, Chapter 12]. In small pores the molecules are interacting more with the wall than with themselves. This process is described by Knudsen diffusion. For ideal gases the validity of this regime depends on the total pressure and the pore diameter.

Fig. 1.13 Micro gas chromatograph. Left: separation column on a silicon chip, covered with a glass plate; right: IC plate and analytical equipment of the micro-GC, courtesy of SLS Microtechnology [80].

For analytical purposes the faster loading results in a fast response time of chromatographic analysis. The loading, detection and cleaning of the chromatographic separation column can be achieved in a very short time, compared to conventional systems. This allows a fast response to signal changes, which is very important for the proper control of modern chemical plants. The fabrication methods and the special system integration of microtechnology are very useful to enhance the performance, as has been demonstrated by Micro Gas Chromatography [80], see Fig. 1.13.

The chromatographic separation column is fabricated with a combined bulk and surface micromachined process on a silicon chip. The silicon-glass chip, on which this separation column with a separation channel length of 0.86 m and diameter of 60 μm is made, has a surface of 1 cm^2. This column is heated to 200 °C within 20 s, using an electric heating power of 5 W, and because of the low thermal capacity cools down to ambient temperature rapidly. This results in a measurement cycle faster than 60 s and a low carrier gas and electrical power consumption.

Pumping In general, microfluidic systems are typical for small flow rates and small holdups in the devices. This is also valid for micropumps, which have now a certain history of development and type variety [81]. Like conventional pumps, the micropumps can be classified into two different types, the displacement pumps with reciprocating or rotating elements, and the dynamic pumps using centrifugal, acoustic or electrical fields [82]. Due to the small dimensions, also the flow rate is small, but also small effects like electrohydrodynamic or magnetohydrodynamic effects can be employed. A pump is mainly characterized by the flow rate and the induced pressure difference for micro process applications. The frequency of displacement pumps is directly coupled with the flow rate up to a certain value, where sealing losses and dynamic effects like fluid acceleration are limiting and decrease the flow rate [81]. The flow rates of a micro diaphragm pump decrease linearly with increasing counterpressure due to the higher forces. Maximal flow rates of about 35 ml/min (0.78 kPa pressure difference) of air and 16 ml/min (4.9 kPa pressure difference) of water have been reported [82]. Flow oscillations caused by pumps are critical for the proper operation of microstructured devices like micro mixers or micro reactors and can be reduced by appropriate flow dampers.

Electro-osmosis employs the negative charge of the wall and corresponding positive charge of ions in the fluid. An external electrical field generates the motion of the counterions and therefore of the entire fluid in the capillary. The electro-osmotic effect decreases with increasing diameter of the channel, which have a characteristic dimension of about 1 μm. High pressure differences up to 2 MPa of water have been realized for low flow rates of 3.6 μm/min. High flow rates up to 33 ml/min have been measured that indicates that the main applications of electro-osmosis pumps are for dosing and analytical applications [82].

The rotating gear pump of HNP Mikrosysteme GmbH [83] belongs to the displacement pumps for dosing and pumping of liquids in analytical and production applications. The pumps, with an internal volume from 3 to 48 μl, deliver a flow rate up to 17 l/h against a counterpressure of 30 to 50 bar.

Fast Nucleation and Controlled Growth of Nanoparticles The small capacity and inertia with small system volume allows high change rates for heating, cooling, and mass transfer, see also Section 1.5.1 Characteristic Times. The fast mixing of two nearly saturated mixtures leads often to a high supersaturation of the mixture and particle generation, see Fig. 1.14, left side. The saturation S is measured with ratio of the partial pressure p_V to the saturation pressure p_S. In gaseous flow the particles (liquid or solid phase) are called aerosols, like fog in humid air. In liquid flow the particulate flow is called a suspension and is often combined with crystallization. Besides the exceeding of solubility limits in liquids, often also chemical reactions are involved in the precipitation process. Two solvable components in solution react to a third component with low solubility, which precipitates after the reaction.

The precipitation process is determined by the nucleation of the dissolved component and the following growth of the particles. A high supersaturation generates many nuclei in the mixture, see Fig. 1.14 right. The faster the mixing process, the more nuclei are formed. The homogeneous nucleation of the dissolved component is followed by the growth of the particles, which is determined by the mass transfer from the bulk to the nucleus.

Experimental investigations of the precipitation of barium sulfate in aqueous solution have shown that the particle size distribution strongly depends on the mixing intensity [86, 86a]. The investigated T-mixer has circular cross sections with an inlet diameter of 0.5 mm and an outlet diameter of 1 mm. The mean particle diameter decreases from about 200 nm for Re numbers of 255 to about 50 nm for a Re number of 6360. The authors set up a calculation model for the nucleation and growth of the particles depending on the flow regimes and interfacial energy within the T-mixer.

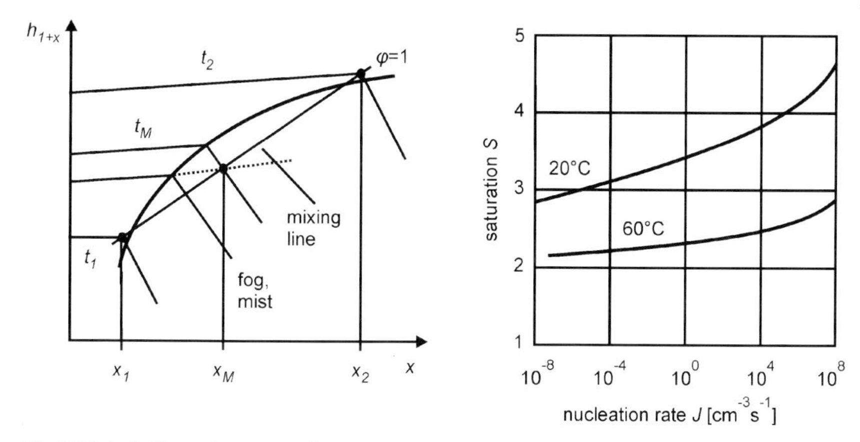

Fig. 1.14 Left: Saturation curve of a gas–vapor system, e.g. humid air system (fog generation); right: nucleation rate for the saturation at various temperatures [adopted from 84 and 85, Chapter Je].

In general, fast mixing in microstructures can be used for the production of nanoparticles in droplet flow, emulsions, and with fast chemical reactions and successful precipitation in bulk flow. The main problem is the sticking and adhesion of particles at the wall, see the fouling topic in Chapter 3, Section 4.

1.5.4 Other Topics

The microfabrication technology allows, under certain conditions, a cost-effective mass production and a dense integration of various elements. The integration of microstructured devices into the environment also needs conventional fabrication technologies and has to be considered during the device design. Many microfabrication techniques allow very complex geometries, a switching and a proper arrangement of various process steps.

The main drawbacks of microsystems for production applications and industrial use are the low mass flow rates and the sensibility to fouling and blocking. A high flow rate is achieved with internal and external numbering up. Fouling considerations can be found in Chapter 3.

Mechanical unit operations often include mechanical gears and motions with friction and wear that cause many problems in microsystems. Therefore, pure mechanical processes like pressing or crushing are not feasible and reasonable in micro process technology.

1.6 Conclusions and Outlook

Only a few physical principles and their potential for miniaturization have been demonstrated in this chapter. Schubert [87] gives a comprehensive collection of all known physical principles. Many of these principles are translated for design purposes by Koller [88] and illustrated by examples. This could be a starting point for further investigations. The aim should be the fruitful combination and connection between microsystem technology and process engineering.

The picture in this chapter and from the entire book is certainly not entirely complete, but is one of the first steps for a deeper understanding and a help for further applications of microsystem technology in process engineering and technology. Both sides can learn from each other and contribute to the progress. One motivation for process engineering can be summarized under the concept of process intensification. Additionally, processes in microstructured devices open new routes in mixing, transformation and separation with the help of system integration.

An emerging market is governed by a technology push from new possibilities and by a market pull from the consumer and application needs. In chemical technology and production a long-term business and complex production plants with decades of lifetime allow no easy change. The role of unit operations should be extended to integrated processes in microstructures to provide a comprehensive toolbox for treating various tasks from process engineering. Innovative reactor design adjusted to chemical reactions, optimized fluid flow, parallel and high throughput testing, new reaction routes, closer to the molecule and the transformation (time, rate, concentration) are the new applications of microstructured equipment in chemical industry, in energy conversion and fuel cells, and life science.

For the future research, many aspects need to be addressed like fluid flow and transport phenomena or two phase flow in capillary structures, like using small effects, fluctuations, surface effects and coupled processes, high throughput devices like mixers, premixers, reactors, heat exchangers, application of cost-effective fabrication technologies, and integrated processes like mixers with chemical reactions and heat exchangers. The integration of micro structures into macro equipment has to be carried out to use the benefits and opportunities of both fields and additional emergent effects.

References

1 W. R. A. Vauck, H. A. Müller, *Grundoperationen chemischer Verfahrenstechnik*, 11. Aufl., Deutscher Verlag für Grundstoffindustrie, Stuttgart, **2000**.

2 M. Gad-el-Hak (Ed.), *The MEMS Handbook*, CRC Press, Boca Raton, **2003**.

3 D. J. Young, C. A. Zormann, M. Mehregany, *MEMS/NEMS Devices and Applications*, in B. Bushan (Ed.), Springer Handbook of Nanotechnology, Springer, Berlin, **2004**, Chap. 8.

4 H. Fujita (Ed.), *Micromachines as Tools for Nanotechnology*, Springer, Berlin, **2002**.

4a H. Behrens, *Mikrochemische Technik*, Voss, Hamburg, **1900**.

5 F. Emich, *Lehrbuch der Mikrochemie*, Bergmann, Wiesbaden, **1911**.

6 R. Schütte, *Diffusionstrennverfahren*, in Ullmanns Encyklopädie der technischen Chemie, E. Bartholomé (Ed.), Vol. 2, Verfahrenstechnik I: (Grundoperationen), **1972**.

7 M. Baerns, H. Hofmann, A. Renken, *Chemische Reaktionstechnik*, Thieme, Stuttgart, **1992**.

8 E. Klemm, M. Rudek, G. Markowz, R. Schütte, *Mikroverfahrenstechnik*, in: Winnacker–Küchler: Chemische Technik – Prozesse und Produkte, Vol. 2, Wiley-VCH, Weinheim, **2004**, Chap. 8.

9 M. Hoffmann, M. Schlüter, N. Räbiger, *Experimental Investigation of Mixing in a Rectangular Cross-section Micromixer*, AIChE Annual Meeting, **2004**, paper 330b.

10 E. Blass, *Entwicklung verfahrenstechnischer Prozesse – Methoden, Zielsuche, Lösungssuche, Lösungsauswahl*, Springer, Berlin, **1997**.

11 T. Melin, R. Rautenbach, *Membranverfahren – Grundlagen der Modul- und Anlagenauslegung*, Springer, Berlin, **2004**.

12 E. U. Schlünder, *Einführung in die Stoffübertragung*, Vieweg, Braunschweig, **1997**.

13 R. H. Perry, D. W. Green, J. O. Maloney, *Perry's Chemical Engineers' Handbook*, McGraw-Hill, New York, **1998**.

14 P. A. Schweitzer (Ed.), *Handbook of Separation Techniques for Chemical Engineers*, McGraw-Hill, New York, **1988**.

15 J. Benitez, *Principles and Modern Applications of Mass Transfer Operations*, John Wiley, New York, **2002**.

16 P. G. de Gennes, F. Brochart-Wyart, D. Quéré, *Capillary and Wetting Phenomena Drops, Bubbles, Pearls, Waves*, Springer, New York, **2004**, Chapter 9.

17 V. Hessel, S. Hardt, H. Löwe, *Chemical Micro Process Engineering*, Wiley-VCH, Weinheim, **2004**.

18 V. Hessel, H. Löwe, A. Müller, G. Kolb, *Chemical Micro Process Engineering*, Wiley-VCH, Weinheim, **2005**.

19 A. Sommerfeld, *Vorlesungen über Theoretische Physik*, Band V, Thermodynamik und Statistik, Verlag Harry Deutsch, Thun, **1988**.

20 L. D. Landau, E. M. Lifschitz, *Lehrbuch der theoretischen Physik*, Band X, Physikalische Kinetik, Akademie-Verlag, Berlin, **1983**.

21 C. Cercignani, *The Boltzmann Equation and its Applications*, Springer, New York, **1987**.

22 J. H. Ferziger, H. G. Kaper, *Mathematical theory of transport processes in gases*, Elsevier, Amsterdam, **1972**.

23 M. Krafzyk, *Gitter-Boltzmann-Methoden: Von der Theorie zur Anwendung*, Habilitationsschrift, Technical University of Munich, **2001**.

24 Y. Sone, *Kinetic Theory and Fluid Dynamics*, Birkhäuser, Boston, **2002**.

25 L. D. Landau, E. M. Lifschitz, *Lehrbuch der theoretischen Physik*, Band VI, Hydrodynamik, Akademie-Verlag, Berlin, **1981**.

26 K. F. Knoche, F. Bŏsnjakovič: *Technische Thermodynamik, Teil II*, Steinkopff, Darmstadt, **1997**.

27 H. D. Baehr, K. Stephan, *Wärme- und Stoffübertragung*, Springer, Berlin, **2004**.

28 A. Bejan, *Advanced Engineering Thermodynamics*, Wiley, New York, **1997**.

29 R. Haase, *Transportvorgänge*, Steinkopff, Darmstadt, **1973**.

30 A. Schönbucher, *Thermische Verfahrenstechnik*, Springer, Berlin, **2002**.

31 D. Jou, J. Casas-Vazquez, G. Lebon, *Extended Irreversible Thermodynamics*, Springer, Berlin, **2001**.

32 J. Zierep, *Ähnlichkeitsgesetze und Modellregeln der Strömungslehre*, Braun, Karlsruhe, **1972**.

33 H. Wetzler, *Mathematisches System für die universelle Ableitung der Kennzahlen*, Hüthig, Heidelberg, **1987**.

34 A. Bejan, *Convection Heat Transfer*, Wiley, New York, **2004**.

35 M. Kaspar, *Mikrosystementwurf, Entwurf und Simulation von Mikrosystemen*, Springer, Berlin, **2000**.

36 S. D. Senturia, *Microsystem Design*, Kluwer, Boston, **2000**.

37 G. Gerlach, W. Dötzel, *Grundlagen der Mikrosystemtechnik*, Hanser, München, **1997**.

38 N.T. Nguyen, *Mikrofluidik – Entwurf, Herstellung und Charakterisierung*, Teubner, Stuttgart, **2004**.

39 N.T. Nguyen, S.T. Werely, *Fundamentals and Applications of Microfluidics*, Artech House, Boston, **2002**.

40 G.E. Karniadakis, A. Beskok, *Microflows – Fundamentals and Simulations*, Springer, New York, **2002**.

41 M. Faghri, B. Sunden (Eds.), *Heat and Fluid Flow in Microscale and Nanoscale Structures*, WIT Press, Southampton, **2004**.

42 S. Succi, *The Lattice Boltzmann Equation for Fluid Dynamics and Beyond*, Oxford University Press, New York, **2001**.

43 M. Krafzcyk, *Gitter-Boltzmann-Methoden: Von der Theorie zur Anwendung*, Habilitationsschrift TU München, **2001**, URL: www.inf.bauwesen.tu-muenchen.de/

44 H.K. Versteeg, W. Malalasekera, *An Introduction to Computational Fluid Dynamics – The Finite Volume Method*, Prentice Hall, Harlow, **1995**.

45 J.H. Ferziger, M. Peric, *Computational Methods for Fluid Dynamics*, Springer, Berlin, **1999**.

46 H.D. Bockhardt, P. Güntzschel, A. Poetschukat, *Grundlagen der Verfahrenstechnik für Ingenieure*, Deutscher Verlag für Grundstoffindustrie, Stuttgart, **1997**.

47 T. Fischer, Wavelet-Transformation von instationären Wirbeln und turbulenten Strömungsvorgängen, Diploma work University of Stuttgart, **1997**, URL: http://www.csv.ica.uni-stuttgart.de/homes/tf/diplom/all.html

48 N. Kockmann, P. Woias, Separation Principles in Micro Process Engineering, IMRET8, TK129a, **2005**.

49 S. Ookawara, D. Street, K. Ogawa, *Quantitative Prediction of Separation Efficiency of a Micro-Separator/Classifier by an Euler-Granular Model*, IMRET8, TK130c, **2005**.

50 S. Ookawara, N. Oozeki, K. Owara, *Experimental Benchmark of a Metallic Micro-Separator/Classifier with Representative Hydrocyclone*, IMRET8, TK129g, **2005**.

51 Little Things Factory, *Product catalogue*, **2005**. URL: *www.ltf-gmbh.de/deu/produkt/mikrofluidik/separieren/mikrozyklon.htm*

52 R. Jaeckel, G.W. Oetjen, *Molekulardestillation*, Chemie-Ingenieur-Technik, **1949**, 21, 169-208.

53 W. Jorisch, *Vakuumtechnik in der chemischen Industrie*, Wiley-VCH, Weinheim **1999**.

54 H. London (Ed.), *Separation of Isotopes*, Georges Newnes Ltd., **1961**.

55 G. Veser, *Chemical Engineering Science*, **2001**, 56, 1265–1273.

56 C.M. Miesse, C.J. Jensen, R.I. Masel, M.A. Shannon, M. Short, *Sub-millimeter Scale Combustion*, AIChE Journal, **2004**, 50, 3206.

57 S. Middelhoek, S.A. *Audet, Silicon Sensors*, TU Delft, Dep. of Electrical Engineering, **1994**.

58 G. Kluge, G. Neugebauer, *Grundlagen der Thermodynamik*, Spektrum Akademischer Verlag, Heidelberg, **1994**.

59 E.W. Becker, W. Ehrfeld, P. Hagmann, A. Maner, D. Münchmeyer, *Microelectronic Engineering*, **1986**, 4, 35–56.

60 MicroChemTec, URL: *www.microchemtec.de*

61 R. Barron, *Cryogenic Systems*, McGraw-Hill, New York, **1966**.

62 R.B. Peterson, *Microscale Thermophysical Eng.* **1998**, 2, 121–131.

63 A. Pais, *Raffiniert ist der Herrgott... Albert Einstein – Eine wissenschaftliche Biographie*, Spektrum Akademischer Verlag, Heidelberg, **2000**, Chapters 4 and 5.

64 R. Schmidt, *Challenges in Electronic Cooling*, ICMM2003-1001, 951–959, **2003**; S. Billat, H. Glosch, M. Kunze, F. Hedrich, J. Frech, J. Auber, H. Sandmaier, W. Wimmer, W. Lang, *SensAct A*, **2002**, 97–98, 125–130.

65 C.J. Geankoplis, *Transport Processes and Separation Process Principles*, Prentice Hall, Upper Saddle River, **2003**.

66 K.I. Sowata, K. Kusakabe, *Design of microchannels for use in distillation devices*, IMRET7, **2003**.

67 H. Fink, M.J. Hampe, *Designing and Constructing Microplants*, IMRET3, **1999**, 664-673.

68 S. Häberle, H.P. Schlosser, R. Zengerle, J. Ducree, *A Centrifuge-Based Microreactor*, IMRET8, 129f, **2005**.

69 E. U. Schlünder, F. Thurner, *Destillation, Absorption, Extraktion*, Thieme, Stuttgart, **1986**.

72 S.H. Cypes, J.R. Engstrom, *Chem. Eng. J.* **2004**, 101, 49–56.

71 T.A. Ameel, I. Papautsky, R.O. Warrington, R.S. Wegeng, M.K. Drost, *J. Propulsion Power*, **2000**, 16, 577–582.

72 V.S. Stenkamp, W. teGrotenhuis, *Microchannel Absorption for Portable Heat Pumps*, IMRET8, 129e, **2005**.

73 J.D. Tice, H. Song, A.D. Lyon, R.F. Ismagilov, *Langmuir*, **2003**, 19, 9127–9133.

74 A, Günther, S.A. Khan, M. Thalmann, F. Trachsel, K.F. Jensen, *Lab-on-a-Chip*, **2004**, 4, 278–286.

75 W. Ehrfeld, V. Hessel, H. Löwe, *Microreactors*, Wiley-VCH, Weinheim, **2000**.

76 Y. Okubo, M. Toma, H. Ueda, T. Maki, K. Mae, *Chem. Eng. J.* **2004**, 101, 39–48.

77 W.E. teGrotenhuis, R. Cameron, M.G. Butcher, P.M. Martin, R.S. Wegeng, *Micro Channel Devices for Efficient Contacting of Liquids in Solvent Extraction*, Separation Science and Technology, PNNL-SA-28743, **1998**.

78 K. Kusakabe, K.I. Sotowa, H. Katsuragi, *Internal flow in a droplet formed in a microchannel contactor for solvent extraction*, IMRET7, **2003**.

79 A. Tuchbreiter, J. Marquardt, B. Kappler, J. Honerkamp, M.O. Kristen, R. Mülhaupt, *Macromol. Rapid Commun.* **2003**, 24, 48–62.

80 SLS MicroTechnology GmbH, **2005**, URL: www.sls-micro-technology.de

81 P. Woias, *Sens. Act. B*, **2005**, 105, 28–38.

82 D.J. Laser, J.G. Santiago, *J. Micromech. Microeng.* **2004**, 14, R35–R64.

83 HNP Mikrosysteme GmbH, **2005**, URL: www.hnp-Mikrosysteme.de

84 K. Schaber, *Thermodynamik disperser Systeme*, Universtäts-Skriptum, **2005**, URL: www.ttk.uni-karlsruhe.de/scripten/pdf/tds.pdf

85 H. Kraussold (Ed.), *VDI-Wärmeatlas*, VDI-Verlag, Düsseldorf, **1997**.

86 H.C. Schwarzer, W. Peukert, *AIChE J.* **2004**, 50, 3234–3247.

86a H.-C. Schwarzer, *Nanoparticle Precipitation – An Experimental and Numerical Investigation Including Mixing*, Logos, Berlin, **2005**.

87 J. Schubert, *Dictionary of Effects and Phenomena in Physics Descriptions, Applications, Tables*, VCH, Weinheim, **1987**.

88 R. Koller, *Konstruktionslehre für den Maschinenbau*, Springer, Berlin, **1998**.

2
Momentum and Heat Transfer in Microsized Devices

Heinz Herwig, Technische Thermodynamik, Technische Universität Hamburg-Harburg, Hamburg, Germany

Abstract
Microsized devices with characteristic flow passages that are less than 1 mm wide, but not so small that molecular effects would have to be accounted for, are analyzed systematically with respect to the appropriate way of modelling momentum and heat transfer. Based on the dimensional analysis of the underlying continuum equations (Navier–Stokes and first law of thermodynamics) scaling effects can be identified that characterize flow and heat transfer in small, microsized devices compared to the same geometries but on a macrosized scale. As far as heat transfer is concerned variable property effects on the microscale turn out to be more important than in conventional macrosized devices. Special aspects that are dealt with in detail are heat-transfer enhancement in bent conduits and the analysis of momentum and heat transfer by the second law of thermodynamics, from which the entropy generation in these flows can be determined.

Keywords
Microsized devices, dimensional analysis, scaling effects, variable properties, entropy generation

Advanced Micro and Nanosystems Vol. 5. Micro Process Engineering. Edited by N. Kockmann

ISBN: 3-527-31246-3

2.1 Introduction

2.1.1 The Continuum Approach

Within a continuum-theory approach the widely accepted definition of micro-structured devices in which momentum and heat transfer occur due to forced convection is a characteristic flow passage that is less than 1 mm wide. Lower bounds are due to the occurrence of noncontinuum effects like a velocity slip or a temperature jump at the wall that arise when the ratio of a typical length characterizing the molecular fluid structure and a characteristic width of the flow passage exceeds a critical value.

For a gas flow this critical value can be the Knudsen number $\mathrm{Kn}=\Lambda/l$ that should be less than 10^{-3} for a continuum approach to be appropriate to analyze convective heat and momentum transfer problems, see for example [1]. Here Λ is the mean free path length of the molecules and l a characteristic channel height. With $\Lambda \approx 5\times10^{-8}$ m (gas at standard pressure and temperature) and $\mathrm{Kn}=10^{-3}$ the critical length for l is ≈ 50 μm and noncontinuum effects might occur for a characteristic channel height l that is less than this critical value.

Often the first effect of noncontinuum when l is decreased is assumed to be the violation of the no-slip condition at the wall, see for example [2]. Then it is assumed that a modified continuum approach with a so-called slip boundary condition is appropriate for $10^{-3} < \mathrm{Kn} < 10^{-1}$. With $\Lambda \approx 5\times10^{-8}$ m this corresponds to characteristic channel dimensions l between 50 μm and 0.5 μm.

For liquids there is no free mean path length Λ and thus the Knudsen number cannot be used to identify continuum flow regimes. However, a general interaction length of molecules exists for liquids being roughly ten times smaller than Λ for gases (since the density of liquids is about 10^3 times larger than that of gases). Thus, as a first approximation it can be assumed that the critical length l for which noncontinuum effects occur is ten times smaller for flows of liquids compared to that of gases.

Typical diameters or channel heights in microsized devices are of the order of 100–500 μm, and thus are well in the range for which a continuum approach is appropriate.

2.1.2
From Macro to Micro Dimensions

When the characteristic length scale l of a certain problem is decreased below 1 mm one by definition leaves the "macroworld" and enters the "microworld" of momentum and heat transfer. If, however, nondimensional quantities are introduced by referring all lengths to l and all other quantities like velocities and temperature to w_c and ΔT, respectively, the question arises: What is "macro", what is "micro"? Since now all nondimensional quantities are of order one, irrespective of whether l is in the macro- or the microrange, there is no formal distinction between "small" (=micro) and "large" (=macro) values of the same quantity.

On that background the crucial question arises: Are the physics of a certain problem the same for macro and micro dimensions, i.e. do they have the same representation in a nondimensional form? Or, asked in a similar way: Is everything in the microworld the same as we know it from the macroworld, only just smaller?

The answer is: No, because there may be scaling effects. These are effects that can be neglected in macrosized channels but have a strong influence in microchannels. Each of them is related to a certain nondimensional group, with a different magnitude of this group for macro- and microflow situations.

These scaling effects can be identified after a problem has been subjected to a dimensional analysis, revealing which nondimensional groups govern the physics of this problem.

2.2
Dimensional Analysis and Scaling Effects

2.2.1
Dimensional Analysis of a Problem

Physical problems can and should be described in nondimensional variables because then more general solutions are possible compared to solving a problem in its dimensional form.

When a mathematical model of the physical problem, for example a set of differential equations, exists, nondimensionalization is straightforward. With a set of reference quantities that are characteristic with respect to the physics of the problem the basic equations can be cast into their nondimensional form. As part of this procedure combinations of the reference quantities and physical properties appear as nondimensional groups in the equations and/or the boundary conditions. These groups are often named after famous researchers like the Reynolds, Mach and Nusselt numbers.

The important step in this procedure is to select characteristic reference quantities. They should characterize the geometry as well as the process within it, when one and/or the other is changed by orders of magnitude. If, for example, a flow through a channel is nondimensionalized, then $w_c = \dot{m}/\rho A$, i.e. the mean velocity in the cross section A, would be a characteristic velocity since it vanishes when the mass flux $\dot{m}$ goes to zero. Alternatively cw_c, i.e. some fixed multiple of w_c might be used as reference velocity with the consequence that then the factor c would appear in the equations. However, what should not be used is for example $w_c = \sqrt{p_0/\rho}$ with p_0 being some typical pressure in the channel and ρ the density. Though the units of $\sqrt{p_0/\rho}$ are m/s, i.e. that of a velocity, this combination is not characteristic for velocities that occur in the flow for different process situations, i.e. for different mass fluxes in the channel.

When the mathematical model of a physical problem is not known, however, one can nevertheless determine the nondimensional groups of the problem. Then, one has to decide which physical quantities will affect the process. Collecting all relevant quantities corresponds to deciding which mathematical/physical model would be appropriate to describe the problem, even though its exact mathematical form does not yet exist. Finding this *list of relevant physical quantities* is the crucial step of the procedure based on Buckingham's so-called Pi-theorem, for example described in [3].

According to the Pi-theorem a number of $(m-n)$ dimensionless groups exist for a problem. Here, m is the number of relevant quantities and n the number of basic dimensions that appear in them, like length, mass, temperature, ... The dimensionless groups can either be found by a formal but very laborious procedure or simply by try and error. Since their number is known, $(m-n)$, one only has to combine the relevant quantities such that dimensionless groups appear. Once it is checked that the $(m-n)$ groups are independent of each other (i.e. not multiples of two other groups, for example), one has found a set of nondimensional groups that are the nondimensional parameters of the solution.

2.2.2 Scaling Effects in the Dimensional Analysis

When a physical/mathematical model for a certain problem exists, either in terms of a set of differential equations or in terms of a number of dimensionless groups (with a yet unknown relation between these groups) the question arises, whether this model is appropriate for the problem on very different geometrical scales, i.e. for example for macro as well as for micro dimensions. If a model has been developed in a macroscale situation, for example, it is not automatically appropriate for microscales. The reason is that certain effects might well have been neglected in the macro situation but might be of importance for microscales. If one had started on the microscale level one would have taken these effects as relevant effects into consideration (when collecting all relevant quantities for finding dimensionless groups of a problem). They then would turn out to be irrelevant when applying the model to macroflows. From these considerations scaling effects can be defined as those *effects that gain or lose their relevance when the geometrical length scales of the problem are changed by orders of magnitude.*

With respect to dimensionless groups two kinds of scaling effects may occur:

- Dimensionless groups gain very different numbers when geometrical length scales are changed by orders of magnitude.
- New dimensionless groups come in or existing groups drop out of a physical/mathematical model, when the geometrical length scales are changed by orders of magnitude.

Note that scaling effects always refer to a change in geometrical length scales within an existing physical/mathematical model.

2.2.3 Identifying Scaling Effects when Changing from Macro to Micro Dimensions

In order to be more specific a particular flow and heat-transfer problem can be compared for a macro- and a microflow situation. When certain process and boundary conditions are the same for both (macro and micro), one immediately gets the ratio "micro/macro" for all other quantities. What is kept constant is somehow arbitrary, however. From such numbers it follows which dimensionless groups may change their order of magnitude, come in or drop out, when length scales change from macro to micro dimensions.

In Fig. 2.1 typical numbers are given for the parameters of a simple channel flow with heat transfer for a macro and a micro situation. These numbers can be inserted into the nondimensional groups to identify scaling effects when the geometrical scale of the channel is changed from the macro to the micro size.

In Table 2.1 the dimensionless groups are collected that determine nondimensional heat transfer results in a *macrostandard analysis*. In addition, those groups are listed that might represent scaling effects with respect to this macrostandard analysis.

From the typical numbers in Fig. 2.1 we now can conclude how the dimensionless groups behave when scales are changed from the macro to the micro size. Together with the information for which limit ($\rightarrow 0$ or $\rightarrow \infty$) the physics associated with a certain nondimensional group have "no effect", we can immediately decide whether it is a scaling effect or not.

According to this procedure there are six definite scaling effects:

1. $\mathrm{Re} = O(1)$: Different from macroscale channel flows the Reynolds number is small, i.e. $\mathrm{Re} = O(1)$ instead of $\mathrm{Re} \rightarrow \infty$.
2. $\mathrm{Pe} \rightarrow 0$: The Peclet number is associated with axial heat conduction effects that thus may be important in microscale flows.

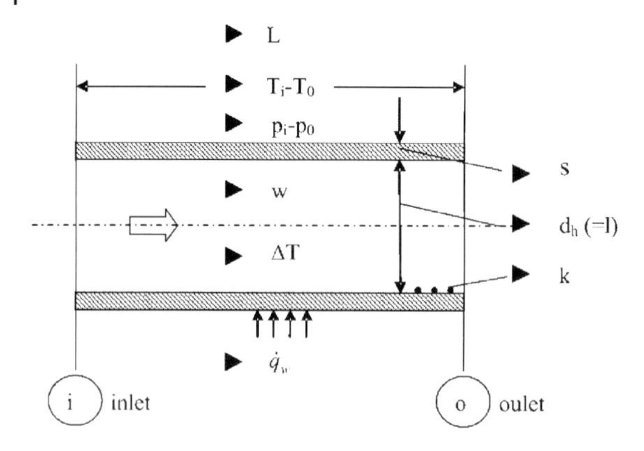

Parameter	Macro	Micro	Micro/macro
Pipe length L	10^{-1} m	10^{-2} m	0.1
Hydraulic diameter D_h (=l)	10^{-2} m	10^{-4} m	0.01
Wall thickness s	10^{-2} m	10^{-2} m	1
Wall roughness height k	10^{-4} m	10^{-6} m	0.01
Mean velocity w	1 m/s	1 m/s	1
Heat flux density $\dot{q}_w$	10 W/m^2	10 W/m^2	1
Pressure drop p_i–p_O	(10^{-2} mbar)	(10 mbar)	1000 [1)]
Temperature difference ΔT (wall–bulk)	(1 °C)	(10^{-2} °C)	0.01 [2)]
Temperature increase T_O–T_i	(1 °C)	(10 °C)	10
$\Delta T/(T_O$–$T_i)$	1	10^{-3}	0.001 [3)]

Fig. 2.1 Typical numbers for the dimensional parameters of macro and micro pipe flow with heat transfer. Note: Numbers in this table are only orders of magnitude; (...) typical for gases, 1) based on $\Delta p = \frac{64}{\mathrm{Re}} \frac{L}{l} \frac{\rho}{2} w^2$, 2) based on $\mathrm{Nu} = \frac{\dot{q}_w l}{\lambda \Delta T} = \text{const.}$, 3) based on $\dot{m} c_p (T_O - T_i) = \dot{q}_w \pi d_h L$.

3. $\mathrm{Ec} \to \infty$: The Eckert number is associated with viscous dissipation effects that thus may not be neglected in microscale flows.
4. $\Lambda_s \to \infty$: Compared to the characteristic channel size $l (= d_h)$ the wall thickness often is large so that a heat-transfer problem is almost always a conjugate problem (convective heat transfer plus heat conduction in the wall).
5. $K_{zTa} \to \infty$: Since often there is a strong streamwise increase in temperature, variable property effects due to this rise in bulk temperature have to be taken into account.
6. $K_{zp\rho} \to \infty$: For gases the density ρ strongly depends on pressure (and temperature, see (5)). Since in microsized flow situations the pressure

Table 2.1 Nondimensional heat transfer correlation $\mathrm{Nu} \equiv \dot{q}_w l/(\lambda \Delta T) = \mathrm{Nu}\,(\ldots)$, a: general physical property, $a = \rho, \eta, \lambda, c_p$; ref. Eqs. (2.7), (2.22).

$$\mathbf{Nu} = \mathbf{Nu}\,(\underbrace{x^*, \mathbf{Re}, \mathbf{Gr}, \mathbf{Pr}}_{\text{macrostandard}}, \underbrace{\mathbf{Pe}, \mathbf{Ec}, \mathbf{Ma}, \Lambda_k, \Lambda_s, K_{nij}, K_{zij}}_{\text{possible scaling effects}})$$

Dimensionless group	Definition	Micro/macro	
Dimensionless coordinate	$x^* = \frac{x}{l}$	$\to \infty$	Macro standard analysis
Reynolds number	$\mathrm{Re} = \frac{wl}{\nu}$	$\to 0$	
Grashof number	$\mathrm{Gr} = \frac{g\beta\Delta T l^3}{\nu^2}$	$\to 0$	
Prandtl number	$\mathrm{Pr} = \frac{\nu}{a}$	$O(1)$	

Dimensionless group	Definition	Micro/macro	No effect for	Scaling effect?
Peclet number	$\mathrm{Pe} = \frac{wl}{a}$	$\to 0$	$\mathrm{Pe} \to \infty$	Yes
Eckert number	$\mathrm{Ec} = \frac{w^2}{c_p \Delta T}$	$\to \infty$	$\mathrm{Ec} \to 0$	Yes
Mach number	$\mathrm{Ma} = \frac{w}{c}$	$O(1)$	$\mathrm{Ma} \to 0$	No
Wall roughness	$\Lambda_k = \frac{k}{l}$	$O(1)$	$\Lambda_k \to 0$	?
Wall thickness	$\Lambda_s = \frac{s}{l}$	$\to \infty$	$\Lambda_s \to 0$	Yes
Normal temperature coefficient K_{nTa}	$= \frac{\Delta T}{T}\left[\frac{\partial a}{\partial T}\frac{T}{a}\right]$	$\to 0$	$K_{nTa} \to 0$	No
Axial temperature coefficient K_{zTa}	$= \frac{T_0 - T_i}{T}\left[\frac{\partial a}{\partial T}\frac{T}{a}\right]$	$\to \infty$	$K_{zTa} \to 0$	Yes
Axial pressure coefficient $K_{zp\rho}$	$= \frac{p_i - p_0}{p}\left[\frac{\partial \rho}{\partial p}\frac{p}{\rho}\right]$	$\to \infty$	$K_{zpp} \to 0$	Yes

drop can be strong the density/pressure relation has to be accounted for. This may result in a strong downstream increase in Mach number, whereas otherwise Ma often is small ($\mathrm{Ma} \to 0$).

In Table 2.1 the only open question is left with the influence of wall roughness since it may not be appropriate to associate with it only one single parameter k (mean roughness height), neglecting its distribution and the specific form of the roughness elements.

2.3
Basic Equations in Microsized Devices

2.3.1
Continuum Model

Within a continuum approach it is assumed that the physics of a problem can be modelled by differential equations for the field variables. This implies that the derivatives of all dependent variables exist in space and time. Therefore local and instantaneous properties such as the density, pressure and velocity are defined as averages over spatial elements and time spans large compared with characteristic molecular length and time scales but small enough in comparison to the corresponding scales for the macroscopic phenomena to permit the use of differential equations. For a gas, for example, the length and time scales on the molecular level may be the mean free path length and the inverse of the molecular collision frequency. Macroscopic scales may be $a/|\partial a/\partial z|$ and $a/|\partial a/\partial t|$ for any dependent macroscopic property a.

As mentioned in Section 2.1 already, a continuum model is appropriate for Knudsen numbers $\mathrm{Kn} < 10^{-3}$ when the fluid is a gas and for even smaller macroscopic scales when the fluid is a liquid.

The macroscopic differential equations for mass, momentum and energy will be presented here without derivation. For details the reader is referred to advanced textbooks in fluid mechanics like [4–6].

The subsequent equations are based on the conservation of mass, momentum and energy in infinitesimal control volume elements. These balance equations only form a determinate set of equations (number of unknowns = number of equations) after certain constitutive and thermodynamic equations of state have been introduced. These are:

- thermal and caloric equations of state for the fluid under consideration i.e. $\rho = \rho(T, p)$, $h = h(T, p)$.
- stress/strain and heat flux/temperature gradient relations as so-called constitutive equations. When they are chosen as linear isotropic relations with constant coefficients η (viscosity) and λ (thermal conductivity) this implies *Newtonian fluid* behavior and thermodynamic processes not too far away from thermodynamic equilibrium.

2.3.2
Navier–Stokes and Energy Equations

With the above made assumptions the basic equations are the so-called *Navier–Stokes equations* supplemented by the *energy equation*. They are given in vector notation here and for special cases in Cartesian and cylindrical coordinates afterwards.

Continuity equation:

$$\frac{D\rho}{Dt} + \rho \operatorname{div} \vec{w} = 0 \tag{2.1}$$

Momentum equation (Newtons law):

$$\rho \frac{D\vec{w}}{Dt} = \rho\vec{g} - \operatorname{grad} p + \operatorname{Div}\left[\eta\left(2\nabla\vec{\mathbf{w}} - \frac{2}{3}\vec{\boldsymbol{\delta}} \operatorname{div} \vec{w}\right)\right] \tag{2.2}$$

Energy equation (first law of thermodynamics):

$$\rho \frac{Dh^{+}}{Dt} = -\operatorname{div} \vec{q} + \rho \vec{w} \cdot \vec{g} + \hat{D} + \frac{\partial p}{\partial t} \tag{2.3}$$

Here $\nabla \vec{\mathbf{w}}$ is the velocity gradient tensor and $\vec{\boldsymbol{\delta}}$ the Kronecker unit tensor. In the energy equation $h^{+} = h + \rho \vec{w}^2/2$ is the total enthalpy and $\hat{D}$ is a group of diffusion terms.

Equations (2.1)–(2.3) from an Eulerian viewpoint hold in a fixed frame of reference (coordinate system) so that the substantial derivatives on the left-hand sides comprise local and convective derivatives $(D\ldots/Dt = \partial\ldots/\partial t + \vec{w} \cdot \operatorname{grad}\ldots)$.

Equation (2.3) expresses the conservation of the total energy that is the sum of its mechanical and thermal parts. Only when both parts are balanced individually a dissipation term Φ occurs explicitly. Then Φ describes the redistribution of energy between its mechanical and thermal (internal) parts. The individual energy equations (the sum of which is (2.3)) are:

mechanical energy:

$$\rho \frac{D\vec{w}^2/2}{Dt} = \rho \vec{w} \cdot \vec{g} + \hat{D} + \frac{\partial p}{\partial t} - \frac{Dp}{Dt} - \Phi \tag{2.3 a}$$

thermal (internal) energy:

$$\rho \frac{Dh}{Dt} = -\operatorname{div} \vec{q} + \frac{Dp}{Dt} + \Phi \tag{2.3 b}$$

Here Φ is the dissipation of mechanical energy that is accompanied by entropy production, see Section 2.7. It reads

$$\Phi = \vec{\boldsymbol{\tau}} : \nabla\vec{\mathbf{w}} \quad \text{with} \quad \vec{\tau} = \left[\eta\left(2\nabla\vec{\mathbf{w}} - \frac{2}{3}\vec{\boldsymbol{\delta}} \operatorname{div} \vec{w}\right)\right] \tag{2.4}$$

and is the inner product of the two tensors $\vec{\boldsymbol{\tau}}$ (viscous stress tensor) and $\nabla\vec{\mathbf{w}}$ (velocity gradient tensor).

The boundary conditions for the set of differential equations are

- no slip; zero normal velocity at the wall (if there is neither suction nor blowing)
- no temperature jump; temperature or/and temperature gradients prescribed at the wall.

2.3.3
The Role of Density

In the above equations density is assumed to be a variable quantity, which sometimes is associated with the term *compressibility*. In order to get a clear picture in connection with variable-density effects the total differential of $\rho(T, p)$, i.e.

$$d\rho = \left(\frac{\partial \rho}{\partial T}\right) dT + \left(\frac{\partial \rho}{\partial p}\right) dp \tag{2.5}$$

should be analyzed. Nondimensionalized with reference values $\rho_r = \rho_r(T_r, p_r)$ and introducing nondimensional temperature and pressure *differences* it reads ($\rho^* = \rho/\rho_r$; $\hat{T}^* = (T - T_r)/T_r$; $\hat{p}^* = (p - p_r)/p_r$):

$$d\rho^* = K_{\rho T}\, d\hat{T}^* + K_{\rho p}\, d\hat{p}^* . \tag{2.6}$$

Here

$$K_{\rho T} = \left[\frac{T}{\rho}\frac{\partial \rho}{\partial T}\right]_r ; \quad K_{\rho p}\left[\frac{p}{\rho}\frac{\partial \rho}{\partial p}\right]_r \tag{2.7}$$

are properties of the fluid and indicate, how strongly the density of a fluid is affected by changing temperature or pressure.

From Eq. (2.6) it follows immediately under which conditions non-negligible density variations ($d\rho^* \neq 0$) occur. Here, $\neq 0$ means "non-negligible":

- $K_{\rho T}\, d\hat{T}^* \neq 0$: When the fluid can change density due to a change in temperature ($K_{\rho T} \neq 0$; ideal gas: $K_{\rho T} = -1$) then $dT^* \neq 0$ results in $d\rho^* \neq 0$. This, for example, is the case for strong heat transfer with gas as a fluid.
- $K_{\rho p}\, d\hat{p}^* \neq 0$: When the fluid can change density due to a change in pressure $K_{\rho p} \neq 0$; ideal gas: $K_{\rho p} = 1$) then $dp^* \neq 0$ results in $d\rho^* \neq 0$. When the equations that model the flow situation can be nondimensionalized with $(\rho w^2)_r$, i.e. fluid inertia effects are important, $\hat{p}^*$ can be rewritten:

$$\hat{p}^* = C\,\mathrm{Ma}^2 p^*; \; p^* = \frac{p - p_r}{(\rho w^2)_r} \tag{2.8}$$

with the Mach number $\mathrm{Ma} = w_r/c_r$ and $C = \rho_r c_r^2/p_r$ which is the isentropic exponent $C = \kappa_r$ when the fluid is an ideal gas.

When p^* is the appropriate nondimensionalization density changes $\neq 0$ occur for Ma $\neq 0$ according to Eq. (2.8). Often, Ma $= 0.3$ is assumed as a threshold above which a flow should be treated as a *compressible* flow, i.e. as a flow that is non-negligibly affected by variable density effects. That means, however, that Ma < 0.3 is only a necessary condition for treating a flow as incompressible ($\rho =$ const.), since density may also vary due to temperature effects.

2.4 Basic Equations in Microsized Slender Channels

Often, geometries in microsized devices are of the so-called slender-channel type, i.e. their streamwise dimensions are much larger than the dimensions perpendicular to it. A typical geometry of that kind is a tube or a plane channel in a microsized device.

For macrosized channels this situation in addition to the geometrical particularity of slenderness is accompanied by high Reynolds numbers. This should be kept in mind when the slender-channel transformation, often used in macroflow situations, is applied next. This transformation takes into account that streamwise and normal coordinates due to the geometry of slender channels gain very different numbers.

Table 2.2 shows the nondimensional and retransformed variables for slender-channel geometries that are assumed to be either 2-dimensional or rotationally symmetric. In addition, a steady and incompressible flow is assumed for which viscosity and thermal conductivity may change with temperature.

The nondimensional equations for these flows follow from Eqs. (2.1)–(2.3). They read with $i=0$ for the 2-D case (Cartesian coordinates) and $i=1$ for rotational symmetry (cylindrical coordinates), for details see ref. [7]:

continuity equation:

$$\frac{\partial w^*}{\partial z^*} + \frac{1}{n^{*i}} \frac{\partial (n^{*i} v^*)}{\partial n^*} = 0 \tag{2.9}$$

z-momentum equation:

$$w^* \frac{\partial w^*}{\partial z^*} + v^* \frac{\partial w^*}{\partial n^*} = -\frac{\partial p^*}{\partial z^*} + \frac{1}{n^{*i}} \frac{\partial}{\partial n^*} \left[n^{*i} \eta^* \frac{\partial w^*}{\partial n^*} \right] + \underline{\frac{1}{\mathrm{Re}^2} \frac{\partial}{\partial z^*} \left[\eta^* \frac{\partial w^*}{\partial z^*} \right]} \tag{2.10}$$

n-momentum equation:

$$\underline{\frac{1}{\mathrm{Re}^2} \left[w^* \frac{\partial v^*}{\partial z^*} + v^* \frac{\partial v^*}{\partial n^*} \right]} = -\frac{\partial p^*}{\partial n^*} + \underline{\frac{1}{\mathrm{Re}^2} \left[\frac{\partial}{\partial n^*} \left(\frac{\eta^*}{n^{*i}} \frac{\partial}{\partial n^*} (n^{*i} v^*) \right) + \frac{\partial}{\partial z^*} \left(\eta^* \frac{\partial v^*}{\partial z^*} \right) \right]} \tag{2.11}$$

Table 2.2 Nondimensional variables for slender geometries two-dimensional: $n = y$, $l = H$; rotationally sym.: $n = r$, $l = R$.

z^*	n^*	w^*	v^*	p^*	T^*	η^*	λ^*	Φ^*
$\frac{z}{l\mathrm{Re}}$	$\frac{n}{l}$	$\frac{w}{w_\mathrm{r}}$	$\frac{v\mathrm{Re}}{w_\mathrm{r}}$	$\frac{p-p_\mathrm{r}}{\rho w_\mathrm{r}^2}$	$\frac{T-T_\mathrm{r}}{\Delta T}$	$\frac{\eta}{\eta_\mathrm{r}}$	$\frac{\lambda}{\lambda_\mathrm{r}}$	$\frac{\Phi l^2 \mathrm{Re}^2}{w_\mathrm{r}^2}$

thermal energy:

$$w^* \frac{\partial T^*}{\partial z^*} + v^* \frac{\partial T^*}{\partial n^*} = \frac{1}{\mathrm{Pr}} \frac{1}{n^{*i}} \frac{\partial}{\partial n^*} \left[n^{*i} \lambda^* \frac{\partial T^*}{\partial n^*} \right] + \underline{\frac{1}{\mathrm{Pr}\mathrm{Re}^2} \frac{\partial}{\partial z^*} \left[\lambda^* \frac{\partial T^*}{\partial z^*} \right] + \frac{\mathrm{Ec}}{\mathrm{Re}^2} \Phi^*} \tag{2.12}$$

with the nondimensional groups Re, Pr and Ec from Table 2.1 that here are

$$\mathrm{Re} = \frac{\rho w_\mathrm{r} l}{\eta_\mathrm{r}}; \ \mathrm{Pr} = \frac{\eta_\mathrm{r}}{c_\mathrm{p} \lambda_\mathrm{r}}; \ \mathrm{Ec} = \frac{w_\mathrm{r}^2}{c_\mathrm{p} \Delta T} \tag{2.13}$$

For macrosized channels the Reynolds number often is high so that then all terms in Eqs. (2.9)–(2.12) that are multiplied by $(1/\mathrm{Re}^2)$ are asymptotically small and can be neglected for $\mathrm{Re} \to \infty$. In particular, the n-momentum equation then reduces to $\partial p^*/\partial n^* = 0$ and is no longer accounted for.

For microflows, however, the Reynolds number is an order one quantity and $1/\mathrm{Re}^2$ is not small. Therefore, if $\mathrm{Re} \approx 1$ and the flow is not a priori assumed to be fully developed, the full Navier–Stokes equations have to be considered. Also, the v-velocity component is not negligibly small! Only when Reynolds numbers are well above $\mathrm{Re} \approx 1$ might one take the slender-channel equations into consideration for flows in microsized devices. As far as the energy equation is concerned viscous dissipation cannot be neglected. Not only is $1/\mathrm{Re}^2 = O(1)$ in the viscous dissipation term, it is also multiplied by the Eckert number Ec, which for microflows is much larger than for macroflows, see Table 2.1.

The basic equations only have been provided for laminar flow assuming the Reynolds number always to be well below the critical Reynolds number (transition to turbulence).

For internal flows critical Reynolds numbers are of the order of 10^3 when Re is defined with a mean velocity and a characteristic wall-to-wall distance. The critical Reynolds number $\mathrm{Re}_\mathrm{c} = w_\mathrm{m} l/v$ with w_m as cross section averaged velocity is for a:

- circular pipe flow: $\mathrm{Re}_\mathrm{c} \approx 2300$ (l: diameter)
- plane channel flow: $\mathrm{Re}_\mathrm{c} \approx 2000$ (l: distance of the walls)
- plane Couette flow: $\mathrm{Re}_\mathrm{c} \approx 1800$ (l: distance of the walls)

There is some controversy whether critical Reynolds numbers for microdevices are smaller than those for macrogeometries, see for example [8], the order of magnitude ($O(\mathrm{Re}_c) = 10^3$), however, is generally accepted.

2.5 Streamtube Approximations (1-D)

An approximate method to calculate momentum and heat transfer in conduits is the so-called streamtube approach. A streamtube is a slender control volume with an entrance cross section 1, an exit cross section 2, and stream surfaces as side walls. Therefore, by definition (of a streamline or a stream surface) there is no mass flux across the sidewalls.

In the streamtube approach it is also assumed that all quantities are constant across each cross section of the streamtube. They may, however, change in the streamwise direction. When the flow in a conduit (tube or channel) is approximately described by the streamtube approach, the conduit is assumed to be the streamtube, since impermeable walls have the same property as stream surfaces have (though at walls there are no streamlines since due to the no-slip condition velocity is zero at the walls).

When a coordinate is introduced that follows the centerline of the streamtube the flow is one-dimensional in that coordinate. The conservation equations then no longer are differential equations though they can systematically be deduced from the basic equations (2.1)–(2.3), see for example [6] for details of this derivation.

The constant value of a fluid or flow property at a cross section can be interpreted as the mean value of that property with respect to the cross section. A velocity w=const. in a certain cross section, for example, in that sense is either

- the mean value of a velocity profile, which in the real flow is not constant (due to the no-slip condition at the wall, for example), or
- a plug flow profile that is assumed to exist in the cross section.

2.5.1 Streamtube Equations for Constant Properties

With the coordinate z directed along the conduit the basic equations for flow and heat transfer between two cross sections 1 and 2 in a dimensional form and for constant properties, i.e. also for constant density, are:

continuity equation:

$$\rho w_2 A_2 = \rho w_1 A_1 \tag{2.14}$$

momentum equation/mechanical energy equation:

$$\frac{w_2^2}{2} + \frac{p_2}{\rho} + g y_2 = \frac{w_1^2}{2} + \frac{p_1}{\rho} + g y_1 + w_{t12} - \varphi_{12} \tag{2.15}$$

thermal energy equation:

$$e_2 = e_1 + q_{12} + \varphi_{12} \tag{2.16}$$

Equation (2.15) is an extended (by w_{t12} and φ_{12}) version of the so-called *Bernoulli equation*. It is a momentum as well as a mechanical energy balance since the mechanical energy equation emerges when the momentum equation is multiplied by the flow velocity (scalar product in the original vector form).

The three process quantities that appear in the conservation equations are:

- w_{t12}: specific technical work, increasing ($w_{t12} > 0$, pump) or decreasing ($w_{t12} < 0$, turbine) the mechanical energy between 1 and 2
- q_{12}: specific heat, increasing ($q_{12} > 0$, heating) or decreasing ($q_{12} < 0$, cooling) the internal energy e between 1 and 2
- φ_{12}: specific viscous dissipation, equally decreasing the mechanical energy and increasing the internal energy

For microsized pumps or turbines the occurrence of w_{t12} is as in the macrosized case. Heating or cooling microdevices also is a standard situation in which, however, one carefully has to make sure that q_{12} is definitely transferred between 1 and 2. Often heat conduction in solid walls occurs and should be accounted for when the effective heat flux is determined that enters or leaves the flow between the cross sections 1 and 2.

Viscous dissipation is directly linked to a pressure drop that may be very high in narrow conduits. With a constant cross section A and for constant flow levels y a streamwise pressure drop is due to viscous dissipation and $\varphi_{12} = (p_1 - p_2)/\rho$ holds. In macroflow situations φ_{12} often is approximately determined by accounting for individual contributions of device elements like pipes, bends, valves and fittings that are forming the flow passage. Then φ is given by

$$\varphi = \sum_i \varphi_i = \sum_i K_i \frac{\rho}{2} w_i^2 \tag{2.17}$$

where K_i is the loss coefficient of element i and w_i a characteristic velocity in it. Here the question arises, whether the same way of accounting for flow losses can be used on the microscale. In Eq. (2.17) loss coefficients are constant numbers (and tabulated for various geometries of the flow elements) when the flow is fully turbulent. Only then does $\varphi_i \sim w_i^2$ hold. In laminar flow, however, $\varphi_i \sim w_i$ and K_i is not a constant when it is introduced according to Eq. (2.17). This may be illustrated for the loss coefficient of a straight pipe. It is defined

$$K_{\text{pipe}} = \lambda_{\text{R}} L/d_{\text{h}} \tag{2.18}$$

with the friction factor λ_R. For fully turbulent flow (at rough walls) λ_R is a constant number (depending on the relative roughness height k/d_h). For laminar flow, however, $\lambda_R \sim \text{Re}^{-1}$ and thus $\lambda_R \sim w^{-1}$ holds.

When λ_R for laminar pipe flow is known, the loss coefficient K_{pipe} can be determined. For other flow elements, however, loss coefficients are unknown (and cannot be taken from tables of loss coefficients that are for turbulent flows only!).

Thus Eq. (2.17), well established for fully turbulent macroflows is inappropriate for microflows since they are laminar in most cases!

2.5.2
Hydrodynamic Entrance Length

A flow in a pipe or channel is hydrodynamically fully developed when its velocity profile does not change in the (downstream) z-direction. This only can occur when the cross section is z-independent. Since the fully developed flow profile is reached asymptotically for $z/d_h \to \infty$ a certain criterion is needed for the definition of the fully developed state.

The hydrodynamic entrance length L_{hyd} is defined as the distance from the pipe or channel entrance for which the centerline velocity differs by less than 1% from the value for $z/d_h \to \infty$. For laminar incompressible flows this length is approximately

$$\frac{L_{\text{hyd}}}{L_{\text{R}}} = \frac{C_1}{1 + C_2 \text{Re}/C_1} + C_2 \text{Re} \tag{2.19}$$

with $\text{Re} = \rho w L_R / \eta$ and

- $L_R = R$, $C_1 = 1.2$, $C_2 = 0.224$ for pipe flow
- $L_R = H$, $C_1 = 0.89$, $C_2 = 0.164$ for channel flow

with the radius R and half the plate distance H, respectively.

For Reynolds numbers $\text{Re} \approx 1$, typical for microflow situations, the entrance length is very short ($L_{\text{hyd}} \approx R, H$); for higher Reynolds numbers, however, it takes several characteristic lengths (R, H) before the flow is fully developed. Though these are numbers for the flow entrance length they are also characteristic lengths after which a disturbance of the flow is damped out on its way downstream. Again, this happens on a short length scale for low Reynolds number flows.

2.5.3 Streamtube Approach for Compressible Flows

In a compressible flow, density is a function of pressure and temperature so that the sum of both parts of the energy equation (mechanical and thermal energy) is needed. Adding Eqs. (2.15) and (2.16) and introducing the specific enthalpy $h = e + p/\rho$ leads to (first law of thermodynamics, total energy balance)

$$h_2 + \frac{w_2^2}{2} + gy_2 = h_1 + \frac{w_1^2}{2} + gy_1 + w_{t12} + q_{12} \tag{2.20}$$

Here φ_{12} no longer appears explicitly since it is an internal redistribution of mechanical into thermal energy.

In which situations microflows become compressible can best be seen when the continuity equation (2.14) is rewritten (assuming $A_1 = A_2$ for this purpose) by introducing $w = \text{Ma}\, c$, $c = \sqrt{\kappa RT}$ and $\rho = p/(RT)$. Here, Ma is the Mach number and ideal gas behavior is assumed for the speed of sound, c, and the thermal equation of state. From $w_2\rho_2 = w_1\rho_1$ we thus get

$$\text{Ma}_2 = \text{Ma}_1 \frac{p_1}{p_2} \sqrt{\frac{T_2}{T_1}} \tag{2.21}$$

When Ma_1 is assumed to be small ($\text{Ma}_1 \approx 0.1$, for example) Ma_2 may increase appreciably mainly due to a strong pressure drop $p_1/p_2 \gg 1$. A rise in temperature will also lead to an increase in Ma_2, however, $\sqrt{T_2/T_1}$ will be close to 1 since T is the absolute temperature in Kelvin.

When Ma_2 exceeds 0.3 the flow has to be treated as compressible, see Section 2.3.3. In macrodimensions three standard compressible flow situations are often encountered and their theoretical description can be found in advanced textbooks. They are

- inviscid compressible flow in conduits with varying cross sections without heat transfer (isentropic flow),
- inviscid compressible pipe or channel flow with heat transfer (Rayleigh flow),
- viscous compressible pipe or channel flow with adiabatic walls (Fanno flow).

In microflow situations, viscous effects are important so that out of the three flows only Fanno flows are of interest in this geometrical range. In very long pipes or channels the Mach number can increase up to $\text{Ma}_2 = 1$, which is the so-called viscosity choked flow situation, for details see ref. [9].

2.6
Heat Transfer in Microsized Devices

2.6.1
Scaling Considerations

Heat transfer on small scales compared to that in macrosized devices is special with respect to several scaling effects, identified in Table 2.1. These are

- axial heat conduction ($\mathrm{Pe} \to 0$),
- conjugate heat transfer ($\Lambda_s \to \infty$),
- variable fluid property effects ($K_{zTa} \to \infty$),
- low Reynolds number flow ($\mathrm{Re} \to 0$).

As shown in Section 2.4 already, the theoretical model is different when Reynolds numbers are $\mathrm{Re} \approx 1$ or smaller or when they are relatively high. For small Reynolds numbers and a flow that is not fully developed the full Navier–Stokes equations are needed and in the energy equations neither axial heat conduction nor viscous dissipation can be neglected.

In general, axial heat conduction in a fluid cannot be neglected unless the Peclet number is high (i.e. for $\mathrm{Pe} \to \infty$). Since, however, in microdevices $\mathrm{Pe} = \mathrm{RePr}$ often is small, axial heat conduction may be important. Equation (2.12) shows that it actually is the *axial change* of axial conduction, i.e. the change of, $\dot{q}_z = -\lambda \partial T / \partial z$, i.e. $\partial \dot{q}_z / \partial z$. Only when a fluid element loses more or less energy by axial conduction on its downstream side than it gains on its upstream side, is the energy balance affected by axial heat conduction. Thus, whenever a nonlinear axial temperature distribution occurs this effect appears. Predominately that will be close to the beginning or the end of a heat-transfer section.

Axial heat conduction may also be important in the solid walls of the micro conduit that, compared to the characteristic length of the cross section, $l = d_h$, may be very thick. When, for example, a constant wall heat flux density $\dot{q}_w$ is not imposed at the inner wall surface but at the outer wall, then almost always strong conjugate effects occur, i.e. heat conduction in the wall and convective heat flux to the fluid are both and simultaneously present.

2.6.2
Variable Property Effects

For the analysis of the influence of variable properties in microflow situations in contrast to those for macroflows (i.e. when going from macro- to microscales) three aspects are important.

1. Reynolds numbers are small. That is why microflows almost always are laminar. Conventional methods to account for variable property effects assume large Reynolds numbers, however.

2. Axial temperature gradients are large. Consequently a constant property solution cannot be a good approximation over an appreciable downstream length L.
3. Cross-sectional temperature differences ΔT are small. In macroflow situations there is an appreciable temperature difference over the cross section and a very small one over finite axial distances L. In microflow situations, this is vice versa, or at least both are of equal importance, see Fig. 2.1, last line.

For an incompressible flow, property variations may occur due to the temperature and pressure dependence of η (viscosity), λ (thermal conductivity) and c_p(specific heat capacity). Taking water as a typical fluid it turns out that the pressure dependence for all three properties is very small and that the temperature dependence of c_p is almost negligible. This follows from the sensitivity coefficients

$$K_{aT} = \left(\frac{T}{a}\frac{\partial a}{\partial T}\right)_{T_r}, \quad K_{ap}\left(\frac{p}{a}\frac{\partial a}{\partial p}\right)_{p_r} \tag{2.22}$$

For $a = \eta, \lambda, c_p$ their numerical values at the reference temperature $T_r = 293$ K and the reference pressure $p_r = 1$ bar are shown in Table 2.3.

As far as variable property effects are concerned there are various degrees to which they can be accounted for in conventional macroflows. The different levels of accuracy, i.e. the different models, are

I. *constant properties*: There is assumed to be one single value for a property in the entire flow field.

II. *quasiconstant properties*: The problem is treated like the constant-property case, but with property values according to the mean temperature at position z. Then, e.g., large but slow changes of the mean temperature can be accounted for, like in a situation when $\dot{q}_w = \text{const.}$ acts on very long pipes.

III. *weakly variable properties*: In a perturbation approach the dominating influence of radial property variations can be accounted for and can be combined with the axial variations by assuming the quasiconstant property case as the basic case that has to be perturbed, see ref. [10] for details.

IV. *strongly variable properties*: If properties change equally strongly in both directions, radially and axially, and these changes are "not small", the fully coupled equations have to be solved. Here "not small" means that the

Table 2.3 Sensitivity coefficients for water at $T_r = 293$ K, $p_r = 1$ bar (data from [7]).

$K_{\eta T}$	$K_{\lambda T}$	K_{cT}	$K_{\eta p}$	$K_{\lambda p}$	K_{cp}
–7.13	0.82	–0.05	$-3 \cdot 10^{-4}$	$8 \cdot 10^{-5}$	$-6 \cdot 10^{-5}$

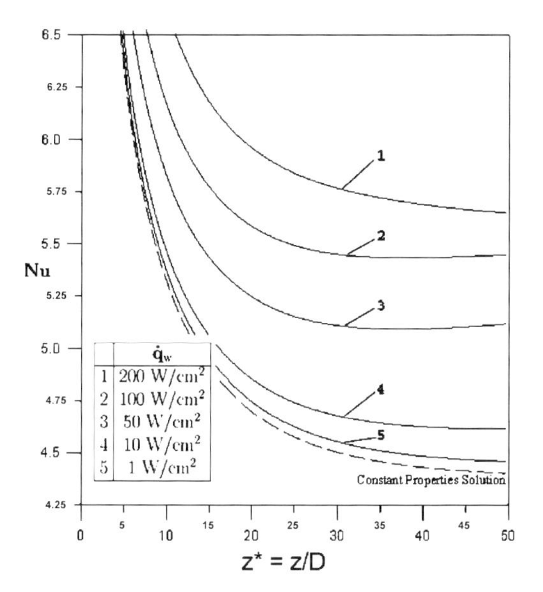

Fig. 2.2 Nusselt number results from Navier–Stokes solutions for different heating rates $\dot{q}_w$ in a hydrodynamically and thermally developing pipe flow of water. Re $= 75$, $T_0 = 278$ K (5 °C). - - - constant properties (model I); ___ variable properties (model IV).

temperature field is influenced not only by the variation of those properties that appear in the energy equation (i.e. by λ and c_p), but also by the modifications of the flow field due to variations of η and ρ (which are determined by the comprehensive forms of the momentum and continuity equations, respectively).

As an example, Fig. 2.2 shows how the Nusselt number $\mathrm{Nu} = \dot{q}_w D/(\lambda(T_w - T_b))$ is affected by the influence of variable properties $\eta(T)$ and $\lambda(T)$, for details see refs. [11–13]. Here T_b is the local bulk (caloric mean) temperature. T_0 in Fig. 2.2 is T_b at $z^* = 0$.

2.6.3 Viscous Heating

The order of magnitude considerations in Fig. 2.1 show that pressure drop in microsized devices can be large and as a consequence viscous dissipation is strong. The redistribution from mechanical to thermal energy in association with this dissipation (of mechanical energy) is called *viscous heating* since the fluid increases its thermal (internal) energy like in a process that involves heat transfer. This effect is covered by φ_{12} in Eq. (2.16) and by the term $\mathrm{Ec}\Phi^*/\mathrm{Re}^2$ in Eq. (2.12).

What happens to the temperature of the fluid under the influence of viscous heating depends on the thermal boundary conditions. Two limiting cases can be considered and will be shown for a pipe flow:

- a fully developed viscous dissipation situation in which the temperature profile due to dissipation is independent of the axial coordinate z. Then a constant heat flux density exists at the pipe wall, which transfers the thermal energy due to viscous heating out of the pipe. This heat flux density for a circular pipe is

$$\dot{q}_{\mathrm{w}} = \frac{8\eta w^2}{D} \tag{2.23}$$

and can be large when the diameter D becomes small, like for microflows.

- an adiabatic wall so that the increase in thermal energy due to viscous heating increases the bulk temperature of the fluid downstream. The adiabatic wall temperature (neglecting entrance effects) then is

$$T_{\mathrm{ad}} = T_0 + \frac{\eta w^2}{\lambda}\left(16\frac{z/D}{\mathrm{Pe}} + 1\right) \tag{2.24}$$

with T_0 the bulk temperature at $z = 0$, and Pe the Peclet number of the flow (which is small for microflows).

For details, see [7]. Even if practical situations do not have one of the two special thermal boundary conditions, Eqs. (2.23) and (2.24) show that viscous dissipation effects become important for $D \to 0$, $z/D \to \infty$ and $\mathrm{Pe} \to 0$, respectively. This, however, is typical for microflows.

2.6.4 Heat-transfer Enhancement in Bent Conduits

Since microflows are laminar, turbulent mixing of flow particles is absent and as a consequence heat transfer (as well as mass transfer in multicomponent fluids) is poor.

In order to enhance mixing of fluid particles of different temperatures and thus enhance heat-transfer flow separation, recirculation and/or vortices may be implemented in the flow field. This can either be done by geometrical inserts like twisted tapes (see [14] for turbulent pipe flow as an example) or by sharp or curved bends of the pipe or channel geometry.

The physics in these flows can be characterized by the nondimensional Dean number $\mathrm{Dn} = \mathrm{Re}(d/R_{\mathrm{c}})^{1/2}$, first introduced by [15] in helical flows, i.e. for flows in coiled pipes. Here d is the diameter of a pipe and R_{c} the radius of curvature (either of the helicalicity or of a single bend).

The flow through a single bend of a pipe is characterized by the formation of a single pair of so-called Dean vortices, which sometimes is interpreted as a secondary flow (see [16] for a discussion of this interpretation).

These vortices produce significant heat-transfer enhancement with only moderate extra pressure drop. The alignment of the flow with the vorticity in these structures allows significant mixing of the fluid without creating large pressure-drop penalties, as for example shown by [17] in serpentine geometries, which are a combination of successive 90° bends.

For similar results in sharp bends and T-joints see [18], for example.

2.7 Entropy Generation

For predicting an efficient use of energy in thermal systems like heat exchangers the second law of thermodynamics is very useful since the amount of entropy produced is directly linked to the loss of energy quality.

Basically, there are two ways in which to determine the overall entropy production in a system in which viscous dissipation and an increase of entropy due to heat transfer occur:

- the *direct method*: Entropy production due to dissipation and heat conduction is determined locally. Then an integration over the entire flow domain leads to the overall entropy production. This can be done in the postprocessing phase of a CFD-calculation, for example, since all information about the local entropy production is known, once the velocity and temperature fields in the flow domain have been determined. For laminar flows this approach is straightforward.
- the *indirect method*: From the global entropy balance the overall entropy production is determined by integration over the control volume surface. For an arbitrary control volume V shown in Fig. 2.3 the balance is

$$\underbrace{\iiint_V S_{\mathrm{PRO}}\,\mathrm{d}V}_{\text{PRODUCTION},0} = \underbrace{\iint_{A_1} \rho w s\,\mathrm{d}A}_{\text{CONVECTION in},1} - \underbrace{\iint_{A_2} \rho w s\,\mathrm{d}A}_{\text{CONVECTION out},2} - \underbrace{\iint_{A_0+A_1+A_2} (\vec{q}/T)\,\mathrm{d}A}_{\text{MOLECULAR FLUX},3} \tag{2.25}$$

  where s is the specific entropy and w a velocity perpendicular to the surface. The left-hand side, i.e. the overall entropy production is known, once all surface integrals on the right-hand side have been determined. This, again, for laminar flows is straightforward, refs. [19] and [14].

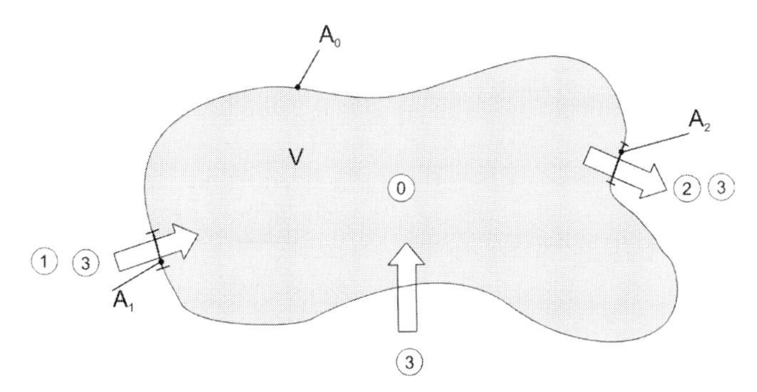

Fig. 2.3 Flow domain with volume V, impermeable wall A_0, inlet A_1 and outlet A_2. The numbers refer to the terms in Eq. (2.25).

As an example the direct method will be applied to a fully developed pipe flow (Hagen–Poiseuille) with constant wall heat flux density. For this simple case the flow and temperature fields are

$$w = 2w_\mathrm{m}\left[1 - \left(\frac{r}{R}\right)^2\right]; \quad v = 0 \tag{2.26}$$

$$T - T_\mathrm{w} = \frac{-\dot{q}_\mathrm{w} R}{\lambda}\left[\frac{3}{4} + \frac{1}{4}\left(\frac{r}{R}\right)^4 - \left(\frac{r}{R}\right)^2\right] \tag{2.27}$$

Here w_m is the cross section averaged flow velocity, T_w the wall temperature and $\dot{q}_\mathrm{w}$ is the constant heat flux density in W/m^2.

There are two entropy production rates, one by viscous dissipation ($\dot{S}_\mathrm{PRO,D}$) and one by heat-conduction effects ($\dot{S}_\mathrm{PRO,C}$). From the analytical velocity and temperature results (2.26), (2.27) they can be determined in their nondimensional form ($r^* = r/R$):

$$\dot{S}^*_\mathrm{PRO,D} \equiv \frac{\dot{S}_\mathrm{PRO,D}}{\lambda T_\mathrm{b}^2/\dot{q}_\mathrm{w}^2} = \underbrace{16\frac{\mathrm{Pr}\hat{\mathrm{Ec}}}{\hat{\mathrm{Nu}}^2}}_{\Pi} r^{*2} \tag{2.28}$$

$$\dot{S}^*_\mathrm{PRO,C} \equiv \frac{\dot{S}_\mathrm{PRO,C}}{\lambda T_\mathrm{b}^2/\dot{q}_\mathrm{w}^2} = (2r^* - r^{*3})^2 \tag{2.29}$$

Here, T_b is the local bulk temperature (caloric mean temperature) and $\hat{\mathrm{Ec}}$ and $\hat{\mathrm{Nu}}$ are defined according to

$$\hat{\mathrm{Ec}} \equiv \frac{w_\mathrm{m}^2}{c_\mathrm{p} T_\mathrm{b}} = \mathrm{Ec}\frac{\Delta T}{T_\mathrm{b}}; \quad \hat{\mathrm{Nu}} \equiv \frac{\dot{q}_\mathrm{w} R}{\lambda T_\mathrm{b}} = \mathrm{Nu}\frac{\Delta T}{T_\mathrm{b}} \tag{2.30}$$

In Fig. 2.4 the nondimensional entropy production rates are shown for three different numbers of the parameter $\Pi = 16\,\mathrm{Pr}\hat{\mathrm{Ec}}/\hat{\mathrm{Nu}}$ (see Eq. (2.28)) of the problem. This parameter only affects the nondimensional $\dot{S}^*_\mathrm{PRO,D}$ and is implicitly contained in $\dot{S}^*_\mathrm{PRO,C}$.

Both rates are zero on the centerline of the pipe ($r^* = 0$) and increase towards the wall ($r^* = 1$). Different from turbulent flows, where entropy production is concentrated in regions very close to the wall, see [14], in laminar pipe flow appreciable parts of the overall production rates occur in the inner part of the flow region. This happens because the velocity and temperature gradients in laminar flow are not so concentrated at the near-wall regions as is the case for turbulent flows.

With increasing $\hat{\mathrm{Ec}}$ numbers, i.e. with increasing numbers for Π, entropy production due to dissipation more and more dominates the overall dissipation rate ($\dot{S}^*_\mathrm{PRO,D} + \dot{S}^*_\mathrm{PRO,C}$). For $\Pi = 1/16$ both contributions are equal, for larger values of Π dissipation effects are stronger than heat-conduction effects with respect to entropy production, for details see [20].

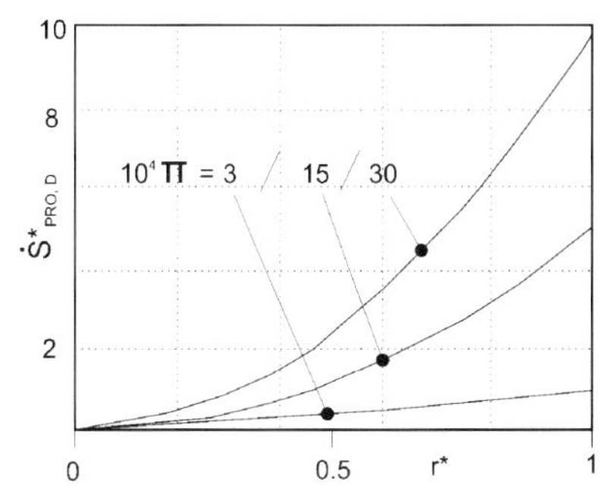

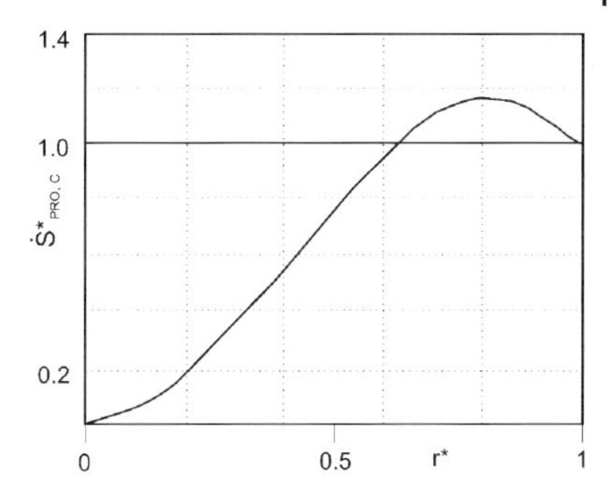

Fig. 2.4 Entropy production rates in a fully developed pipe flow for three different values of $\Pi = 16\,\mathrm{Pr}\hat{\mathrm{Ec}}/\hat{\mathrm{Nu}}$.

References

1 Gad-el Hak, M. (1999): The fluid mechanics of micro-devices – the Freeman Scholar Lecturer, Journal of Fluids Engineeering, **121**, 5–33

2 Kardiadakis, G. E. K.; Beskok, A. (2002): Microflows: Fundamentals and Simulation, Springer, New York

3 Szirtes, T. (1998): Applied Dimensional Analysis and Modeling, McGraw-Hill, New York

4 Batchelor, G. K. (2000): An Introduction to Fluid Dynamic, 2nd edn, Cambridge University Press, Cambridge, U.K.

5 Panton, R. L. (1996): Incompressible Flow, 2nd edn, John Wiley & Sons, Inc., New York

6 Herwig, H. (2002): Strömungsmechanik, Springer, Berlin, Heidelberg, New York

7 Gersten, K.; Herwig, H. (1992): Strömungsmechanik, Vieweg, Braunschweig

8 Guo, Z. Y. (2000): Size effect on flow and heat transfer characteristics in MEMS. In: Heat transfer and transport phenomena in microscale. Banff, Canada, Oct. 2000, 24–31

9 Munson, B. R., Young, D. F.; Okiiski, T. H. (2001): Fundamentals of Fluid Mechanics, 4th edn, John Wiley & Sons, Inc., New York

10 Herwig, H. (1985): The effect of variable properties on momentum and heat transfer in a tube with constant heat flux across the wall, Int. J. Heat Mass Transfer, **28**, 423–431

11 Herwig, H.; Mahulikar, S. P. (2005): Variable Property Effects in Single-Phase Incompressible Flows through Micro Tubes, paper ICMM 2005-75082, Proc. of the 3rd Int. conf. on Microchannels and Minichannels, June 13–15, 2005, Toronto, Canada

12 Mahulikar, S. P.; Herwig, H. (2005): Theoretical investigation of scaling effects from macro-to-microscale convection due to variations in incompressible fluid properties, Appl. Phys. Lett., **86**, 104–105

13 Mahulikar, S. P.; Herwig, H.; Hausner, O.; Kock, F. (2004): Laminar gas microflow convection characteristics due to steep density gradients, Europhys. Lett., **86**, 811–817

14 Herwig, H.; Kock, F. (2005): Direct and Indirect Methods of Calculating Entropy Generation Rates in Turbulent Convective Heat Transfer Problems, to be published in Heat and Mass Transfer

15 Dean, W. R. (1927): Note on the motion of a fluid in a curved pipe, Philos. Mag. **7**, 208–223

16 Herwig, H.; Hölling, M.; Eisfeld, T. (2005): Sind Sekundärströmungen noch zeitgemäß?, Forschung im Ingenieurwesen, **69**, 115–119.

17 Rosaguti, N. R.; Fletcher, D. F.; Haynes, B. S. (2005): Laminar Flow and Heat Transfer in a Periodic Serpentine channel, Chem. Eng. Techn., **28**, 353–361

18 Kockmann, N.; Engler, M.; Haller, D.; Woias, P. (2004): Fluid dynamics and transfer processes in bended microchannels, Proc. Microchannels and Minichannels Conf., Rochester 2004, paper ICMM 2004-2331

19 Kock, F.; Herwig, H. (2005): Entropy production calculation for turbulent shear flows and their implementation in CFD codes, Int. J. Heat and Fluid Flow, **26**, 672–680.

20 Kock, F. (2004): Bestimmung der lokalen Entropieproduktion in turbulenten Strömungen und deren Nutzung zur Bewertung konvektiver Transportprozesse, Dissertation, TU Hamburg-Harburg

3
Transport Processes and Exchange Equipment

Norbert Kockmann, Laboratory for Design of Microsystems, Department of Microsystem Engineering (IMTEK), University of Freiburg, Germany

Abstract

With miniaturization of process equipment the characteristic lengths are reaching into the order of magnitude of boundary layers or transport lengths of microprocesses at interfaces. Almost all transport processes dealt with are in the continuum range and can be described by linear correlations, but the limits and exceptions are given for heat and mass transfer. Steady-state and transient processes in diffusion and heat conduction are described as well as the convective single-phase flow in exchange equipment. Dimensionless parameters are derived to design and evaluate the exchange equipment. Mixing characteristics and short-length heat transfer phenomena are addressed to describe special features of microstructured devices. Important coupled processes are shortly introduced with their actual application and potential for future microsystems and process equipment. The integration of microstructured devices or elements will be a key issue for the future development in process intensification.

Keywords
Miniaturization benefits, diffusion, convection, mass and heat transfer, coupled processes

Advanced Micro and Nanosystems Vol. 5. Micro Process Engineering. Edited by N. Kockmann

ISBN: 3-527-31246-3

3.1
Transport Properties of Pure Substances and Mixtures

The fundamental transport equations, described in Chapter 1, include transport properties for diffusion, heat conduction, or viscous motion of a fluid, which have been regarded as a linear relationship between the force and the corresponding flux. In small structures, these linear coefficients may be limited due to the influence of the surface. This plays a major role for gases, see Chapter 2 and [1] for additional reading. For liquids, the limits of the linear correlation are reached in far smaller scales, which are beyond the scope of this book and the most application of microprocess technology. Interested readers are referred to [2, 3].

Surface effects become more important in microstructures due to the higher surface-to-volume ratio. The viscosity of a liquid is reported to be higher in the vicinity of the surface. Viscoelastic effects due to the surface influence are described in [2, Chapter 6]. A solid surface induces due to its electrical charge an adsorbed layer of the liquid, the Kelvin–Helmholtz double layer EDL (electrokinetic double layer). This layer reaches about three molecular layers from the wall into the bulk fluid. In very small channels ($d_h < \sim 100$ nm) and for relatively large molecules, this layer may influence the flow and the transport processes at the wall. The wall friction can be increased. For the special cases, an order of magnitude estimation will help to quantify the influence of this wall effect. In most cases in this book, the influence of the EDL might be neglected [4, Chapter 32]. The EDL influences the local Nu number, too, in micropolar flow, for example, the Nu number is about 7% smaller than in laminar flow.

Capillary forces may block microchannels but may also be used for filling and directed transport of liquids in microstructures. Many aspects of interfacial transport processes are under current investigation, a good overview is given by [5]. Important for process engineering, the vapor pressure depends not only on the temperature but also on the surface shape. At curved surfaces (bubbles with a concave surface, droplets with a convex surface, capillary fillings) [6, Chapter 1] the vapor pressure p_s depends on the surface curvature, see Fig. 3.1.

The vapor pressure is determined with the Gibbs–Thompson equation at strongly curved surfaces

$$\ln \frac{(p_s)_r}{p_s} = \frac{2\sigma}{RT\rho_L r} \tag{3.1}$$

For concave surfaces (bubbles), the vapor pressure is smaller, at convex surfaces (droplets), the vapor pressure is higher than at a plane surface. Hence, during evaporation small bubbles are formed at first; during condensation small droplets occur at first. For water and most other organic liquids, the pressure difference is smaller than 1% for bubbles or droplets smaller than 100 nm in diameter. For example, a vapor bubble in water with a diameter of 40 nm exhibits a 5% lower vapor pressure.

For high force gradients the linear correlations must be enlarged by further terms, which include relaxation times of fast processes. This is important for high energy densities like in laser processing or fast chemical reactions (detonations). For example, the Fourier equation has to be expanded by nonlinear terms for heat transfer processes with high-energy fluxes or high temperature gradients. The linear relationship is valid for temperature gradients up to a few thousand Kelvin per cm, which is also 0.1 K/μm in microsystems. Here, the more accurate Maxwell–Catteneo equation is used to describe the heat flux [7].

$$q = -\lambda \nabla T - t_q \frac{\partial q}{\partial t} \tag{3.2}$$

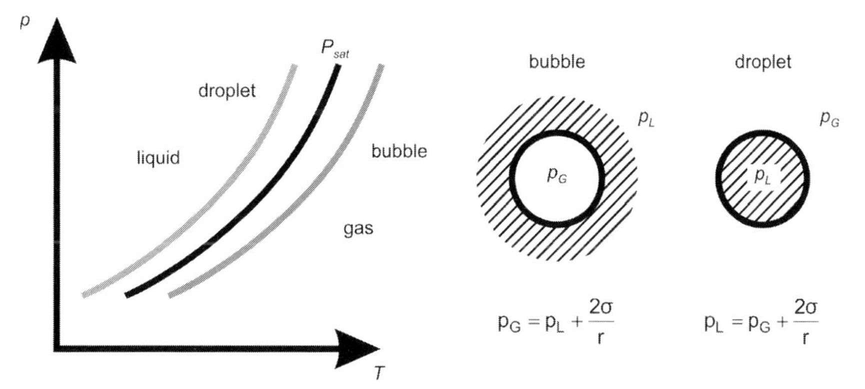

Fig. 3.1 Relation between the vapor pressure and the surface curvature. In small bubbles the gas dissolves earlier into the surrounding liquid. From a droplet the liquid evaporates earlier into the surrounding vapor [6].

The measurement of the timely propagation of heat waves gives the order of magnitude of the relaxation time t_q, which is 10^{-12} s for metals, about 10^{-9} s for gases, and 10^{-11} s for liquids.

The anisotropic conduction in crystals gets important in microstructures smaller than 1 μm. These processes are described by rational irreversible thermodynamics [8] or by the extended irreversible thermodynamics [9], where further application limits of the linear correlations can be found.

3.2 Diffusion, Mixing, and Mass-transfer Equipment

The mixing process is a fundamental unit operation that influences many other transport processes like heat transfer, chemical reactions, or separation processes. Two major parts, the diffusion and the convection transport determine the mixing process. In microfluidic systems with predominant laminar flow, the diffusion is the prevailing process. However, for many microprocess applications, the convection part is systematically used to enhance the transport properties besides the reduction of the diffusion length.

As already described in Chapter 1, the diffusive mass transport of the component i is described by the Fick's first law (one-dimensional form) [10, Chapter 17]

$$\dot{n}_i = \frac{\partial n_i}{\partial t} = -\rho D_{ij} \frac{\partial c_i}{\partial x} \tag{3.3}$$

Here, the flow rate depends on the concentration gradient and the binary diffusion coefficient. Typical values for the binary diffusion coefficients are for gases 10^{-4}–10^{-5} m^2/s, for self-diffusion in liquids 10^{-8}–10^{-9} m^2/s, for small particles in liquids 10^{-9}–10^{-11} m^2/s, and for solids 10^{-14}–10^{-19} m^2/s. Other important diffusion forms like multicomponent diffusion or diffusion in microporous systems are described in [10, Chapter 19; 11, Chapter 3]. Fick's second law describes the unsteady behavior of the concentration profile, here in the one-dimensional form,

$$\frac{\partial c}{\partial t} = D \frac{\partial^2 c}{\partial x^2} \tag{3.4}$$

A more detailed discussion and some analytic solutions may also be found in [12, 13]. An order of magnitude estimation from the above equation gives the characteristic mean diffusion length in one direction. The factor 2 comes from the uniform diffusion in both directions.

$$\bar{x}^2 = 2Dt \tag{3.5}$$

This equation was derived by Einstein in 1905 with the examination of molecular fluctuations [14]. The typical diffusion length for gases and liquids is shown in Fig. 3.8. Thermal fluctuations restrict the sedimentation of small particles in liquids, which can also be described with Eq. (3.5). Particles smaller than 1 μm in diameter hardly settle down in water, because their sedimentation velocity is in the range of the fluctuation velocity.

3.2.1
Diffusive Mass Transport and Concentration Distribution in Fluids

To give an impression of the concentration field development with time, some analytical and numerical solutions of the diffusion equations are given in the following. Starting from a concentration peak c_0, the concentration profile is determined from the one-dimensional distribution of a concentration peak, see Fig. 3.2, left [20, Section 1.5].

$$c(x,t) = \frac{n_0}{\sqrt{4\pi Dt}} \exp\left(-\frac{x^2}{4Dt}\right) \tag{3.6}$$

with the total amount of the species n_0 that diffuses into the surrounding area. The time-dependent diffusion in a plate is determined by an infinite Fourier series. The solution of this series expansion is time-consuming; charts and short cuts for various geometries and boundary conditions are proposed in [15, Chapter 7].

Derived from the source integral, Eq. (3.6), the one-dimensional diffusion between two semi-infinite bodies can be described by

$$c(x,t) = \frac{c_0}{2}\left(1 - \text{erf}\left(\frac{y}{\sqrt{4Dt}}\right)\right) = \frac{c_0}{2}\,\text{erfc}\left(\frac{y}{\sqrt{4Dt}}\right) \tag{3.7}$$

with the Gaussian error function erf and the complementary error function erfc, see Fig. 3.2, right.

An important diffusion setup in micro process technology is the concentration homogenization between small fluid lamellae. These fluid lamellae are pro-

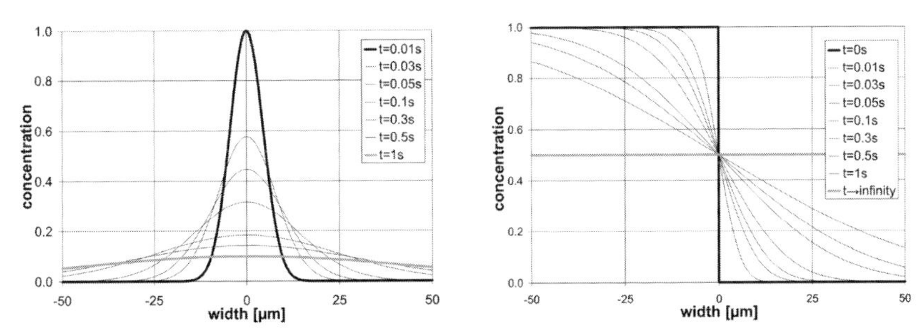

Fig. 3.2 Left: Diffusion from a source peak n_0=1/89206.2, D=10^{-9}; Right: Diffusion between two semi-infinite bodies, according [11, Chapter 3; 12, p. 36].

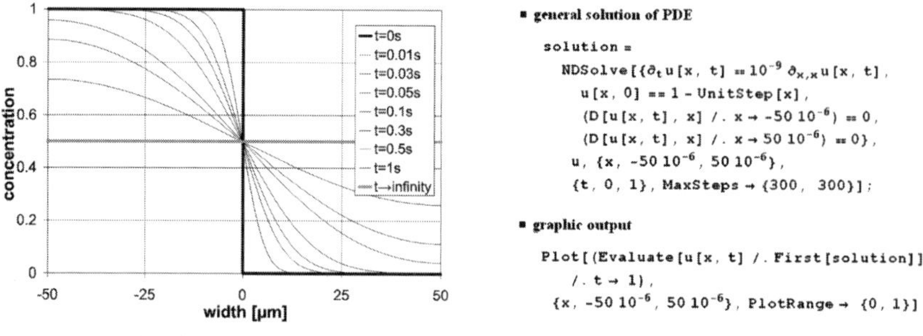

Fig. 3.3 Time development of the concentration profile in a symmetric fluid lamella; Mathematica® routine to solve numerically and plot the concentration profile.

duced in lamination micromixers, the serial lamination in split and recombine mixers (SAR) or the parallel lamination in interdigital or multilamination mixers. Small lamellae lead to a short diffusion length. The time development of the concentration profile may be determined in various ways. In [16], a harmonic series expansion is used to determine the solution of the diffusion equation (3.4) for a SAR micromixer. The coupling with an exponential diffusion leads to the calculation of an integral mixing residual as a measure of the mixing quality.

The numerical solution of the diffusion equation (3.4) with symmetrical boundary conditions is shown in Fig. 3.3. With this routine, different parameters, properties, and geometrical setups can be simulated and compared. An analytical solution with a Fourier series is given by Kluge and Neugebauer [17]. For high mass flux conditions, which may occur in microprocesses, correction factors are proposed by [15, Chapter 7].

3.2.2
Convective Mass Transport

The species transport can be described by the following mechanisms: diffusion (Eqs. (3.3) and (3.4)), convection in the flow direction (Eq. (3.8)), and transport in radial direction (Eq. (3.9)):

$$\dot{n}_i = -\rho D_{ij} \Delta c_i \tag{3.8}$$

$$\dot{n} = \rho c_i A w \tag{3.9}$$

$$\dot{n} = \beta_i \Delta c_i = \beta_i (c_{i,w} - c_{i,\text{bulk}}) \tag{3.10}$$

There are no generally valid solutions of the transport equation, but solutions exist for special cases for the momentum, heat, and mass transfer, see for example [18]. For a given process some simplifications will help to find an analytical solution, see also [19, p. 82]. Numerical solutions are suited for more complex systems, but they also need simplifications.

The analogy between heat and mass transfer with similar fundamental equations gives, for similar geometrical and process conditions, the same solutions. For laminar flow in a straight channel the transfer of species and heat from the wall into the bulk flow can be solved analytically [19, p. 82; 20, Chapter 3.8] and described by dimensionless parameters. The dimensionless heat transfer coefficient, the Nu number, is determined for laminar flow in three cases: a) in a pipe with constant heat flux; b) in a pipe with constant wall temperature; c) over a plate.

$$\mathrm{Nu_q} = \left.\frac{\alpha_z z}{\lambda}\right|_{\mathrm{q=const}} = \frac{48}{11} = 4.36 \tag{3.11}$$

$$\mathrm{Nu_T} = \left.\frac{\alpha_z z}{\lambda}\right|_{\mathrm{T=const}} = 3.6568 \tag{3.12}$$

$$\mathrm{Nu_z} = \frac{\alpha_z z}{\lambda} = 0.664\,\mathrm{Re}^{1/2}\mathrm{Pr}^{1/3} \tag{3.13}$$

These equations are valid for thermal and hydrodynamic developed flow. For the mass transfer from the wall to the bulk of the fluid the dimensionless mass-transfer coefficient is determined in an analogous way, for example the plate.

$$\mathrm{Sh_z} = \frac{\beta_{i,z} z}{D_{ij}} = 0.664\,\mathrm{Re}^{1/2}\mathrm{Sc}^{1/3} \tag{3.14}$$

By elimination of the Re number, the analogy between the heat and mass transfer can be expressed by

$$\mathrm{Sh_z} = \mathrm{Nu}\left(\frac{\mathrm{Sc}}{\mathrm{Pr}}\right)^{1/3} = \mathrm{NuLe}^{1/3} \tag{3.15}$$

The turbulent transport in mixing can be expressed by mixing length theory or engulfment theory, see for example [21]. The mixing from the wall into the bulk of the fluid is important for catalytic wall reactions, for adsorption processes at the wall, or for precipitation at the wall.

3.2.3
Convection and Mixing of Two or More Components

To compare various mixing states and mixing equipment some fundamental definitions are helpful. The Danckwerts segregation intensity I is defined with the mean square deviation of the concentration profile in a cross section [22, Chapter 1].

$$I = \sigma^2/\sigma^2_{\max} \quad \text{with} \quad \sigma^2 = \int (c_i - \bar{c}_i)^2, \; \sigma^2_{\max} = \bar{c}_i(c_{i,\max} - \bar{c}_i) \tag{3.16}$$

The segregation intensity can be transformed to a value between 0 (completely segregated) and 1 (completely mixed), the so-called mixing quality α.

$$\alpha = 1 - \sqrt{\sigma^2/\sigma^2_{\max}} \tag{3.17}$$

For determination and interpretation of the proposed mixing quality, the grid scale, on which the concentration has been measured, is important and should also be given. Unfortunately, the grid scale or measuring scale of the mixing quality can often not be found. A further mixing characterization, which is still under discussion, was presented by Bothe et al. [23] and describes the potential of diffusive mixing. Schlüter et al. [24] show a first application with experimental data.

The convection in microchannels plays a major role for highly efficient micromixers with high mass flow rates, see [25]. The convective effects are induced by bent, curved, or otherwise structured complete channel progression. Additional elements in microchannels also induce secondary flow structure, but the main channel keeps a straight direction. These secondary flow structures are also called advection flow.

In Fig. 3.4, the mixing of two components with an equal fraction (1 : 1 mixture) is numerically calculated for a channel with a 90° bend and square cross section. The numerical diffusion becomes visible in the broad concentration belt and relative low concentration gradient directly behind the bend. This numerical effect results from the insufficient grid distance (too wide) for the concentration distribution, but does not change the trend of the mixing quality with the Reynolds number [25a]. After a constant mixing channel length of 1000 μm, the mixing quality is determined [26]. Besides the diffusion, the two components

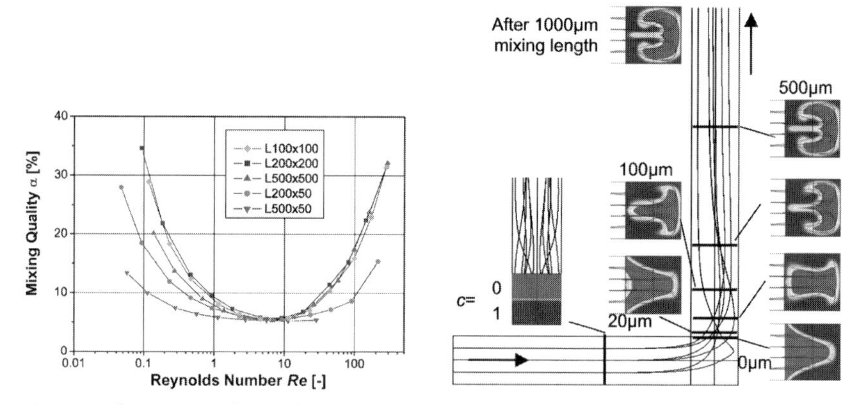

Fig. 3.4 Left: Mixing quality α after the 90° bend of five various mixers (L-shaped, channel width × depth) over the Re number; right: Mixer geometry (cross section 100×100 μm^2) and concentration profiles at various channel locations, Re = 99, w = 0.85 m/s [27].

mix by convective effects for Re numbers higher than about 10. For lower Re numbers, the straight laminar flow allows only diffusive mixing. The mixing quality α is proportional to the inverse of the Re number due to the residence time of the fluid. Smaller channels show a higher mixing quality because of their short diffusion length, see Fig. 3.4, left side.

The mixing quality α increases with the Re number due to a convective enlargement of the component interface for Re numbers larger than 10. The centrifugal force from the curved flow presses the faster fluid parts from the middle to the outer wall of the bend and causes a stretching of the component interface. The mushroom-like interface structure is clearly visible in Fig. 3.4, right side. The influence of the relevant forces is described in [27]. The situation can be compared with the flow in curved channels, first studied by Dean and described in Chapter 2 of this book.

For Re numbers larger than 10, the mixing quality increases with the Re number. The Re number and the aspect ratio of the channel are important for the mixing quality, the channel width is now represented by the Re number. This means that the convection in microchannels is scale invariant. The mixing in curved microchannels shows a similar behavior, as shown in [28].

The flow in a T-shaped micromixer shows a comparable behavior with the formation of secondary vortices, as displayed in Fig. 3.5, left side. With increasing Re number, the flow starts to form secondary vortices at the T-junction at the beginning of the mixing channel. Unfortunately, the component interface lies directly on the symmetry plane of the two components. Hence, the mixing of the two components is not affected by the vortex formation, as can be seen in the development of the mixing quality over the Re number in Fig. 3.6, left side.

With further increasing Re number, the symmetry of the flow structure breaks up and fluid from one side swaps to the opposite side. This effect can be seen in the middle of Fig. 3.5, where the concentration distribution in the mixing channel is shown. The two components are intertwined in two counter-rotating vortices that enhance the mixing quality dramatically. This is shown in Fig. 3.6, left side. Like the mixing in 90° bends for higher Re numbers, the mixing process is scale invariant and depends only on the Re number and on the

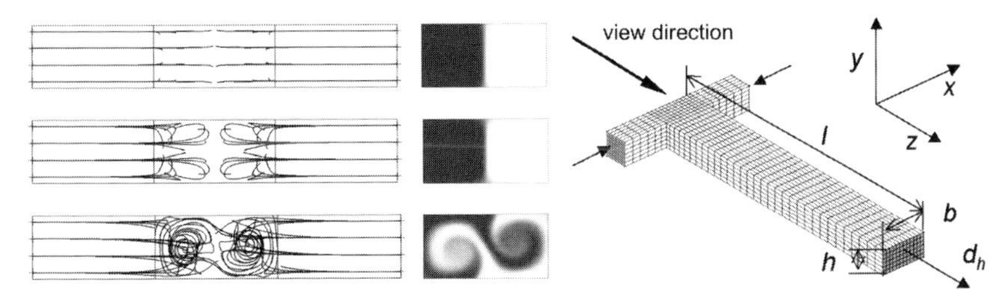

Fig. 3.5 Left: Flow regimes in a rectangular cross section T-shaped micromixer; middle: regarding concentration profiles of the laminar diffusive and the engulfment mixing; right: geometrical setup of the T-mixer [29].

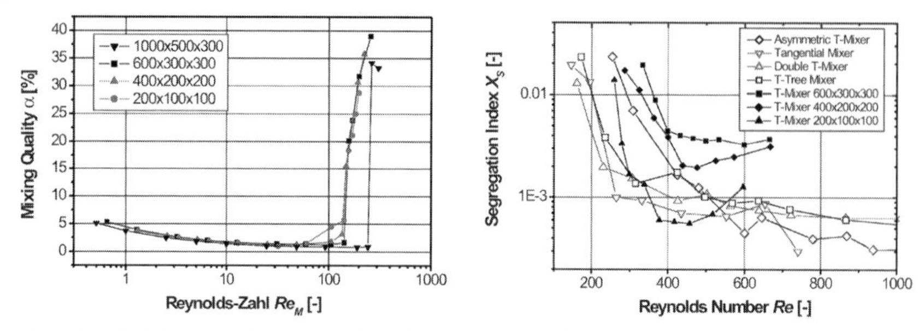

Fig. 3.6 Left: Mixing quality in a T-shaped micromixer with 4 different geometries over the Re number in the mixing channel; right: Experimental data from the iodide-iodate-reaction (Villermaux–Dushman). The segregation index is inversely proportional to the mixing quality α [25].

aspect ratio. The inlet Re number, which represents the incoming momentum, is the main parameter of the symmetry breakup of the flow instability.

Due to further flow instabilities, the steady-flow simulations are only valid for Re numbers lower than about 250. For higher Re numbers, only transient simulations and experimental data display the physical picture of the flow in T-shaped micromixers, see Fig. 3.6, right.

A mixer can generally be divided into two parts, the contacting elements like T- or Y-shaped junctions and the pure mixing elements like bends, curves or internals of the channel. More complex contacting elements for micromixers and their arrangement on a silicon chip can be found in [25]. The optimized combination of contacting elements, meandering channels, and their arrangement on a single silicon chip was investigated by [30]. A total volume flow of about 25 kg/h of an aqueous solution was completely mixed in a single silicon chip (20×20 mm^2 footprint with 16 parallel mixing elements) within a mixing time below 1 ms causing a pressure loss below 1 bar. An overview of this work is given in [31].

The experimental determination of the mixing quality is still a subject under investigation, but colored liquids [16] and some chemical reactions give a good result. For visualization pH indicators with color change are used for flow indication [32]. For slow mixing processes like in SAR mixers, the metal complex reaction of iron rhodanide is used for the flow visualization [16].

The mixing in T-junctions at sufficiently high Re numbers (> about 150, depending on the geometry) is a combination of convection plus diffusion effects. The interplay of the vortex generation with the interface between the components can be described by characteristic points, which also indicate the mixing process. Ottino [33] defines two characteristic points, the elliptic with converging flow and the hyperbolic point with diverging flow. Both can be found in the concentration profile of the T-shaped micromixer. The elliptic point is located in the center of the vortex (2 points in the engulfment flow) and indicates a stable flow with contracting fluid lamellae (mixing by stretching and thin-

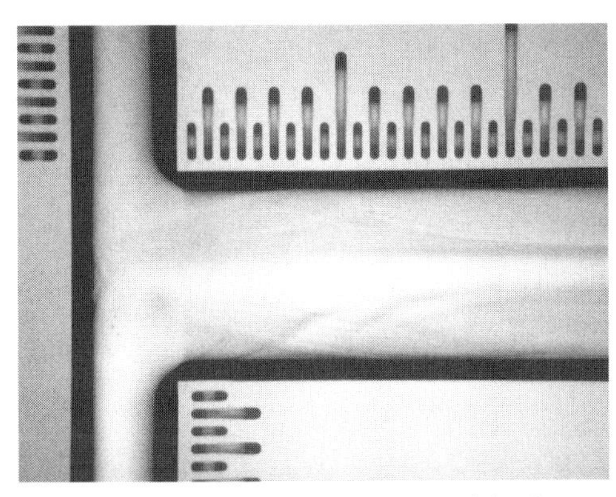

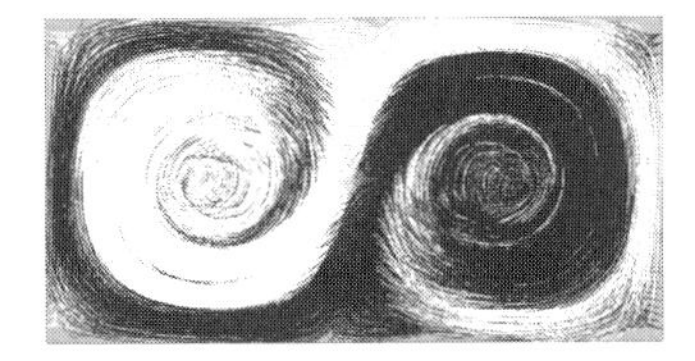

Fig. 3.7 Left: Experimental visualization of the flow structure in engulfment flow, Bromothymole Blue redox-reaction in a micromixer; right: simulations of the flow situation, illustrated by 7000 streamlines each at two different Reynolds numbers, $Re_M = 179$ (top) and $Re_M = 249$ (bottom) [25].

ning). The hyperbolic point is located in the center of the mixing channel between the two vortices. It is combined with converging flow and a stretching effect of the fluid lamellae (mixing by stretching). The location of these characteristic points and the location of interface of the components are very important for mixing process.

A large class of micromixers uses various internals in straight channels (passive mixers) and external forces (active mixers) to induce secondary flow structures. A good overview of the flow principles of chaotic advection and the geometrical arrangement of the internals can be found in [34]. In the context of microprocess engineering chaotic advection is the splitting, stretching, folding, and breaking the flow by internal elements at low or intermediate Re numbers. Nguyen and Wu [34] distinguish three different ranges of Re number: $Re > 100$ as high, $10 < Re < 100$ as intermediate, and $Re < 10$ as low, which also agrees very well with the above mentioned effects, see Figs. 3.4 and 3.6. For high Re numbers the internals are obstacles on the wall or in the channel as well as meandering or zig-zag-shaped channels. The pressure loss caused by the internals is one important design issue for determining the geometrical shape.

For intermediate Re numbers, mainly twisted microchannels and 3-dimensional L-shaped microchannels in various forms cause chaotic advection. The modification of the channel wall with ribs and grooves brings chaotic advection to low Re number flow. A prominent version of the grooved channel walls is the herringbone mixer [35], which shows good mixing characteristics. These mixers are often used for lab-on-chip application, where a good mixing at low Re numbers should be achieved and the pressure loss is not the main issue. Further mixing geometries of passive micromixers and examples of active micromixers are given in [36].

Some research groups are working on multiphase flow mixing, either the mass transfer between two almost immiscible phases (extraction) or the mixing within a bubble or a meniscus in a two-phase flow. The mass transfer between two immiscible phases can be described by the *two-film* theory, see also Chapter 1, Fig. 1.7. Other models are the penetration and the surface renewal model [20, Chapter 1.5]. The slug flow in microchannels enhances the mixing process in one phase due to the Taylor dispersion. The effect was experimentally described by [37] for the mixture of aqueous solutions in oil. Gavriilidis et al. [38] proposed a CFD model for Taylor flow characteristics and axial mixing. Most applications of this mixing process can be found in analysis systems, but the extraction may be based on this mechanism. The research group of K. F. Jensen at MIT describes the production of monodisperse silica nanoparticles in a multiphase flow system [39].

3.2.4 Mixing Time Scales and Chemical Reactions

Besides the mixing time scale the flow through a device and the transformation in a device possess their own time characteristics. The channel length l divided by the mean velocity determines the mean residence time t_P of a fluid in a straight channel:

$$t_P = \frac{l}{\bar{w}} \tag{3.18}$$

With the equipment miniaturization the channels become shorter and the mean residence time t_P decreases. Considering laminar flow with mixing solely by diffusion, the mixing time t_D also decreases with the channel width

$$t_D = \frac{(b/2)^2}{2D} = \frac{b^2}{8D} \tag{3.19}$$

The ratio of the mixing time t_D to the residence time t_P indicates the completeness of mixing in a channel by diffusion. If the ratio is smaller than 1, the diffusion has enough time to complete in the channel.

$$\frac{t_D}{t_P} = \frac{b^2}{8D}\frac{\bar{w}}{l} = \frac{b}{8l}\mathrm{Re}\cdot\mathrm{Sc} = \frac{\mathrm{Pe}}{8}\frac{b}{l} \tag{3.20}$$

The introduction of the Reynolds number Re (with the channel width as the characteristic length), of the Schmidt number Sc, and their combination, the Peclet number Pe gives the conclusion, that high Pe numbers leads to an incomplete mixing in straight channels. For increasing velocities, and therefore higher Re or Pe numbers, the mixing quality a decreases proportional to 1/Re, see also Fig. 3.4. A smaller channel with constant length will increase the mixing quality due to the smaller diffusion length.

The typical diffusion length and times are shown in Fig. 3.8, where also characteristic length and time scales of mixing equipment are integrated for compar-

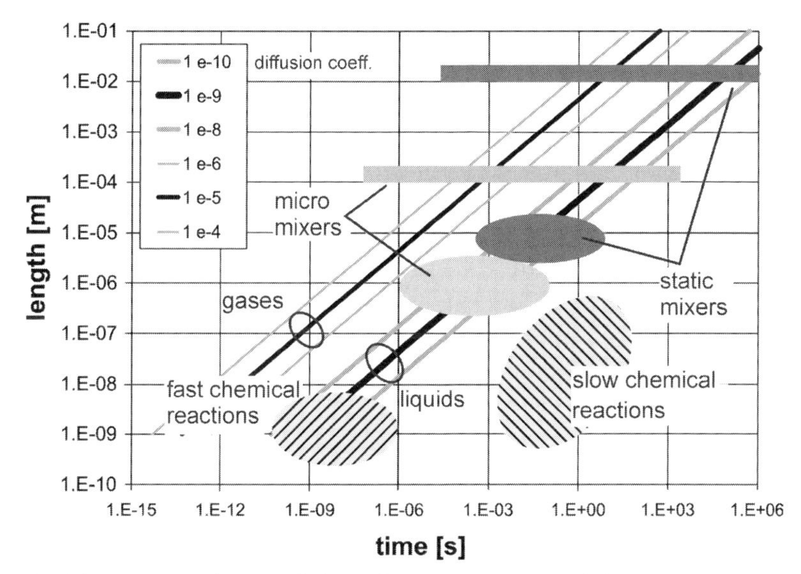

Fig. 3.8 Migration paths (straight lines) for different fluids: gas with 10^{-4}–10^{-5} m^2/s, liquid self diffusion with 10^{-8}–10^{-9} m^2/s, and a liquid with small particles 10^{-9}–10^{-11} m^2/s together with typical length scales for mixing equipment and chemical reactions; straight bar: equipment structure length of conventional static and micromixers; area: typical length of fluid elements during mixing.

ison. In convective micromixers, small fluid structures down to a few micrometers are produced in relatively large channel structures for high throughput and low fouling potential.

In microreactors the time scale of the chemical reaction has also to be considered. The concentration development of an educt or a product over the time depends on the reaction order and is described by the reaction kinetics. For an mth-order reaction the reaction rate is determined with the specific reaction rate k:

$$\frac{\mathrm{d}n}{\mathrm{d}t} \propto r = k \cdot c_j^m \tag{3.21}$$

Other reaction types can be found in the literature, for example [40, Chapter 7]. The characteristic time t_R for a reaction with constant volume can be determined with an order of magnitude estimation to

$$t_\mathrm{R} \propto \frac{1}{k \cdot c_j^{m-1}} \tag{3.22}$$

Comparing the time scales and the corresponding dimensionless ratios, the Damköhler number Da is very important for the design of microreactors, see Chapter 6. The design of mass transfer equipment requires a close look onto the local transport processes, the fluid distribution, and the pressure loss [41].

The entire apparatus can be described by integral parameters of an effective diffusion coefficient D_{eff} and a logarithmic concentration difference dc_{log}, comparable to heat transfer equipment. The method to derive integral parameters is described in Section 3.3.

3.2.5
Mixing Evaluation

The characterization and comparison of different micromixers is a complex task due to the wide application range of the equipment. The Reynolds number Re is one basic parameter to describe the flow characteristics. Low Re numbers are typical for analytical applications with small fluid amounts where diffusion is the dominating process. Microreactors for high-throughput screening or for production purposes show higher Re numbers and often use convective effects for mixing enhancement [34].

Another important parameter to characterize the mixing process is the mixing time. Mixing is a transient process with asymptotic behavior of the mixing quality. For long mixing times, the mixing quality α tends to reach 1. This asymptotic behavior can be described by a characteristic time, which is given by the slope at the mixing start or at a certain value of the mixing quality (for example 95%). A short mixing time gives a high mass-transfer rate and a high selectivity and yield for complex chemical reactions. This general statement has to be adjusted to the actual reaction type, where other conditions may be important.

In process engineering, inline static mixers are characterized by the pressure loss, which occurs over the channel length with inserts [42, Chapter 8]. In general the pressure loss in a channel with mixing elements is described by

$$\Delta p = \left(\zeta + \lambda_R \frac{l}{d_h} \right) \frac{\rho}{2} \bar{w}^2 \tag{3.23}$$

with the pressure-loss coefficient of a straight channel λ and the pressure-loss coefficient ζ of channel elements like bends or junctions. The terms in the bracket can also be expressed as the sum of all elements in the mixing channel. The pressure loss is formed into a dimensionless number Eu, the Euler number

$$\mathrm{Eu} = \frac{\Delta p}{\rho \cdot \bar{w}^2} = \frac{1}{2} \left(\zeta + \lambda_R \frac{l}{d_h} \right) \tag{3.24}$$

The pressure loss in the mixing channel is the effort one has to make to achieve a proper mixing. The performance of the mixing device is given by a short mixing length and a high mass flow rate, represented by the Reynolds number. To describe an effective mixing device with a single number the ratio of the performance to the effort gives a new dimensionless group, which was called mixer effectiveness ME_{II} by Kockmann et al. [31].

$$\mathrm{ME_{II}} = \frac{\text{performance}}{\text{effort}} = \frac{\mathrm{Re}}{\mathrm{Eu}} \frac{d_\mathrm{h}}{l_\mathrm{m}} \tag{3.25}$$

A high value of the mixer effectiveness $\mathrm{ME_{II}}$ indicates a mixing device with a short mixing length regarding the hydraulic diameter and a high flow rate generating a low pressure loss. With the mixing length $l_\mathrm{m} = \bar{w} t_\mathrm{m}$ and nearly square channel cross sections $A_\mathrm{m} = d_\mathrm{h}^2$ the mixer effectiveness can be written as

$$\mathrm{ME_{II}} = \frac{\dot{m}^2}{\eta d_\mathrm{h}^2 \Delta p t_\mathrm{m}} \tag{3.26}$$

Unfortunately, the micromixers known in the literature are often not characterized in such detail that they can be compared with the mixer effectiveness. In Fig. 3.9, the effectiveness of five micromixers is displayed. It can clearly be seen, that the diffusive mixers are less effective than the convection mixing devices. Therein, the dissipated energy is mainly used for the generation of small fluid lamellae, which shortens the diffusion length.

In Eq. (3.25), the Re number represents the flow rate through a mixing device. To characterize the mixing process itself, it is sufficient to consider only the pressure loss and the mixing length.

$$\mathrm{ME_{I}} = \frac{\text{performance}}{\text{effort}} = \frac{1}{\mathrm{Eu}} \frac{d_\mathrm{h}}{l_\mathrm{m}} \tag{3.27}$$

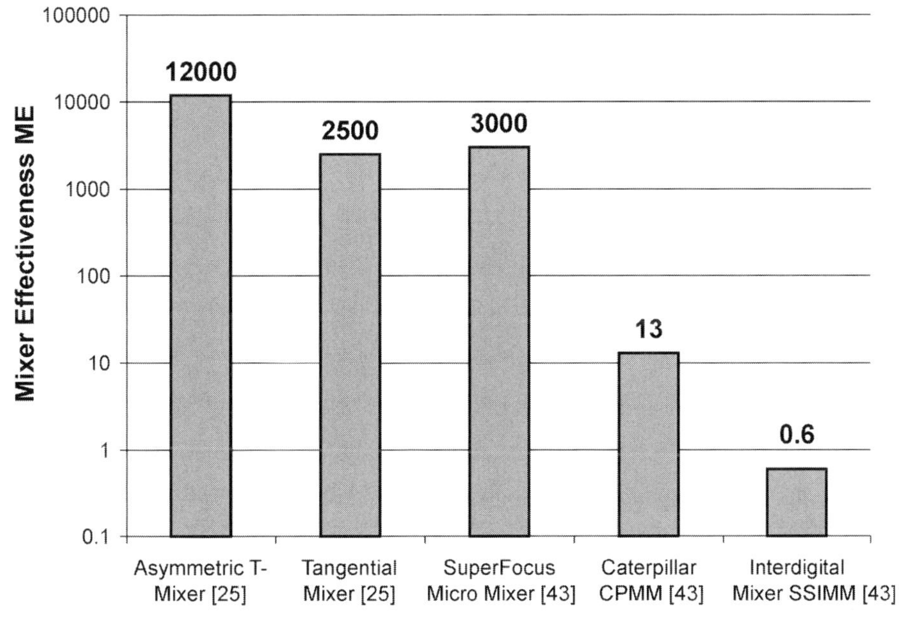

Fig. 3.9 Comparison of the mixing effectiveness for various micromixers, data from [25] and from IMM, Mainz, Germany [43].

With the above simplifications the mixing effectiveness can be written as

$$\mathrm{ME_I} = \frac{\dot{m}}{d_\mathrm{h} \Delta p t_\mathrm{m}} \tag{3.28}$$

The mixing effectiveness of various mixing principles is displayed in Fig. 3.10, left side. The mixing effectiveness $\mathrm{ME_I}$ consists of primary parameters and combines the requirements that were given above: It is dimensionless and increases with increasing mass flow, decreasing pressure loss and decreasing mixing time. It is therefore suitable to serve as a tool to compare different micromixers for use in chemical production processes.

With the Re number and the mixing time t_m, the mixing effectiveness can be written as

$$\mathrm{ME_I} = \mathrm{Re} \frac{\eta}{\Delta p t_\mathrm{m}} \tag{3.29}$$

Introducing the pressure-loss coefficients from Eq. (3.23), the mixing effectiveness can be written as

$$\mathrm{ME_I} = \frac{2}{\left(\zeta + \lambda_\mathrm{R} \dfrac{l}{d_\mathrm{h}}\right)} \frac{d_\mathrm{h}}{l_\mathrm{m}} \tag{3.30}$$

If the pressure loss of the mixing device can be described with convective effects only ($\lambda_\mathrm{R} = 0$), Eq. (3.30) is simplified to

$$\mathrm{ME_I} = \frac{2}{\zeta} \frac{d_\mathrm{h}}{l_\mathrm{m}} \tag{3.31}$$

With Eq. (3.31) the pressure-loss coefficient, the hydraulic diameter and the mixing length readily describe the mixing effectiveness. This gives the opportu-

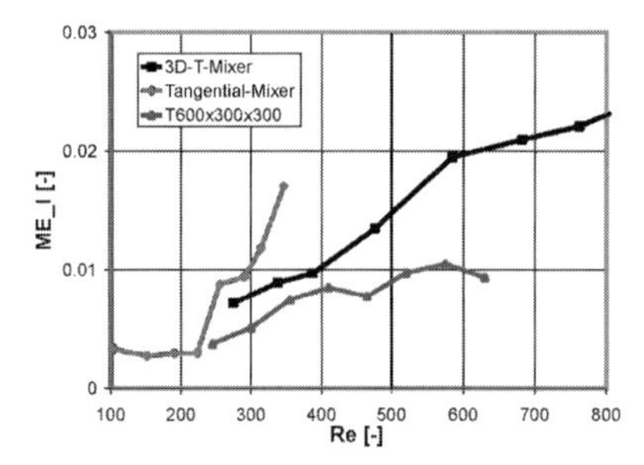

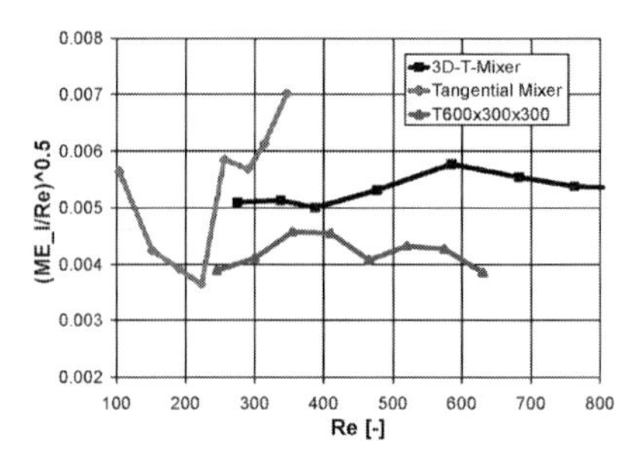

Fig. 3.10 Left: Mixing effectiveness of various mixing principles, see also [31]; right: dimensionless group to compare mixing times.

nity to compare different mixing processes and find optimum arrangements for the particular application.

An important parameter for mixers in general is represented by the dissipated energy ε during the mixing process. In stirred vessels, the dissipated energy is the stirring power divided by the stirred mass in the vessel. For static mixers, the dissipated energy is calculated from the pressure loss, the volume flow and the mass in the static mixer [42, Chapter 8].

$$\varepsilon \overset{\text{stirred vessel}}{=} \frac{P}{m} \overset{\text{static mixer}}{=} \frac{\Delta p_{\mathrm{V}} \cdot \dot{V}}{m} = \frac{\Delta p_{\mathrm{V}} \cdot \bar{w}}{\rho \cdot l_{\mathrm{m}}} \tag{3.32}$$

With $l_{\mathrm{m}} = \bar{w} t_{\mathrm{m}}$ and substituting the pressure loss from Eq. (3.32) into Eq. (3.29) results in

$$\mathrm{ME_I} = \mathrm{Re} \frac{\nu}{\varepsilon t_{\mathrm{m}}^2} \tag{3.33}$$

According to Zlokarnik [42, Chapter 8] there are three different types of micromixing within liquids: molecular diffusion, laminar deformation below the Kolmogorov length scale, and mutual enclosing of fluid lamellae with different composition that lead to the growth of a micromixed volume. The third process, the enclosing of relatively large volumes ("engulfment" according to [21]) is the limiting process with the characteristic time t_{E}

$$t_{\mathrm{E}} = 17.3 \left(\frac{\nu}{\varepsilon} \right)^{0.5} \tag{3.34}$$

Introducing Eq. (3.34) into Eq. (3.33) and rearranging leads to the comparison of two characteristic times.

$$\frac{t_{\mathrm{E}}}{t_{\mathrm{m}}} = 17.3 \sqrt{\frac{\mathrm{ME_I}}{\mathrm{Re}}} \tag{3.35}$$

The "engulfment" time t_{E} serves as an independent time scale to compare different mixing times of various mixers. A high ratio means a short mixing time and good mixing compared to static, turbulent mixers, see Fig. 3.10, right side. The small values of the mixing effectiveness and of the time ratio in Fig. 3.10 indicate a relatively high energy input into micro mixers compared to turbulent mixing in vessels or static mixers mainly due to wall friction losses. Nevertheless, micro mixers are very useful for very short mixing and residence times as well as for a high heat transfer load, see also Chapter 6.

The characterization of mixing devices with the mixing effectiveness was recently proposed and is still under discussion. Open points to discuss are certainly the definition and measurement of the mixing length, which is combined with a proper definition of mixing quality or segregation intensity. Additionally, the definition and measurement of the pressure loss length has to be addressed in future experimental and simulation investigations.

3.3 Heat Transfer and Heat Exchangers

The heat transfer topic is very diversified and covered by many textbooks, journals, and conferences. A comprehensive picture is given by [20, 44] as well as in [45, 46].

3.3.1 Heat Conduction in Small Systems

Fourier's first law describes the correlation between the steady heat flow and the driving temperature difference, here for one-dimensional heat conduction:

$$\dot{Q} = -\lambda A(r)\frac{\mathrm{d}T}{\mathrm{d}r} \tag{3.36}$$

The integration over the coordinate r leads for constant heat conductivity λ to

$$T_1 - T_2 = \dot{Q}\frac{1}{\lambda}\int_{r_1}^{r_2} \frac{1}{A(r)}\mathrm{d}r \tag{3.37}$$

The integral divided by the thermal conductivity is often called the thermal resistance R_{th} in analogy to the electrical resistance. For the geometrical forms of a plate, a cylinder, and a sphere the thermal resistance R_{th} in Eq. (3.37) is determined to the following expressions in Table 3.1. In analogy to the electrical resistance the thermal resistances of composed systems are arranged in series by simple adding and in parallel by adding the inverse.

The heat conductivity in microsystems is influenced by the microstructure of the material. Grain boundaries and crystal lattices form additional resistances to the heat transfer. In regular crystals, the heat transfer coefficient depends on the crystal orientation and the Fourier equation has to be enlarged to the tensor notation

$$q = -\lambda\frac{\mathrm{d}T}{\mathrm{d}x} \quad \rightarrow \quad \vec{q} = -\Lambda \text{ grad } T \tag{3.38}$$

Table 3.1 Thermal resistance R_{th} of simple geometrical elements of a plate, a cylinder, and a sphere.

Plate or slab	Cylindrical element	Spherical element
A = const., $n = 0$	$A = 2r\pi L$, $n = 1$	$A = 4r^2\pi$, $n = 2$
$R_{th} = \frac{1}{\lambda}\frac{r_2 - r_1}{A}$	$R_{th} = \frac{1}{\lambda}\frac{\ln r_2/r_1}{2\pi L}$	$R_{th} = \frac{1}{\lambda}\frac{(1/r_1 - 1/r_2)}{4\pi}$

with the heat conductivity tensor

$$\Lambda = \begin{pmatrix} \lambda_{11} & \lambda_{12} & \lambda_{13} \\ \lambda_{21} & \lambda_{22} & \lambda_{23} \\ \lambda_{31} & \lambda_{32} & \lambda_{33} \end{pmatrix} \tag{3.39}$$

The solution of the three-dimensional heat conduction is often only possible with numerical methods.

The unsteady heat transfer in small systems is very fast and can often be approximated with a fast asymptotic behavior to the steady-state solution. The time-dependent second Fourier law is derived from a differential element with the balance of the heat capacity and the heat conduction

$$\rho A(r) c_p \frac{\partial T}{\partial t} = \frac{\partial}{\partial r}\left(\lambda A(r) \frac{\partial T}{\partial r}\right) \tag{3.40}$$

Constant material properties and mathematical simplification lead to

$$\frac{\partial T}{\partial t} = a\left(\frac{\partial^2 T}{\partial r^2} + \frac{n}{r}\frac{\partial T}{\partial r}\right) \tag{3.41}$$

with the temperature conductivity $a = \lambda/\rho c_p$ and the geometrical factor n, see Table 3.1. The transient temperature development in a semi-infinite body is given in the one-dimensional form:

$$\frac{\partial T}{\partial t} = a\frac{\partial^2 T}{\partial x^2};\ t \geq 0,\ x \geq 0 \tag{3.42}$$

The introduction of the dimensionless temperature Θ leads to:

$$\frac{\partial \Theta}{\partial t} = a\frac{\partial^2 \Theta}{\partial x^2} \quad \text{with} \quad \Theta = \frac{T - T_0}{T_w - T_0} \tag{3.43}$$

The solution of Eq. (3.43) depends on the boundary conditions at the wall, see also Section 3.2.1. With constant wall temperature (T_W=const.) the dimensionless temperature is determined by the error function.

$$\frac{T - T_0}{T_w - T_0} = \text{erf}(x^*) \quad \text{with} \quad x^* = \frac{x}{2\sqrt{at}}\ . \tag{3.44}$$

The wall heat flux can be calculated from Eq. (3.44). For a constant heat transfer coefficient α at the wall the solution of Eq. (3.43) is derived with the help of two dimensionless numbers, the Fourier number $\text{Fo} = at/x^2$ and the Biot number $\text{Bi} = \alpha x/\lambda$. The temperature development during cooling of the body is given by:

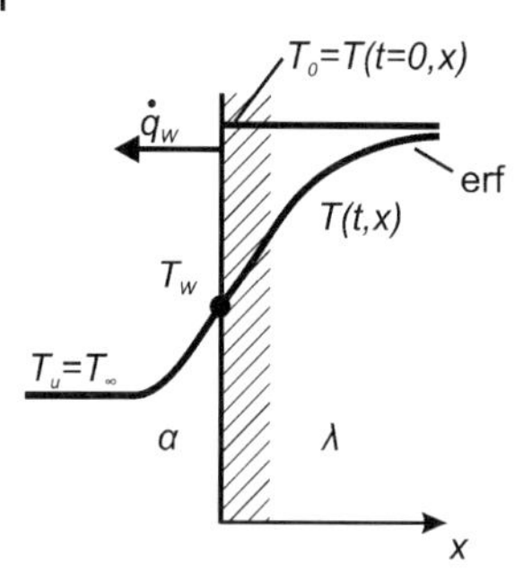

Fig. 3.11 Temperature distribution in a semi-infinite body during cooling.

$$\Theta_{\rm c} = \frac{T - T_\infty}{T_0 - T_\infty} = \operatorname{erf}(x^*) - e^{\rm FoBi^2+Bi}\operatorname{erfc}\left(\sqrt{\rm Fo}{\rm Bi} + x^*\right) \tag{3.45}$$

and for heating

$$\Theta_{\rm H} = 1 - \Theta_{\rm C} \tag{3.46}$$

The characteristic time of the heating or cooling process is given by the combination of the Fo and Bi numbers and is independent of the length (semi-infinite body). A miniaturization will not influence the temperature development and the heat flux from a semi-infinite body.

$$t_{\rm c} = at\left(\frac{\alpha}{\lambda}\right)^2 = {\rm FoBi}^2 \tag{3.47}$$

For small bodies (high Fo number and low Bi number), the temperature development can be approximated by asymptotic solutions. The temperature inside the small body is only a function of time

$$\Theta = \frac{T - T_\infty}{T_0 - T_\infty} = e^{-\frac{\alpha A t}{m c_{\rm p}}} \quad (\text{for} \quad {\rm Fo} > 0.3 \quad \text{and} \quad {\rm Bi} < 0.2) \tag{3.48}$$

or more exactly for the three basic geometries (plate, cylinder, sphere):

$$\Theta = e^{-(n+1){\rm BiFo}} \tag{3.49}$$

The characteristic time in Eq. (3.48) is given by

$$t_{\rm c} = \frac{m c_{\rm p}}{\alpha A} \tag{3.50}$$

Other solutions of more complex geometries are given in [47, Section 1.4].

3.3.2
Convective Heat Transfer in Microchannels

For convective flow in channels the pressure loss is given by Eq. (3.23). The friction factor of fully developed laminar flow in straight channels is proportional to 1/Re, the constant factor C is determined by the geometry. For circular pipes the constant factor is $C=64$, for a square cross section $C=56.92$. More data for various channel cross sections may be found in [4, Chapter 6]. For straight laminar flow, the dimensionless heat transfer coefficient, the Nusselt number Nu, is constant, depending on the boundary conditions,

$$\mathrm{Nu} = \frac{\alpha d_\mathrm{h}}{\lambda} \tag{3.51}$$

For constant wall heat flux, the Nu number is $\mathrm{Nu_q}=4.3$, for constant wall temperature the Nu number is $\mathrm{Nu_T}=3.66$. In a wide gap or narrow slit, the Nu number is 7.54 (q=const) and 8.24 (T=const) for double-sided heat transfer and 4.86 (q=const) and 5.39 (T=const) for single-sided heat transfer. For smaller channels, the heat transfer coefficient increases due to the constant Nu number. With decreasing channel dimensions, the transfer area and the mass flow through the channel is decreased, too. The transported heat is limited by these circumstances. To maintain a high heat transfer coefficient with a high transport rate, an optimum channel dimension has to be found [48], which is also oriented on the fabrication process. This process is described in Chapter 8 in more detail.

At the entrance of a channel or behind channel elements like joints, expansions or contractions, the disturbed flow enhances the radial transport in the channel. This results in an increased pressure loss as well as an increased heat transfer. For a local description of the heat transfer in entrance flow the dimensionless channel length X^* from the entrance is defined according to

$$X^* = \frac{L}{d_\mathrm{h}\mathrm{Pe}} = \frac{L}{d_\mathrm{h}\mathrm{Re}\cdot\mathrm{Pr}} \tag{3.52}$$

The mean Nu number in the entrance flow is calculated according the following approach [20]:

$$\mathrm{Nu_{me}} = \frac{\mathrm{Nu_m}}{\tanh(2.432\,\mathrm{Pr}^{1/6}X^{*1/6})} \tag{3.53}$$

This equation is valid for the entire channel length X^* and $\mathrm{Pr} > 0.1$. The analytical values for the entrance flow do agree well with the numerical values of the flow and heat transfer simulations for two volumes flows [27]. After a certain length, the velocity and temperature profile no longer change. This so-called fully developed flow is described in Chapter 2 and in [4, Chapter 32].

In turbulent flow (for Re > Re_{crit}=2300 in channel flow) the pressure loss is proportional to the square mean velocity and the heat transfer can be calculated according to Gnielinski [44, Chapter Ga]

$$\mathrm{Nu} = \frac{\xi/8(\mathrm{Re}-1000)\mathrm{Pr}}{1+12.7\sqrt{\xi/8}(\mathrm{Pr}^{2/3}-1)}\left(1+\left(\frac{d_\mathrm{h}}{l}\right)^{2/3}\right)K_\mathrm{Pr} \tag{3.54}$$

with $\xi = (1.8\log_{10}\mathrm{Re} - 1.5)^{-2}$ and $K_\mathrm{Pr} = (\mathrm{Pr}_\mathrm{fluid}/\mathrm{Pr}_\mathrm{wall})^{0.11}$.

The above correlation is valid for $0.5 < \mathrm{Pr} < 2000$, $2300 < \mathrm{Re} < 5 \times 10^6$ and $1 < L/d_\mathrm{h} < \infty$. Turbulent flows do not often occur in microchannels, but the manifolds or inlet and outlet headers may have turbulent conditions. A review of the heat transfer in various kinds of microstructured heat exchanger is given in [49], for heat sinks and electronic cooling [48] and for circular micropipes in [50]. Multiphase flow heat transfer is covered in Chapter 4, for film flow see [44, Chapter Md] or [45, Chapter 8.5].

Rarefied Gases The above equations are valid for liquids and dense gas above the continuum limit (Kn < 0.001). For rarefied gases the transfer processes at a wall change due to the low number of gas particles. Experiments with rarefied gas flow through a capillary indicated a linearly decreasing volume flow with decreasing pressure down to a certain value [51]. For even lower pressures, the volume flow remains constant and subsequently even increasing volume flow. The reduced flow resistance was explained with a slip velocity at the wall, which results from the molecular motion and the insufficient momentum transfer from the wall into the bulk fluid. In the same way, the molecular motion in the gas influences the energy transfer into the fluid. A temperature jump occurs at the wall, which increases the heat transfer resistance, see Fig. 3.12.

For rarefied gas flow (0.001 < Kn < 0.1) the boundary condition of the gas velocity at the wall is described by

$$u(x=0) = \zeta\left(\frac{\partial u}{\partial x}\right)_{x=0} \tag{3.55}$$

The slip length ζ can be calculated with the accommodation coefficient β of the tangential velocity and the mean free path Λ of the molecules. Experimental data for β can be found in [52].

$$\zeta \approx \frac{2-\beta}{\beta}\Lambda \tag{3.56}$$

The determination of the friction factor and the resulting volume flow in rarefied gas flow can be found in [44, 52, 53, Chapter Mo]. The temperature jump at the wall is described in a similar way with the temperature jump coefficient g.

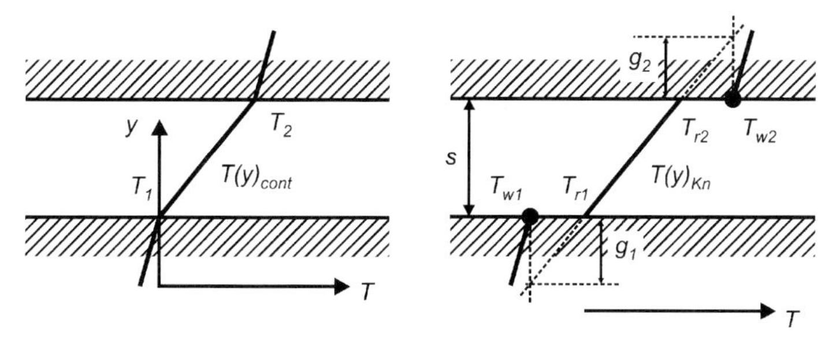

Fig. 3.12 Temperature gradient in a gap; left: linear development for dense gases (Kn <0.001); right: temperature jump for rarefied gases (Kn >0.001).

$$T(x=0) = T_r = T_w + g\left(\frac{\partial T}{\partial x}\right)_{x=0} \tag{3.57}$$

The temperature-jump coefficients g can be determined from kinetic theory from the thermal accommodation coefficient γ, a material parameter f and the mean free path Λ.

$$g = \frac{2-\gamma}{\gamma}\frac{15}{8}f\Lambda \tag{3.58}$$

The material parameter f is calculated from the linearized Boltzmann transport equation (see Chapter 1) and is given by Frohn [53]

$$f = \frac{16}{15}\frac{\lambda}{\eta c_v}\frac{1}{\kappa+1} = \frac{16}{15}\frac{1}{\mathrm{Pr}}\frac{\kappa}{\kappa+1} \tag{3.59}$$

For flow processes the ratio of the temperature jump coefficient and the characteristic length is important and is derived from Eq. (3.58).

$$\frac{g}{l_{\mathrm{char}}} = \frac{2-\gamma}{\gamma}\frac{15}{8}f\,\mathrm{Kn} \tag{3.60}$$

For monatomic gases with $\kappa=5/3$ and $\mathrm{Pr}=2/3$ and complete accommodation ($\gamma=1$) the length ratio reduces to

$$\frac{g}{l_{\mathrm{char}}} = \frac{15}{8}\mathrm{Kn} \tag{3.61}$$

The linear temperature distribution in a gap is determined with the jump coefficient g and Eq. (3.57) to

$$\frac{T(y) - T_2}{T_1 - T_2} = \left(1 - \frac{y + g_1}{s + g_1 + g_2}\right). \tag{3.62}$$

The temperature jump coefficient g can be regarded as an additional distance of the gap. For $0.1 < \mathrm{Kn} < 10$, monatomic gases, and complete accommodation, the temperature gradient is only a function of the Kn number.

$$\frac{T(y/s) - T_2}{T_1 - T_2} = \left(1 - \frac{y/s + (15/8)\,\mathrm{Kn}}{1 + (15/4)\,\mathrm{Kn}}\right) \tag{3.63}$$

The derivation of the temperature profile and comparison of the coefficients gives the heat flux for $0.1 < \mathrm{Kn} < 10$, which is given as the ratio with the continuum heat flux,

$$\frac{\dot{q}}{\dot{q}_{\mathrm{cont}}} = \frac{1}{1 + (15/4)\,\mathrm{Kn}} \tag{3.64}$$

For $\mathrm{Kn} > 10$ (free molecular flow) the ratio of the heat fluxes in a gap also depends only on the Kn number,

$$\frac{\dot{q}_{\mathrm{FM}}}{\dot{q}_{\mathrm{cont}}} = \frac{4}{15\,\mathrm{Kn}} \tag{3.65}$$

The distance of the plates no longer plays a role; the molecules only hit the walls and not their partners. The unsteady heat transfer of rarefied gases is treated in the same manner, see [44, Chapter Mo].

Assuming a round capillary with the outer radius r_A and constant wall heat flux q, the heat transfer of rarefied gas flow (for $\mathrm{Kn} < 0.1$) is expressed with the dimensionless heat transfer coefficient, the Nu_q number.

$$\mathrm{Nu}_q = \frac{\bar{q}}{(T_w - \bar{T})}\frac{r_A}{\lambda} = \frac{\alpha r_A}{\lambda} = 24\left(11 - 6\,\Delta\bar{w} + (\Delta\bar{w})^2 + 24\frac{g}{r_A}\right) \tag{3.66}$$

The dimensionless slip velocity Δw is determined with the slip length from Eq. (3.56).

$$\Delta w = \frac{w(r_A)}{\bar{w}} = \left(1 + \frac{r_A}{4\zeta}\right)^{-1} \tag{3.67}$$

For noncircular cross section, the half of the hydraulic diameter d_h can be taken for r_A. The Nu number for constant wall temperature is about 5% higher than for constant wall heat flux. A numerical study with the Monte-Carlo method [54] indicates that the slip flow model covers very well the convective heat transfer between continuum and molecular flow. The influence of the axial heat conduction has to be considered together with the viscous heat dissipation, but expansion cooling and thermal creep can be neglected.

3.3.3
Micro Heat Exchangers

The basic configuration of micro heat exchangers does not differ from conventional equipment. Heat is transferred from one fluid to the other (for two-flow heat exchanger) with a certain driving temperature difference ΔT. The balance equations for cooling and heating as well as the transfer correlation are independently valid from the scale. In the next paragraph some special issues of micro heat exchangers are addressed.

The energy conservation for cooling and heating both streams entering the heat exchanger is described for a differential element and for the entire apparatus,

$$\text{cooling:}\quad d\dot{Q}_1 = -\dot{m}_1 c_{p1} dT_1;\ \dot{Q}_1 = -\dot{m}_1 c_{p1}(T_1 - T_1') \tag{3.68}$$

$$\text{heating:}\quad d\dot{Q}_2 = \dot{m}_2 c_{p2} dT_2;\ \dot{Q}_2 = \dot{m}_2 c_{p2}(T_2 - T_2') \tag{3.69}$$

If no heat losses to the environment have to be considered (good thermal insulation), both heat amounts are equal and have to be transferred across the channel wall. The heat transfer correlation is described by the kinetic equation for heat transport between the two streams,

$$d\dot{Q}_1 = kb(T_1 - T_2)dz = kA dT_{\text{tot}} \tag{3.70}$$

The heat transfer area A is given by the integration of the channel width over the channel length,

$$A = \int_0^L b dz \tag{3.71}$$

In micro heat exchangers the determination of the heat transfer area is difficult due to the small channels and relatively thick walls. For compact heat exchangers with rectangular cross section channels, Harris et al. [55] proposed a "wall efficiency" comparable to a fin efficiency η_{sw} to include the heat conduc-

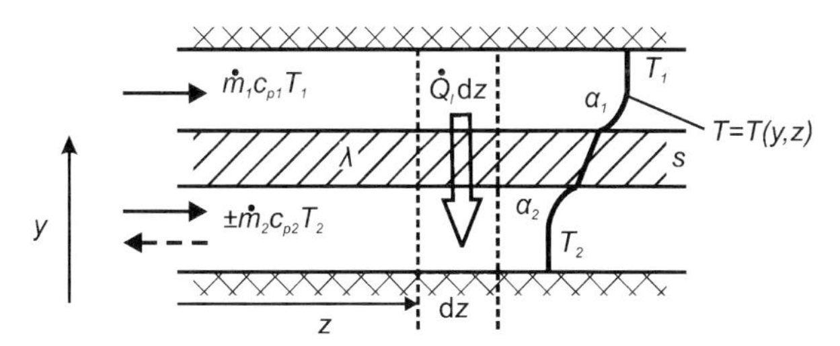

Fig. 3.13 Setup of a heat exchanger channel element.

tion in the side walls (thickness d_{sw}) of the channels. The heat exchanger area A is determined from the channel length l, the channel width d, and the channel height h multiplied by the wall efficiency η_{sw}

$$A = (w + \eta_{sw} h) \cdot l \tag{3.72}$$

with the fin efficiency from the inner heat-transfer coefficient a_i and the heat conductivity λ_w of the wall,

$$\eta_{sw} = \frac{\tanh\left(\sqrt{\frac{2a_i}{d_{sw}\lambda_w}} \frac{h}{2}\right)}{\sqrt{\frac{2a_i}{d_{sw}\lambda_w}} \frac{h}{2}} \tag{3.73}$$

The addition of the single heat-transfer resistances determines the overall heat-transfer coefficient k, here given for a plate

$$k = \left(\frac{1}{a_1} + \frac{s}{\lambda_w} + \frac{1}{a_2}\right)^{-1} \tag{3.74}$$

and for arbitrary cross sections with a reference area A_k

$$\frac{k}{A_k} = \left(\frac{1}{a_1 \cdot A_1} + \frac{s}{\lambda_w \cdot A_w} + \frac{1}{a_2 \cdot A_2}\right)^{-1} \tag{3.75}$$

Typical overall heat transfer coefficients of micro heat exchangers range from 2600 W/m^2 K for gas/liquid flow to 26 000 W/m^2 K for liquid/liquid flow [56]. Conventional plate heat exchangers, which exhibit a very good transfer characteristic, range from 200 to 2500 W/m^2 K for gas/liquid and liquid/liquid flow, respectively [44].

The integration of Eq. (3.70), combined with Eqs. (3.68) and (3.69), over the entire heat exchanger or channel length leads to the mean logarithmic temperature difference with the temperature differences at point z_1 and z_2 (i.e. inlet and outlet of the heat exchanger)

$$\dot{Q}_{ges} = kA\Delta T_{log} = kA \frac{\Delta T_{z1} - \Delta T_{z2}}{\ln\left|\frac{\Delta T_{z1}}{\Delta T_{z2}}\right|} \tag{3.76}$$

The mean logarithmic temperature difference does not depend on the flow direction inside the single-pass heat exchanger with co-current or counter-current flow. For more complex flow inside the heat exchanger, the mean temperature difference is determined for the individual parts, see [44]. The temperature development along the channel length in a single-pass heat exchanger with co-current (left side) and counter-current flow (right side) is sketched in Fig. 3.14.

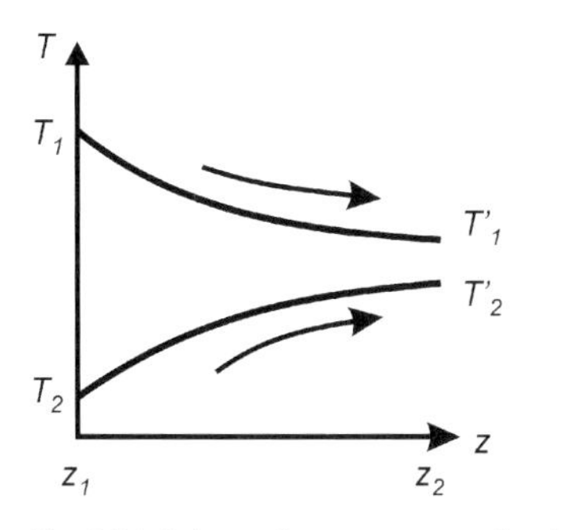

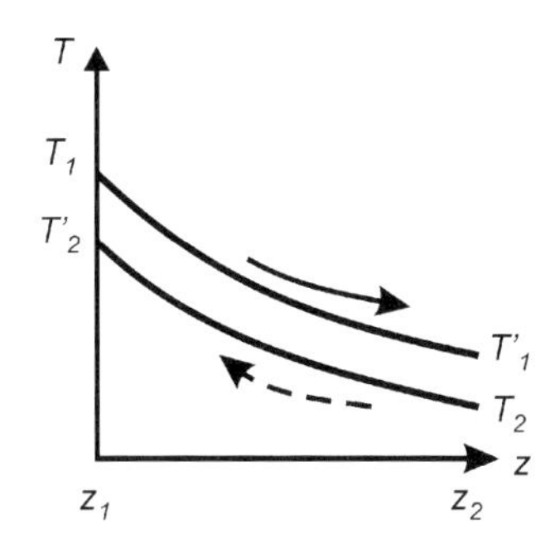

Fig. 3.14 Schematic temperature development along the channel length. Left: co-current flow; right: counter-current flow.

To determine the mean temperature difference from Eq. (3.76) the outlet temperatures of both streams have to be known. They have to be calculated iteratively by numerical outlet-temperature estimation for strongly varying fluid properties with the temperature or for complex flow regimes in the heat-exchanger passages. With a constant overall heat transfer coefficient and fluid properties the differential heat transfer equations can be solved to determine the outlet temperatures [57]. The heat exchange in a differential channel element (3.70) with the driving temperature difference (T_1–T_2) leads to the cooling (3.68) of mass flow 1 or heating (3.69) of mass flow 2,

$$\frac{\mathrm{d}T_1}{\mathrm{d}z} = -\frac{bk}{\dot{m}_1 c_{\mathrm{p}1}}(T_1 - T_2)\,; \tag{3.77}$$

$$\frac{\mathrm{d}T_2}{\mathrm{d}z} = \pm\frac{bk}{\dot{m}_2 c_{\mathrm{p}2}}(T_1 - T_2)\begin{cases} +: & \text{co-current} \\ -: & \text{counter-current} \end{cases} \tag{3.78}$$

The addition of both equations leads to the differential equation for the driving temperature difference.

$$\frac{\mathrm{d}(T_1 - T_2)}{T_1 - T_2} = -kb\mathrm{d}z\left(\frac{1}{\dot{m}_1 c_{\mathrm{p}1}} \pm \frac{1}{\dot{m}_2 c_{\mathrm{p}2}}\right)\begin{cases} +: & \text{co-current} \\ -: & \text{counter-current} \end{cases} \tag{3.79}$$

To solve this ordinary differential equation, the dimensionless temperatures θ_i and the dimensionless ratio N_i of the heat transfer capacity to each fluid flow heat capacity are introduced:

$$\theta_1 = \frac{T_1 - T_1'}{T_1 - T_2},\ \theta_2 = \frac{T_2' - T_2}{T_1 - T_2},\ N_i = \frac{kA}{(\dot{m}c_{\mathrm{p}})_i} \tag{3.80}$$

The dimensionless temperatures are also called the heat exchanger efficiency, which denotes the actual temperature change of one flow compared to the maximal possible temperature change. The capacity ratio is called the number of transfer units first introduced by Chilton [57]. The physical meaning of number of transfer units is threefold:

- the temperature change of the fluid compared to a mean temperature difference $\Delta T_{\log}$,
- with the residence time t_P from Eq. (3.18) the transfer units can be expressed as

$$N_i = \frac{kA}{(mc_p)_i} t_{Pi} \tag{3.81}$$

  A short residence time, which often occurs in micro heat exchangers, leads to a low temperature change of the fluid and a high mean temperature difference.
- with Eq. (3.50), which denotes the characteristic time of unsteady heat transfer or the relaxation time t_{rel} the number of transfer units can also be regarded as the ratio of two characteristic times, where the local heat transfer coefficient a is replaced by the total heat transfer coefficient k.

$$N_i = \frac{t_P}{t_{rel}} \tag{3.82}$$

The ratio of the heat capacity fluxes is often used for further calculations.

$$C_1 = \frac{(\dot{m}c_p)_1}{(\dot{m}c_p)_2} = \frac{N_2}{N_1}; \; C_2 = \frac{N_1}{N_2} = 1/C_1 \tag{3.83}$$

To develop the dimensionless temperature profile along the dimensionless channel length (heat exchanger length) z^*, Eq. (3.78) represents the energy balance of a channel element of fluid 2 in counter-current flow.

$$N_2(T_2 - T_1) + \frac{dT_2}{dz^*} = 0 \tag{3.84}$$

The algebraic transformation and integration over the heat exchanger length leads to the correlation for the outlet temperatures, here for counter-current flow. The correlations for other flow situations can be derived in a similar way and are given in [44]

$$\theta_i = -\frac{1 - \exp((C_i - 1)N_i)}{1 - C_i \exp((C_i - 1)N_i)} \tag{3.85}$$

The above equation is not determined for equal heat capacity flows on both sides (C=1), but this case is given by

$$\theta = \frac{N}{N+1} \tag{3.86}$$

The local temperatures over the heat exchanger length can be derived from the energy balance and are given here for counter-current flow,

$$\theta_1(z^*) = \frac{T_1(z^*) - T_1'}{T_2 - T_1'} = -\frac{N_1}{N_2 - N_1}(1 - \exp(-(N_2 - N_1)z^*)) \tag{3.87}$$

$$\theta_2(z^*) = \frac{T_2(z^*) - T_2}{T_1' - T_2} = \frac{N_2}{N_2 - N_1}(1 - \exp(-(N_2 - N_1)z^*)) \tag{3.88}$$

The co-current flow is similar and can be found in the literature [44, 45, 57].

With this set of equations, the temperatures and heat fluxes can be determined for a given heat exchanger. The heat transfer area, cross sections and wall dimensions can be determined for a heat exchanger layout. The above equations are also valid for mass transfer equipment, if the same physical processes occur.

The design task of a heat exchanger can be twofold: determine the heat transfer area and number of channels from a given process task (temperatures and flow rates) or determine the outlet conditions and characteristics of a given heat exchanger (A, k, geometry). For further information see [44, 46].

3.3.4
Design Issues for Exchange Equipment

Microstructured heat exchangers exhibit not only high transfer rates, but also have drawbacks or points to consider during design and operation. The axial heat conduction in the relatively thick walls has to be considered. The equal distribution on a large number of channels O(1000) is a major task for the manifold design as well as an operating problem to avoid fouling or blocking of single passages or complete parts.

The axial wall conduction in microstructured heat exchangers with gaseous flow was addressed by Stief et al. [58] in a numerical investigation of gas-phase micro heat exchanger. The order of magnitude estimation gave an optimal heat conductivity of the wall material of about 2.5 W/m K, which corresponds to glass or ceramics. Besides this particular example of gas-phase heat exchange some other hints were given for the design of microsystems. This order of magnitude estimation is a good application of the correlations for the heat exchanger.

The temperature profile of the fluids in a counter-current heat exchanger is schematically given in Fig. 3.14, right, and in more detail in Fig. 3.15. Three cases are displayed, low axial heat conduction with good radial heat conduction (thick line), low axial heat conduction with also low radial heat conduction (thick dotted line), and high axial heat conduction (fine dotted line). It is clearly visible that the first case exhibits the best heat exchanger efficiency θ. The last case with high axial heat transfer also exhibits high heat transfer rates at both ends of the heat exchanger, but in the middle of the apparatus, there is only marginal heat transferred.

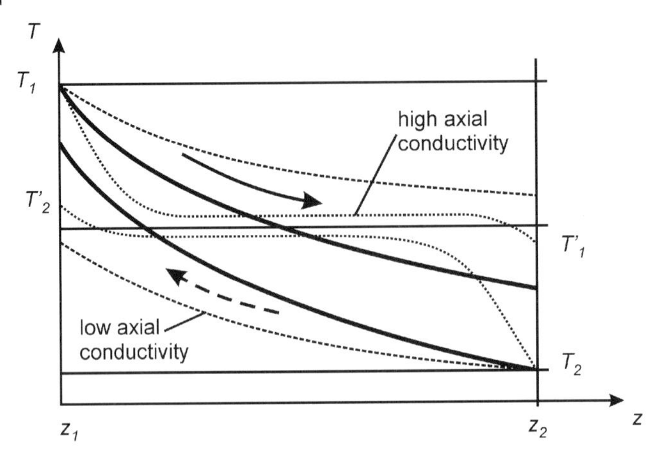

Fig. 3.15 Schematic temperature profile in a counter-current heat exchanger with high axial heat conduction in the wall (thick dotted lines for heat exchanger with low axial heat conduction), from [58].

To estimate the influence of the axial heat conduction in the wall we start with the heat transfer equation similar to Eq. (3.70), where the transported heat depends on three factors.

$$\dot{Q}_i = k_i \cdot A_i \cdot \Delta T_{\mathrm{W}i} \tag{3.89}$$

The axial heat conduction along the heat exchanger length imposes an additional heat flux Q_z on the wall, which increases the temperature difference over the wall. The original temperature difference over the wall in the y-direction comes from the heat transfer. The temperature difference over a small wall element in the channel direction depends on the mean local temperature difference of the temperature change of the flows.

The heat transfer in the z-direction, regarding one channel, depends on the following factors.

$$k_z = \frac{\lambda_{\mathrm{W}}}{\mathrm{d}z};\ A_z = bs;\ \Delta T_{\mathrm{W}z} = \frac{s}{l}\left(\frac{\theta_1 + \theta_2}{2}\right)(T_1 - T_2) \tag{3.90}$$

The same wall element dz transports the heat in the radial direction depending on following factors

$$k_y = \left(\frac{1}{\alpha_1} + \frac{s}{\lambda} + \frac{1}{\alpha_2}\right)^{-1};\ A_y = b \cdot f_{\mathrm{A}} \cdot l;\ \Delta T_{\mathrm{W}y} = \frac{ks}{\lambda_{\mathrm{W}}}\Delta T_{\log} \tag{3.91}$$

with the geometrical factor f_{A} that represents the wall cross section compared to the channel cross section. This ratio equals 1 for thin side walls and is small for very thick side walls that enable an additional axial heat transfer. The comparison of axial heat fluxes with the radial heat flux results in the following dimensionless groups.

$$\frac{\dot{Q}_z}{\dot{Q}_y} \rightarrow \text{a)}\ \frac{k_z}{k_y} = 1 + \frac{\lambda_W}{s}\left(\frac{1}{\alpha_1} + \frac{1}{\alpha_2}\right)$$
$$\text{b)}\ \frac{A_z}{A_y} = \frac{s}{l f_A}$$
$$\text{c)}\ \frac{\Delta T_{Wz}}{\Delta T_{Wy}} = \frac{\lambda_W}{lk}\left(\frac{\theta_1 + \theta_2}{2}\right)\frac{(T_1 - T_2)}{\Delta T_{\log}} = \frac{\lambda_W}{k}\frac{(T_1 - T_2)}{l \cdot \Delta T_{\log}}\left(\frac{\theta_1 + \theta_2}{2}\right) \tag{3.92}$$

The ratios in Eq. (3.92) indicate a high influence of the axial heat conduction not only for a good wall heat conductivity in a gas-phase heat exchanger (low α). A short heat exchanger ($s \approx l$) with thick walls ($f_A \ll 1$) is sensitive to the axial heat conduction as well as a heat exchanger with a high temperature load ($T_1 – T_2 \gg \Delta T_{\log}$). Equation (3.92) gives a first hint of the relevance of axial heat conduction, but the actual temperature profile for different flow regimes, geometrical arrangements, heat capacity flows, and wall heat conductivities is quite complex. Some more information can be found in [58], but for the actual design of a microstructured heat exchanger, a more detailed analysis with numerical methods should be carried out.

Micro heat exchangers possess a large number of channels to provide a high flow rate. The headers and manifolds at the entrance and at the outlet are very important for the almost even distribution of the fluids. The effect of fluid maldistribution can be described by the mass flow difference $\Delta\dot{m}$ between the single channels [57].

Channels with low flow rates undergo a high temperature change, while the temperature change is low in channels with high flow rates. Behind the passages the fluid streams mix and this results in a mean temperature that is lower than the maximum reachable temperature for cooling. Therefore, the efficiency of the heat exchanger is decreased due to the maldistribution in the channels. A rough estimation in [57] shows for the asymptotic efficiency $\theta = 1$ the following behavior:

$$\max\theta = 1 - \frac{\Delta\dot{m}}{\dot{m}} \tag{3.93}$$

Therefore, the mass flux deviation reduces directly the efficiency of the heat exchanger. Particularly in micro heat exchangers with the huge number of passages and potentially high transfer rates and equipment loading, the even distribution of the fluid streams is very important for the equipment performance.

For the entire residence time within the equipment, the flow in the manifolds and fluidic connection has to be considered, which cannot to be neglected in micro devices. The dead volumes in the headers have to avoid by proper design.

Fouling and blocking of the small passages with geometrical dimensions from 10 micrometers up to the millimeter size is the major problem during testing and operation of these devices, see [59, Chapter 1]. Although the problem is omnipresent in microsystem technology, fouling still attracts too little attention in the technical-scientific investigation of microsystems. Hessel et al. [60, Chapters 1 and 2] present two fouling-mitigation techniques in more detail. A separation layer of a flow component is placed between the wall and the particulate or reacting flow and hinders the attachment of solids to the wall. The impingement of two liquid jets and the fast reaction prohibits the fast contact of the reacting flow with the wall (impinging jet micromixer [43]). Disposable devices for singular use often avoid the fouling problem in analytical applications where the device costs are not important.

Experimental results from Wengeler et al. [61] indicate two major fouling effects in microchannels with particulate, laminar flow. Very small particles (<30 nm) migrate to the wall by self-diffusion in the carrier fluid. Larger particles (>150 nm) accumulate at the wall behind bends, nozzles, or expansions or parallel to vortex flow, where a flow direction perpendicular to the wall exists. Smaller particles follow the flow and do not contact the wall. To mitigate particulate fouling, a filtering of the inlet flows is recommended if it is possible. The filtering is only reasonable for particles larger than about 150 nm, smaller particles will pass through the equipment.

A high specific surface energy of a rough surface facilitates the attachment of particles [62]. Construction materials consisting of fine powder (metal or ceramic sintering process) produce a rough surface with a high specific energy. Due to the high surface-to-volume ratio various surface activities have to be considered, e.g. Ni as a catalyst for hydrocarbons from organic liquids. Ni atoms at the surface induce the growth of carbon nanotubes and carbon black, which can block the channels. Other metals like Pt or Cu may have similar effects, hence, the application of alloys has to be considered very carefully. Material experience from larger equipment may not be valid for microstructured devices.

Other mechanisms of fouling in microstructures like corrosion, chemical reactions and precipitation on the surface, adhesion, etc. are not addressed here. Many experimental and operational data have to be collected to give a broad picture of the fouling problem.

The major issue to control fouling in microchannels is the proper design of the channel, the junctions, the expansions, the nozzles, or the elimination of dead zones [63]. A careful design of slender channels without rough changes in cross sections helps to prevent precipitation and attachment of particles. Larger channel diameters allow high flow velocities, and smooth surfaces also help to mitigate precipitation. High flow velocities in microchannels will hinder particles attaching to the wall and already attached particles may be washed away by the high shear gradient. The rule of thumb says "as small as necessary and beneficial, not as small as possible". For elements with high fouling risk one should think about cleanable or disposable elements. Due to the complex flow and great variety of different applications, a fouling-sensitive design and the collection of more experimental data is essential for the successful operation of microstructured devices.

The microfabrication technologies allow a broad variety of geometrical shapes and elements and their combinations. A low pressure loss and a low mean temperature difference are essential for an optimal operation and good heat exchanger performance. These effects can be summarized and generalized in the minimum entropy production [64]. The theory is presented under the concept "constructal theory" [65, 66] and leads to a proper geometrical optimization. An example of optimal spacing for heat transfer in microchannels is given in [67].

3.4 Coupled Processes

This section gives a short overview of the thermodynamic description of coupled transport processes and their implication on microtechnology. The methodology of thermodynamics of irreversible processes is not wide spread in engineering thermodynamics, which deals more with equilibrium thermodynamics and linear efficiency calculations. Any real process is connected with irreversible effects and should be treated from the beginning with that emphasis [68]. For a short introduction and the implication on engineering application see also [69]. For further textbook reading on the general topic see [17, 70], for more detailed information please refer to the following books [71–74]. The matrix in Fig. 3.16 gives an overview about the coupling of momentum, diffusion, thermal, and electrical currents.

The phenomenological description of the correlation between the forces and the related fluxes is given by

$$J_i = L_{ij} \cdot X_i \tag{3.94}$$

The phenomenological coefficients L_{ij} are experimentally determined and depend on the material. Examples for the linear correlations are found on the diagonal of the matrix, the Fourier law of heat transfer

$$Q = -\lambda \frac{A}{l} \Delta T \quad \text{or} \quad \dot{q} = -\lambda \Delta T \tag{3.95}$$

flux \ force	momentum	temperature	concentration	voltage
momentum	$\tau = -\eta \frac{du}{dx}$	thermo-osmosis Knudsen pump	concentration pressure	electroosmosis
heat	mech.-caloric effect thermo-molec. pressure	$\dot{q} = -\lambda \frac{dT}{dx}$	diffus.-thermo effect **Dufour** effect	electrotherm. effect **Peltier** effect
mass	pressure diffusion	thermodiffusion **Soret** effect	$\dot{n} = -D \frac{dc}{dx}$	electrophoresis
el. current	electrical current by mass flow	thermoelectricity **Seebeck** effect	electrical current by concentration diff.	$I = -\sigma \frac{dU}{dx}$

Fig. 3.16 Coupling matrix of transport processes.

or the description of species flow by Fick's law. Additionally, Newton's law of shear stress and Ohm's law for electrical current are given in Fig. 3.16.

The product of the flux J with the generalized force X is called the local entropy production in a system, the dissipation function $\dot{\sigma}$,

$$\dot{\sigma} = J_i \cdot X_i \tag{3.96}$$

For real processes the entropy production is always positive due to the irreversible nature of transport processes.

In microsystem technology coupled phenomena are becoming important, which has been shown in Section 1.5 as a consequence of the gradient increase. In Fig. 3.16, the coupling effects are given for momentum, heat and mass fluxes, and electrical current with the corresponding forces. The description of other coupled effects can be found in [71, 72]. The description of coupled flow phenomena can be described with a system of equations or in matrices, here for two coupled transport processes.

$$\begin{aligned} J_1 &= L_{11}X_1 + L_{12}X_2 \\ J_2 &= L_{21}X_1 + L_{22}X_2 \end{aligned} \quad \text{or} \quad \begin{bmatrix} J_1 \\ J_2 \end{bmatrix} = \begin{bmatrix} L_{11} & L_{12} \\ L_{21} & L_{22} \end{bmatrix} \cdot \begin{bmatrix} X_1 \\ X_2 \end{bmatrix} \tag{3.97}$$

The Onsager reciprocal relation gives the relation between the nondiagonal elements of the symmetric matrix and is based on the microscopic reversibility of the fundamental processes [71],

$$L_{12} = L_{21} \quad \text{or} \quad \left.\frac{J_1}{X_2}\right|_{X_1} = \left.\frac{J_2}{X_1}\right|_{X_2} \tag{3.98}$$

The vertical line indicates a constant value for the displayed parameter. The Onsager reciprocal relation reduces the number of parameters to three independent coefficients. With three experimental test series, the complete matrix can be determined. Additionally to the first-order correlations, there are another six combination possibilities for coupling of two transport processes,

$$\left.\frac{J_1}{J_2}\right|_{X_1} = -\left.\frac{X_2}{X_1}\right|_{J_2} = -\frac{L_{12}}{L_{22}} \tag{3.99}$$

$$\left.\frac{J_1}{J_2}\right|_{X_2} = -\left.\frac{X_2}{X_1}\right|_{J_1} = -\frac{L_{11}}{L_{12}} \tag{3.100}$$

$$\left.\frac{J_1}{X_2}\right|_{J_2} = -\left.\frac{J_2}{X_1}\right|_{J_1} = L_{12} - \frac{L_{11}L_{22}}{L_{12}} \tag{3.101}$$

Some examples of the electrokinetic coupling are given by Hasselbrink [75]. As a further result from the entropy production in Eq. (3.96) the sign of the coefficients can be determined [76].

$$L_{12} > 0; \; L_{11}L_{22} - L_{12}^2 > 0 \tag{3.102}$$

The second equation gives an estimation of the order of magnitude for the coefficients of the coupled processes. For example, the thermodiffusion coefficient in an aqueous system can be estimated with $O(L_{12}) = 10^{-9}$ m^2/s using the order of the heat conductivity $O(L_{11}) = 0.6$ W/m K and the order of the diffusivity $O(L_{22}) = 10^{-9}$ m^2/s [71, p. 406].

The Curie symmetry principle allows only the coupling of the same tensor systems with identical transformation behavior due to rotation in the space [71, 72]. This principle admits only combinations with the same tensor character: a scalar can be combined with a scalar process (chemical reactions, pressure diffusion), a vector with a vector process (heat transfer, diffusion, electrical conduction) and a tensor with a tensor process (pressure and shear tensor for complex media) [72, p. 57].

Many microelectromechanical systems (MEMS) employ coupled processes for a special purpose like sensing (magnetoelectro coupling in a Hall sensor [77]) or cooling (thermo electric coupling in a Peltier or Seebeck element [78]). Further applications with special interest for microprocess engineering are given in the following.

3.4.1
Thermoelectric Energy Conversion

The coupled process between electrical transport and heat transport is described by the Seebeck effect, a temperature difference ΔT induces an electrical potential drop $\Delta\varphi$, and the Peltier effect, an electrical potential difference $\Delta\varphi$ induces a heat flux. The phenomenological coefficients give the correlation between the fluxes and the generalized forces,

$$\begin{aligned} J_e &= \sigma_e \Delta\varphi + L_{12}\Delta T \\ J_Q &= L_{21}\Delta\varphi + \lambda\Delta T \end{aligned} \tag{3.103}$$

The coefficients in Eq. (3.103) are the electrical and thermal conductivity, σ_e and λ, respectively. The coupling coefficients, the Peltier coefficient $\pi_{AB} = L_{21}$ and the Seebeck coefficient $\alpha_{AB} = L_{12}$, depend on the material combination A and B. With the Onsager relation, both coefficients can be combined to

$$\pi_{AB} = \alpha_{AB} \cdot T \tag{3.104}$$

With this relation, a voltage measurement of a material combination under a temperature gradient gives the thermoelectric activity of the combination. In this circumstance a third effect has to be mentioned: the Thomson heat, which occurs when an electrical current flows through a conductor with a temperature gradient. The Thomson effect depends on the Peltier coefficient and on its derivation with the temperature.

An overview of the fabrication techniques, the appropriate materials and the variety of applications can be found in [79]. Besides measuring applications, the thermoelectric effect is often used for the generation of electrical energy from a given temperature difference. A wristwatch is powered by heat released from the human body [80] or the energy supply of deep space missions like "Voyager" with radioisotope thermoelectric generators (RTGs). The miniaturization of thermoelectric devices is very promising due to integration of a high number of material pairs in a small device, which results in a high voltage, and due to the flexible application of small energy converters [81]. Recent studies show a good thermal efficiency of an optimized geometrical arrangement to yield a low thermal conductivity and a high electrical conductivity [78].

An interesting application is the skilful combination of a microreactor with the temperature control of Peltier elements [82]. A microreactor array on a chip is covered on both sides with a Peltier element and submerged in a heat bath to control the ambient temperature. The heat release from an exothermic reaction in the microreactor is adsorbed by one Peltier element, the other element serves for the temperature control of the calorimeter. The calorimeter exhibits a time constant of about 3 s for a volume of about 50 µl. Two example reactions of high energetic reactions are given by Schifferdecker et al. [83] that match quite well with data from literature. An open question is certainly the local distribution of the reaction progress and local inhomogeneities.

3.4.2 Electro-osmotic Flow and Electrokinetic Pumping

The correlation between the electrical current and the fluid flow is called the electrokinetic flow. A related kind of electrically induced flow is given by the electrophoretic flow of species through a matrix or gel. This effect is often used for separation and detection of larger molecules and biological reagents on microfluidic chips.

Kirby and Hasselbrink [84] describe in two review articles the typical surface characteristics and Zeta potential of silicon, glass, and polymers for microfluidic applications. They recommend controlling precisely the temperature and the pH value of the solution to achieve reliable experimental data. To measure the streaming potential and the electro-osmotic mobility are the most effective methods to determine the Zeta potential. The thermodynamics of irreversible processes and the Onsager reciprocity relations are helpful for minimizing experimental effort [85]. In a system with two coupled effects, here the electrical and hydrodynamic flux, only three measurements are necessary to determine all coefficients [75]. Thermodynamics can also be used for finding the possible applications, for example, the energy conversion to produce an electrical current from hydrodynamic motion.

Electro-osmotic flow is often used for molecular separation on microfluidic chips [86]. Pfeiffer et al. [87] give an overview and some heuristic and numerical optimization design techniques. Starting from geometrical channel elements like straight channels, 180° turns, meandering channels, and spirals, complete separation chips are designed and compared to optimize the channel arrangement. The aim of the study is a minimized area consumption and an optimized topology, where electronic tools like VLSI techniques can be useful.

Electro-osmotic pumping, also called electrokinetic pumping, employs the negative charge of the solid wall and the corresponding positive charge of ions in the fluid. An external electrical field generates the motion of the counterions and therefore of the entire fluid in the capillary. The electro-osmotic effect decreases with increasing channel diameter, which have characteristic dimension of about 1 μm. High pressure differences up to 2 MPa of water have been realized for low flow rates of 3.6 μm/min. High flow rates up to 33 ml/min have been measured that indicates that the main application of electroosmosis pumps are for dosing and analytical applications [88]. The efficiency of electrokinetic pumping is mainly determined by the channel width compared to the EDL length. An optimal ratio of 2 to 5 was found by Min et al. [89] in an analytical and numerical study for silica and a pH-value larger than 8, which gives an EDL thickness of about 10 nm.

3.4.3 Thermodiffusion

The thermodiffusion describes the effect of different migration velocities of molecules with different size or mass in a thermal gradient field and the generation of a concentration gradient. Assisted by natural convection the thermodiffusion was employed for the separation of isotopes in the Clusius–Dickel column, see London [90]. While the Fourier law describes the coupling between the heat flux and temperature gradient, Fick's first law gives the linear relationship between the mass or species flow and the concentration gradient. The cross coupling between the heat flux and concentration gradient (Dufour effect L_{qD}) as well as the cross coupling between the species flux and a temperature gradient (Soret effect L_{Dq}) takes the following form,

$$\begin{aligned} J_{\mathrm{q}} &= -\lambda \Delta T + L_{\mathrm{qD}} \Delta c \\ J_{\mathrm{n}} &= L_{\mathrm{Dq}} \Delta T - D \Delta c \end{aligned} \tag{3.105}$$

The Soret coefficient L_{Dq} is 3 to 5 magnitudes smaller than the first order effects of heat conduction L_{qq} or diffusion L_{DD} [91]. The thermodiffusion is a single-phase separation process for gaseous and liquid mixtures without unwanted interfacial forces. The fluid mixture is flowing through a number of parallel microchannels with a superimposed temperature gradient. In a gas, smaller molecules tend to diffuse to the hotter side, larger ones diffuse to the colder side. At the end of the channel, the flow is split up and rearranged into the next channels. By cascading the parallel channels, the concentration of the smaller molecules is gradually enriched at the hot side.

Kockmann et al. [92] investigated the separation of a gaseous mixture (gas, 50% H_2, 50% CO_2) in a microchannel array with a given temperature profile at the walls and a resulting temperature profile over the channels. The numerical simulations of an array of 5×3 channels (200 μm wide and 400 μm deep) were performed with the commercial FVM-code CFD-ACE+ by the ESI group. Based on the simulations and the available fabrication techniques, the structures were designed and fabricated in silicon to demonstrate the separation in liquid mixtures, see Fig. 3.17. A silicon wafer was structured with successive DRIE and KOH etching and finally covered by two Pyrex lids. A difficult point was the drilling or grinding of the fluid connection with 1 μm diameter into the bottom glass lid. The first fluidic tests gave indications of how to handle microfluidic devices in a test environment.

A similar coupled process with a thermal gradient is used in vacuum technology to pump rarefied gases, the so-called Knudsen effect or thermal transpiration. The effect is based on the correlation between the pressure and temperature of two connected systems in the free molecular flow regime ($Kn > 10$)

$$\frac{p_1}{p_2} = \sqrt{\frac{T_1}{T_2}} \tag{3.106}$$

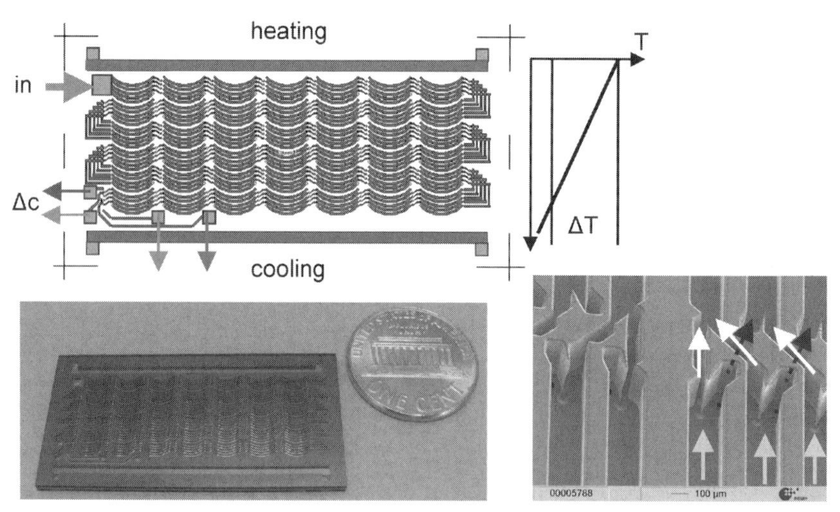

Fig. 3.17 Top: chip layout of a 40 × 20 mm^2 separation chip with 42 blocks of 5 bent channels each. The inlet and the four outlet connections for the mixture are arranged on the left side, the heating and cooling channel on the top and bottom side, respectively. Bottom: left: Fabricated silicon chip with glass lids; right: SEM image of the RIE fabricated crossing structure, 400 μm deep. The arrows indicate the crossing direction of the divided flow between the channel blocks.

For a silicon micropump, McNamara and Gianchandani [93] give a short explanation of the underlying theory and show the microfabrication of the 1.5×2.0 mm silicon chip. The pump consists of five large channels, 10 μm deep and 30 μm wide, connected with narrow channels, 10 μm wide and 100 nm deep. Two of the large channels are heated with a polysilicon heater up to 600 °C. With an on-chip pressure sensor, a pressure difference of about 3 kPa was measured for a power input of 60 mW. Multiple stages may be cascaded to get a pump with a higher pressure difference for applications in gas sampling, pneumatic actuation, or vacuum encapsulation. Alexeenko et al. [94] describe the theoretical modelling of the flow in a Knudsen pump with the Monte-Carlo method (DSMC).

Interesting effects, which are not covered here, are connected with rarefied gas flow, surface effects and thermal gradients. Sone [95] reported on a "ghost effect", which describes a flow phenomena of rarefied gases under heat transfer.

3.4.4
Pressure Diffusion

An advanced process of the microcyclone, the pressure diffusion of a flow, causes a species flow and a concentration gradient. Within a separation nozzle, see Fig. 3.18, the pressure diffusion is assisted by centrifugal forces to separate uranium isotopes. An early application for isotope separation of chemically etched microstructures is described by Schütte [91]. The high energy demand compared to ultracentrifuges was the limiting factor for the application of sepa-

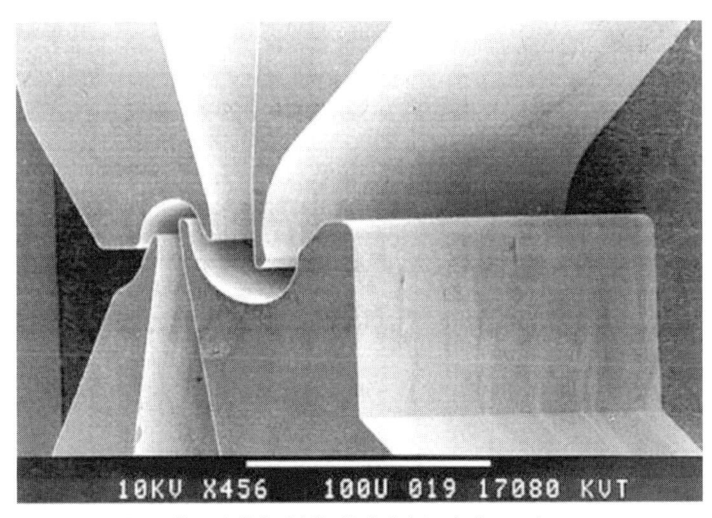

Fig. 3.18 Separation nozzle for pressure diffusion, LIGA fabricated, courtesy of Forschungszentrum, Karlsruhe, Germany.

ration nozzles. Modern processes are working with ultracentrifuges, which cause a higher centrifugal force, an easy staging, and higher separation factors [91]. A few years later, Becker et al. [96] reported upon experimental investigations with separation nozzle devices, which show only slightly higher energy consumption than contemporary gas-diffusion plants. For microstructured devices fabricated with the help of synchrotron radiation, lithography, and galvanoplastics, Becker et al. [96, 97] reported an increase by the factor of three for the gas pressure and "considerable savings" for the enrichment of uranium isotopes. The nozzle channels were fabricated with a channel height of about 400 μm and with structural details of about 0.1 μm.

The succeeding development of more appropriate and advanced geometries led to the development of the LIGA technology [98, 99, Chapter 7] and is one of the birthplaces in Germany for microsystem technology and micro process engineering.

3.5 Conclusions and Outlook

Transport processes and equipment hold the best promise for the successful application in various fields, where high transfer rates, intelligent and prolific combinations of microstructures in macrodevices and new process routes are needed. With the fundamentals and some basic applications of heat and mass transfer, the benefits of miniaturization have been shown. High gradients and a high specific surface in devices with various construction materials lead to fast equilibrium. The characteristic dimensions of the microstructure internals are in the range of boundary layers or entrance lengths, where high gradients enforce the transfer processes. Additionally, the cascading of effects and the integration of the various elements enhance and guide the entire process. This enables new possibilities for difficult processes and opens new opportunities for the future.

References

1 J.H. Ferziger, H.G. Kaper, *Mathematical Theory of Transport Processes in Gases*, North Holland, Elsevier, **1972**.
2 J. Frenkel, *Kinetic theory of liquids*, Dover, New York, **1955**.
3 D.J. Evans, G.P. Morriss, *Statistical Mechanics of Nonequilibrium Liquids*, Academic Press, London, **1990**.
4 M. Gad-el-Hak (Ed.), *The MEMS Handbook*, CRC Press, Boca Raton, **2003**.
5 R. Narayanan, D. Schwage (Eds.), *Interfacial Fluid Dynamics and Transport Processes*, Springer, Berlin, **2003**.
6 A. Mersmann, *Thermische Verfahrenstechnik – Grundlagen und Methoden*, Springer, Berlin, **1980**.
7 D.D. Joseph, L. Preziosi, *Rev. Mod. Phys.*, **1989**, *61*, 41–73.
8 I. Müller, T. Ruggeri, *Rational Extended Thermodynamics*, Springer, New York, **1998**.

9 D. Jou, J. Casas-Vazquez, G. Lebon, *Extended Irreversible Thermodynamics*, Springer, Berlin, **2001**.

10 R.B. Bird, W.E. Stuart, E.N. Lightfoot, *Transport Phenomena*, Wiley, New York, **2002**.

11 A. Schönbucher, *Thermische Verfahrenstechnik*, Springer, Berlin, **2002**.

12 W. Jost, *Diffusion – Methoden der Messung und Auswertung*, Steinkopff, Darmstadt, **1957**.

13 E.L. Cussler, *Diffusion – Mass Transfer in Fluid Systems*, Cambridge University Press, Cambridge, **1997**.

14 J. Stachel, *Einsteins Annus mirabilis – Fünf Schriften, die die Welt der Physik revolutionierten, Zweiter Teil*, Rowohlt, Reinbek, **2001**.

15 C.J. Geankoplis, *Transport Processes and Separation Process Principles*, Prentice Hall, Upper Saddle River, **2003**.

16 F. Schönfeld, V. Hessel, C. Hofmann, *Lab Chip*, **2004**, *4*, 65–69.

17 G. Kluge, G. Neugebauer, *Grundlagen der Thermodynamik*, Spektrum Akademischer Verlag, Heidelberg, **1994**.

18 B. Weigand, *Analytical Methods for Heat Transfer and Fluid Flow Problems*, Springer, Berlin, **2004**.

19 M. Baerns, H. Hofmann, A. Renken, *Chemische Reaktionstechnik*, Thieme, Stuttgart, **1992**.

20 H.D. Baehr, K. Stephan, *Wärme- und Stoffübertragung*, Springer, Berlin, **2004**.

21 J.R. Bourne, *Org. Proc. Res. Dev.* **2003**, *7*, 471–508.

22 M. Kraume (Ed.), *Mischen und Rühren*, Wiley-VCH, Weinheim, **2003**.

23 D. Bothe, C. Stemich, H.J. Warnecke, *Chem.-Ing.-Techn.* **2004**, *76*, 1480–1484.

24 M. Schlüter, M. Hoffmann, N. Räbiger, *Chem.-Ing.-Techn.* **2004**, *76*, 1682–1688.

25 M. Engler, T. Kiefer, N. Kockmann, P. Woias, *IMRET8*, TK128d, **2005**.

25a N. Kockmann, T. Kiefer, M. Engler, P. Woias, Silicon microstructures for high throughput mixing devices, *J. Microfluids Nanofluidics*, acc. paper, **2005**.

26 N. Kockmann, M. Engler, D. Haller, P. Woias, ICMM2004-2331, **2004**, 165–171.

27 N. Kockmann, M. Engler, D. Haller, P. Woias, *Heat Trans. Eng.* **2005**, *26*, 71–78.

28 F. Jiang, K. Drese, S. Hardt, M. Küpper, F. Schönfeld, *AIChE J.* **2004**, *50*, 2297–2305.

29 N. Kockmann, M. Engler, P. Woias, *Chem.-Ing.-Techn.* **2004**, *76*, 1777–1783.

30 N. Kockmann, T. Kiefer, M. Engler, P. Woias, ICMM2005-75125, **2005**.

31 N. Kockmann, T. Kiefer, M. Engler, P. Woias, *Transducers 05*, EA363, **2005**; and acc. contribution to *Sens. Act. B*.

32 M. Engler, N. Kockmann, T. Kiefer, P. Woias, ICMM04-2412, **2004**, 781–788.

33 J.M. Ottino, *The Kinematics of Mixing: Stretching, Chaos, and Transport*, Cambridge University Press, Cambridge, **1997**.

34 N.T. Nguyen, Z. Wu, *J. Micromech. Microeng.* **2005**, *15*, R1–R16.

35 A.D. Stroock, S.K.W. Dertinger, A. Ajdari, I. Mesic, H.A. Stone, G.M. Whitesides, *Science*, **2002**, *295*, 647–651.

36 V. Hessel, H. Löwe, F. Schönfeld, *Chem. Eng. Sci.* **2005**, *60*, 2479–2501.

37 J.D. Tice, H. Song, A.D. Lyon, R.F. Ismagilov, *Langmuir*, **2003**, *19*, 9127–9133.

38 A. Gavriliidis, P. Agneli, W. Salman, *IMRET8*, TK128c, **2005**.

39 S.A. Khan, A. Günther, M.A. Schmidt, K.F. Jensen, *Langmuir*, **2004**, *20*, 8604–8611.

40 R.H. Perry, D.W. Green, *Perry's Chemical Engineers' Handbook*, McGraw, Hill, New York, **1998**.

41 N. Aoki, S. Hasebe, K. Mae, *IMRET8*, TK132a, **2005**.

42 M. Zlokarnik, *Rührtechnik – Theorie und Praxis*, Springer, Berlin, **1999**, or English translation: *Stirring, Theory and Practice*, Wiley-VCH, Weinheim, **2001**.

43 IMM Institut für Mikrotechnik Mainz GmbH, *Process Technology of Tomorrow – The Catalogue*, Mainz, **2003**.

44 H. Kraussold, Ed., *VDI-Wärmeatlas*, VDI-Verlag, Düsseldorf, **1997**.

45 J.H. Lienhardt IV, J.H. Lienhardt V, *A Heat Transfer Textbook*, Phlogiston Press, Cambridge, **2003**, URL: web.mit.edu/lienhard/www/ahtt.html.

46 G.F. Hewitt (Ed.), *Heat Exchanger Design Handbook HEDH*, Begell House, New York, **2002**.

47 F. Bosnjakovic, K.F. Knoche, *Technische Thermodynamik Teil II*, Steinkopff, Darmstadt, **1997**.

48 S.V. Garimella, V. Singhal, ICMM2003-1018, **2003**, 159–169.

49 B. Palm, *Microscale Thermophys. Eng.* **2001**, *5*, 155–175.

50 G.P. Celata, ICMM2003-1019, **2003**, 171–179.

51 M. Knudsen, *Kinetic Theory of Gases*, Wiley, London, **1950**.

52 S. Chapman, T.G. Cowling, D. Burnett, *The Mathematical Theory of Non-Uniform Gases*, University Press, Cambridge, **1970**.

53 A. Frohn, *Einführung in die Kinetische Gastheorie*, Aula-Verlag, Wiesbaden, **1988**.

54 N.G. Hadjiconstantinou, O. Simek, *J. Heat Transfer*, **2002**, *124*, 356–364.

55 C. Harris, M. Despa, K. Kelly, *J. MEMS*, **2000**, 4, 502–508.

56 E. Klemm, M. Rudek, G. Markowz, R. Schütte, *Mikroverfahrenstechnik*, in: Winnacker-Küchler: Chemische Technik – Prozesse und Produkte, Vol. 2, Wiley-VCH, Weinheim, **2004**, Chapter 8.

57 E.U. Schlünder, *Einführung in die Wärmeübertragung*, Vieweg, Braunschweig, **1991**.

58 T. Stief, O.U. Langer, K. Schubert, *Chem.-Ing.-Techn.* **1998**, *70*, 1539–1544, also in *Chem.-Eng.-Tech.* **1999**, *21*, 297–302.

59 V. Hessel, S. Hardt, H. Löwe, *Chemical Micro Process Engineering – Fundamentals, Modeling and Reactions*, Wiley-VCH, Weinheim, **2004**.

60 V. Hessel, H. Löwe, A. Müller, G. Kolb, *Chemical Micro Process Engineering – Processing and Plants*, Wiley-VCH, Weinheim, **2005**.

61 R. Wengeler, M. Heim, M. Wild, H. Nirschl, G. Kaspar, N. Kockmann, M. Engler, P. Woias, ICMM2005-75199, **2005**.

62 R.J. Adrian, E. Yamaguchi, P. Vanka, T. Plattner, W. Lai, *12th Int. Symp. Appl. Laser Techniques to Fluid Mechanics*, Lisbon, **2004**.

63 N. Kockmann, M. Engler, P. Woias, *ECI Heat Exchanger Fouling and Cleaning: Challenges and Opportunities*, Kloster Irsee, **2005**.

64 A. Bejan, *Convection Heat Transfer*, Wiley, New York, **2004**.

65 A. Bejan, *Shape and Structure, From Engineering to Nature*, Cambridge University Press, **2000**.

66 J. Lewins, *Int. J. Heat Mass Transfer*, **2003**, *46*, 1541–1543.

67 M. Favre-Marinet, S. Le Person, A. Bejan, *Microscale Thermophys. Eng.* **2004**, *8*, 225–237.

68 Y. Demirel, S.I. Sandler, *J. Phys. Chem. B*, **2004**, *108*, 31–43.

69 I. Müller, *Chem.-Ing.-Techn.* **2000**, 72, 194–202.

70 A. Bejan, *Advanced Engineering Thermodynamics*, John Wiley, New York, **1997**.

71 R. Haase, *Thermodynamik der irreversiblen Prozesse*, Steinkopff, Darmstadt, **1963**.

72 S.R. de Groot, P. Mazur, *Non-Equilibrium Thermodynamics*, Dover Publications, New York, **1984**.

73 D. Kondepudi, I. Prigogine, *Modern Thermodynamics – From Heat Engines to Dissipative Structures*, Wiley, Chichester, **1998**.

74 Y. Demirel, *Nonequilibrium Thermodynamics – Transport and Rate Processes in Physical and Biological Systems*, Elsevier, Amsterdam, **2002**.

75 E.F. Hasselbrink, *Electrokinetic Pumping and Power Generation: Onsager Reciprocal Processes*, ICMM2004, panel session, **2004**.

76 Y. Demirel, S.I. Sandler, *Int. J. Heat Mass Transfer*, **2001**, 44, 2439–2451.

77 S. Middelhoek, S.A. *Audet, Silicon Sensors*, TU Delft, Dep. of Electrical Engineering, **1994**.

78 M. Strasser, *Entwicklung und Charakterisierung mikrostrukturierter thermoelektrischer Generatoren in Silizium-Halbleitertechnologie*, Shaker, Aachen, **2004**.

79 G.S. Nolas, J. Sharp, H.J. Goldsmid, *Thermoelectrics – Basic Principles and New Material Developments*, Springer, Berlin, **2001**.

80 M. Kishi, H. Nemoto, M. Yamamoto, S. Sudou, M. Mandai, S. Yamamoto, *Microthermoelectric modules and their application to wrist-watches as an energy source*, IEEE Proc. ICT **1999**, 301–307.

81 H. Böttner, *Thermoelectric Micro Devices: Current State, Recent Developments and Future Aspects for Technological Progress and Applications*, IEEE Proc. ICT **2002**, 511–518.

82 J. Antes, D. Schifferdecker, H. Krause, S. Loebbecke, *Chem.-Ing.-Tech.* **2004**, *76*, 1332–1333.

83 D. Schifferdecker, J. Antes, S. Loebbecke, *IMRET8*, 134b, **2005**.

84 J. Kirby, E. F. Hasselbrink, *Electrophoresis*, **2004**, *25*, 187–202.

85 E. Brunet, A. Ajdari, *Phys. Rev. E*, **2004**, *69*, 016306.

86 J. Khandurina, A. Guttman, *Curr. Opin. Chem. Biol.* **2002**, *6*, 359–366.

87 A. J. Pfeiffer, T. Mukherjee, S. Hauan, *Ind. Eng. Chem. Res.* **2004**, *43*, 3539–3553.

88 D. J. Laser, J. G. Santiago, *J. Micromech. Microeng.* **2004**, *14*, R35–R64.

89 J. Y. Min, E. F. Hasselbrink, S. J. Kim, *Sens. Actuators B*, **2004**, *98*, 368–377.

90 H. London (Ed.), *Separation of Isotopes*, Georges Newnes Ltd., **1961**.

91 R. Schütte, *Diffusionstrennverfahren*, in *Ullmanns Encyklopädie der technischen Chemie*, E. Bartholomé (Ed.), Vol. 2, Verfahrenstechnik I (Grundoperationen), **1972**.

92 N. Kockmann, P. Woias, *IMRET8*, TK129a, **2005**.

93 S. McNamara, Y. Gianchandani, *Transducers '03*, **2003**, 1919–1922.

94 A. A. Alexeenko, S. F. Gimelshein, E. P. Muntz, A. D. Ketsdever, *Modeling of Thermal Transpiration Flows for Knudsen Compressor Optimization*, AIAA Paper **2005**, 963.

95 Y. Sone, Kinetic Theory and Fluid Dynamics, Birkhäuser, Boston, **2002**.

96 E. W. Becker, W. Bier, W. Ehrfeld, K. Schubert, R. Schütte, D. Seidel, *Naturwissenschaften*, **1976**, *63*, 407–411.

97 E. W. Becker, W. Ehrfeld, D. Münchmeyer, H. Betz, A. Heuberger, S. Pongratz, W. Glashauser, H. J. Michel, R. v. Siemens, *Naturwissenschaften*, **1982**, *69*, 520–523.

98 E. W. Becker, W. Ehrfeld, P. Hagmann, A. Maner, D. Münchmeyer, *Microelectronic Eng.* **1986**, *4*, 35–56.

99 W. Menz, J. Mohr, *Mikrosystemtechnik für Ingenieure*, VCH, Weinheim, **1997**.

4
Multiphase Flow, Evaporation, and Condensation at the Microscale

Michael K. Jensen, Yoav Peles, and Theodorian Borca-Tasciuc,
Rensselaer Polytechnic Institute, Troy, USA
Satish G. Kandlikar, Rochester Institute of Technology, Rochester, USA

Abstract
A review of boiling and condensation heat transfer, fluid flow, and pressure drop characteristics in mini- and microchannels is presented. Because this topic of research is relatively new, the literature is not as extensive as with conventional-sized channels. Nevertheless, flow patterns, heat-transfer coefficients (boiling and condensing), the critical heat flux condition, onset of nucleate boiling, and two-phase pressure drop data have been acquired and quantified to a limited extent. Likewise, engineering correlations and analytic/numerical approaches have been developed to predict behavior. Flow- and heat-transfer similarities and differences compared to large-scale systems have been observed, but further investigations are needed to fully assess and predict these thermo-hydraulic phenomena at the microscale.

Keywords
Boiling, condensation, two-phase flow, flow patterns, pressure drop, critical heat flux

Advanced Micro and Nanosystems Vol. 5. Micro Process Engineering. Edited by N. Kockmann

ISBN: 3-527-31246-3

4.1 Introduction

Boiling/evaporation and condensation are very effective heat-transfer mechanisms with much higher heat-transfer coefficients compared to single-phase flows; with change of phase, flow rates are reduced because of use of the enthalpy of vaporization, and more uniform temperatures are obtained than with single-phase flows. Reductions in channel hydraulic diameters cause heat-transfer coefficients to increase dramatically; in addition, small channel diameters can result in very high area densities (heat transfer area per unit volume), smaller volumes of working fluids in heat-transfer systems, and faster response times. Hence, when change of phase heat transfer is combined with small channels, the result is a very efficient, compact engineering system for heat transfer.

The microelectronics industry is searching for new cooling concepts that will enable the use of ultra high heat flux IC chips, and ultracompact heat exchangers are being developed for the automotive, process, aerospace, and other industries and applications. Other applications include use in microrefrigerators, microchemical reactors, and microrockets. Forced convection boiling and condensation in microchannels may be the cooling/heating mechanisms of choice for these ultra high heat flux applications, as well as for other microelectromechanical systems (MEMS) applications.

Despite the many attractive characteristics of microscale phase change heat transfer, there also are some issues that need to be addressed. In particular, there has been a lack of in-depth knowledge of heat transfer mechanisms at the microscale, uncertainties about applying correlations for conventional-sized channels to microchannels, and concern about possible operational differences compared to conventional scale (e.g., flow oscillations, low critical heat flux, etc.).

Because of the potential benefits of and interest in using small channels in a variety of heat transfer applications, in the past several years two-phase flow boiling and condensation in channels with cross-sectional dimensions on the order of 100 μm to 5 mm (with more experimental investigations using channels toward the large end of this range, which is still smaller than conventional-sized channels) have been topics of active research in academia as well as by the microelectronics and other industries. Reviews have been presented on boiling (e.g., Palm [1], Mehendale et al. [2], and Kandlikar [3, 4]) and condensation (e.g., Cavallini et al. [5]). These research thrusts have provided knowledge and insight into heat transfer and fluid flow mechanisms during boiling and condensation and have resulted in heat transfer correlations, flow pattern identification, and knowledge about other fluid flow and heat transfer characteristics in microchannels.

This chapter presents a survey of present understanding of multiphase flow, and evaporation and condensation heat transfer in mini- and micro-channels.

4.2 Two-phase Flow Patterns

The intent of study of two-phase flow patterns in boiling heat transfer is to define the boundaries of the liquid and vapor domains in space and, to some respect, in time. Since boiling and condensation heat transfer is related to the flow structure, identification of the two-phase flow patterns provides means to evaluate the heat-transfer mechanisms and the hydrodynamic flow field. This allows proper development of correlations and models to predict the pressure drop, heat-transfer coefficient, and the critical heat flux condition. It should be noted, however, that information on the flow pattern does not guarantee comprehensive knowledge of the heat-transfer mechanism. In fact, considerable effort to link flow patterns to boiling mechanisms have been conducted over the last half century, and a set of qualitative traits linking flow patterns to boiling mechanism, heat-transfer coefficient and pressure drop in conventional scale systems has been established. However, because most of these correlations are ad hoc, adaptation of large-scale knowledge to microchannels should be carefully exercised. Since the flow structure during forced boiling in channels is complex and varies considerably along the flow path, often it is desirable to study the morphology of the two-phase flow under adiabatic flow conditions. Thus, a review of adiabatic two-phase flows in microchannels is included.

4.2.1 Adiabatic Flow Patterns

Adiabatic two-phase flow in tubes with hydraulic diameters in the order of 1 mm have been extensively investigated in the last decade (e.g., Lee and Lee [6]; Zhao and Bi [7]; Yang and Shieh [8]; Triplett et al. [9]; Xu et al. [10]; Lin et al. [11]; Ide et al. [12]; Mishima and Hibiki [13]; Keska and Fernando [14]). Kawaji [15] summarized the major differences in flow patterns, void fraction, and pressure drop between adiabatic two-phase flows in channels with hydraulic diameters of $O(1\ \mathrm{mm})$ and of $O(10\ \mathrm{mm})$. Two-phase flows in minichannels (~ 1 mm) are characterized by diminishing gravitation effects, the disappearance of a stratified flow pattern, and a shift in some of the flow-pattern boundaries. However, generally the two-phase flow patterns in conventional scale channels and minichannels appear to be morphologically similar (Kawaji and Chung [16]).

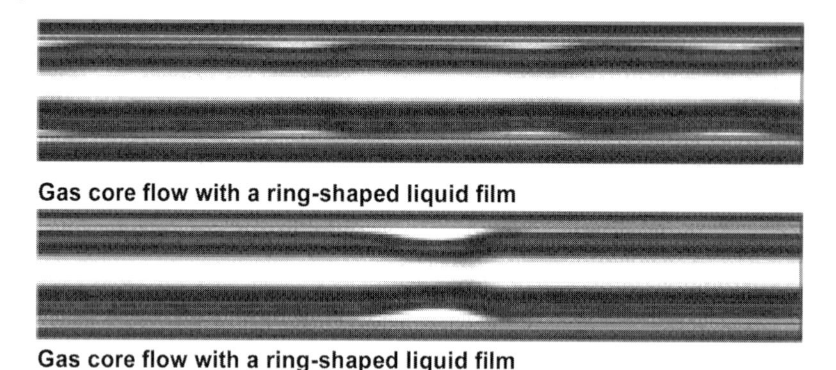

Fig. 4.1 Liquid-ring flow pattern (from Kawahara et al. [21], used with permission).

Although relatively little is known about two-phase flow patterns in channels with hydraulic diameter in the 100-μm range, it appears that for microchannels the flow morphology is modified in addition to other alterations in the flow characteristics. Kawaji and Chung [16] defined the boundaries between minichannels and microchannels for adiabatic flow to be between 100 μm and 250 μm, based on unique flow patterns and strong deviations in void fraction from that found in minichannels. The *liquid-ring* flow pattern reported by several studies (Feng and Serizawa [17, 18]; Serizawa and Feng [19,20]; Kawahara et al. [21]) to be unique to microchannels is observed at low gas flow rates and is characterized by axi-symmetrical distribution around the channel inner wall as shown in Fig. 4.1.

In the work by Serizawa and coworkers [17–20] on air–water flow inside 25-μm and 100-μm capillary tubes, five flow patterns were identified: dispersed bubbly flow, gas slug flow, liquid ring flow, liquid lumped flow, and liquid droplet flow. Kawaji, Chung and coworkers [20–23] observed slug flow, liquid-ring flow, gas core flow with a serpentine liquid film, and semi-annular flow patterns in a 100-μm circular tube with the boundaries shown in Fig. 4.2. Kawahara et al. [21] compared the flow map depicted in Fig. 4.2 with maps obtained in ~1 mm diameter channels by Damianides and Westwater [24], Fukano and Kariyasaki [25], Triplett et al. [9], and Zhao and Bi [7] and delineated the differences and similarities. Intermittent flow patterns such as plug and slug flow, and an annular (or semi-annular) flow pattern were all characteristics of minichannels as well as microchannels, while bubbly and churn flow patterns were absent in channels with hydraulic diameters in the 100-μm range. Differences in the location of the transition boundaries were also found, but due to variations among the flow-pattern maps for ~1-mm diameter channels, no discussion was presented.

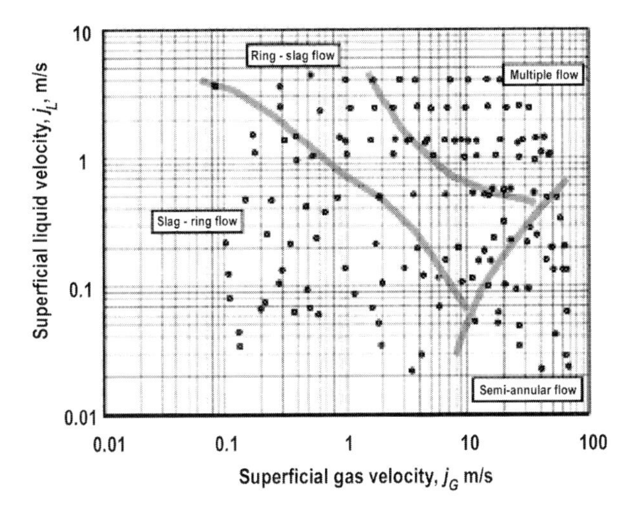

Fig. 4.2 Two-phase flow regime map for a 100-μm microchannel (from Kawahara et al. [21], used with permission).

4.2.2
Boiling Flow Patterns

Similar to conventional scale, various flow patterns have been reported to develop during boiling in microchannels including bubbly, slug, churn, annular, mist, and dry-out. Several authors also reported unique flow patterns commonly referred as *rapid bubble growth* (Kandlikar [3, 4]; Wu and Cheng [26, 27]), which is related to significant flow and pressure oscillations, occasionally observed in single tubes (Kamidis and Ravigururajan [28]) but more often in multichannel arrangements (Jiang et al. [29, 30]; Cornwell and Kew [31, 32]; Mertz et al. [33]; Kennedy et al. [34]; Kandlikar [35]). In the initial stage of the ebullition cycle during the *rapid bubble growth* a spherical bubble grows until it attains a size comparable to the channel hydraulic diameter. Constrained by the channel walls the bubble can not expand radially, and thus grows rapidly in the longitudinal direction (downstream as well as upstream) causing flow reversal. Although the *rapid bubble growth* is frequently considered along other flow patterns, it is perhaps better to categorize this phenomenon as flow instability. It should be noted that several remedies are available to suppress this deleterious flow oscillation, but a comprehensive discussion on boiling instabilities is left to Section 4.5.

Early studies (Peng et al. [36–38]; Hu et al. [39]) on boiling in microchannels have reported radically different flow phenomena from those in conventional scale channels, particularly the existence of what has been referred to as "fictitious boiling" by Hu et al. [39]. "Fictitious boiling" was used to imply that the liquid has reached conventional nucleate boiling conditions, but internal evaporation and bubble growth have not yet been realized, or there exist countless microbubbles within the liquid that cannot be visualized by ordinary means.

However, later studies performed by various independent researchers suggest that (when ignoring flow instabilities) there are no unique flow patterns during flow boiling in microchannels that fundamentally diverge from the typical patterns exhibit in conventional scale channels, and it appears that the bubble-nucleation process is part of the boiling environment in microchannels as visualized by several investigators (Zhang et al. [40]; Wu and Cheng [26, 27]; Lee et al. [41, 42]; Zhang et al. [43]; Jiang et al. [44]; Balasubramanian and Kandlikar [45, 46]). Nevertheless, currently, no general criteria exist that categorize the conditions for various boiling flow patterns in microchannels.

Jiang et al. [44] performed a visualization study on 34 and 35 parallel triangular microchannels having hydraulic diameters of 26 μm and 53 μm to identify the flow patterns in flow boiling in microchannels and observed three stable boiling patterns depending on the heat flux. At low heat fluxes, bubble formation, growth, and collapse was observed to occur on several nucleation sites in the channel. At moderate heat fluxes large bubbles were generated mainly in the inlet/outlet common passage, while at moderate to high heat fluxes an unstable intermitted annular flow with liquid droplets entrained inside the vapor core was detected. At even higher heat fluxes stable annular flow predominated and no liquid droplets within the vapor core were detected. Similar to results obtained by Zhang et al. [43] bubbly flow was not detected in the microchannels under any circumstances. Zhang et al. [43] also observed mostly annular flow in their 25–60-μm microchannels. However, when analytical models were applied to predict the pressure drop and temperature the homogeneous model (which is by conventional criteria more appropriate to model bubbly flows) predicted the experimental results better than the annular model.

Several other studies reported broader range of flow patterns. Wu and Cheng [26] observed bubbly flow, slug flow, and churn flow during flow boiling in parallel trapezoidal 158.8-μm and 82.8-μm hydraulic diameter silicon microchannels. Steinke and Kandlikar [47] also observed bubbly flow alongside slug flow, churn flow, annular flow, annular flow with nucleation in the thin film, and dry-out in a set of six parallel square microchannels of 214 μm × 200 μm. In the bubbly flow the bubble sizes are seen to be from 15 μm to 193 μm (comparable to the entire channel width), and the smaller bubbles tend to flow near the wall. Balasubramanian and Kandlikar [45] compared flow patterns between single minichannels and parallel microchannels and concluded that flow reversal (herein classified as flow instability) in parallel channels is more pronounced than in single channels, and the occurrence of churn flow depends on a combination of channel size and aspect ratio. They also observed that in the large aspect ratio channel bubbles were attached to the corner of the channel.

Most studies reporting the absence of bubbly flow and the existence of annular flow do, in fact, indicate the appearance of bubble nucleation process, which seemingly contradicts the premise on which the annular flow assumption is made. In order to better understand this contradiction it is useful to examine the manner in which flow patterns are identified. There is a general agreement that the bubbly flow pattern predominates when bubbles, each occupying a relatively small fraction of the channel cross section, flow as a two-phase mixture. The definition for a single bubble originating from a single nucleation site, occupying a considerable fraction of the channel (herein defined as a *sizeable bubble*), and periodically detaching is less clear. Some investigators have used the term bubble/slug flow, while others have used periodic annular or simple annular flow.

The importance of carefully defining the flow pattern is not purely semantic since the definition encompasses a set of attributes that are used to identify heat-transfer mechanisms, at least by conventional scale practice. For instance, if one defines the *sizeable bubble* as the annular flow pattern convective boiling-type correlations need to be employed when attempting to predict heat-transfer coefficients, while nucleate boiling-type correlations should be selected when defining the pattern as bubble/slug flow. Furthermore, heat-transfer coefficient trends will deviate considerably depending on the flow-pattern definition used. The question regarding the appropriate definition is still a question of an ongoing debate within the heat-transfer community. Some advocate for annular (convective boiling predominance), some for bubbly/slug (nucleate boiling predominance), while others argue for both (depending on the thermo-hydraulic conditions). Perhaps the traditional linkage between flow patterns and boiling mechanism employed in conventional-scale apparatus is not as appropriate for micrometer-sized channels, and the old concepts need to be revisited. However, before any conclusive arguments can be made, the community has to acknowledge that the state-of-the-art experimental instrumentation capabilities have to be considerably improved in order to obtain more accurate and reliable local measurements.

4.2.3 Condensation Flow Patterns

Condensation flow patterns in mini- and microchannels have similarities and differences with those patterns observed in larger size channels. The driver for the differences is the changed relationship (relative magnitudes) among shear, gravitational and surface tension forces as channel diameter decreases and when the channel is noncircular. The literature is limited for studies in true microchannels; minichannel (in the order of 1 to 3 mm) studies are more common. Accurate flow-pattern prediction (with variation in quality) is needed for effective heat-transfer and pressure-drop predictions.

Begg et al. [48] experimentally and numerically observed a lessening of stratification as tube diameter decreased; with increasing mass flux, stratification produced significant film thicknesses between the top and bottom of the tubes. Mederic et al. [49] experimentally identified three flow pattern regions (annular/stratified, plug/Taylor bubble, and spherical bubble) and used the capillary length, $l_c = [\sigma/(\rho_l - \rho_v)g]^{1/2}$, to indicate the relative influences of capillary and gravitational forces. Wu and coworkers [50, 51] performed a study of water condensation in an 82.8-μm trapezoidal channel and observed that with decreasing mass flux, different flow patterns progressed along the channel: fully droplet flow; droplet/annular/injection/slug-bubbly; annular/injection/slug-bubbly; or fully slug-bubbly flow. Depending on the flow pattern, the wall temperature had significant variations with time.

Coleman and Garimella [52] and Garimella [53] present the most comprehensive results from condensation experiments in micro- and minichannels for circular and noncircular tubes with hydraulic diameters ranging between 0.4 and 4.91 mm. Four different flow regimes were identified (Fig. 4.3): intermittent, wavy, annular, and dispersed. Each of these regimes was subdivided into flow patterns. Flow-pattern maps were developed. They concluded that the size of the intermittent regime increases with decreasing hydraulic diameter due to larger surface tension forces compared to gravitational forces (Fig. 4.4 a); this enhances the development of plug and slug flows. As hydraulic diameter decreases, the size of the wavy flow regime decreases and is replaced by annular flow whose overall range expands (Fig. 4.4 b). Sharp corners tended to retain liquid due to surface tension effects, but tube shape had less influence on flow patterns than hydraulic diameter. A transition criterion from intermittent to other flow regimes was proposed by Garimella et al. [54]:

$$x \leq \frac{a}{G+b} \quad \text{and} \quad \begin{aligned} a &= 69.57 + 22.60 \exp(0.259 D_\mathrm{h}) \\ b &= -59.99 + 178.8 \exp(0.383 D_\mathrm{h}) \end{aligned} \tag{4.1}$$

where G is the total mass flux (kg/m^{2}s) and D_h is the hydraulic diameter in mm. When plotted on a figure, the area to the left of this line is in the intermittent flow regime; to the right are in either the wavy or annular flow regime.

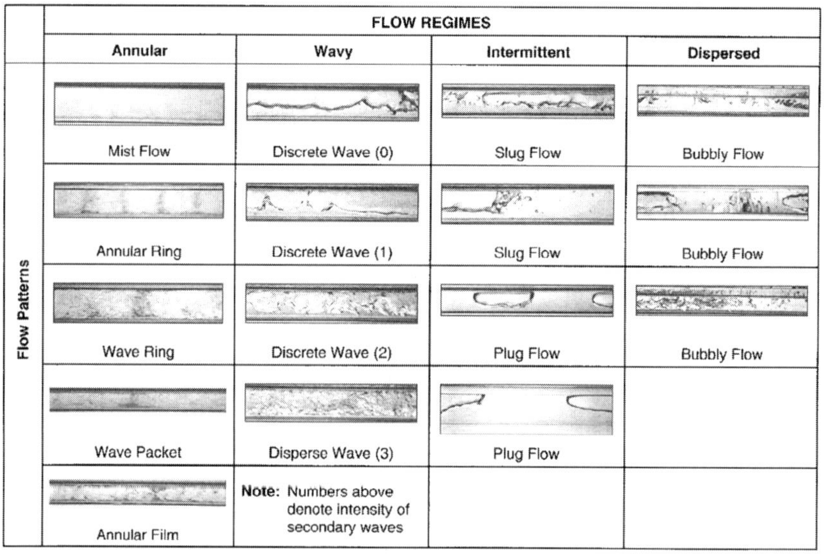

	FLOW REGIMES			
	Annular	Wavy	Intermittent	Dispersed
Flow Patterns	Mist Flow	Discrete Wave (0)	Slug Flow	Bubbly Flow
	Annular Ring	Discrete Wave (1)	Slug Flow	Bubbly Flow
	Wave Ring	Discrete Wave (2)	Plug Flow	Bubbly Flow
	Wave Packet	Disperse Wave (3)	Plug Flow	
	Annular Film	Note: Numbers above denote intensity of secondary waves		

Fig. 4.3 Descriptions of two-phase flow patterns in condensation (Coleman and Garimella [52], used with permission).

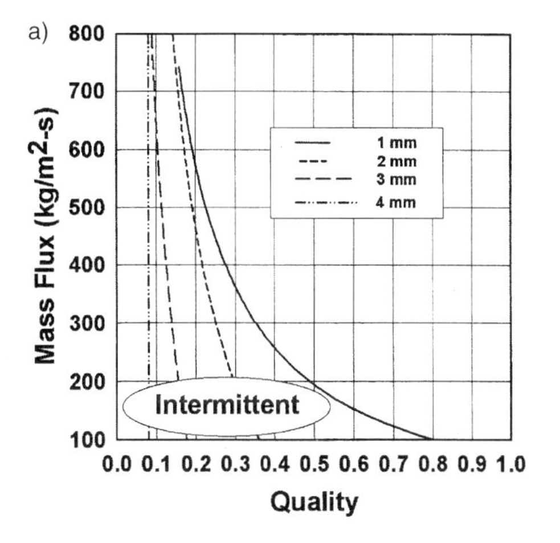

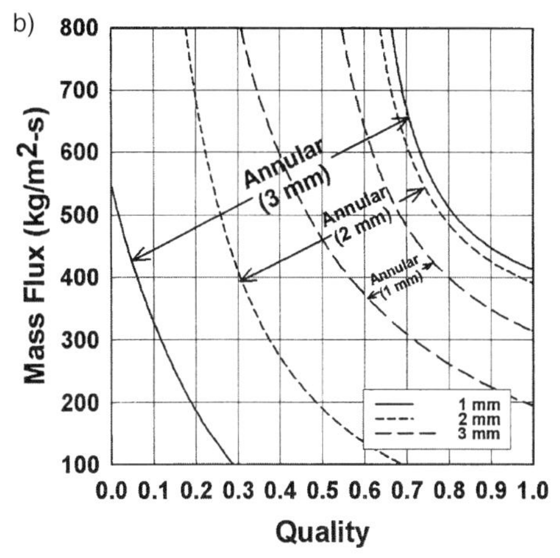

Fig. 4.4 Effect of hydraulic diameter on (a) intermittent flow regime and (b) annular flow regime (Garimella [53], used with permission).

4.3 Pressure Drop

Two-phase flow in channels results in a characteristic pressure drop, which is most often used as a criterion to help design heat exchangers. The determination of the pressure drop is important to size the circulation pump and to calculate the decrease in the saturation temperature. Generally, the pressure drop is composed of friction, acceleration, and hydrostatic components. To calculate the hydrostatic and acceleration pressure drop, the primary quantity needed is the void fraction, whereas for the friction component the two-phase friction factor (or the friction multiplier) is required. Two basic modeling approaches have been widely used for gas-liquid two-phase flows in channels (as well as other flow configurations such as flow over tube bundle (Jensen [55])) – the "homogeneous" model and the "separated flow" model. The "homogeneous" model considers the two phases to flow as a single phase possessing mean fluid properties, while the separated flow model considers the phases to be artificially segregated into two streams, one for the liquid and one for the vapor. The separated flow models can be further classified into two classes: models that do not consider the gas/liquid interface, and models that include interface exchange. Models that do not account for interface exchange are semi-empirical and are experimentally based. In order to maintain the experimental flow parameters constant throughout the test section, most studies have been conducted under adiabatic flow conditions, and validated under diabatic tests. The widely used Lockhart–Martinelli [56] correlation employs a two-phase friction multiplier, ϕ_L^2 or ϕ_G^2, defined as:

$$\left(\frac{\Delta P_f}{\Delta Z}\right)_{TP} = \phi_L^2 \left(\frac{\Delta P_f}{\Delta Z}\right)_L \text{ or } \left(\frac{\Delta P_f}{\Delta Z}\right)_{TP} = \phi_G^2 \left(\frac{\Delta P_f}{\Delta Z}\right)_G \tag{4.2}$$

where $(\Delta P_f/\Delta Z)_{TP}$ is the two-phase frictional pressure drop, and $(\Delta P_f/\Delta Z)_L$ and $(\Delta P_f/\Delta Z)_G$ are the pressure drop when liquid and gas are assumed to flow alone in the channel, respectively. The frictional multiplier is correlated in terms of the Lockhart–Martinelli parameter, X, given by:

$$X = \frac{(\Delta P_f/\Delta Z)_L}{(\Delta P_f/\Delta Z)_G} \tag{4.3}$$

A widely used curve fit formula to correlate the two-phase friction multiplier with the Lockhart–Martinelli parameter was first suggested by Chisholm and Laird [57]:

$$\phi_L^2 = 1 + \frac{C}{X} + \frac{1}{X^2} \tag{4.4}$$

where C is an empirical adjustment factor.

4.3.1 Boiling and Adiabatic Pressure Drop

Modeling aimed at predicting two-phase pressure drop in channels with hydraulic diameter smaller than 200 μm are quite limited. However, attempts to develop or use homogeneous models (Zhang et al. [43]; Kawahara et al. [21]), Lockhart–Martinelli-type friction factor models (Kawahara et al. [21]; Chung et al. [23]), and separated flow models that include interface exchange (Qu and Mudawar [58]; Zhang et al. [43]; Chung et al. [23]) are documented. The homogeneous model developed by Zhang et al. [43] and Koo et al. [59] predicted better the experimental results of flow boiling in 25- to 60-μm rectangular channels data than did an annular flow model. However, the results did not provide sufficient evidence to select between annular and homogeneous models. On the other hand, Kawahara et al. [21] reported poor prediction of homogeneous models of adiabatic nitrogen–water two-phase flow in a 100-μm circular tube, and developed Lockhart–Martinelli-type correlations. For conventional scale systems a value of $C=5$ in the two-phase friction multiplier should be used for liquid laminar and vapor laminar flow according to the original work of Chisholm and Laird [57]. However, based on previous results reviewed by Kawaji and Chung [16] and obtained by Mishima and Hibiki [13] and Lee and Lee [6] a functional dependency of C on the hydraulic diameter (Mishima and Hibiki [13]) and other dimensionless parameters proposed by Lee and Lee [6] was used to correlate the experimental results. It was concluded that the data can be correlated well using the Lockhart–Martinelli type parameter, but a modified expression for the C value has to be included in order to predict the results favorably, as can be seen from Fig. 4.5. In a later study, Chung and coworkers [22, 23] extended a mechanistic separated flow model developed by Garimella et al. [54, 60] to estimate the two-phase pressure drop. Good agreement to experimental results was obtained when accurate void fraction data are used.

A detailed model development and comprehensive comparison to large-scale pressure drop models for a heated microchannel was given in Qu and Mudawar [58, 61] based on microchannel heat sink containing 21 parallel 231×713 μm channels. Comparison of their pressure-drop data to six macrochannels correlations and four correlations developed for mini/microchannels resulted in moderate to poor agreements. From the large-scale correlations the homogeneous flow models resulted in improved prediction to the experimental data in com-

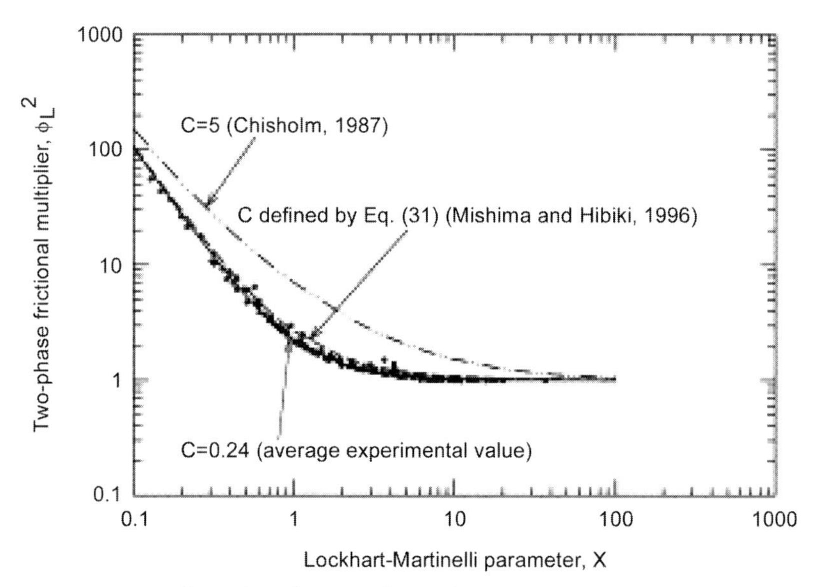

Fig. 4.5 Variation of two-phase friction multiplier data with the Lockhart–Martinelli parameter (from Kawahara et al. [21], used with permission).

parison to separated flow models. The authors hypothesized that the poor prediction was a result of the turbulent-flow assumption used to develop the separated-flow models, while for microchannels the flow regime is predominantly laminar. Except for one model the correlations developed for mini/microchannels predicted the experimental results adequately. A model developed in a previous study (Qu and Mudawar [58]) was also employed successfully.

4.3.2 Condensation Pressure Drop

Pressure drop in condensing systems has been investigated experimentally and analytically. Both complete and partial condensation systems have been studied in circular and noncircular channels. Comparisons have been made of experimental data against prediction techniques (e.g., correlations) for conventional-size channels and new predictive approaches have been developed.

Koyama et al. [62] compared several correlations against R134a condensation in 1-mm multiport (rectangular) tubes with and without microfins. Reasonable agreement was obtained, but they concluded that tube diameter should be taken into account in the frictional pressure drop. They modified the Mishima–Hibiki correlation [13] to include surface tension and kinematic viscosity. Baird et al. [63] condensed R123 and R11 in 0.92- and 1.95-mm circular tubes and determined that the Lockhart–Martinelli correlation reproduced the data reasonably well. They also developed an analytic shear stress model for annular flow that tended to underpredict the pressure drop.

Garimella and coworkers [53, 54, 60, 64] performed a series of pressure-drop experiments with condensation of R134a in circular and noncircular tubes for hydraulic diameters from 0.5 to 5 mm. One conclusion was that the Lockhart–Martinelli [56], Chisholm [65], and Friedel [66] correlations were not able to predict the data with acceptable accuracy. Hence, Garimella and coworkers developed a new multiple-flow regime pressure-drop model that was phenomenologically based. They separated the data into only three flow regimes: mist/annular/disperse, intermittent/discrete wave, and an overlap zone. This broad characterization was sufficient to accurately capture experimental trends. For intermittent/discrete wave flow, model development assumed, among other items, that: vapor flowed in long solitary bubbles separated from the wall and other bubbles with liquid; the bubble traveled faster than the liquid slug ahead of the bubble; and the length/frequency/speed of bubbles was constant. A correlation for slug frequency was developed that was the same for circular and noncircular tubes (except triangular). For annular/mist/disperse flow, a liquid film was assumed to uniformly cover the tube wall, and interfacial friction factors were developed, depending on whether the liquid-phase Reynolds number was laminar or turbulent. For the overlap zone a four-point interpolation scheme was used.

The model predictions matched the experimental data well. Figure 4.6 a illustrates parametric trends for circular tubes for different qualities, mass fluxes, and tube diameters. As can be seen, hydraulic diameter does have a significant

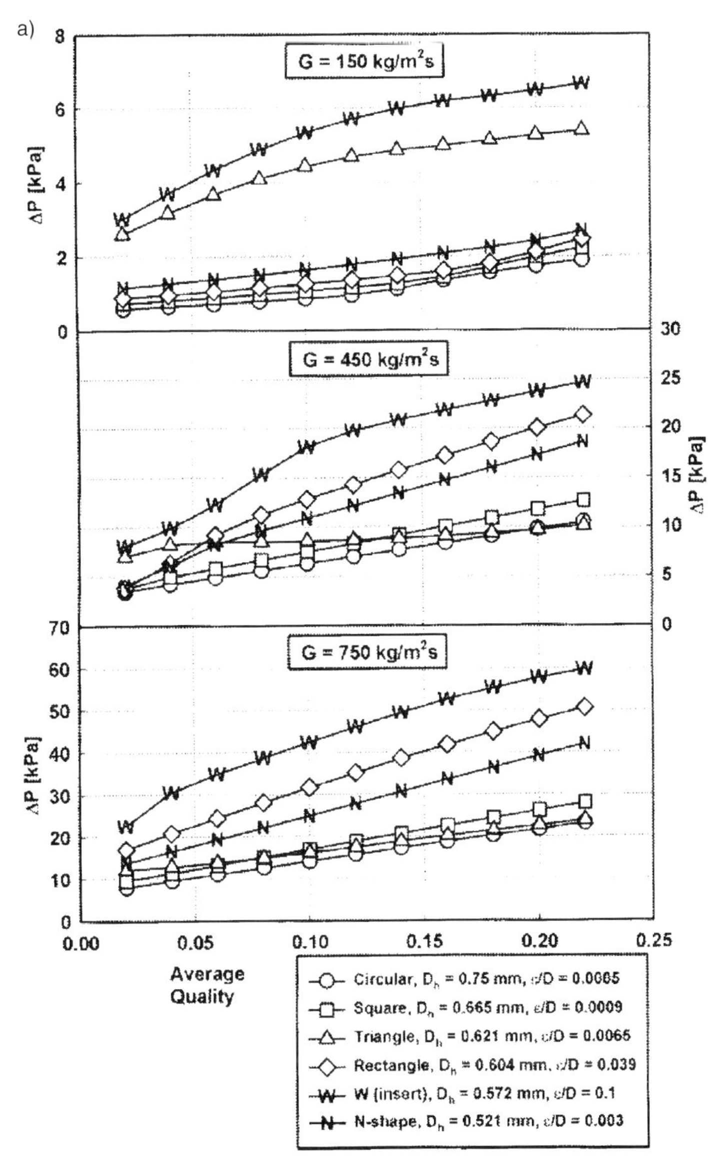

Fig. 4.6 Effect of channel diameter (a) and (b) shape on condensation pressure drop (Garimella et al. [60, 64], used with permission).

influence otherwise there would have been only one line on each graph. The influence of tube shape and surface roughness is shown in Fig. 4.6 b. For a nominal diameter of 0.75 mm, the pressure drop can vary by a factor of three depending on the tube; note that triangular tubes behaved significantly differently from the other tubes at low mass fluxes.

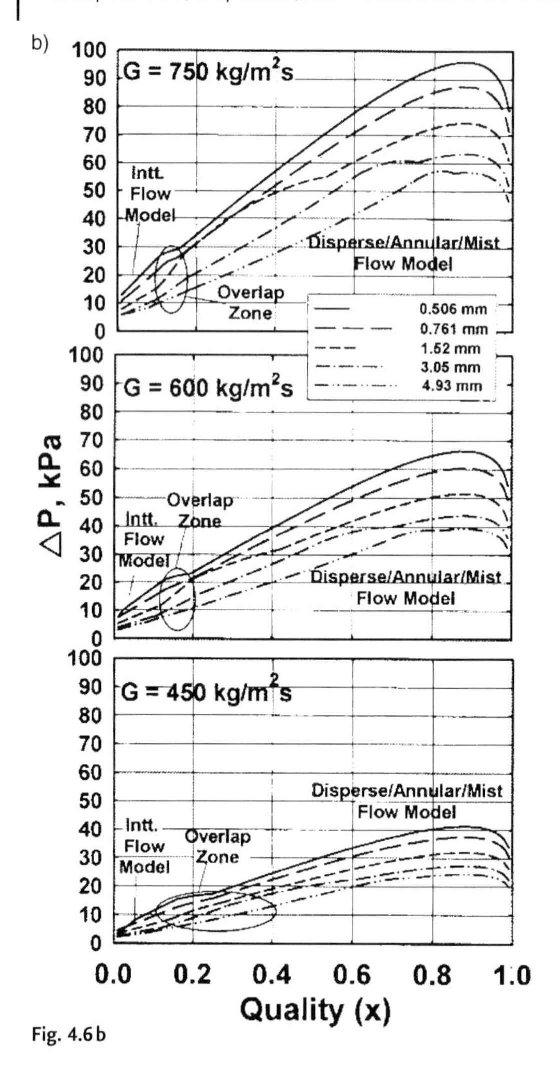

Fig. 4.6 b

4.4
Boiling/Evaporation

The catalyst for exploring boiling phenomena in microchannels is the large heat-transfer coefficients. However, flow boiling in conduits is perhaps the most complex convective heat-transfer process, and several factors must be carefully considered before this heat transfer mode can be successfully employed in high heat-flux applications. For boiling to occur, the superheat, or the difference between the temperature of the wall and the saturation temperature of the fluid, must be large enough to initiate nucleation of some of the cavities on the surface. While the criterion for the onset of nucleate boiling in small-diameter channels is expected to be similar to the one obtained for larger-diameter channels, the heat-transfer coefficients may be different because of the significantly different channel sizes. New correlations have been developed for these cases. Unfortunately, there is much less understanding of the boiling limit, or the critical heat-flux condition, at the microscale. This section summarizes the current understanding of boiling heat transfer in microchannels.

4.4.1 Onset of Nucleate Boiling

Evaporation in microchannels and minichannels can be accomplished with either subcooled liquid entry or with a two-phase entry resulting from a throttle valve in a refrigeration system. Operation with subcooled liquid entry allows a more uniform liquid distribution in parallel channels and is therefore preferred in practical systems. As liquid flows through these small hydraulic diameter channels, it experiences a very high heat-transfer coefficient with the single-phase liquid flow. Eventually, the conditions for onset of nucleation are met and nucleating bubbles are formed in the flow.

The criterion for onset of nucleate boiling in small-diameter channels is expected to be similar to those obtained for larger-diameter channels. The nucleation criterion of Hsu [67] can be applied. Using this criterion, the onset condition is expressed in terms of the local wall superheat $\Delta T_{Sat} = (T_W - T_{Sat})$. The range of nucleating cavity radii under a given set of conditions is expressed as:

$$\{r_{c,min}, r_{c,max}\} = \frac{\delta_t}{4}\left(\frac{\Delta T_{Sat}}{\Delta T_{Sat} + \Delta T_{Sub}}\right)\left[1 \mp \sqrt{1 - \frac{12.8\,\sigma T_{Sat}(\Delta T_{Sat} + \Delta T_{Sub})}{\rho_V h_{LV} \delta_t \Delta T_{Sat}^2}}\right] \tag{4.5}$$

where $\Delta T_{Sat} = (T_{Sat} - T_{bulk})$, σ – surface tension, ρ_V – density of vapor, and h_{LV} – latent heat of vaporization. The thickness of the local thermal boundary layer is obtained as $\delta_t = k_L/h_L$, where h_L is the local heat-transfer coefficient prior to nucleation and k_L ist the liquid thermal conductivity.

In the above expression, Hsu used a contact angle of 53.1 degrees. Bergles and Rohsenow [68] considered a hemispherical bubble with a contact angle of 90 degrees and obtained a different set of numerical constants in the above expression. Kandlikar et al. [69] numerically analyzed the flow around a bubble and observed the liquid streamlines around a bubble to yield a point location located at a distance of 1.1 times the bubble radius above the heater surface. The resulting range of nucleating cavity radii (truncated-stagnation model) is given by:

$$\{r_{c,min}, r_{c,max}\} = \frac{\delta_t \sin\theta_r}{2.2}\left(\frac{\Delta T_{Sat}}{\Delta T_{Sat} + \Delta T_{Sub}}\right)\left[1 \mp \sqrt{1 - \frac{8.8\,\sigma T_{Sat}(\Delta T_{Sat} + \Delta T_{Sub})}{\rho_V h_{LV} \delta_t \Delta T_{Sat}^2}}\right] \tag{4.6}$$

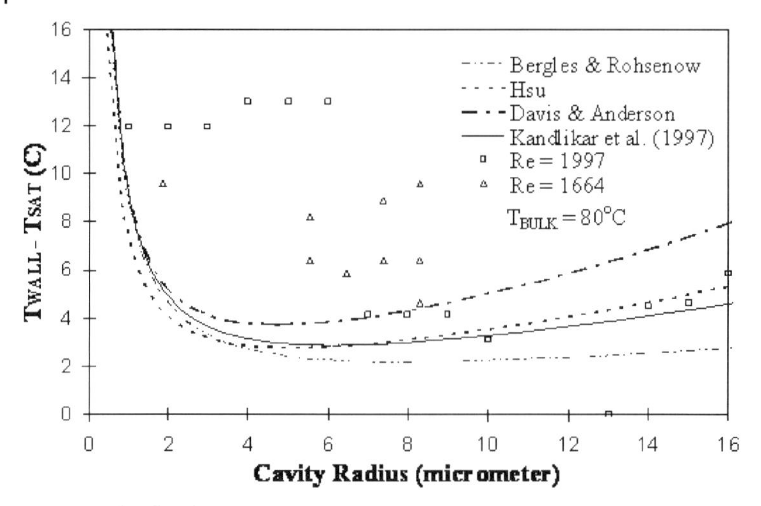

Fig. 4.7 Radii of nucleation cavities.

where θ_r is the receding contact angle. Kandlikar et al. [69] conducted experiments with water boiling in a 1-mm high and 40-mm deep channel and observed the nucleation over a heater surface using a high-speed camera. The radii of nucleation cavities for a given wall superheat are plotted in Fig. 4.7. These cavity radii are expected to fall between the respective range predicted by a nucleation criterion. Also shown in the figure are nucleation criteria by Hsu [67], Bergles and Rohsenow [68], Davis and Anderson [70], and Kandlikar et al. [69] using the truncated-stagnation model. The Bergles and Rohsenow model is seen to predict the widest range of nucleation cavities while Davis and Anderson's model is seen to provide the narrowest range.

Assuming the availability of all cavity sizes on the heater surface, the condition for onset of nucleate boiling can be obtained by setting the minimum and maximum cavity radii to be equal in Eq. (4.6):

$$\Delta T_{\mathrm{Sat,ONB}} = \sqrt{8.8\,\sigma T_{\mathrm{Sat}} q''/(\rho_{\mathrm{V}} h_{\mathrm{LV}} k_{\mathrm{L}})} \tag{4.7}$$

4.4.2 Heat-transfer Coefficients

Some of the experimental data in the literature showed heat-flux dependence and was correlated by the all-nucleate boiling-type correlations, such as Lazarek and Black [71] and Tran et al. [72]. However, the presence of both nucleate boiling and convective boiling is reported by Kandlikar and Balasubramanian [73]. Since the flow Reynolds number is low, they recommended the single-phase all-liquid heat-transfer coefficient in the Kandlikar [74] correlation.

For $Re_{LO} > 100$:

$$h_{TP} = \text{larger of} \begin{cases} h_{TP,NBD} \\ h_{TP,CBD} \end{cases} \tag{4.8}$$

$$h_{TP,NBD} = 0.6683\,Co^{-0.2}(1-x)^{0.8}h_{LO} + 1058.0\,Bo^{0.7}(1-x)^{0.8}F_{Fl}h_{LO} \tag{4.9}$$

$$h_{TP,CBD} = 1.136\,Co^{-0.9}(1-x)^{0.8}h_{LO} + 667.2\,Bo^{0.7}(1-x)^{0.8}F_{Fl}h_{LO} \tag{4.10}$$

where the Convection number is given by $Co = [(1-x)/x]^{0.8}[\rho_V/\rho_L]^{0.5}$, the boiling number Bo is given by $Bo = q''/(Gh_{LV})$, q'' – heat flux, x – quality, $Re_{LO} = GD(1-x)/\mu$.

The single-phase all-liquid flow heat-transfer coefficient h_{LO} is given by

$$\text{for } 10^4 \le Re_{LO} \le 5\times 10^6 \quad h_{LO} = \frac{Re_{LO}Pr_L(f/2)(k_L/D)}{1+12.7(Pr_L^{2/3}-1)(f/2)^{0.5}} \tag{4.11}$$

$$\text{for } 3000 \le Re_{LO} \le 10^4 \quad h_{LO} = \frac{(Re_{LO}-1000)Pr_L(f/2)(k_L/D)}{1+12.7(Pr_L^{2/3}-1)(f/2)^{0.5}} \tag{4.12}$$

$$\text{for } 100 \le Re_{LO} \le 1600 \quad h_{LO} = \frac{Nu k}{D_h} \tag{4.13}$$

where f is the friction factor.

In the transition region between Reynolds numbers of 1600 and 3000, a linear interpolation is suggested for h_{LO}.

For Reynolds number below and equal to 100 ($Re \le 100$) the nucleate boiling mechanism governs, and the following modification in the Kandlikar correlation is proposed:

Table 4.1 Fluid–surface parameter F_{Fl} in flow boiling correlation, Eqs. (4.9)–(4.14).

Fluid	F_{Fl}
Water	1.00
R-11	1.30
R-12	1.50
R-13B1	1.31
R-22	2.20
R-113	1.30
R-114	1.24
R-134a	1.63
R-152a	1.10
R-32/R-132	3.30
R-141b	1.80
R-124	1.00
Kerosene	0.488

For $\mathrm{Re_{LO}} \leq 100$,

$$h_{\mathrm{TP}} = h_{\mathrm{TP,NBD}} = 0.6683\,\mathrm{Co}^{-0.2}(1-x)^{0.8}h_{\mathrm{LO}} + 1058.0\,\mathrm{Bo}^{0.7}(1-x)^{0.8}F_{\mathrm{FL}}h_{\mathrm{LO}} \tag{4.14}$$

The single-phase all-liquid flow heat-transfer coefficient h_{LO} in Eq. (4.13) is found from Eq. (4.12).

The fluid-surface parameter F_{FL} in Eqs. (4.9), (4.10) and (4.14) for different fluid-surface combinations is given in Table 4.1. These values are for copper or brass surfaces. For stainless steel surfaces, use $F_{\mathrm{FL}}=1$ for all fluids.

4.4.3
Critical Heat-flux Condition

The critical heat-flux (CHF) condition refers to a sudden, sharp decrease in the heat-transfer coefficient that sets the upper limit on boiling heat transfer. In heat-flux-controlled situations the CHF condition can manifest itself by a rapid and sometimes catastrophic rise in the wall temperature (burn-out). In temperature-controlled situations, the CHF can cause a large decrease in the heat-transfer rate. The CHF condition varies with channel geometry and material properties, type of fluid, and operating conditions (flow rate, pressure level, inlet temperature, etc.). An example of boiling curves revealing the CHF condition is shown in Fig. 4.8. The ability to predict the critical heat flux is of vital importance for the operation of two-phase microsystems. However, the critical heat flux (CHF) condition in microchannels has attracted much less attention than the boiling heat-transfer coefficient. As a result, the mechanism of critical heat flux is not as well understood. An examination of the limited CHF data indicates (Bergles and Kandlikar [75]) that CHF in parallel microchannels seems to be the result of either an upstream compressible volume instability or an excursive instability rather than the conventional dry-out mechanism. Section 4.5 discusses these mechanisms in more detail.

This section presents a survey of the current understanding of CHF in mini- and microchannels. Most of the existing data are for channels with diameters in the mm range, while only a few studies have been conducted in microchannels below 200 μm hydraulic diameters. Additional information on heat transfer and fluid flow in microchannels can be found in surveys by Palm [1], Mehendale et al. [2], and Kandlikar [3, 4].

Some of the earliest studies of CHF were conducted by Bergles [76, 77] and Bergles and Rohsenow [78]. They studied subcooled CHF with water at very high mass fluxes in tubes as small as 0.5 mm. Lazarek and Black [71] investigated 3.1-mm diameter tubes (length-to-diameter ratio $L/D=39.7$ and 79.4) and found no differences between the CHF in vertical upflow and downflow of R113. Comparison against a CHF correlation for larger-diameter tubes suggested that the minichannel had a greater critical quality. In other CHF studies Kamidis and Ravigururajan [28] used R113 and 1.59, 2.78, 3.97, and 4.62 mm tubes (L/D ranged from 24 to 11). CHF increased with flow rate. Yu et al. [79]

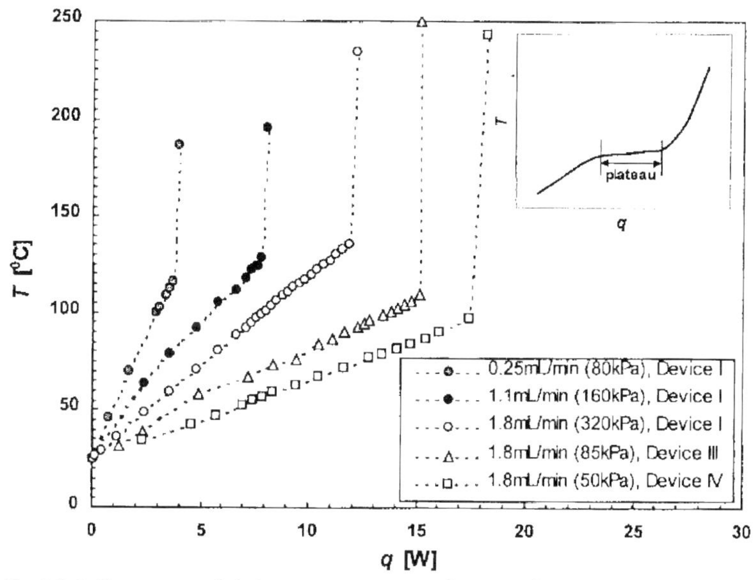

Fig. 4.8 Boiling curves of device temperature as a function of dissipated power for various conditions. The critical heat flux is indicated by the sharp increase in temperature at a given flow rate (dotted lines) (from Jiang et al. [30], used with permission).

studied the CHF with water in a 2.98 mm tube ($L/D=305$). The CHF increased with mass flux and was dependent on the inlet liquid subcooling.

Bowers and Mudawar [80] studied the CHF with R113 in 17 parallel 2.54-mm and 0.51-mm diameter channels formed in nickel blocks heated over the central 10-mm section. The tests were carried out at low mass fluxes typical of microchannels (7.0–28.2 kg/m^2 s). The inlet pressure was 1.38 bar, and the inlet subcooling ranged from 10 to 32 °C. Good conduction around the circumference of each channel was assumed, so that the heat flux was determined by heat input divided by the channel wall area. Boiling curves were generated that terminated in well-defined critical heat fluxes. It was found that inlet subcooling had no effect on the CHF condition at low flows and the CHF increased almost directly proportional to mass flux. For the same mass flux, inlet temperature and pressure, the minichannel ($L/D=3.94$) had a CHF up to 2.5 times that of the microchannel ($L/D=19.6$). The authors attributed the high CHF to conduction removing heat from downstream areas undergoing partial or total liquid film dry-out. The bulk fluid condition at the end of the heated section was high quality or even superheated.

An example of CHF studies in silicon microfabricated test sections is provided by Jiang et al. [29, 30]. The authors report a microchannel heat sink integrated with a heater and an array of implanted temperature sensors. There were up to 58 or 34 channels of rhombic shape, having a hydraulic diameter of 40 or

80 μm ($L/D \approx 500$ and 250, respectively). Due to the fabrication method, there was no transparent cover plate to view the two-phase flow. CHF data were taken for water entering at 20 °C. It appears that the CHF condition was characterized by a rapid rise in the average of all the temperature sensors (Fig. 4.8). The critical power was found to be a linear function of the total volume flow rate, which ranged from 0.25 to 5.5 ml/min. Since no boiling plateau could be identified, the authors speculated that in such small channels, bubble formation may be suppressed and recommended flow-visualization studies to determine the governing heat-transfer mechanism.

In one of the most comprehensive studies to date, Qu and Mudawar [81] report the CHF in 21 separate parallel micro-channels that were 215×821 μm. The heat flux was based on the heated three sides of the channel. Deionized, deaerated water was supplied over the mass-flux range of 86–368 kg/m^{2}s, with an inlet temperature of 30 or 60 °C and an outlet pressure of 1.13 bar. In order to eliminate upstream compressible volume instabilities it was necessary to install a throttle valve upstream of the test section. As CHF was approached vapor backflow from all of the channels into the inlet plenum was observed. A correlation for circumferential heating conditions was developed based on this study, as well as R-113 data from a previous study

$$\frac{q''_{\mathrm{p,m}}}{Gh_{\mathrm{LV}}} = 33.43\left(\frac{\rho_{\mathrm{g}}}{\rho_{\mathrm{f}}}\right)^{1.11}\left(\frac{G^2 L}{\sigma\,\rho_{\mathrm{f}}}\right)^{-0.21}\left(\frac{L}{d_{\mathrm{e}}}\right)^{-0.36} \tag{4.15}$$

where $q''_{\mathrm{p,m}}$ is the critical heat flux based on the channel heated inside area, G is the mass-velocity, h_{fg} is the latent heat of vaporization, ρ_{l} and ρ_{g} are (respectively) liquid and vapor densities, L is the heated length of the microchannel, and d_{e} is heated equivalent diameter of rectangular channel; d_{e} should be set equal to d for circular mini/microchannel heat sinks. Figure 4.9 shows a comparison of the CHF data for water and R-113 in mini/microchannel heat sinks with predictions from the above correlation. The experimental data for the two fluids compares very well with this correlation.

There are three important points that should be noted with the above CHF studies. First, in the multichannel flows, flow distribution was assumed to be uniform in all channels; however, flow-maldistribution effects (different flow rates in the parallel channels) may have been present. Since the CHF condition is dependent on flow rate, data from experiments in which the flows in individual channels are unknown may be unreliable.

Second, flow oscillations and their effects were not mentioned in most of these microchannel CHF studies. In conventional (single) channels, flow oscillations have a significant negative effect on the CHF condition. For example, Mayinger et al. [82] observed stable and unstable CHF in upflow, with unstable flow CHF values as much as 20 to 50% lower than the stable values. Ruan et al. [83] recorded CHF values in oscillatory downward flows that were only 20% of the stable values. For low flows and pressures (probably typical of what it

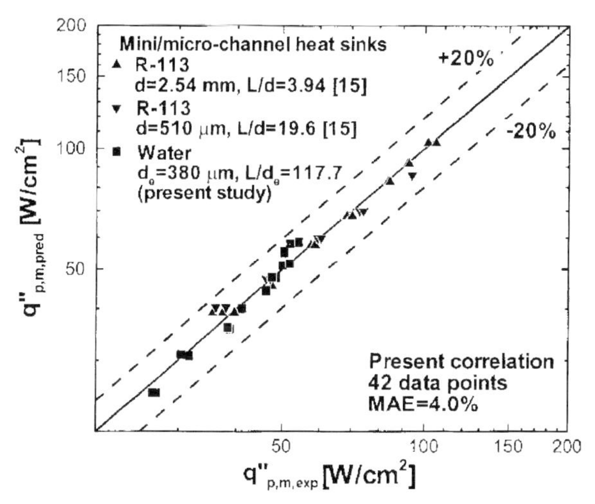

Fig. 4.9 Comparison of CHF data for water and R-113 in mini/microchannel heat sinks with predictions from Eq. (4.15) (from Qu and Mudawar [81] used with permission).

might be expect in microchannels), Ozawa et al. [84] found that the CHF condition is a decreasing function of amplitude and period of the flow oscillation and were as low as 40% of the steady-state value. Therefore, the effect of flow instabilities on the CHF condition in parallel microchannels must be assessed.

Third, in the microchannel studies reviewed above, wall thicknesses were the same order of magnitude as or larger than the channel diameters. Conjugate effects may lead to nonuniform temperatures around the circumference of the microchannel and presumably influence the CHF condition (sometimes increasing and sometimes decreasing the CHF). Bowers and Mudawar [80] pointed out (but did not quantify) that conduction effects delayed the CHF condition in their multichannel test section manufactured in a large metallic block. In addition, Jiang et al. [29, 30], Kamidis and Ravigururajan [28], Ammerman and You [85], and Zhang et al. [43] all used very large blocks compared to the small channel size, but did not comment on any possible conduction effects. In single-tube tests, Larazek and Black [71] used tubes with I.D./O.D. of 3.1 mm/3.9 mm and noted that the CHF had critical qualities greater than in larger-diameter tubes. Roach et al. [86] used 1.17 mm/2.39 mm tubes and observed that the CHF condition initiated at very high qualities. However, the effect of conduction on the CHF condition has to be assessed in a quantitative manner.

4.5
Two-phase Instabilities

The nucleation criteria described in Section 4.4.1 provide the condition for nucleation over cavities of certain radii. If the nucleation cavities are not available in the range of radii given by the nucleation criterion (e.g. Eq. (4.5) or Eq. (4.6)), nucleation will not occur at the location. The liquid flows down the channel until it encounters cavities that locally satisfy the nucleation criterion.

As nucleation is delayed downstream, the liquid is further heated and attains a superheated liquid state. Under these conditions, the resulting nucleation will cause an explosive growth of a bubble as the superheated liquid surrounding the bubble evaporates rapidly in the channel. The rapid bubble growth introduces a major instability during flow boiling in microchannels, sometimes leading to reversed flow of the liquid–vapor mixture all the way into the inlet manifold. Such reversed flow has been observed by a number of investigators, including Cornwell and Kew [31, 32] and Peles et al. [87]. Figure 4.10 shows a high-speed photographic sequence obtained by Kandlikar and Balasubramanian [88].

The six parallel minichannels are connected to a common header at the inlet and outlet ends. The rapid bubble movement and reverse flow causes large pressure fluctuations as reported by a number of researchers. Figure 4.11 shows a plot of instantaneous pressure drop reported by Kandlikar and Balasubramanian [88]. The large pressure fluctuations and the extended periods of dry operation following the reverse flow condition are detrimental to the heat-transfer process and leads to low values of critical heat flux.

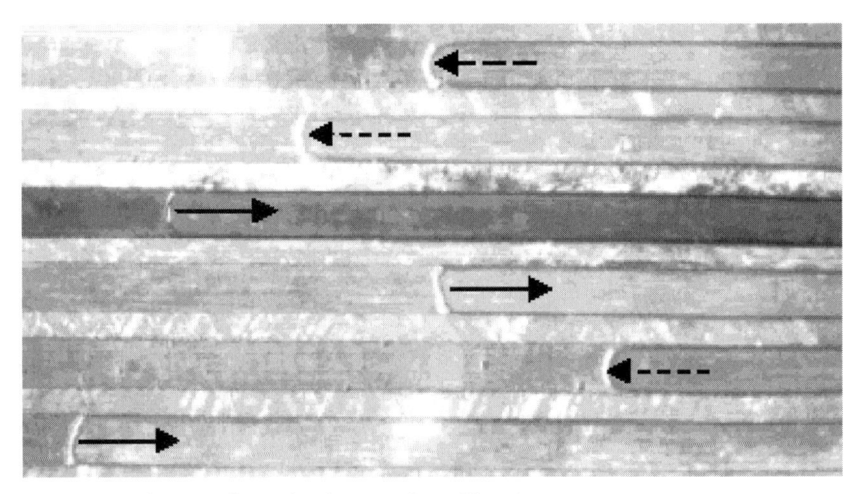

Fig. 4.10 Simultaneous forward and reverse flow of liquid/vapor interfaces during flow boiling of water in 200 μm deep×1054 μm wide minichannels (Kandlikar and Balasubramanian [88]).

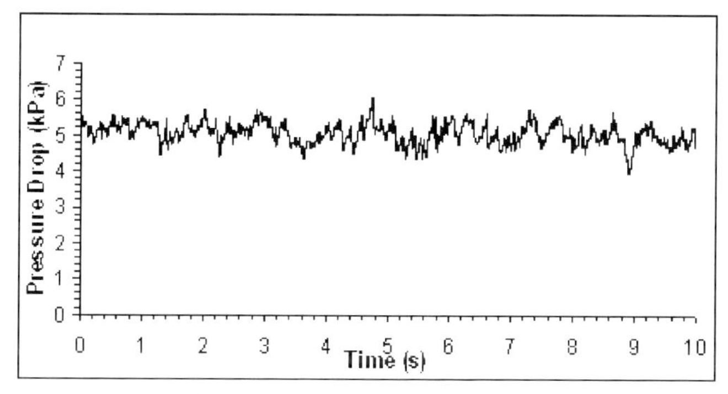

Fig. 4.11 Pressure fluctuations observed during flow boiling of water in 1054×197-μm parallel minichannels (Balasubramanian and Kandlikar [88]).

The pressure fluctuations and instabilities can be reduced by introducing pressure-drop elements at the inlet of the channels and incorporating artificial nucleation sites on the channel walls to initiate nucleation at relatively low values of wall superheat.

4.6 Condensation Heat Transfer

Experimental and theoretical investigations have been used to study condensation heat transfer in micro- and minichannels. Not surprisingly, differences with condensation in conventional-sized channels have been observed. In particular, in noncircular channels as the hydraulic diameter decreases, surface tension should have an increasing influence (the so-called "Gregorig effect" [89]); surface tension and condensate film curvature lead to variations in liquid-film thickness around a perimeter and, hence, increased heat transfer. The influence of surface tension, as well as the other two major forces of gravity and shear, has been described and the dependence of heat transfer with hydraulic diameter has been shown.

Perhaps due to inaccurate experimental results reported with early single-phase heat-transfer studies (see, for example, an explanation by Palm [90]), a number of investigators have given considerable thought to the experimental apparatus and procedures used to evaluate heat-transfer performance with condensation in micro- and minichannels. Conventional approaches yield data with very high uncertainties because of the small heat duties, high heat-transfer coefficients (h) and low flow rates. Hence, specific and careful techniques are needed to produce average heat-transfer coefficients over narrow quality ranges because condensation heat-transfer coefficients are strongly dependent on vapor quality since the flow pattern changes along the length of a condenser tube; with fluid-cooled channels, care must be taken to ensure that the condensation heat-transfer resistance is the governing thermal resistance. A number of investigators (Garimella and Bandhauer [91], Garimella [53], Baird et al. [63], Cavallini et al. [92, 93], Shin and Kim [94]) have proposed improved techniques and have presented data using these experimental test rigs. Additional data have been produced, for example, by Yan and Lin [95], Wang et al. [96], Koyama et al. [62], Begg et al. [97], Wu and Cheng [51], Webb and Ermis [98]. Cavallini et al. [5] present a brief review of condensation in micro- and minichannels.

A variety of trends have been observed. Baird et al. [63] showed only small differences between R-123 condensing in 0.92- and 1.95-mm tubes. This is consistent with the results of Yan and Lin [95], who obtained heat-transfer coefficients in 2-mm tubes that were only about 10% higher than those in an 8-mm tube. On the other hand, the measurements by Begg et al. [97] for condensing water in tubes ranging from 1.7 to 4.0 mm indicated a large increase (of the order of 76%) in the heat-transfer coefficient for similar conditions. Likewise, the Shin and Kim [94] study of condensing R134a in circular and rectangular channels with hydraulic diameters ranging from 0.493 to 1.067 mm found definite increases in the heat-transfer coefficient with decreasing hydraulic diameter; for circular channels, $h \propto D_h^{-0.54}$, and for rectangular channels $h \propto D_h^{-0.45}$. Representative data are shown in Fig. 4.12. The circular-tube Nusselt numbers had a stronger dependence on mass flux and quality than the rectangular channels. Wang et al. [96] found that with increasing quality, the Nusselt number became more sensitive to increases in mass flux. This result was attributed to different flow patterns (stratified versus annular). In the smallest channel tested to date,

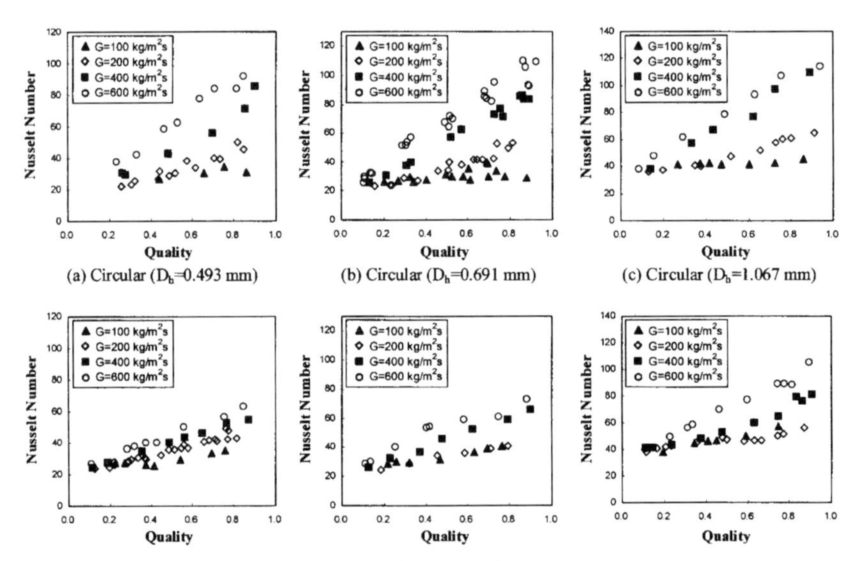

Fig. 4.12 Effect of mass flux and vapor quality on condensation heat transfer (Shin and Kim [94], used with permission).

Wu and Cheng [51] condensed water in an 82.8-μm trapezoidal channel. In the intermittent flow regime, large fluctuations in wall temperatures were measured, but these decreased with increasing mass flux.

Koyama et al. [62] condensed R134a in 1-mm rectangular channels and compared their results to several correlations developed for conventional-sized channels from the literature. Most overpredicted the data, but the data were in relatively good agreement with the Moser et al. [99] correlation. Nevertheless, Koyama et al. [62] developed a new correlation that agreed better with the data. The data of Webb and Ermis [98] also was overpredicted by correlations from conventional-sized channels, but they recommended that the correlation by Akers et al. [100] be used.

Theoretical analyses of condensation in small circular and noncircular channels have been presented by a number of investigators. Because the results of a number of studies suggested the predominance of the annular flow regime over a significant portion of a channel length, the most common assumption is a

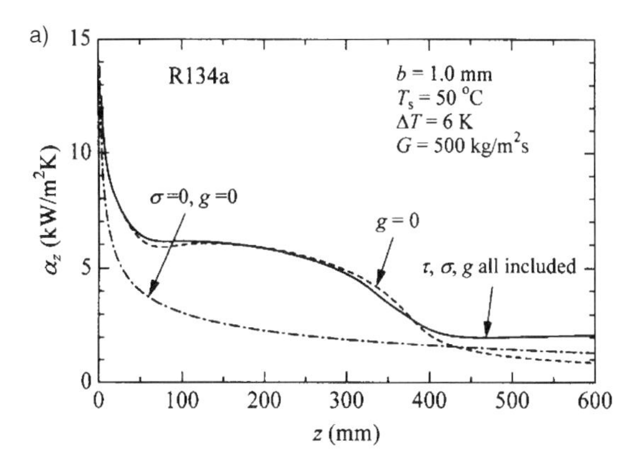

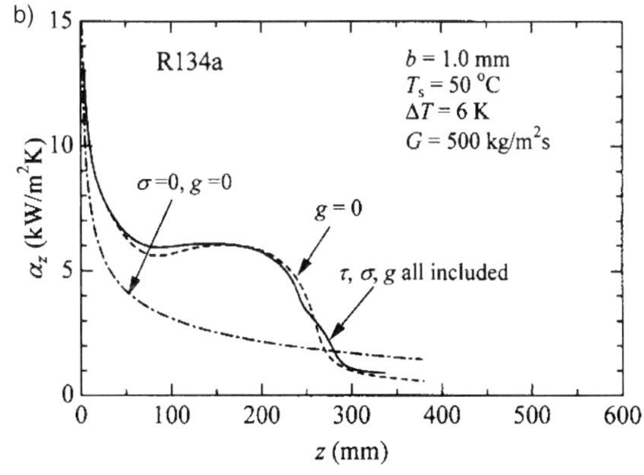

Fig. 4.13 Variation of mean (over perimeter of channel) heat-transfer coefficient with distance (Wang and Rose [102], used with permission).

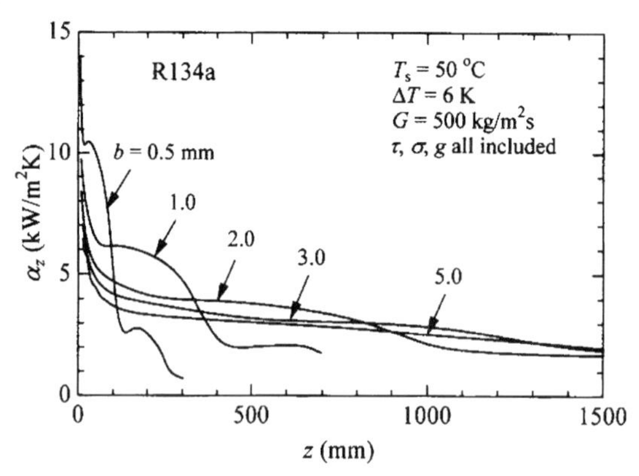

Fig. 4.14 Effect of hydraulic diameter on heat transfer in square channels (Wang and Rose [102], used with permission).

smooth film (uniform and nonuniform thickness) covering the surface of the tube. Both analytic and numerical research has been pursued. Typical of some of the analytic studies are those of Begg et al. [48], Wang et al. [96], and Baird et al. [63]. Begg et al. [48] for condensation of water in 0.5-, 1.0-, and 1.5-mm circular tubes concluded that the length of film condensation was restricted due to surface tension effects, which resulted in liquid blockage ("flooding"), and that condensation is different in small and large tubes. Wang et al. [96] for condensing R134a in 1.46-mm rectangular channels indicated that their experimental results showed surface tension changed the phase distribution in the flow and stabilized annular flow at lower vapor velocities. However, their analysis ignored surface tension and gravity. Baird et al. [63] formulated a shear-dominated annular flow model that had reasonable agreement with data, but this study, too, ignored surface tension.

In a series of papers, Wang and Rose [101–103] performed a detailed numerical analysis of condensing annular flow of R134a, R22, and R410A in noncircular channels (square, triangular) with hydraulic diameters ranging from 0.5 to 5 mm. Surface tension, gravity, and shear were included in the calculations, with results being circumferential film thicknesses and heat-transfer coefficients. As shown in Fig. 4.13, surface tension effects about doubled the peripheral average heat-transfer coefficient over a large portion at the start of the tube; gravity was important at distances further along the tube due to thinning of the film on the upper surface. Higher mass fluxes resulted in enhanced heat transfer over longer lengths of tube. Consistent with the conclusions of Begg et al. [48], Fig. 4.14 indicates that smaller tubes tend to "flood" more rapidly than larger tubes, thus leading to reduced heat-transfer coefficients; tubes greater than 3 mm showed little effect due to surface tension.

4.7 Advanced Test Sections

Experimental investigation plays a major role in understanding two-phase flow and heat transfer in microchannels and microtubes. Figure 4.15 shows an example of an experimental flow loop for boiling heat-transfer studies. Preheaters are used to control the inlet fluid temperature while flowmeters and valves are used to obtain the desired flow rate. The setup allows for measurements of pressure at the inlet and exit ports of the test section, electrical power, and temperature at different locations along the test section. Downstream of the test section the fluid is condensed and recirculated. If condensation experiments are performed, the heater is replaced with a cooling device (such as a thermoelectric cooler) and a liquid/vapor mixture is passed through the test section. In both types of experiments, careful design and instrumentation of the fluidic loop and test sections are critical to obtain reliable data. Briefly discussed below are some of the efforts in the development of advanced test sections for two-phase heat transfer in microchannels.

Fabrication and instrumentation of test sections have become increasingly challenging as the hydraulic diameters decreased from the macroscale to sub-mm and micrometer scale. Fabrication of the test sections is performed typically by: (1) miniaturized fabrication technologies that often use conventional machine tools or tools adapted to operate in the microregime or (2) semiconductor style microfabrication technologies. Kandlikar and Grande [104] have presented a summary of microchannel fabrication technology and its evolution. While the semiconductor fabrication technology is a batch fabrication method that can integrate sensors and actuators on the test section, the conventional fabrication technologies are serial and cannot provide the integration of test-section instrumentation.

An example of conventionally machined layered microchannel heat sink is shown by Qu and Mudawar [81]. The microchannels (0.215 mm wide by 0.821 mm height) were cut into the top surface of an oxygen-free copper block. Heat was supplied to the block from high-power cartridge heaters inserted into bores in the underside of the block. The copper block was placed along the central section of a thermally insulating fiberglass housing. A thermoplastic semi-transparent cover was used to cover the channels and to provide visual access to the flow boiling. Type K thermocouples were used to measure the streamwise temperature distribution of the section, while pressure transducers were used to monitor the inlet and outlet pressure.

An example of employing semiconductor fabrication technologies is provided by Jiang et al. [29, 30] who studied phase change in microchannel heat sinks with integrated heater and temperature sensors. Arrays of channels (with 40–80 μm hydraulic diameters) were fabricated between two bonded silicon wafers using a maskless self-align and self-stopping anisotropic etch. A heater and distributed thermistors were patterned in a polysilicon film deposited on one of the wafers and ion implantation doping was used to control their electrical re-

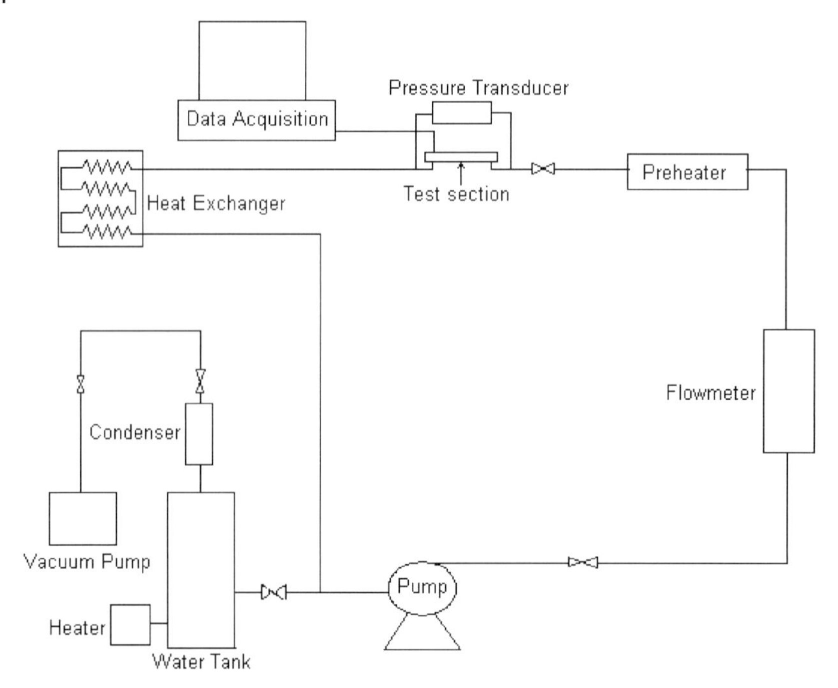

Fig. 4.15 Schematic of a flow loop used for two-phase heat-transfer studies.

sistivity. However, the device did not allow for optical access to study the boiling phenomena and because the heater was localized near the fluid inlet the heat flux along the channels was nonuniform.

A test section with integrated heater and temperature sensors and a nearly constant heat-flux boundary condition was developed and tested by Zhang et al. [43]. The single and multiple channel experimental structures used plasma-etch silicon to pattern the channels and a Pyrex glass cover to allow visualization studies. The channels (20–40 μm width and 50–100 μm depth) bridge between a reservoir and an outlet that are patterned in the wafer. The narrow bridge design for the channels helps to concentrate the heat flux from the heater to the microchannel region. Ion implantation was used to define the resistive heater and the distributed temperature sensors in the back-side of the silicon substrate while the fluidic passages were patterned on the front side. The heater/temperature sensor array was embedded in the bottom wall of the channel. However, the authors report that the silicon heaters/temperature sensors were sensitive to pressure fluctuations in the channels, and that the silicon substrate may have a significant contribution to heat-loss effects.

Recently, Jain et al. [105] have developed an instrumented glass test-section to study the critical heat flux in one side heated uniform heat flux microchannels. The low thermal conductivity (1 W/m K), thin substrate device (2×150 μm) minimizes the conjugate heat-loss effects and can provide optical access for

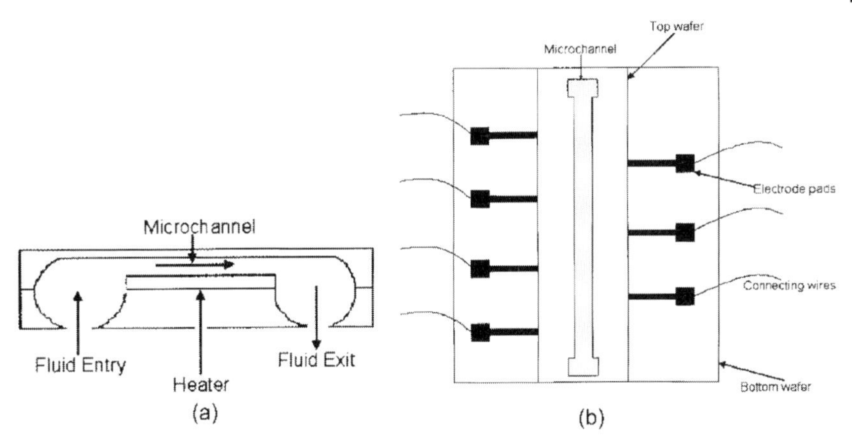

Fig. 4.16 (a) Cross-sectional view and (b) top view of an instrumented glass microchannel device.

front-side and back-side visualization. A cross section and the top view of the device are shown in Fig. 4.16. The device consists of two layers: the bottom layer contains the fluid entry/exit ports, the heater, and the temperature sensors; the top layer contains the microchannel. The channel may have a hydraulic diameter between 50–300 μm, depending on the selected width and depth. The heater and temperature sensors are made of a thin metallic film. The temperature-sensor array monitors the temperature along small heater sections. The temperature of each section of the heater is inferred from the change in resistance.

In the examples discussed above both heater and temperature sensors have been integrated with the micromachined test sections for boiling heat-transfer studies. However, it becomes increasingly important to monitor the spatial and temporal distribution of pressure in the test channels, in order to understand the effect of instabilities on heat transfer and fluid flow. Distributed pressure sensors have been integrated on test sections, but just for fluid-flow studies (Ho and Tai [106]). It is envisioned that advanced heat/transfer and fluid-flow sections with embedded heaters and distributed temperature and pressure sensors could be developed in the near future using MEMS technology.

4.8 Summary

While the last half decade has witnessed significant advances in knowledge about boiling, condensation and adiabatic two-phase flows in mini- and microchannels, this field of study is still in its infancy. Flow patterns, heat-transfer coefficients (boiling and condensing), the critical heat-flux condition, onset of nucleate boiling, and two-phase pressure drop data have been acquired and quantified to a limited extent. Likewise, attempts have been made to develop engineering correlations to accurately predict behavior. Flow and heat-transfer similarities and differences compared to large-scale systems have been observed, but further investigations are needed to fully assess and predict these thermohydraulic phenomena at the microscale.

Microchannel two-phase systems have matured to a degree that the heat-transfer community is ready to take on new challenges and develop more sophisticated second-generation microchannel systems. Recent advancements in microfabrication technologies, coupled with the growing fundamental knowledge of phase change in miniature systems, can enhance the use of microchannel heat transfer as the technique of choice for current and future ultra high power density devices. However, before this can materialize many challenges in boiling and condensation in microchannels need to be overcome; for example, in boiling, knowledge has to be acquired to suppress severe flow oscillations, reduce high superheat temperatures at the onset of nucleate boiling, increase the critical heat flux condition, and reduce pressure drops. Furthermore, improvements in state-of-the-art experimental capabilities are required to advance the thermal and hydraulic knowledge of change of phase heat transfer in channels with length scales on the order of 50 to 500 μm.

References

1. B. Palm, in *Heat Transfer and Transport Phenomena in Microscale*, Proc. Int. Conf., Banff, Canada, Oct. 2000, Begell House, New York, **2000**, 54–64.
2. S.S. Mehendale, A.M. Jacobi, and R.K. Shah, *Appl. Mech. Rev.*, **2000**, *53*, 175–193.
3. S.G. Kandlikar, *Exp. Thermal Fluid Sci.*, **2002a**, *26*, 389–407.
4. S.G. Kandlikar, *Heat Trans. Eng.*, **2002b**, *23*, 5–23.
5. A. Cavallini, G. Censi, D. Del Col, L. Doretti, G.A. Longo, L. Rossetto, *Heat Transfer 2002, Proc. 12th Int. Heat Trans. Conf.*, **2002**, Elsevier, Paris, France.
6. H.J. Lee, S.Y. Lee, *Int. J. Multiphase Flow* **2001**, *27*, 783–796.
7. T.S. Zhao, Q.C. Bi, *Int. J. Heat Mass Trans.*, **2001**, *44*, 2523–2534.
8. C.-Y. Yang, C.-C. Shieh, *Int. J. Multiphase Flow* **2001**, *27*, 1163–1177.
9. K.A. Triplett, S.M. Ghiaasiaan, S.I. Abdel-Khalik, D.L. Sadowski, *Int. J. Multiphase Flow* **1999**, *25*, 377–394.
10. J.L. Xu, P. Cheng, T.S. Zhao, *Int. J. Multiphase Flow* **1999**, *25*, 411–432.
11. S. Lin, P.A. Kew, K. Cornwell, "Characteristics of air/water flow in small tubes," *Int. J. Heat Technology*, **1999**, *17*, 63–70.
12. H. Ide, H. Matsumura, Y. Tanaka, T. Fukano, *Trans. JSME*, **1997**, *63*, 452–460.
13. K. Mishima, T. Hibiki, *Int. J. Multiphase Flow* **1996**, *22*, 703–712.
14. J.K. Keska, R.D. Fernando, *J. Fluids Eng.*, **1994**, *116*, 247–256.
15. M. Kawaji, in *Handbook of Phase Change: Boiling and Condensation* (Eds.: S.G. Kandlikar, M. Shoji, V.K. Dhir Taylor & Francis, Wash., DC, **1999**, chap. 9.2.
16. M. Kawaji, P.M.-Y. Chung, *Microscale Thermophysical Eng.*, **2004**, *8*, 239–257.
17. Z.P. Feng, A. Serizawa, *Proc. 18th Multiphase Flow Symp. Japan*, **1999**, 33–36.
18. Z.P. Feng, A. Serizawa, *Proc. 37th Nat. Heat Tran. Symp. Japan*, **2000**, 351–352.
19. A. Serizawa, Z.P. Feng, Z. Kawara, *Exp. Thermal Fluid Sci.*, **2002**, *26*, 703–714.
20. A. Serizawa, Z.P. Feng, *Proc. US-Japan Seminar on Two-Phase Flow Dynamics*, **2000**, 429–451.

21 A. Kawahara, P.M.-Y. Chung, M. Kawaji, *Int. J. Multiphase Flow,* **2002**, *28,* 1411–1435.

22 P.M.-Y. Chung, M. Kawaji, *Int. J. Multiphase Flow,* **2004**, *30,* 735–761.

23 P.M.-Y. Chung, M. Kawaji, A. Kawahara, Y. Shibata, *J. Fluids Eng.,* **2004**, *126,* 546–552.

24 C.A. Damianides, J.W. Westwater, *Proc. 2nd UK Nat. Conf. Heat trans.,* **1988**, 1257–1268.

25 T. Fukano, A. Kariyasaki, *Nucl. Eng. Des.,* **1993**, *141,* 59–68.

26 H.Y. Wu, P. Cheng, *Int. J. Heat Mass Trans.,* **2003**, *46,* 2603–2614.

27 H.Y. Wu, P. Cheng, *Int. J. Heat Mass Trans.,* **2004**, *47,* 3631–3641.

28 D.E. Kamidis and T.S. Ravigururajan, *Proc. NHTC2000: 33rd Nat. Heat Trans. Conf.,* Albuquerque, NM, Paper No. NHTC2000-12100, **1999**, 1–8.

29 L. Jiang, M. Wong, Y. Zohar, *J. Microelectromechanical Sys.,* **1999**, *8,* 358–365.

30 L. Jiang, Y. Wang, M. Wong, Y. Zohar, *J. Micromech. Microeng.,* **1999**, *9,* 422–428.

31 K. Cornwell, P.A. Kew, *Energy Efficiency in Pro. Tech.,* **1992**, *37,* 624–638.

32 K. Cornwell, P.A. Kew, *Proc. CEC Conf. Energy Efficiency in Process Tech.,* Athens, October 1992, Paper 22, Elsevier Applied Sciences, 624–638.

33 R. Mertz, A. Wein, M. Groll, *Calore e Technologia,* **1996**, *14,* 47–54.

34 J.E. Kennedy, G.M. Jr. Roach, M.F. Dowling, S.I. Abdel-Khalik, S.M. Ghiaasiaan, S.M. Jeter, Z.H. Quershi, *J. Heat Trans.,* **2000**, *122,* 118–125.

35 S.G. Kandlikar, *Multiphase Sci. Technol.,* **2001**, *13,* 207–232.

36 X.F. Peng, B.X. Wang, *Int. J. Heat Mass Trans.,* **1993**, *36,* 3421–3427.

37 X.F. Peng, H.Y. Hu, B.X. Wang, *J. Enhanced Heat Trans.,* **1994**, *4,* 315–326.

38 X.F. Peng, H.Y. Hu, B.X. Wang, *Int. J. Heat Mass Trans.,* **1998**, *41,* 101–106.

39 H.Y. Hu, G.P. Peterson, X.F. Peng, B.X. Wang, *Int. J. Heat Mass Trans.,* **1998**, *41,* 3483–3490.

40 L. Zhang, E.N. Wang, K.E. Goodson, T.W. Kenny, *Int. J. Heat Mass Trans.,* **2005**, *48,* 1572–1582.

41 M. Lee, Y.Y. Wang, M. Wong, Y. Zohar, *J. Micromech. Microeng.,* **2003**, *13,* 155–164.

42 P.C. Lee, F.G. Tseng, C. Pan, *Int. J. Heat Mass Trans.,* **2004**, *47,* 5575–5589.

43 L. Zhang, J.-M. Koo, L. Jiang, M. Asheghi, K.E. Goodson, J. Santiago, T.W. Kenny, *J. Microelectromechanical Sys.,* **2002**, *11,* 12–19.

44 L. Jiang, M. Wong, Y. Zohar, *J. Microelectromechanical Sys.,* **2001**, *10,* 80–87.

45 P. Balasubramanian, S.G. Kandlikar, *Proc. 2003 ASME Intl. Mech. Eng. Cong.,* **2003**, 315–322.

46 P. Balasubramanian, S.G. Kandlikar, *Heat Trans. Eng.,* **2005**, *26,* 20–27.

47 M.E. Steinke, S.G. Kandlikar, *1st Int. Conf. Microchannels and Minichannels,* **2003**, 567–579.

48 E. Begg, D. Khrustalev, A. Faghri, *J. Heat Trans.,* **1999**, *121,* 904–915.

49 B. Mederic, M. Miscevic, V. Platel, P. Lavieille, J.-L. Joly, *Proc. 1st Int. Conf. Microchannels and Minichannels* (Ed.: S. Kandlikar), ASME, New York, USA, **2003**, 707–712.

50 H.Y. Wu, P. Cheng, B.C.-P. Siu, Y.-K. Lee, *Proc. 2nd Int. Conf. Microchannels and Minichannels* (Ed.: S. Kandlikar), ASME, New York, USA, **2004**, 657–660.

51 H.Y. Wu, P. Cheng, *Int. J. Heat Mass Trans.,* **2005**, *47.*

52 J.W. Coleman, S. Garimella, *Int. J. Refrig.,* **2003**, *26,* 117–128.

53 S. Garimella, *Heat Trans. Eng.* **2004**, *25,* 104–116.

54 S. Garimella, J.D. Killion, J.W. Coleman, *J. Fluids Eng.* **2002**, *124,* 205–214.

55 M.K. Jensen, in *Boiling Heat Transfer* (Ed.: R.T. Lahey, Jr.), **1992**, 483–514.

56 R.W. Lockhart, R.C. Martinelli, *Chem. Eng. Prog.,* **1949**, *45,* 39–48.

57 D. Chisholm, A.D.K. Laird, *Trans. ASME,* **1958**, *80,* 276–286.

58 W. Qu, I. Mudawar, *Int. J. Heat Mass Trans.,* **2003a**, *46,* 2737–2753.

59 J.-M. Koo, L. Jiang, L. Zhang, P. Zhou, S.S. Banerjee, T.W. Kenny, J. Santiago, K.E. Goodson, in *Proc. Int. MEMS Workshop,* **2001**, 422–426.

60 S. Garimella, J.D. Killion, J.W. Coleman, *J. Fluids Eng.* **2003**, *125,* 887–894.

61 W. Qu, I. Mudawar, *Int. J. Heat Mass Trans.*, **2003b**, *46*, 2773–2784.

62 S. Koyama, K. Kuwahara, K. Nakashita, *Proc. 1st Int. Conf. Microchannels and Minichannels* (Ed.: S. Kandlikar), ASME, New York, USA, **2003**, 193–205.

63 J.R. Baird, D.F. Fletcher, B.S. Haynes, *Int. J. Heat Mass Trans.*, **2003**, *46*, 4453–4466.

64 S. Garimella, A. Agarwal, J.D. Killion, *Heat Trans. Eng.* **2005**, *26*, 1–8.

65 D. Chisholm, *Two-Phase Flow in Pipelines and Heat Exchangers*, George Godwin, Longman Group, New York, **1983**.

66 L. Friedel, European Two Phase Flow Group Meeting, Ispra, Italy, E2, **1979**.

67 Y.Y. Hsu, *J. Heat Trans.*, **1962**, *84*, 207–216.

68 A.E. Bergles, W.M. Rohsenow, *J. Heat Trans.*, **1964**, *86*, 365–372.

69 S.G. Kandlikar, V.R. Mizo, M.D. Cartwright, Ikenze, HTD-Vol. 342, *ASME Proc. 32nd Nat. Heat Trans. Conf.*, **1997**, *4*, 11–18.

70 E.J. Davis, G.H. Anderson, *AIChE J.*, **1966**, *12*, 774–780.

71 G.M. Lazarek, S.H. Black, *Int. J. Heat Mass Trans.*, **1982**, *25*, 945–960.

72 T.N. Tran, M.W. Wambsganss, D.M. France, *Int. J. Multiphase Flow*, **1996**, *22*, 485–498.

73 S.G. Kandlikar, P. Balasubramanian, *Heat Trans. Eng.*, **2004**, *25*, 86–93.

74 S.G. Kandlikar, *J. Heat Trans.*, **1990**, *112*, 219–228.

75 A.E. Bergles, S.G. Kandlikar, *J. Heat Trans.*, **2005**, *127*, 101–107.

76 A.E. Bergles, *Doctoral Dissertation*, **1962**, Massachusetts Institute of Technology, Cambridge, Massachusetts.

77 A.E. Bergles, ASME Paper No. 63-WA-182, **1963**.

78 A.E. Bergles and W.M. Rohsenow, M.I.T. Engineering Projects Laboratory Report No. DSR 8767-21, **1962**.

79 W. Yu, D.M. France, M.W. Wambsganss, and J.R. Hull, *Int. J. Multiphase Flow*, **2002**, *28*, 927–941.

80 M.B. Bowers and I. Mudawar, *Int. J. Heat Mass Trans.*, **1994**, *37*, 321–334.

81 W. Qu and I. Mudawar, *Int. J. Heat Mass Trans.*, **2004**, *47*, 2045–2059.

82 F. Mayinger, O. Schad, and E. Weiss, *Brennst.-Warme-Kraft*, **1966**, *18*, 288–294.

83 S.W. Ruan, G. Bartsch, and S.M. Yang, *Exp. Thermal and Fluid Sci.*, **1994**, 7.

84 M. Ozawa, H. Umekawa, K. Mishima, T. Hibiki, and Y. Saito, *Chem. Eng Res. Des.*, **2001**, *79*, 389–401.

85 C.N. Ammerman and W.M. You, *J. Heat Trans.*, **2001**, *123*, 976–983.

86 G.M. Roach, S.I. Abdel-Khalik, S.M. Ghiaasiaan, M.F. Dowling, and S.M. Jeter, *Nucl. Sci. Eng.*, **1999**, *131*, 411–425.

87 Y. Peles, L.P. Yarin, G. Hetsroni, *Int. J. Multiphase Flow*, **2001**, *27*, 577–598.

88 S.G. Kandlikar, P. Balasubramanian, ASME Paper No. ICMM 2004-2379, *2nd Int. Conf. Microchannels Minichannels*, June 17–19, 2004, Rochester, NY USA, pp. 539–550, also to appear in *Journal of Heat Transfer*, **2005**.

89 R. Gregorig, *Z. Angew. Math. Phys.*, **1954**, *5*, 36–49.

90 B. Palm, *Proc. 1st Int. Conf. Microchannels and Minichannels* (Ed.: S. Kandlikar), ASME, New York, USA, **2003**, 25–31.

91 S. Garimella, T.M. Bandhauer, *Proc. 2001 ASME Int. Mech. Eng. Congress and Expo.*, ASME, New York, USA, **2001**, IMECE2001/HTD-24221.

92 A. Cavallini, G. Censi, D. Del Col, L. Doretti, G.A. Longo, L. Rossetto, C. Zillio, *Proc. 1st Int. Conf. Microchannels and Minichannels* (Ed.: S. Kandlikar), ASME, New York, USA, **2003**, 691–698.

93 A. Cavallini, D. Del Col, L. Doretti, M. Matkovic, L. Rossetto, C. Zillio, *Proc. 2nd Int. Conf. Microchannels and Minichannels*, (Ed.: S. Kandlikar), ASME, New York, USA, **2004**, 625–632.

94 J.S. Shin, M.H. Kim, *Proc. 2nd Int. Conf. Microchannels and Minichannels* (Ed.: S. Kandlikar), ASME, New York, USA, **2004**, 633–664.

95 Y.Y. Yan, T.F. Lin, *Int. J. Heat Mass Trans.*, **1999**, *42*, 697–708.

96 W.-W.W. Wang, T.D. Radcliff, R.N. Christensen, *Exp. Thermal Fluid Sci.*, **2002**, *26*, 473–485.

97 E. Begg, B. Holley, A. Faghri, *Heat Transfer 2002, Proc. 12th Int. Heat Trans. Conf.*, **2002**, Elsevier, Paris, France.

98 R.L. Webb, K. Ermis, *J. Enhanced Heat Trans.*, **2001**, *8*, 77–90.

99 K.W. Moser, R.L. Webb, B. Na, *J. Heat Trans.*, **1998**, *120*, 410–417.

100 W.W. Akers, H.A. Deans, O.K. Crosser, *Chem. Eng. Prog. Symp. Series*, **1959**, *55*, 171–176.

101 H. Wang, J.W. Rose, *Proc. 2nd Int. Conf. Microchannels and Minichannels* (Ed.: S. Kandlikar), ASME, New York, USA, **2004**, 661–666.

102 H.S. Wang, J.W. Rose, *J. Heat Trans.*, **2005**, 127.

103 H. Wang, J.W. Rose, *Proc. 3rd Int. Conf. Microchannels Minichannels* (Ed.: S. Kandlikar), ASME, New York, USA, **2005**.

104 S.G. Kandlikar and W.J. Grande, *Heat Trans. Eng.*, **2003**, *24*, 3–17.

105 A. Jain, T. Borca-Tasciuc, A.P. Roday, M.K. Jensen, and S.G. Kandlikar, *Proc. 21st Annual IEEE Semiconductor Measurement and Management Symp.*, March 15–17 **2005**, San Jose, CA, 26–30.

106 C.M. Ho, and Y.C. Tai, *Ann. Rev. Fluid Mech.*, **1998**, *30*, 579–612.

5
Generation and Multiphase Flow of Emulsions in Microchannels

Isao Kobayashi and Mitsutoshi Nakajima, Food Engineering Division, National Food Research Institute, Kannondai, Tsukuba, Ibaraki, Japan

Abstract
This chapter will outline features of microfluidic channels used for emulsification, fundamental flow characteristics of the two immiscible phases inside the channels, and emulsification technologies using microfluidic devices. Size and size distribution of the channels microfabricated on a chip are explained, which affects droplet size and size distribution of the resultant emulsions. The effects of the type of channel materials and the composition of two liquid phases on the droplet-generation behavior are discussed based on the results reported in the cited references. The flow conditions and force balance of the to-be-dispersed phase moving forward inside the microfluidic channel are evaluated using an example. Recent emulsification technologies using the microfluidic channels are reviewed, which have enabled direct generation of monodisperse emulsion droplets. Their potential applications as advanced materials and microreactor vessels are also briefly introduced.

Keywords
Emulsification technology, microfluidic channels, monodisperse droplets, droplet generation, immiscible liquid phases, microspace effects

Advanced Micro and Nanosystems Vol. 5. Micro Process Engineering. Edited by N. Kockmann

ISBN: 3-527-31246-3

5.1 Introduction

An emulsion consists of small, spherical droplets of one of two immiscible liquids in the continuous phase of the other. Emulsion droplets usually have an average diameter in the range of 0.1–100 μm. Emulsions become kinetically stable against droplet coalescence for a finite period when surfactants are added prior to emulsification. Figure 5.1 illustrates polydisperse and monodisperse emulsions. A monodisperse emulsion contains uniformly sized droplets with a typical coefficient of variation ((standard deviation/average droplet diameter) ×100) less than 5%. Monodisperse emulsions are useful for measuring, analyzing, and controlling many of their important physical, physicochemical, and organoleptic properties [1–3]. Monodisperse emulsions can also greatly reduce Ostwald ripening because of the very small Laplace pressure difference between droplets. Monodisperse-emulsion-based materials (e.g., monodisperse multiple emulsions, microparticles, and microcapsules) have attracted a great deal of interest for various applications, including foods, cosmetics, pharmaceuticals, chemicals, electronics, and optics.

Emulsions are commonly produced using conventional emulsification devices, such as high-speed blenders, colloid mills, and high-pressure homogenizers [2].

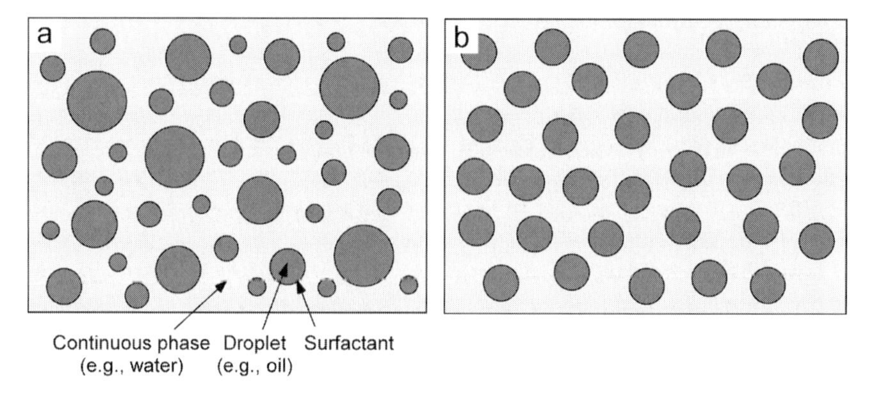

Fig. 5.1 Illustrations of a polydisperse emulsion (a) and a monodisperse emulsion (b).

They apply the large, mechanical shear force to disruption of the to-be-dispersed phase into smaller droplets. However, the emulsions produced by these devices exhibited considerable polydispersity with a typical coefficient of variation over 20%. Their typical droplet diameters were 2–10 μm for high-speed blenders, 1–5 μm for colloid mills, and 0.1–1 μm for high-pressure homogenizers [2]. In addition, the conventional devices require a high energy input for producing emulsions (10^5 to 10^9 J/m^3) [4]. For example in a typical high-pressure homogenizer, the emulsification efficiency, which is the ratio of the energy used for droplet formation to the energy supplied, is less than 0.1% [5].

Several techniques have been proposed for producing quasi-monodisperse emulsions by reducing the size distribution of polydisperse emulsions. Repeated fractionation of polydisperse emulsions can yield a set of quasi-monodisperse emulsions with average droplet diameters of 0.1–2 μm [6], whereas this technique produce quasi-monodisperse emulsions only with significant effect and skill. Shear rupturing of polydisperse emulsion droplets in the thin gap between outer rotating and inner fixed cylinders can produce quasi-monodisperse emulsions with average droplet diameters of 0.1–10 μm and a minimum coefficient of variation of 5%, but their typical coefficients of variation are about 15% [7].

Nakashima et al. [8] proposed direct production of quasi-monodisperse emulsions with a minimum coefficient of variation of 10% using a microporous membrane. This membrane emulsification can produce emulsion droplets by forcing the to-be-dispersed phase in the continuous phase through the membrane pores [8–10]. The resultant droplet size with average values of 0.3–30 μm is primarily controlled by the membrane pore size [8, 11, 12]. This technique is also capable of producing emulsions without strong, mechanical shear stress at a low energy input of 10^4 to 10^6 J/m^3 [13]. While membrane emulsification is a promising technique as described above, it is difficult to produce highly monodisperse emulsions owing to the pore-size distribution of the membranes used.

Microfabrication technology, which originated from semiconductor technology, makes it possible to precisely fabricate micrometer-sized channels on a microfluidic chip. The microfluidic channels have very narrow size distribution (typically $<1\%$), which is a necessary condition for direct production of monodisperse emulsions. Direct production of monodisperse emulsions using microfluidic devices has been a hot research field in the last decade (see Table 5.1). Utilization of the monodisperse droplets generated by these devices is also a promising research field.

In this chapter, we discuss features of microfluidic devices for the flow of liquid phases and review emulsification technologies using microfluidic devices. In Section 5.2, we explain microfluidic channels, their materials, and their surface properties. The forces acting on a liquid/liquid interface moving in the micrometer-sized channel are also evaluated. Section 5.3 first reviews production of emulsions using micromixers with an interdigital channel configuration, and then generation of monodisperse emulsion droplets using several types of microfabricated channel networks. We next review experimental and computational fluid dynamics (CFD) studies of direct production of monodisperse emulsions

Table 5.1 Representative emulsification technologies using microfluidic devices.

Device type	Driving force of droplet generation	Average droplet diameter (μm)	Coefficient of variation (%)	Ref.
Interdigital micromixers	Shear force due to continuous-phase flow	4–60	>10[b]	14, 15
T-shape microchannels	Shear force due to continuous-phase flow	10–380[a]	<5	16, 17
Flow-focusing geometries	Shear force due to continuous-phase flow	9–190[a]	<5	18–20
Tapered capillaries	Shear force due to continuous-phase flow	2–200	<5	21
Microchannel arrays	Spontaneous transformation by interfacial tension	3–90	<5	22–24

a) The generated droplets have disk-like shapes under some device and operation conditions.
b) The value was estimated on the basis of the diameter distribution of the resultant emulsion droplets in ref. 14.

in microchannel (MC) emulsification using microfluidic channel arrays. This section also discusses the effects of the device and process parameters on the emulsification properties for the above-mentioned emulsification technologies. Furthermore, we briefly introduce potential applications of the monodisperse emulsions produced by the above-mentioned emulsification technology.

5.2 Features of Microfluidic Channels for Emulsification

5.2.1 Microfluidic Channels

Microfluidic channels consist of grooves and/or through-holes with a size of 1 to 1000 μm microfabricated on a chip. One can carry out various kinds of liquid operations in the well-designed microfluidic channels (e.g., chemical and biochemical reactions, separation, and mixing), exploiting a micrometer-sized space. Rapid advance in microfabrication technology enables precise fabrication of networks and arrays of the microfluidic channels. Figure 5.2 depicts examples of silicon arrays of the microfluidic channels. They have very narrow size distributions less than 1%. For most emulsification technologies using microfluidic devices, the to-be-dispersed phase that passes through the channel is directly broken up into droplets (see Section 5.3). This droplet-generation process suggests that the resultant droplet size and its distribution greatly depend on

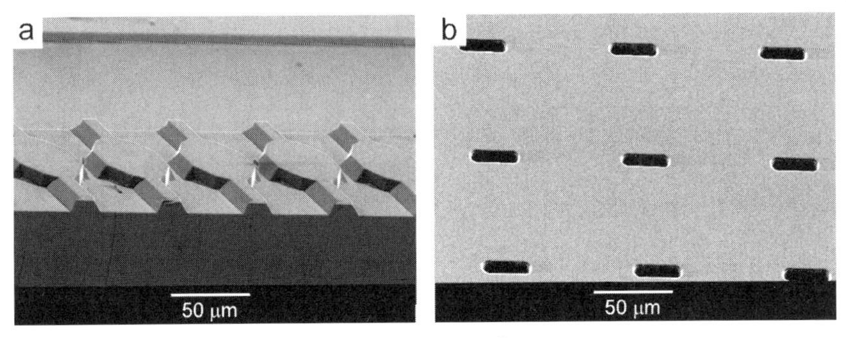

Fig. 5.2 Scanning electron microscopy (SEM) images of silicon arrays of microfluidic channels: (a) a grooved microchannel (MC); (b) an oblong straight-through MC.

the channel size and its distribution. Thus, the microfluidic channels precisely controlled in size and its distribution are useful for generation of monodisperse emulsion droplets with a desired size.

5.2.2
Channel Materials and Their Surface Properties

Microfluidic channels for emulsification are generally made of silicon, silicon compounds, polymers, and metals (Table 5.2). Their typical fabrication techniques, listed in Table 5.2, include chemical wet etching, chemical dry etching, mechanical cutting, injection molding, soft lithography, and electroforming, being described in detail in ref. [25]. The etching techniques enable precise fabrication of the microfluidic channels with a size of over 1 μm. It is also possible to fabricate submicrometer channels by these techniques. The channels precisely fabricated by the other fabrication techniques have a typical size greater than 10 μm.

The inherent surface affinity of the channel materials, such as hydrophilic and hydrophobic, plays an important role in the emulsification behavior. Hydrophilic channel materials include surface-oxidized silicon, silicon compounds, and stainless steel. Hydrophobic channel materials include polymers and nickel. Preventing the to-be-dispersed phase from wetting on the channel surface is a prerequisite for generation of monodisperse emulsion droplets from the channel. The most significant strategy for the successful droplet generation is to select the to-be-dispersed phase liquid with a surface affinity different from that of the channel. Therefore, hydrophilic and hydrophobic channels are commonly used to generate oil-in-water (O/W) and W/O emulsion droplets, respectively. The surface charge of the channel materials is also a key factor affecting the emulsification behavior. For example, the surface-oxidized silicon, which is in contact with the continuous water phase, has a negative surface charge [33]. Tong et al. [34] reported the effect of the surfactant charge on the emulsification

Table 5.2 Materials of microfluidic channels for emulsification.

Materials	Fabrication techniques [25]	Inherent surface affinity	Ref.
Surface-oxidized silicon	Anisotropic wet etching	Hydrophilic	22
	Chemical dry etching	Hydrophilic	19, 24
Quartz glass	Mechanical cutting	Hydrophilic	20
Pyrex glass	Isotropic wet etching	Hydrophilic	26
Photostructurable glass (Foturan™)	Chemical dry etching	Hydrophilic	15
Silicon nitride	Chemical dry etching	Hydrophilic	27
Poly(dimethylsiloxane) (PDMS)	Soft lithography	Hydrophobic	18, 28–30
Poly(methylmethacrylate)	Mechanical cutting	Hydrophobic	17
(PMMA)	Injection molding	Hydrophobic	31
Urethane	Soft lithography	Hydrophobic	16
Stainless steel	Mechanical cutting	Hydrophilic	32
Nickel	LIGA [a)] process	Hydrophobic	33

a) LIGA: *Li*thographie, *G*alvanoformung, *A*bformung.

behavior using a silicon array of the channels, depicted in Fig. 5.2a. The use of anionic surfactants, which have a repulsive interaction with the negatively charged channel surface, led to successful generation of monodisperse O/W emulsion droplets from the channels. In contrast, the use of cationic surfactants, which have an attractive interaction with the negatively charged channel surface, resulted in generation of polydisperse emulsions droplets from the channels and wetting of the to-be-dispersed phase on the channel surface. We thus have to select an appropriate combination of the channel material, the two liquid phases, and the surfactant to achieve successful generation of monodisperse emulsion droplets using the microfluidic channels.

5.2.3
Fundamental Flow Characteristics of the To-be-dispersed Phase Inside the Microfluidic Channels

When the pressure difference applied between the two phases exceeds the following Young–Laplace equation considering the contact angle effect [34, 35], the to-be-dispersed phase moves forward inside a microfluidic channel (Fig. 5.3) and droplet generation from the channel starts.

$$\Delta P = \frac{4\gamma \cos\theta}{d} \tag{5.1}$$

where ΔP is the Young–Laplace pressure between the two phases, γ the interfacial tension, θ the contact angle of the to-be-dispersed phase to the channel surface, and d the channel diameter. The forces acting on the to-be-dispersed phase

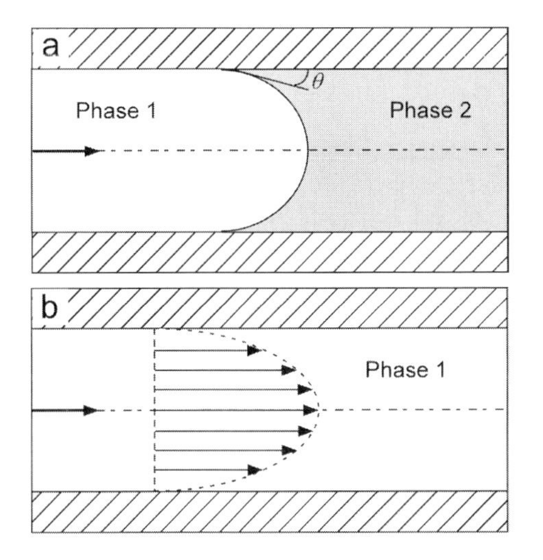

Fig. 5.3 (a) A schematic model of the to-be-dispersed phase moving forward inside a microfluidic channel (e.g., 10 μm in diameter), which is initially filled with the continuous phase. (b) A schematic model explaining the laminar flow of the to-be-dispersed phase inside the channel.

that advances inside the channel can be evaluated using dimensionless numbers, being the Reynolds number (Re), the Bond number (Bo), the Weber number (We), and the Capillary number (Ca) [23]. As an example, the dimensionless numbers are estimated using the following conditions: a channel hydraulic diameter (d_h) of 10^{-5} m, a to-be-dispersed phase density (ρ) of 10^3 kg/m^3, a to-be-dispersed phase viscosity (μ) of 10^{-2} Pa s, a to-be-dispersed phase velocity (U) of 10^{-3} m/s, a gravity force (g) of about 10 m/s^2, and an interfacial tension (γ) of 10^{-2} N/m.

$$\text{Re} = (\text{inertial force/viscous force}) = (\rho d U/\mu) = 10^{-3} \quad (5.2)$$

$$\text{Bo} = (\text{gravitational force/interfacial tension force}) = (\rho g d^2/\gamma) = 10^{-4} \quad (5.3)$$

$$\text{We} = (\text{inertial force/interfacial tension force}) = (\rho d U^2/\gamma) = 10^{-6} \quad (5.4)$$

$$\text{Ca} = (\text{viscous force/interfacial tension force}) = (\mu U/\gamma) = 10^{-3} \quad (5.5)$$

The values of these dimensionless numbers indicate that the interfacial tension force, which is much larger than the other forces, dominates the flow behavior of the to-be-dispersed phase inside the microfluidic channel. The Re value also suggests the laminar flow of the to-be-dispersed phase inside the microfluidic channel, as illustrated in Fig. 5.3. Emulsification technologies using microfluidic devices (see Section 5.3) exploit the aforementioned force balance on a micrometer-sized space.

5.3
Emulsification Technologies Using Microfluidic Devices

5.3.1
Interdigital Micromixers

An interdigital micromixing system was developed by the Institut für Mikrotechnik Mainz in Germany in the late 1990s [36]. Haverkamp et al. [14] applied this interdigital micromixer to the production of silicon oil-in-water emulsions. Figure 5.4 shows SEM micrographs of a mixing element in the interdigital micromixer. The mixing element consists of interdigitated channels with corrugated walls. Typical dimensions of the channels are 25 or 40 μm in width and 300 μm in depth. The liquid phases are introduced into the mixing element as the two counterflows, forming their flow lamellae in interdigitated channels. When they leave the device perpendicular to the direction of the feed flows, the flow lamellae are mixed, and then an emulsion is produced.

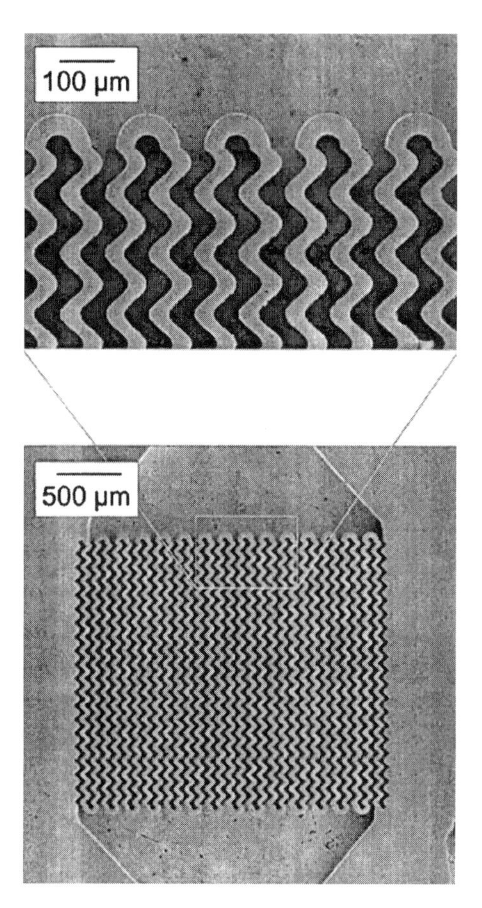

Fig. 5.4 SEM images of a mixing element composed of an interdigital channel configuration with corrugated walls (reproduced from [36] by permission of the American Chemical Society).

The interdigital micromixer yielded emulsions with an average droplet diameter of 4–60 μm by varying the flow rates of the two liquids [14]. The smallest droplets were obtained at the flow rates of 20 ml/h for oil and of 700 ml/h for water. The droplet diameter of the resultant emulsions gradually decreased as increasing the total flow at a fixed ratio (1 : 1) of the flow rates of the two liquids (Fig. 5.5). Their droplet size distribution became narrower at higher total flow. Besides the flow conditions, the channel geometry also significantly affected the droplet diameter of the produced emulsions. For example, decreasing the channel width led to a smaller average droplet diameter [14]. The throughput capacities of the interdigital micromixers are 1.5 l/h for a single mixing unit and 6 l/h for a mixer array [36]. A current problem in the interdigital micromixer is the relatively wide size distribution of the produced emulsions. The interdigital micromixer has been applied to the production of semisolid creams with an average droplet diameter of 1 μm [37].

The use of a glass interdigital micromixer enabled direct observation of the emulsion-droplet generation in the mixing zone [15]. The lamellae of the two phases, which passed through interdigitally arranged channels, flow into the mixing zone in parallel. In the mixing zone, cylinders of the to-be-dispersed phase are broken up into droplets even at high flow velocities of several meters per second (see Fig. 5.6). This droplet generation in the mixing zone takes place within a few milliseconds after the cylinders of the two phases contacted in the flow focusing zone. Hardt et al. [38] carried out three-dimensional computational fluid dynamics (CFD) simulations to analyze droplet generation in the glass interdigital micromixer. Under a fixed total volume flow, the cylinders of the to-be-dispersed phase in the mixing zone became shorter at a high flow ratio of the to-be-dispersed phase, which agreed well with the experimental results.

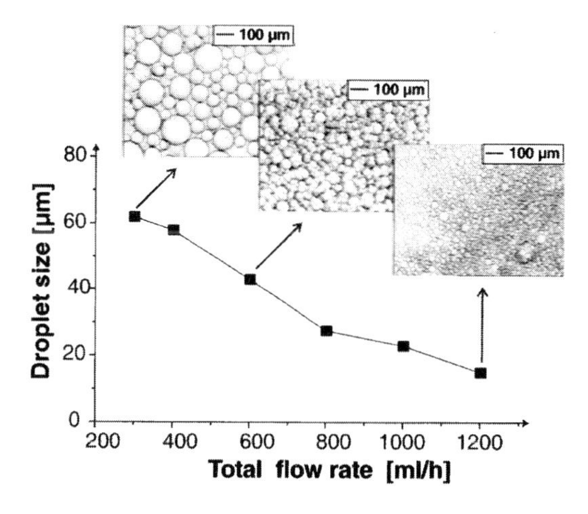

Fig. 5.5 Effect of the total flow rates of the two phases on the droplet size of silicone oil-in-water emulsions generated in a interdigital micromixer (reproduced from [14]). The ratio of the flow rates was kept constant at 1 : 1.

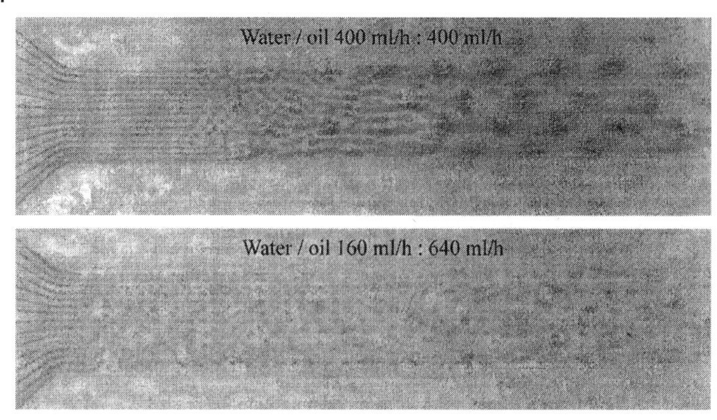

Fig. 5.6 Microscopic images of droplet generation and subsequent coalescence in a rectangular-shaped micromixer for a dyed water-in-silicone oil system (reproduced from [15]).

5.3.2
Emulsion Droplet Generation Using Microfluidic Channel Networks

5.3.2.1 Droplet Generation at a Junction of Microfluidic Channels

Several research groups have proposed the droplet-generation technique using T-shape microfluidic channels fabricated on a chip [16, 17]. The typical droplet-generation process in the channels is depicted in Fig. 5.7. The channels with a typical size over 10 μm consist of a crossflow channel for the continuous phase and a branch channel for the to-be-dispersed phase. The pressurized to-be-dispersed phase, which passed through the channel, begins to flow into the T-junction of the channels, and then is broken up into droplets by the shear force due to the continuous-phase flow at the T-junction. The T-shape channels are capable of generating monodisperse emulsion droplets with average diameters of 10–380 μm and a typical coefficient of variation less than 3% [16, 17, 20, 26]. A hydrophilic chip (e.g., quartz glass) and a hydrophobic chip (e.g., polymethyl methacrylate (PMMA) and urethane) have been used for generation of monodisperse O/W and W/O emulsion droplets, respectively [16, 17, 20].

The flow rates and/or pressures of the two phases in the channels are the most important parameters affecting the droplet diameter. The resultant droplet diameter tends to increase with increasing to-be-dispersed phase pressure [16] and to decrease with increasing continuous phase velocity [17]. The droplet-generation rate at the T-junction, which greatly depends on the flow conditions, achieved a maximum value of 2500 s^{-1} [17]. Generation of monodisperse droplets at the T-junction occurred below a critical flow rate of the to-be-dispersed phase at a certain flow rate of the continuous phase [20]. The T-shape microfluidic channels is advantageous for generation of monodisperse droplets with different average diameters only by varying the flow conditions. Conversely, the droplet breakup sheared by the continuous-phase flow is a sensitive process, requiring skillful control of

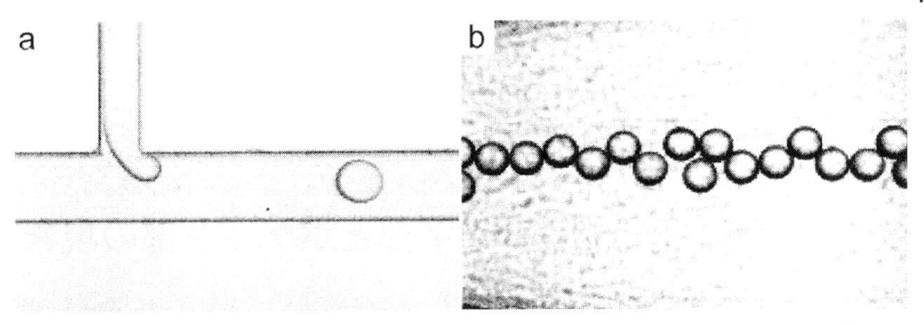

Fig. 5.7 Microscopic images of droplet generation at a T-junction of the microfluidic channels (a) and the generated monodisperse droplets (b) (reproduced from [17] by permission of Elsevier B.V.). The flow rates of the to-be-dispersed phase (1,6-hexanediol diacrylate) and the continuous phase (water (2.0 wt% polyvinyl alcohol)) were 0.1 ml/h and 2.0 ml/h, respectively.

the fluid flow in the channels. This technique has been applied to production of functional polymeric microparticles [20, 39] and multiple emulsions [27]. The multiple emulsions containing a controlled number of internal droplets were produced by two-step droplet breakup at two consecutive T-junctions (see Fig. 5.8).

In addition to the T-shape microfluidic channels, van der Graaf et al. [27] have proposed generation of emulsion droplets using a circular channel (4.8 μm in

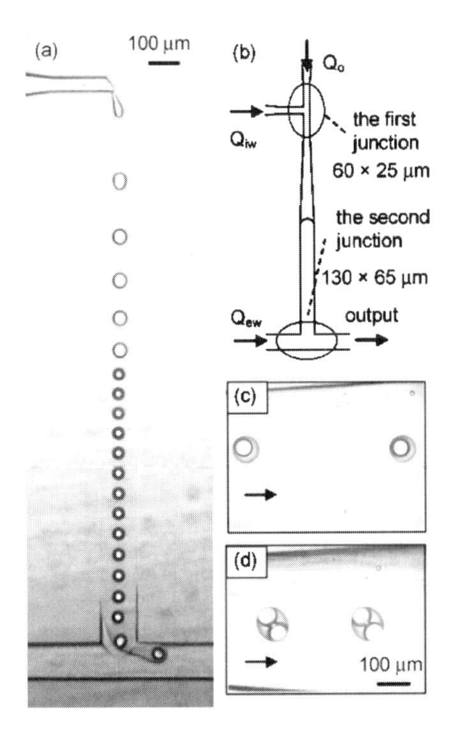

Fig. 5.8 (a) A microscopic image of generation of water-in-oil-water (W/O/W) emulsion droplets via two consecutive T-junctions. (b) Channel configuration (sizes are quoted as width×depth) and flow rates in each channel. Q_{iw}=0.005 ml/h (internal water phase), Q_o=0.02 ml/h (organic phase), Q_{ew}=1.4 ml/h (external water phase). (c, d) Microscopic images of the generated W/O/W droplets flowing inside the channel (reproduced from [27] by permission of the American Chemical Society).

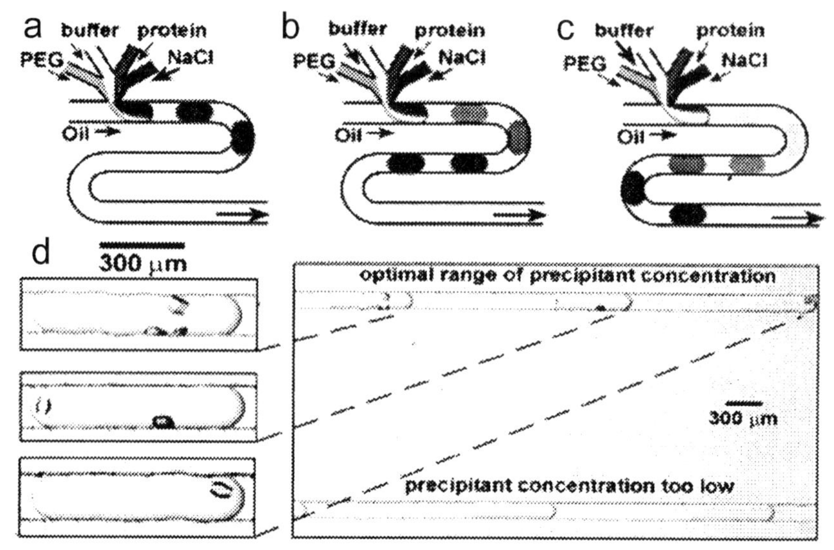

Fig. 5.9 Droplet-based microfluidic system for protein crystallization (reproduced from [30] by permission of the American Chemical Society): (a–c) Schematic illustrations; (d) A polarized microscopic image illustrating crystallization of lysozyme inside droplets of variable composition on a microfluidic chip.

diameter) vertically microfabricated on a silicon chip coated with Si_3N_4. The to-be-dispersed phase that passed through the channel was broken up into droplets by the shear force due to the continuous phase that flows along the chip surface. Emulsion droplets with an average diameter of 10–100 μm were generated by varying the flow conditions using this device.

Monodisperse droplets generated using the first T-junction of microfluidic channels can be passively broken up into precisely controlled daughter droplets at their next T-junction [28]. Monodisperse droplets that flow inside the microfluidic channel are also promising biological and chemical reactors on picoliter and nanoliter scales (Fig. 5.9). Ismagilov's research group has reported reactions inside water droplets in microfluidic systems (e.g., protein crystallization in water droplets) [29, 30].

5.3.2.2 **Droplet Generation Using Flow-focusing Geometry**

Anna et al. [18] proposed a flow-focusing geometry on a microfluidic chip of poly(dimethylsiloxane) (PDMS) for generation of W/O emulsion droplets. Figure 5.10 depicts the typical droplet-generation process using the flow focusing geometry. Three coaxial channels are placed upstream of a small orifice, and a large channel is placed downstream of the orifice. The to-be-dispersed phase flows into the middle channel and the continuous phase flows into the two outside channels. The two phases, which passed through the channels, are then

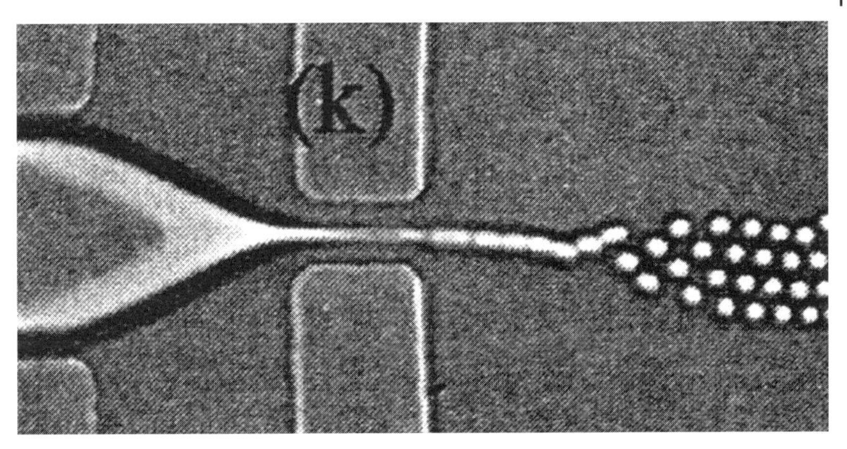

Fig. 5.10 A microscopic image of generation of monodisperse droplets in flow focusing (reproduced from [18] by permission of American Institute of Physics). The flow rates of the to-be-dispersed phase (silicone oil (0.67 wt% Span80)) and the continuous phase (water) were 6.3×10^{-3} ml/h and 0.25 ml/h, respectively.

forced to flow through the orifice. The continuous-phase fluid exerts pressure and viscous stress that force the to-be-dispersed phase fluid into a microthread. The breakup of the microthread then takes place inside or downstream of the orifice. This droplet breakup is a sensitive process and requires skillful control of the fluid flow in the channels, which is analogous to droplet breakup at the T-junction of microfluidic channels. Monodisperse droplets with a wide range of average diameters can be generated using this technique below a critical flow rate of the to-be-dispersed phase [18]. Interestingly, the use of flow focusing, which forms the microthread of the to-be-dispersed phase, allows generation of monodisperse droplets with diameters much smaller than the orifice size.

Xu and Nakajima [19] analyzed the breakup of a microthread focused at a junction of three microfluidic channels (50 μm in depth and 5 μm in width) fabricated on a silicon chip. The width of a microthread at the channel junction decreased with increasing flow rates of the to-be-dispersed phase and the continuous phase. When the microthread achieved a characteristic width of several micrometers, droplet breakup occurred near the channel junction, which is independent of the flow rate of the to-be-dispersed phase. Monodisperse droplets with a diameter of several micrometers to 20 μm were generated at the channel junction. The droplet-generation rate ranged from tens to 200 s^{-1}. Nishisako et al. [20] have also applied a microfluidic device with a flow-focusing geometry to production of bichromal, polymeric microparticles.

Umbanhower et al. [21] have proposed generation of emulsion droplets based on droplet breakup in a coflowing stream, which is similar to droplet generation using the flow-focusing geometry in principle. The to-be-dispersed phase that passed through the outlet of a tapered capillary is broken up into monodisperse droplets (with a minimum coefficient of variation of about 3%) by the coflowing continuous phase in a rotating cup. The resultant droplet diameter, which depends on the capillary diameter and the flow conditions, ranges from 2 to 200 μm.

5.3.3
Microchannel Emulsification

Microchannel (MC) emulsification, proposed by Kawakatsu et al. in 1997 [22], can produce monodisperse emulsions using a microfluidic channel array with a slit-like terrace on a silicon chip (a grooved MC). This channel array was originally developed as a blood-capillary model for measurement of hemorheology [40]. There are two types of MC emulsification devices: a grooved MC and a straight-through MC, which is an array of channels vertically microfabricated on a silicon chip [24]. Figure 5.11 illustrates the emulsification process using a grooved MC and a straight-through MC. This MC emulsification enables production of monodisperse emulsions with average droplet diameters of 3 to 90 μm and coefficients of variation less than 5% [41–43].

A typical experimental setup for MC emulsification is depicted in Fig. 5.12. This emulsification setup consists of an MC plate, a module, apparatuses for supplying the two phases, and a microscope video system [22, 24, 40]. The MC plate is fixed into the module, which has been initially filled with the continuous phase. The pressurized to-be-dispersed phase reaches the channel entrance, and is then pushed out into the continuous phase via the channels to generate emulsion droplets (see Fig. 5.11).

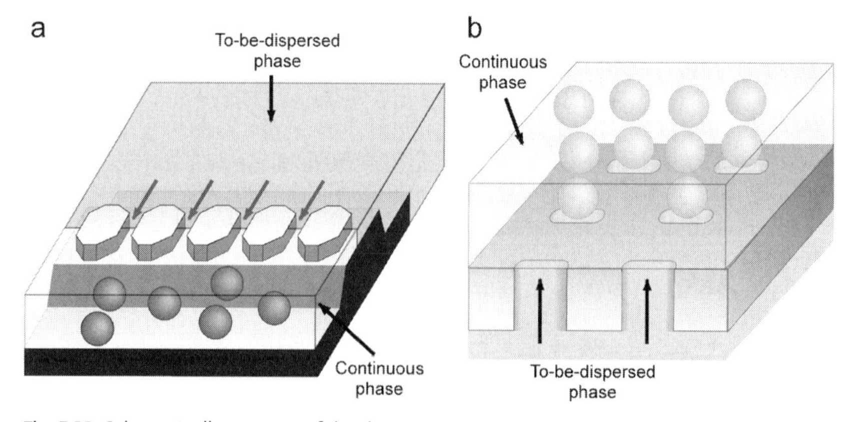

Fig. 5.11 Schematic illustrations of droplet generation in MC emulsification: (a) grooved MC emulsification [22]; (b) straight-through MC emulsification [24].

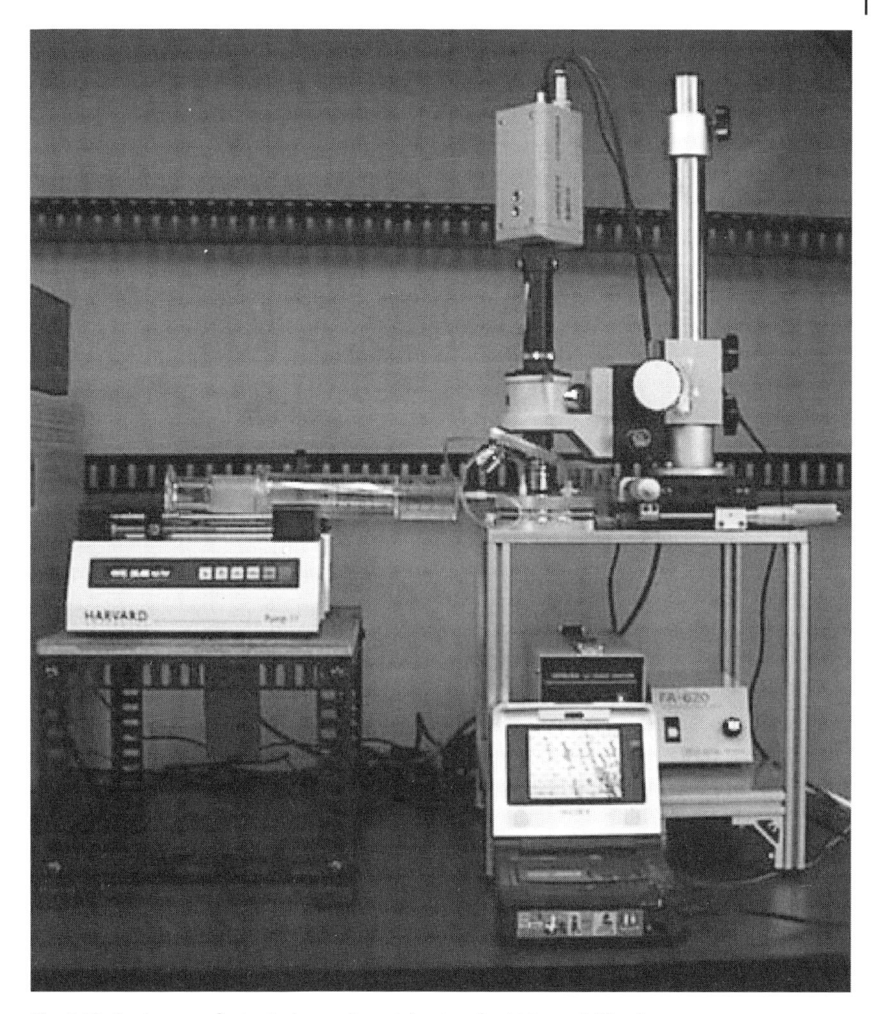

Fig. 5.12 An image of a typical experimental setup for MC emulsification.

5.3.3.1 **Emulsification Characteristics Using Grooved MCs**

A crossflow-type silicon chip with a grooved MC is schematically illustrated in Fig. 5.13. Two terrace lines with a typical height of 100 μm are fabricated on the chip, and a microfluidic channel array of a uniform size is fabricated on the terrace. Figure 5.14 depicts an example of the emulsification process using a grooved MC. The to-be-dispersed phase, which was pressurized exceeding the breakthrough pressure, starts to break through the channels. The breakthrough pressure can be estimated by the Young–Laplace equation considering the contact angle effect (see Eq. (5.1)) [34, 35]. The to-be-dispersed phase that passed through the channels expands on the terrace with a distorted, disk-like shape. The to-be-dispersed phase that passed through the terrace exit then sponta-

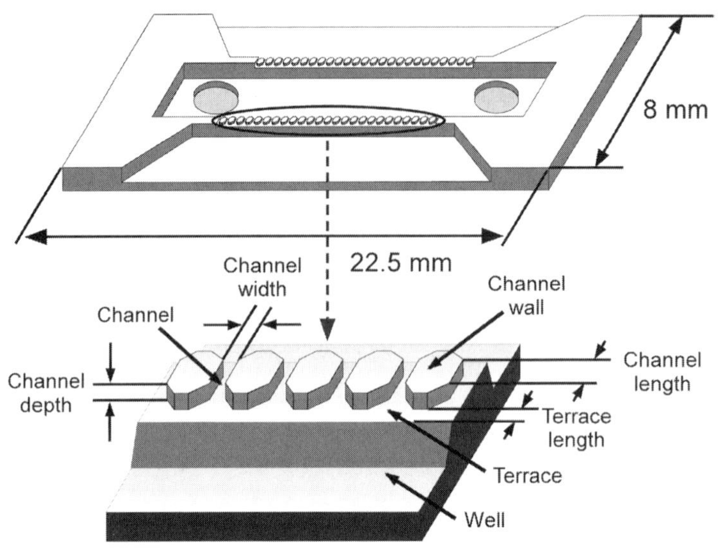

Fig. 5.13 A schematic illustration of a grooved MC plate (top). The grooved MC consists of an array of microfluidic channels and a slit-like terrace (bottom).

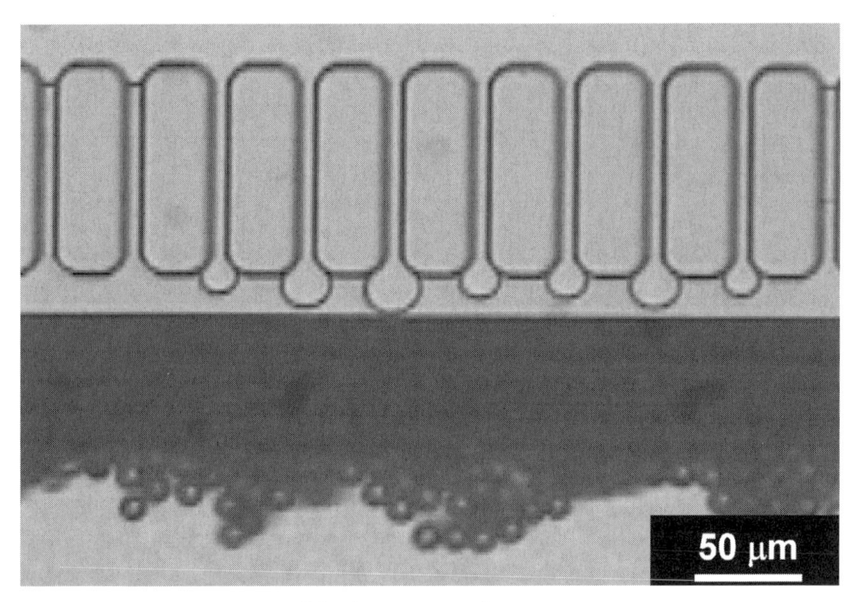

Fig. 5.14 A microscopic image of droplet generation from a grooved MC at the to-be-dispersed phase pressure of 6.3 kPa and the generated monodisperse droplets for soybean oil-in-water (1.0 wt% sodium dodecyl sulfate (SDS)) system.

neously transforms into a spherical droplet by interfacial tension, which is dominant in a microspace [23]. This MC emulsification process makes it possible to generate monodisperse emulsion droplets without applying the external shear force such as the continuous-phase flow, requiring a considerably lower energy input of 10^3 to 10^4 J/m^3 [23]. The reported energy efficiency of MC emulsification had a considerably high value of 65% [23].

The diameter of the droplets generated by MC emulsification is primarily controlled by the channel geometry, depicted in Fig. 5.13. The resultant droplet diameter was greatly influenced by the channel depth and was significantly influenced by the terrace length [44], but was independent of the channel width and channel length [45]. Sugiura et al. [44] proposed an empirical prediction model of the droplet diameter for MC emulsification. For generation of emulsion droplets with a certain diameter, MC structures with narrower and longer channels that cause a large pressure drop in the channel achieved stale droplet generation at a higher droplet-generation rate [45]. The droplet diameter of the resultant monodisperse emulsions was almost constant below the critical flow rate of the to-be-dispersed phase in the channel, which is proportional to the capillary number (Ca [–]) [46]. Ca is defined as the ratio of the viscous force to the interfacial tension force (see Eq. (5.3)). The critical Ca was also independent of the channel size and the interfacial tension. Above the critical Ca, a drastic increase in the droplet diameter was observed, and polydisperse emulsions were produced.

MC emulsification is capable of producing both monodisperse O/W and W/O emulsions [34, 47–49]. Besides the silicon grooved MCs, stainless steel grooved MCs and acrylic grooved MCs were successfully used for production of monodisperse O/W and W/O emulsions, respectively [31, 32]. Monodisperse emulsion droplets generated by MC emulsification have been applied to monodisperse lipid microparticles [35], polymeric microparticles [50, 51], natural polymeric microbeads [52, 53], coacervate microcapsules [54], and W/O/W emulsions [55].

5.3.3.2 Emulsification Characteristics Using Straight-through MCs

A straight-through MC plate, illustrated in Fig. 5.15, has been proposed as a solution to the major problem in a grooved MC plate, which is the low throughput of monodisperse emulsion droplets (typically <0.1 m/h). Uniformly sized channels with a coefficient of variation below 1% are microfabricated in a 10-mm square in the center of the plate. These channels are obtained by deep reactive-ion etching [24].

Figure 5.16 depicts examples of the emulsification process using a circular straight-through MC and an oblong straight-through MC [24]. The cross section of the channels is the most significant device parameter affecting droplet generation in straight-through MC emulsification. The to-be-dispersed phase that passed through the circular channels (10.0 μm in diameter) continuously flowed out in the continuous phase; thus, it was difficult to produce monodisperse emulsions using the circular straight-through MC. In contrast, the to-be-dis-

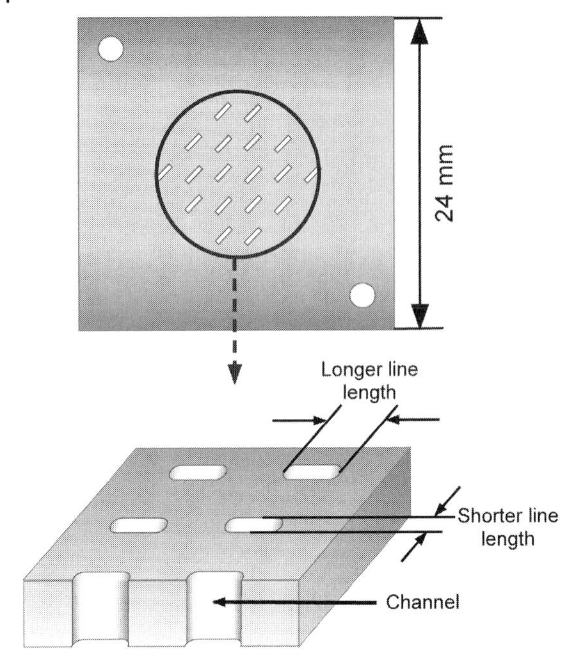

Fig. 5.15 A schematic illustration of a straight-through MC plate (top). The straight-through MC consists of an array of oblong channels vertical to the plate surface (bottom).

persed phase that passed through the oblong channels (9.6 μm in the shorter line and 29.7 μm in the longer line) was stably transformed into monodisperse droplets with an average diameter of 32.5 μm, even without applying the continuous-phase flow. The resultant droplet diameter was independent of the applied flow rates of the continuous phase. The interfacial-tension-based droplet generation, proposed in grooved MC emulsification [23], is applicable to successful droplet generation from the oblong channels. We therefore have reported that the simply elongated cross section of the channels is useful for generation of monodisperse droplets without any external shear force.

Besides the channel cross section, the aspect ratio of the oblong channels considerably affects the droplet-generation process in straight-through MC emulsification. Figure 5.17 depicts droplet generation from oblong channels with different aspect ratios (about 10 μm in the shorter line) [56]. Continuous outflow of the to-be-dispersed phase from the channel exit was observed using the channels with a small aspect ratio of 1.9. On the other hand, monodisperse emulsion droplets were stably generated from the channels with a large aspect ratio of 3.8. The previous studies in straight-through MC emulsification [24, 56] revealed that monodisperse emulsions were successfully produced using the oblong straight-through MCs exceeding a threshold aspect ratio of approximately three.

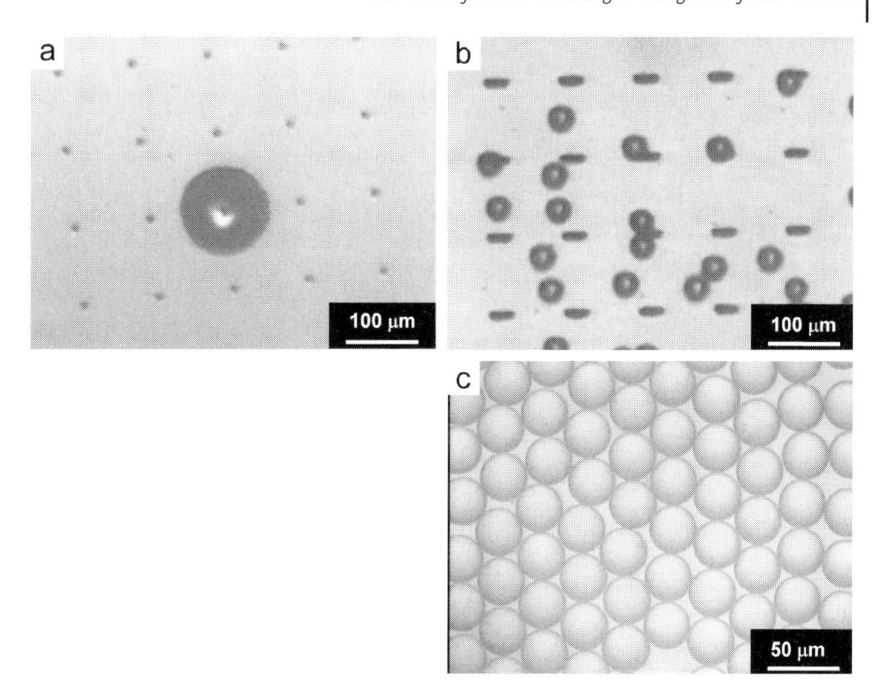

Fig. 5.16 (a) A microscopic image of continuous outflow of the to-be-dispersed phase from a circular straight-through MC at the to-be-dispersed phase pressure of 3.2 kPa. (b, c) Microscopic images of droplet generation from an oblong straight-through MC at the applied to-be-dispersed phase pressure of 1.8 kPa and the generated monodisperse droplets [24].

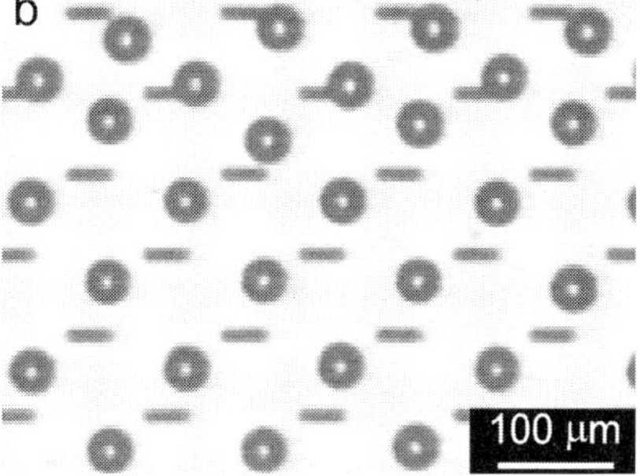

Fig. 5.17 (a) A microscopic image of continuous outflow of the to-be-dispersed phase from an oblong straight-through MC with a channel aspect ratio of 1.9. (b) A microscopic image of generation of monodisperse droplets from an oblong straight-through MC with a channel aspect ratio of 3.8. The oblong straight-through MCs in (a) and (b) have a similar channel size of about 18 μm. The flow rate of the to-be-dispersed phase in (a) and (b) was fixed at 1.0 ml/h [56].

The droplet-generation process using the straight-through MCs was analyzed using the computational fluid dynamics (CFD) method [57]. The to-be-dispersed phase that expands from the channels below a threshold aspect ratio of 3 to 3.5 closed the channel exit, causing continuous outflow of the to-be-dispersed phase into the continuous phase (Fig. 5.18). When the to-be-dispersed phase expanded from the channels exceeding the threshold aspect ratio, sufficient space for water at the channel exit was maintained during droplet generation (Fig. 5.18). This allowed the rapid shrinkage and cutoff of the neck formed inside the chan-

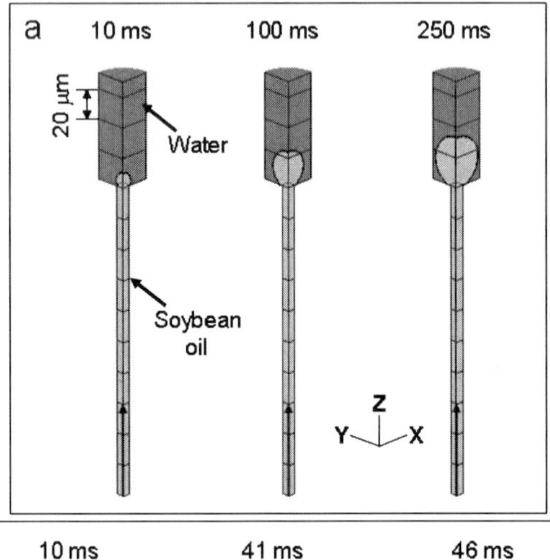

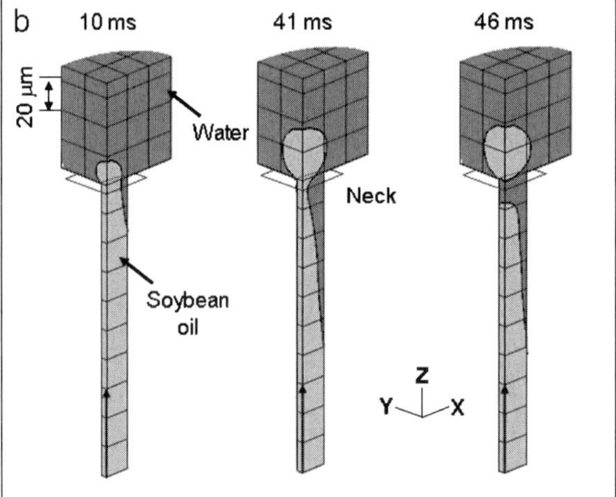

Fig. 5.18 Emulsification behavior using straight-through MCs, calculated by computational fluid dynamics (CFD) [57]. (a) Continuous outflow of the to-be-dispersed phase from a circular straight-through MC (10 μm in diameter). (b) Successful generation of a droplet with a diameter of about 40 μm from an elliptic straight-through MC (40 μm in the major axis diameter and 10 μm in the minor axis diameter).

nel, yielding droplets as small as 40 μm in diameter without applying the continuous-phase flow.

An originally designed straight-through MC plate (Fig. 5.15) had a maximum throughput capacity of monodisperse emulsion droplets of 6 ml/h [58]. The average droplet diameter and coefficient of variation of the produced emulsions were almost independent of the flow rate of the to-be-dispersed phase below the preceding throughput capacity. A large straight-through MC plate with a surface size of 40 mm×40 mm achieved a maximum throughput capacity of monodisperse emulsion droplets of 35 ml/h [59].

5.4 Summary and Outlook

This chapter has outlined recent technologies for direct production of emulsions using microfluidic devices. The microfluidic devices with well-defined channels enable successful production of size-controlled monodisperse emulsions with an average droplet diameter of several micrometers to several hundreds of micrometers by exploiting the flow characteristics and force balance of the two liquid phases inside the microfluidic channels. An appropriate combination of the channel material, the two liquid phases, and the surfactant is also required to achieve successful production of monodisperse emulsions. The resultant monodisperse emulsion droplets have been applied to preparation of monodisperse microparticles and microcapsules and to biological and chemical microreactor vessels. However, the present emulsification techniques using the microfluidic devices do not have throughput capacities sufficient for industrial uses; thus, scale-up of these emulsification devices is an important work in the near future. Further advances in microfluidics are expected to yield a high-throughput production of advanced materials with novel shapes, compositions, functions on the basis of monodisperse emulsion droplets and to establishment sophisticated microfluidic systems, which integrate droplet generation, droplet handling, droplet reaction, and droplet analysis.

Acknowledgements

This work has been supported in part by the Nanotechnology Project of the Ministry of Agriculture, Forestry and Fisheries of Japan.

References

1 E. Dickinson, *Introduction to Food Colloids*, Oxford University Press, UK, **1992**.

2 D.J. McClements, *Food Emulsions: Principles, Practice and Techniques*, CRC Press, Boca Raton, USA, **1999**, pp. 1–16.

3 T.G. Mason, A.H. Krall, H. Gang, J. Bibette, D.A. Weitz in *Encyclopedia of Emulsion Technology, vol. 4* (Ed.: P. Becher), Marcel Dekker, New York, USA, **1996**, Chapter 6.

4 H. Karbstein, H. Schubert, *Chem. Eng. Process.* **1995**, *34*, 205–211.

5 P. Walstra in *Encyclopedia of Emulsion Technology, vol. 1* (Ed.: P. Becher), Marcel Dekker, New York, USA, **1983**, Chapter 2.

6 J. Bibette, *J. Colloid Interface Sci.* **1991**, *147*, 474–478.

7 T.G. Mason, J. Bibette, *Langmuir* **1997**, *13*, 4600–4613.

8 T. Nakashima, M. Shimizu, M. Kukizaki, *Key Eng. Mater.* **1991**, *61/62*, 513–516.

9 S.M. Joscelyne, G. Trägårdh, *J. Membr. Sci.* **2000**, *169*, 107–117.

10 A.J. Abrahamse, A. van der Padt, R.M. Boom, W.B.C. de Heij, *AIChE J.* **2001**, *47*, 1285–1291.

11 R.A. Williams, S.J. Peng, D.A. Wheeler, N.C. Morley, D. Taylor, M. Whalley, D. W. Houldsworth, *Trans IChemE*, **1998**, *76(A)*, 902–910.

12 G.T. Vladisavljevic, H. Schubert, *Desalination* **2002**, *144*, 167–172.

13 V. Schröder, O. Behrend, H. Schubert, *J. Colloid Interface Sci.* **1998**, *202*, 334–340.

14 V. Haverkamp, W. Ehrfeld, K. Gebauer, V. Hessel, H. Löwe, T. Richter, C. Wille, *Fresenius J. Anal. Chem.* **1999**, *364*, 617–624.

15 T. Herweck, S. Hardt, V. Hessel, H. Löwe, C. Hoffmann, F. Weise, T. Dietrich, A. Freitag, *Microreaction Technology (Proceedings of IMRET5)*, Springer, Berlin, Germany, **2001**, pp. 216–229.

16 T. Thorsen, R.W. Roberts, F.H. Arnold, S.R. Quake, *Phys. Rev. Lett.* **2001**, *86*, 4163–4166.

17 T. Nishisako, T. Torii, T. Higuchi, *Lab Chip*, **2002**, *2*, 24–26.

18 S.L. Anna, N. Bontoux, H.A. Stone, *Appl. Phys. Lett.* **2003**, *82*, 364–366.

19 Q. Xu, M. Nakajima, *Appl. Phys. Lett.* **2004**, *85*, 3726–3728.

20 T. Nishisako, T. Torii, T. Higuchi, *Chem. Eng. J.*, **2004**, *101*, 23–29.

21 P.B. Umbanhowar, V. Prasad, D.A. Weitz, *Langmuir* **2000**, *16*, 347–351.

22 T. Kawakatsu, Y. Kikuchi, M. Nakajima, *J. Am. Oil Chem. Soc.* **1997**, *74*, 317–321.

23 S. Sugiura, M. Nakajima, S. Iwamoto, M. Seki, *Langmuir* **2001**, *17*, 5562–5566.

24 I. Kobayashi, M. Nakajima, K. Chun, Y. Kikuchi, H. Fujita, *AIChE J.* **2002**, *48*, 1639–1644.

25 N.-T. Nguyen and S.T. Wereley, *Fundamentals and Applications of Microfluidics*, Artech House, Boston, USA, **2002**, pp. 67–129.

26 S. Okushima, T. Nishisako, T. Torii, T. Higichi, *Langmuir*, **2004**, *20*, 9905–9908.

27 S. van der Graaf, C.G.P.H. Schroën, R.G.M. van der Sman, R.M. Boom, *J. Colloid Interface Sci.*, **2004**, *277*, 456–463.

28 D.R. Link, S.L. Anna, D.A. Weitz, H.A. Stone, *Phys. Rev. Lett.* **2004**, *92*, 0405032–0405035.

29 H. Song, J.D. Tice, R.F. Ismagilov, *Angew. Chem. Int. Ed.*, **2003**, *42*, 768–772.

30 B. Zheng, L.S. Roach, R.F. Ismagilov, *J. Am. Chem. Soc.*, **2003**, *125*, 11170–11171.

31 H. Liu, M. Nakajima, T. Kimura, *J. Am. Oil Chem. Soc.* **2004**, *81*, 705–711.

32 J. Tong, M. Nakajima, H. Nabetani, Y. Kikuchi, Y. Maruta, *J. Colloid Interface Sci.* **2001**, *237*, 239–248.

33 Y. Gu, D. Li, *J. Colloid Interface Sci.*, **2000**, *226*, 328–339.

34 J. Tong, M. Nakajima, H. Nabetani, Y. Kikuchi, *J. Surfactant Deterg.* **2000**, *3*, 285–293.

35 S. Sugiura, M. Nakajima, J. Tong, H. Nabetani, M. Seki, *J. Colloid Interface Sci.* **2000**, *227*, 95–103.

36 W. Ehrfeld, K. Golbig, V. Hessel, H. Löwe, T. Richter, *Ind. Eng. Chem. Res.* **1999**, *38*, 1075–1082.

37 W. Ehrfeld, V. Hessel, H. Löwe, *Microreactors*, VCH, Weinheim, **2000**, p. 70.

38 S. Hardt, F. Schönfeld, F. Weise, C. Hoffmann, V. Hessel, W. Ehrfeld, *Proceedings of the 4th International Conference on Modeling and Simulations of Microsys-*

tems 2001, Hilton Head Island, SC, **2001**, pp. 223–226.

39 D. Dendukuri, K. Tsoi, T. A. Hatton, P. S. Doyle, *Langmuir*, **2005**, *21*, 2113–2116.

40 Y. Kikuchi, K. Sato, T. Kaneko, *Microvasc. Res.* **1992**, *44*, 226–240.

41 I. Kobayashi, M. Nakajima, J. Tong, T. Kawakatsu, H. Nabetani, Y. Kikuchi, A. Shohno, K. Satoh, *Food Sci. Technol. Res.* **1999**, *5*, 350–355.

42 I. Kobayashi, M. Nakajima, H. Nabetani, Y. Kikuchi, A. Shohno, K. Satoh, *J. Am. Oil Chem. Soc.* **2001**, *78*, 797.

43 S. Sugiura, M. Nakajima, M. Seki, *J. Am. Oil Chem. Soc.* **2002**, *79*, 515–519.

44 S. Sugiura, M. Nakajima, M. Seki, *Langmuir*, **2002**, *18*, 3854–3859.

45 S. Sugiura, M. Nakajima, M. Seki, *Langmuir*, **2002**, *18*, 5708–5712.

46 S. Sugiura, M. Nakajima, N. Kumazawa, S. Iwamoto, M. Seki, *Phys. Chem. B*, **2002**, *106*, 9405–9409.

47 J. Tong, M. Nakajima, H. Nabetani, *Eur. J. Lipid Sci. Technol.* **2002**, *104*, 216–221.

48 T. Kawakatsu, G. Trägårdh, Ch. Trägårdh, M. Nakajima, N. Oda, N. T. Yonemoto, *Colloids Surfaces A*, **2001**, *179*, 29–37.

49 S. Sugiura, M. Nakajima, H. Ushijima, K. Yamamoto, M. Seki, *J. Chem. Eng. Japan*, **2001**, *34*, 757–765.

50 S. Sugiura, M. Nakajima, H. Itoh, M. Seki, *Macromol. Rapid Commun.*, **2001**, *22*, 773–778.

51 S. Sugiura, M. Nakajima, M. Seki, *Ind. Eng. Chem. Res.*, **2002**, *41*, 4043–4047.

52 T. Kawakatsu, N. Oda, T. Yonemoto, M. Nakajima, *Kagaku Kogaku Ronbunshu* **2000**, *26*, 122–125.

53 S. Iwamoto, K. Nakagawa, S. Sugiura, M. Nakajima, *AAPS PharmSciTech* **2002**, *3(3)*, article 25.

54 K. Nakagawa, S. Iwamoto, M. Nakajima, A. Shono, K. Satoh, *J. Colloid Interface Sci.* **2004**, *278*, 198–205.

55 S. Sugiura, M. Nakajima, K. Yamamoto, S. Iwamoto, T. Oda, M. Satake, M. Seki, *J. Colloid Interface Sci.* **2004**, *270*, 221–228.

56 I. Kobayashi, S. Mukataka, M. Nakajima, *J. Colloid Interface Sci.* **2004**, *279*, 277–280.

57 I. Kobayashi, S. Mukataka, M. Nakajima, *Langmuir* **2004**, *20*, 9868–9877.

58 I. Kobayashi, M. Nakajima, S. Mukataka, *Colloids Surfaces A*, **2003**, *229*, 33–41.

59 I. Kobayashi, M. Nakajima, S. Mukataka, *Proceedings of International Congress on Engineering and Food 9*, Montpellier, France, **2004**, No. 610.

6
Chemical Reactions in Continuous-flow Microstructured Reactors

Albert Renken and Lioubov Kiwi-Minsker, Swiss Federal Institute of Technology (EPFL), Institute of Chemical Sciences and Engineering, Lausanne, Switzerland

Abstract
The manufacture of chemicals in microreactors and miniaturized equipments has recently become a new branch of chemical reaction engineering. This development is focusing on process intensification and safety. Microreactors, as compared to the conventional equipment, also allow improved environmental protection. Nevertheless, the microreactor technology as a chemical engineering discipline is still in its infancy and the basic approach to the design of microreactors has to be rationalized via specific criteria. Simply miniaturizing a reactor will not guarantee added benefits. In this chapter the background criteria for the beneficial use of reactor miniaturization are described based on the following considerations: i) backmixing and its effect on the residence-time distribution; ii) heat transfer for safe reactor operation; iii) improved mass transfer in microreactors for increased selectivity with lower energy consumption as compared to conventional reactors.

Some examples of the use of microreactors leading to process intensification are presented for selected chemical transformations.

Keywords
Multichannel microreactors, heat and mass transfer, reactor safety, process intensification by miniaturization, pressure drop

Advanced Micro and Nanosystems Vol. 5. Micro Process Engineering. Edited by N. Kockmann

ISBN: 3-527-31246-3

6.1 Introduction

The design of any chemical reactor is based on material and energy balance. A material balance has to be set up for all species participating in the reactions within the reactor. The material balance for a component A_i can be formulated in the following manner:

$$\left\{\begin{array}{c}\text{accumulation}\\ \text{of } A_i \text{ in}\\ \text{the system}\end{array}\right\} = \left\{\begin{array}{c}\text{rate of flow}\\ \text{of } A_i \text{ into}\\ \text{the system}\end{array}\right\} - \left\{\begin{array}{c}\text{rate of flow}\\ \text{of } A_i \text{ out of}\\ \text{the system}\end{array}\right\} + \left\{\begin{array}{c}\text{rate of production}\\ \text{of } A_i \text{ within}\\ \text{the system}\end{array}\right\}$$

$$\frac{\mathrm{d}n_i}{\mathrm{d}t} = \dot{n}_{i,0} - \dot{n}_i + L_{\mathrm{p},i} \qquad (6.1)$$

where n_i represents the number of moles of species A_i in the system at time t, $\dot{n}_i$ is the molar flow rate, and $L_{\mathrm{p},i}$ is the rate of A_i production. If temperature and concentrations of the chemical species are constant, the rate of A_i production corresponds to the product of the reaction volume, V, and the rate of the A_i transformation, R_i,

$$L_{\mathrm{p},i} = R_i \cdot V \qquad (6.2)$$

The rate of A_i transformation (R_i) is the sum of the rates (r_j) of the reactions in which A_i participates:

$$R_i = \sum_j \nu_{i,j} r_j \qquad (6.3)$$

Let us assume the following two reactions occurring in the reactor:

$$\begin{aligned} j = 1:&\ \nu_{1,1}A_1 + \nu_{2,1}A_2 = \nu_{3,1}A_3 + \nu_{4,1}A_4 \\ j = 2:&\ \nu_{3,2}A_3 + \nu_{4,2}A_4 = \nu_{1,2}A_1 + \nu_{2,2}A_2 \end{aligned} \qquad (6.4)$$

The transformation rates for the species A_1, A_3: are given by:

$$R_1 = \nu_{1,1}r_1 + \nu_{1,2}r_2;\ R_3 = \nu_{3,1}r_1 + \nu_{3,2}r_2 \qquad (6.5)$$

The reaction rates depend on the concentrations of the reacting species and can be described by a power law [1]:

$$r_j = k_j \cdot \prod_i c_i^{m_i} \tag{6.6}$$

where c_i is the concentration of reactant A_i and m_i is called the reaction order with respect to A_i. The stoichiometric coefficients, ν_i, in Eq. (6.5) are negative for the reacting species (on the left-hand side of the equations (6.4)) and positive for the species formed (on the right-hand side of the equations (6.4)). If $|\nu_i| = 1$, and first order is assumed for the reactants ($m_i = 1$), Eq. (6.5) can be rewritten:

$$R_1 = -1 \cdot k_1 c_1 c_2 + 1 \cdot k_2 c_3 c_4; \; R_3 = +1 \cdot k_1 c_1 c_2 - 1 \cdot k_2 c_3 c_4 \tag{6.7}$$

A chemical reaction proceeds in the direction in which the free Gibbs enthalpy, G, of the reaction mixture diminishes. When equilibrium is reached ($R_1 = R_3 = 0$) the Gibbs enthalpy has a minimum value. It follows from Eq. (6.7):

$$k_1 c_{1,\mathrm{eq}} c_{2,\mathrm{eq}} = k_2 c_{3,\mathrm{eq}} c_{4,\mathrm{eq}}; \; K_c = \frac{k_1}{k_2} = \frac{c_{3,\mathrm{eq}} c_{4,\mathrm{eq}}}{c_{1,\mathrm{eq}} c_{2,\mathrm{eq}}} \tag{6.8}$$

If the activities of the reactants correspond to their concentrations, the equilibrium constant K can be determined from the second law of thermodynamics:

$$K_c(T) \cong K(T) = \exp\left(\frac{-\Delta G^0}{RT}\right) \tag{6.9}$$

Taking the derivative of Eq. (6.9), where $\Delta G^0 = \Delta H^0 - T\Delta S^0$, the van't Hoff equation can be obtained:

$$\frac{\mathrm{d}\ln K}{\mathrm{d}T} = \frac{\mathrm{d}}{\mathrm{d}T}\left(\frac{-\Delta G^0}{RT}\right) = \frac{\Delta H^0}{RT^2} \tag{6.10}$$

By integrating from the standard temperature (298 K) to the desired temperature T we obtain for constant reaction enthalpy [2]:

$$\ln K = \ln K(298) + \int_{298}^{T} \frac{\Delta H^0}{RT^2} \mathrm{d}T \tag{6.11}$$

The superscript "0" indicates standard conditions.

The reaction rate constant, k_j, in Eq. (6.6) is independent of the composition of the reaction mixture, but is strongly influenced by the temperature as described by the Arrhenius law:

$$k_j = k_{0,j} \cdot \exp\left(\frac{-E_{a,j}}{RT}\right) \tag{6.12}$$

where $k_{0,j}$ is the pre-exponential or frequency factor and $E_{a,j}$ is the apparent activation energy of the reaction. For most reactions, the activation energy lies in the range of 40–300 kJ/mol, resulting in an increase of k_j by a factor of 2 to 50 for a temperature rise of 10 K.

Application of the principle of conservation of energy leads to the energy balance that can be described as follows:

$$\left\{\begin{array}{c}\text{rate of energy}\\\text{accumulation}\\\text{within}\\\text{the system}\end{array}\right\} = \left\{\begin{array}{c}\text{rate of flow}\\\text{of heat}\\\text{to the system}\\\text{from the}\\\text{surroundings}\end{array}\right\} - \left\{\begin{array}{c}\text{rate or work}\\\text{done by}\\\text{the system}\\\text{on the}\\\text{surroundings}\end{array}\right\} + \left\{\begin{array}{c}\text{rate of energy}\\\text{added to}\\\text{the system}\\\text{by mass flow}\\\text{into the system}\end{array}\right\} - \left\{\begin{array}{c}\text{rate of energy}\\\text{leaving the}\\\text{system by}\\\text{mass flow out of}\\\text{the system}\end{array}\right\}$$

$$\frac{dE_{sys}}{dt} = \dot{Q} - \dot{W} + \dot{n}_{in} E_{in} - \dot{n}_{out} E_{out} \tag{6.13}$$

The work term $\dot{W}$ is normally separated into flow work and shaft work, $\dot{W}_s$. Shaft work is, e.g., from the stirrer in a stirred tank or a turbine in a tubular reactor. When shear stress can be neglected, the work term is:

$$\dot{W} = -\sum_i \dot{n}_i p \hat{V}_i \Big|_{in} + \sum_i \dot{n}_i p \hat{V}_i \Big|_{out} + \dot{W}_s \tag{6.14}$$

with $\hat{V}_i$, the molar volume of the reactant A_i.

Introducing in Eq. (6.13) results in

$$\frac{dE_{sys}}{dt} = \dot{Q} - \dot{W}_s - \sum_i \dot{n}_i (E_i + p\hat{V})_i \Big|_{in} + \sum_i \dot{n}_i (E_i + p\hat{V})_i \Big|_{out} \tag{6.15}$$

The energy E_i is the sum of internal energy (U_i), the kinetic energy, the potential energy and all other energies, like electric, magnetic energy or light. For the description of the majority of the chemical reactors the kinetic, potential and other energies can be neglected, resulting in

$$E_i \cong \hat{U}_i \tag{6.16}$$

Introducing the enthalpy

$$\hat{H}_i = \hat{U}_i + p\hat{V}_i \tag{6.17}$$

we obtain finally:

$$\frac{dE_{sys}}{dt} = \dot{Q} - \dot{W}_s + \sum_i \dot{n}_{i,0}\hat{H}_{i,0} - \sum_i \dot{n}_i\hat{H}_i \tag{6.18}$$

The subscript "0" indicates inlet conditions, and $\hat{H}_i$, $\hat{U}_i$, are the molar enthalpy and molar energy, respectively.

6.2 Characteristics of Continuous-flow Reactors

One of the most important parameters to characterize continuous-flow reactors is the degree of backmixing [1]. In the ideal mixed reactor the concentrations and the temperature within the reactor volume are uniform. Complete mixing can be achieved in a continuously operated stirred tank (CSTR) or a loop reactor with a high recycling rate. After a transient period of approximately five mean space-times, the CSTR attains the steady state ($dn_i/dt = 0$) and the material balance, Eq. (6.1), simplifies to

$$0 = \dot{n}_{i,0} - \dot{n}_i + L_{P,i} = \dot{n}_{i,0} - \dot{n}_i + R_i \cdot V \tag{6.19}$$

As concentrations and temperature do not vary within the reactor, the transformation rate, R_i, is constant and the volume of the reaction mixture is given by

$$V = \frac{\dot{n}_{i,0} - \dot{n}_i}{-R_i} \tag{6.20}$$

In contrast to the ideal CSTR, backmixing is excluded in an ideal tubular reactor, characterized by a plug flow of the fluid, and uniform radial composition and temperature. In the plug-flow reactor (PR), the reactants are continuously consumed as they flow down the length to the reactor outlet. The material balance is established for a small volume element (ΔV) at steady state as illustrated in Fig. 6.1.

Taking the limits as ΔV approaches zero, we obtain:

$$-\frac{d\dot{n}_i}{dV} + R_i = 0 \tag{6.21}$$

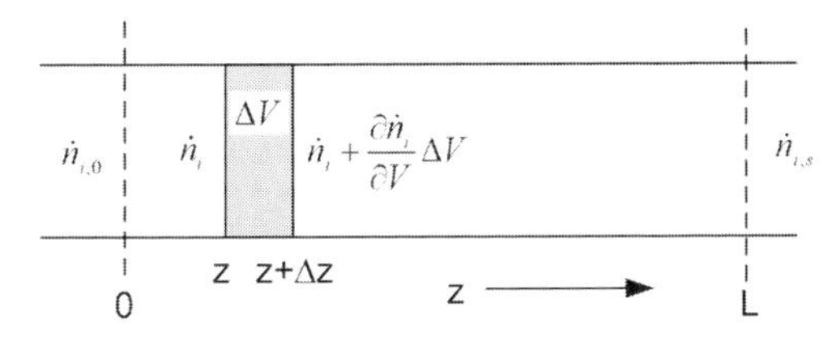

Fig. 6.1 Ideal tubular plug-flow reactor (PR).

The reactor volume is obtained by integration:

$$V = V_{\mathrm{R}} = \int_{\dot{n}_{i0}}^{\dot{n}_{is}} \frac{\mathrm{d}\dot{n}_i}{R_i} \tag{6.22}$$

Introducing the conversion, X, for the key reactant A_1 as

$$X = \frac{\dot{n}_{1,0} - \dot{n}_{1,\mathrm{s}}}{\dot{n}_{1,0}} \tag{6.23}$$

Eq. (6.22) becomes

$$V_{\mathrm{R}} = \dot{n}_{1,0} \int_{0}^{X_{\mathrm{s}}} \frac{\mathrm{d}X}{-R_1} = \dot{V}_0 c_{1,0} \int_{0}^{X_{\mathrm{s}}} \frac{\mathrm{d}X}{-R_1} \tag{6.24}$$

In an ideal plug-flow reactor, all volume elements, ΔV, leaving the reactor have been inside it for exactly the same time. In practice, the volume elements entering the reactor together will spend different times inside the reactor. So, there is a distribution of residence times of the material around the mean value. The distribution of residence times can significantly affect the reactor performance, the product selectivity, and the product yield. The residence-time distribution (RTD) gives important hints for the type of mixing occurring within the reactor. This is one of the key parameters of the reactor. RTD depends on axial mixing and different hydrodynamic effects.

The axial mixing in a tubular reactor can be described by the dispersion model. This model suggests that the RTD may be considered as the result of piston flow with a superposition of longitudinal dispersion. The latter is taken into account by means of a constant effective axial dispersion coefficient, D_{ax}, which has the same dimension as the molecular diffusion coefficient, D_{m}. Usually D_{ax} is much larger than the molecular diffusion coefficient because it incorporates all effects that cause deviation from plug-flow like the differences in radial velocity, eddies and vortices. The mass flow caused by dispersion is described by Fick's law, which leads in the one-dimensional case to

$$J = -D_{\mathrm{ax}} \frac{\partial c}{\partial z} \tag{6.25}$$

The RTD as described by the dispersion model can be derived from the mass balance of a nonreacting species (tracer) over a volume element $\Delta V = S\Delta z$, with S the section of the tube. For constant fluid density and the superficial velocity w we obtain:

$$S\Delta z \frac{\partial c}{\partial t} = w(c_z - c_{z+\Delta z})S + \left(-D_{ax} \frac{\partial c}{\partial z}\bigg|_z + -D_{ax} \frac{\partial c}{\partial z}\bigg|_{z+\Delta z} \right) S$$

and with $\Delta z \rightarrow 0$

$$\frac{\partial c}{\partial t} = -w \frac{\partial c}{\partial z} + D_{ax} \frac{\partial^2 c}{\partial z^2} \tag{6.26}$$

In dimensionless form Eq. (6.26) becomes

$$\frac{\partial C}{\partial \theta} = -\frac{\partial C}{\partial Z} + \frac{1}{\mathrm{Bo}} \frac{\partial^2 C}{\partial Z^2} \tag{6.27}$$

with

$$\theta = \frac{t}{\tau};\ \tau = \frac{L}{w};\ Z = \frac{z}{L};\ C = \frac{c}{\bar{c}_0};\ \bar{c}_0 = \frac{n_{inj}}{V_R};\ \mathrm{Bo} = \frac{w \times L}{D_{ax}}$$

The total amount of a nonreactive tracer injected as a Dirac pulse at the reactor entrance is given by n_{inj}. The Bodenstein number, Bo, is considered as the ratio between the axial dispersion time, $t_{ax} = L^2/D_{ax}$, and the mean residence time, $\bar{t} = \tau = L/w$, which is identical to the space time for reaction mixtures with constant density. For $\mathrm{Bo} = t_{ax}/\tau \rightarrow 0$ the axial dispersion time is short compared to the mean residence time resulting in complete backmixing in the reactor. For $\mathrm{Bo} \rightarrow \infty$ no dispersion occurs. In practice, axial dispersion can be neglected for $\mathrm{Bo} \geq 100$. If a system is open for dispersion at the reactor inlet and outlet [1], the response to a tracer pulse can be predicted with Eq. (6.28) and is shown in Fig. 6.2 (top).

$$C(\theta) = \frac{1}{2} \sqrt{\frac{\mathrm{Bo}}{\pi \cdot \theta}} \exp\left(\frac{-(1-\theta)^2 \mathrm{Bo}}{4\theta} \right) \tag{6.28}$$

The cumulative dispersion, $F(\theta)$, can be obtained by integration of Eq. (6.28) (Fig. 6.2 (bottom)).

For reactions with positive order a broad RTD diminishes the performance of chemical reactors. The loss in performance can become considerable at high conversion for reactions with high reaction order. Even more sensitive to backmixing is the selectivity and yield of the target product, if it is an intermediate in a complex reaction network.

The dispersion in tubular reactors depends on the flow regime, characterized by the Reynolds number, Re, and the physical properties of the fluid, characterized by the Schmidt number, Sc.

Often the axial dispersion in the reactor is correlated on the basis of the axial Péclet number that has as a characteristic length the tube diameter, or the particle diameter in reactors with a packed bed. Correlations and definitions are summarized in Table 6.1. The Bodenstein number characterizing the dispersion in the chemical reactor becomes

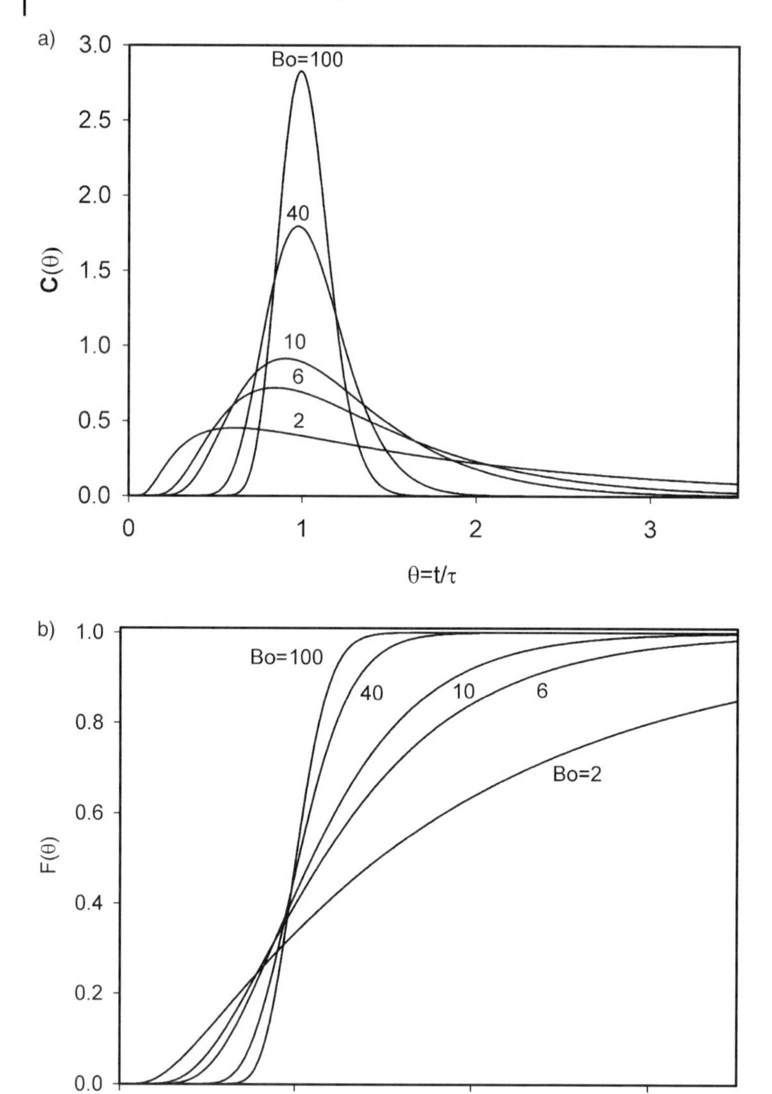

Fig. 6.2 Residence time distribution (top) and cumulative distribution (bottom) predicted by the dispersion model. Parameter: Bodenstein number (Bo).

$$\mathrm{Bo} = \mathrm{Pe}_{\mathrm{ax}} \frac{L}{d} \tag{6.29}$$

Flow in microchannels with diameters between 10 and 500 μm is mostly laminar and has a parabolic velocity profile. Therefore, the molecular diffusion in axial and radial directions plays an important role in RTD. The diffusion in the radial direction tends to diminish the spreading effect of the parabolic velocity profile, while in the axial direction the molecular diffusion increases the dispersion [3, 4].

Table 6.1 Estimation of axial dispersion.

Definitions: $\mathrm{Pe}_{\mathrm{ax}} = \frac{w \cdot d_{\mathrm{p}}}{\varepsilon \cdot D_{\mathrm{ax}}}$; $\mathrm{Re}_{\mathrm{p}} = \frac{w \cdot d_{\mathrm{p}}}{\nu}$; $d_{\mathrm{p}} = 6\frac{V_{\mathrm{p}}}{A_{\mathrm{p}}}$; $\mathrm{Re} = \frac{w \cdot d_{\mathrm{t}}}{\nu}$; $\mathrm{Sc} = \frac{\nu}{D_{\mathrm{m}}}$;

Empty tube, laminar flow:

$$\left.\begin{aligned} D_{\mathrm{ax}} &= D_{\mathrm{m}} + \chi \frac{w^2 d_{\mathrm{t}}^2}{D_{\mathrm{m}}} \\ \frac{1}{\mathrm{Pe}_{\mathrm{ax}}} &= \frac{1}{\mathrm{Re} \cdot \mathrm{Sc}} + \frac{\mathrm{Re} \cdot \mathrm{Sc}}{\chi} \end{aligned}\right. ; \quad \frac{L}{d_{\mathrm{t}}} > 0.04 \frac{w \cdot d_{\mathrm{t}}}{D_{\mathrm{m}}}; \ \chi = \frac{1}{192} \quad \text{for circular tubes} \tag{6.30}$$

Empty tube, turbulent flow:

$$\frac{1}{\mathrm{Pe}_{\mathrm{ax}}} = \frac{3 \times 10^7}{\mathrm{Re}^{2.1}} + \frac{1.35}{\mathrm{Re}^{1/8}}; \ \mathrm{Pe}_{\mathrm{ax}} = \frac{w \cdot d_{\mathrm{t}}}{D_{\mathrm{ax}}} \tag{6.31}$$

Packed bed, gas flow:

$$\frac{1}{\mathrm{Pe}_{\mathrm{ax}}} = \frac{0.3}{\mathrm{Re}_{\mathrm{p}} \cdot \mathrm{Sc}} + \frac{0.5}{1 + \frac{3.8}{\mathrm{Re}_{\mathrm{p}} \cdot \mathrm{Sc}}}; \ \frac{d_{\mathrm{p}}}{d_{\mathrm{t}}} > 15; \ \begin{aligned} 0.008 &< \mathrm{Re}_{\mathrm{p}} < 400 \\ 0.28 &< \mathrm{Sc} < 2.2 \end{aligned} \tag{6.32}$$

Packed bed, liquid flow:

$$\varepsilon \cdot \mathrm{Pe}_{\mathrm{ax}} = 0.2 + 0.011 \cdot \mathrm{Re}_{\mathrm{p}}^{0.48}; \ \frac{d_{\mathrm{p}}}{d_{\mathrm{t}}} > 15; \ 10^{-3} < \mathrm{Re}_{\mathrm{p}} < 10^3 \tag{6.33}$$

To compare microstructured multichannel reactors with conventional fixed-bed reactors, the correct criteria must be defined [5, 6]. Let us assume a fixed-bed reactor of cross section, S_{bed}, and height, L_{bed}, filled with n_{p} spherical particles of diameter, d_{p}. This packed bed is compared to a multichannel reactor with channel diameter, d_{t}, of the same cross-section and height ($S_{\mathrm{bed}} = S_{\mathrm{struc}}$, $L_{\mathrm{bed}} = L_{\mathrm{struc}}$). The geometries of the reactors are presented in Fig. 6.3.

Both reactors are designed to provide identical porosities, ε, resulting in identical space times, τ, and the same (outer, catalytic) surface area per reactor volume, a:

$$\varepsilon_{\mathrm{bed}} = \varepsilon_{\mathrm{struc}} = \varepsilon; \ \tau_{\mathrm{bed}} = \tau_{\mathrm{struc}} = \tau; \ a_{\mathrm{bed}} = a_{\mathrm{struc}} = a \tag{6.34}$$

Under these conditions the following relationship between particle diameter, d_{p}, and the cylindrical channel diameter, d_{t}, is obtained:

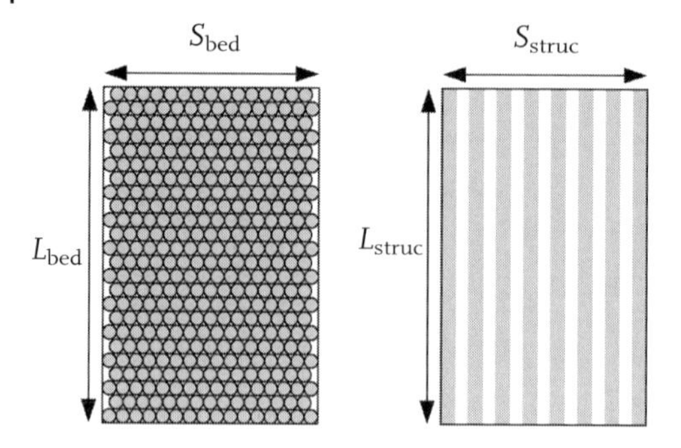

Fig. 6.3 Scheme of the fixed bed (left) and the microstructured reactor (right).

$$d_p = 1.5 \frac{1-\varepsilon}{\varepsilon} d_t \tag{6.35}$$

The residence-time distribution in fixed beds and tubular reactors with laminar flow can be estimated with Eq. (6.32) and Eq. (6.30), respectively. The axial Péclet number for a packed bed and a structured microchannel reactor as function of Re · Sc is shown in Fig. 6.4. For laminar flow in microchannels the axial Pe number as function of Re · Sc passes through a maximum, corresponding to minimal axial dispersion. This phenomenon can be explained as follows: At relatively short radial diffusion time compared to the space time, $t_{D,rad} = d_t^2/D_m \ll L/w = \tau$, axial dispersion is mainly determined by the second term in Eq. (6.30):

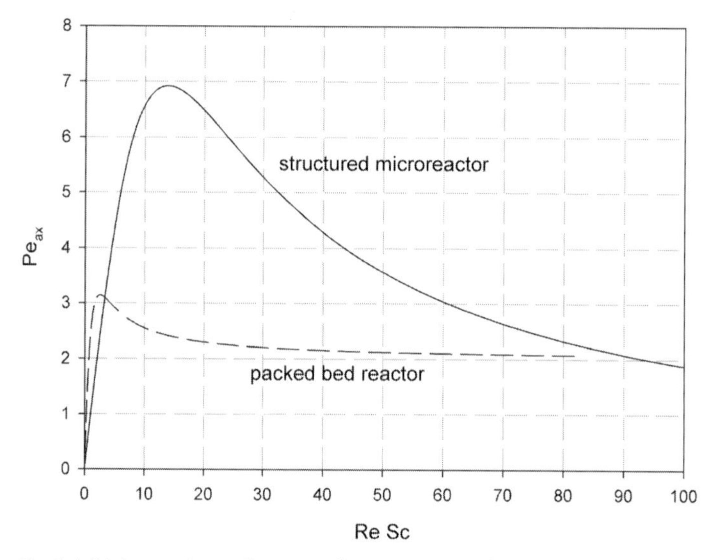

Fig. 6.4 Péclet number as function of Re · Sc in microchannel reactor and in a packed-bed reactor for $\varepsilon = 0.45$.

$$\mathrm{Pe_{ax}} \cong \frac{\chi \cdot D_\mathrm{m}}{w \cdot d_\mathrm{t}} = \frac{\chi}{\mathrm{Re} \cdot \mathrm{Sc}}$$

The axial dispersion decreases as the molecular diffusivity increases because the exchange of material between streamlines with different velocity counteracts the effect of broadening RTD caused by the laminar velocity profile. The effect of axial molecular diffusion becomes important and determines the RTD at very low mean velocities when $D_\mathrm{ax} \cong D_\mathrm{m}$. By differentiation of Eq. (6.30) we obtain the minimal dispersion (maximum $\mathrm{Pe_{ax}}$) at $\mathrm{Re} \cdot \mathrm{Sc} = \sqrt{\chi}$.

In packed-bed reactors, $\mathrm{Pe_{ax}}$ passes also through a small maximum and becomes constant for $\mathrm{Re} \cdot \mathrm{Sc} > 20$.

As the reactor length and the linear velocities in both reactors are identical (Eq. (6.34)), broadening of the RTD depends only on the axial dispersion coefficients, D_ax. The ratio of the estimated dispersion coefficients is shown in Fig. 6.5. This ratio was calculated for two different porosities ($\varepsilon = 0.45$ and $\varepsilon = 0.36$) respecting the relation between d_p and d_t presented by Eq. (6.35).

As is seen from Fig. 6.5, for values of $\mathrm{Re} \cdot \mathrm{Sc}$ between ca. 3.5 and 80 the RTD is narrower in microstructured channel reactors compared to fixed beds. The broadening is reduced by a factor of approximately 3 for $\mathrm{Re} \cdot \mathrm{Sc} \cong 14$. The increased dispersion in fixed beds can be due to the successive constriction and broadening, in addition to the tortuosity in the packing. Therefore, microstructuring the flow in parallel cylindrical channels leads to an improved reactor performance and product selectivity. For $3.5 > \mathrm{Re} \cdot \mathrm{Sc} > 80$, a packed bed ensures narrower RTD and therefore, the use of microstructured reactors will not give any benefits in this respect.

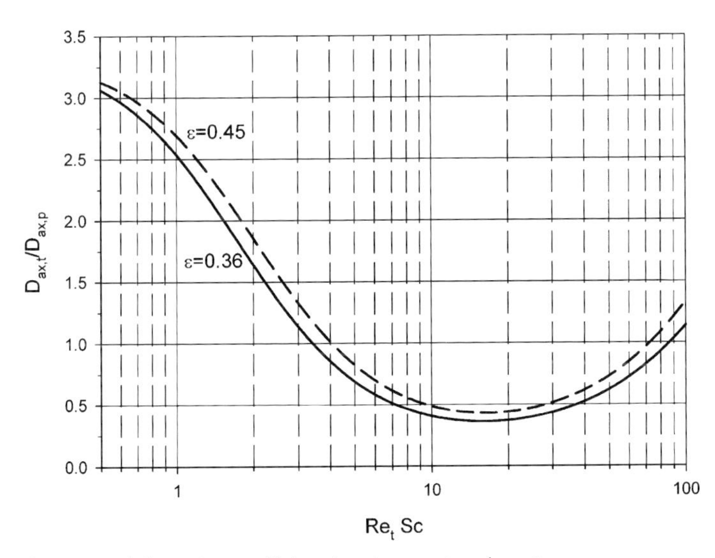

Fig. 6.5 Axial dispersion coefficient in microstructured reactor related to dispersion coefficient in fixed bed.

6.3 Temperature Control in Microchannel Reactors

The knowledge of temperature and concentration profiles along nonisothermal tubular reactors is of primary importance for the design and operation of such reactors. If the locally exchanged heat does not correspond to the heat produced or consumed by chemical reaction, hot, respectively cold spots are formed. As the reaction rates depend exponentially on the temperature, the reactor performance and product yield/selectivity are strongly influenced by the system nonisothermicity. In the case of exothermic reactions, the temperature may exhibit a sudden sharp increase along the reactor length. When such a phenomenon occurs, the reactor is said to operate under "runaway" conditions. In addition, the temperature profile and the hot-spot formation in the reactor may be sensitive to the operational parameters. This is known as a high "parametric sensitivity" of the reactor [7]. The small channel diameters of microstructured reactors improve heat transfer and avoid the former problems enhancing process safety.

The domain of "runaway" and "parametric sensitivity" can be estimated from the reaction kinetics, the reaction enthalpy and the heat-transfer performance of the reactor [8]. For quantitative analysis let us assume that the temperature and composition of the reaction mixture are uniform over the tube cross section and depend only on the axial distance from the feed point. Concerning heat transfer, several studies have shown that for most microstructured multichannel reactors constructed from heat conducting materials the axial wall temperature, T_w, is constant [9]. In addition, laminar flow and a fully developed temperature profile will be assumed. Under these conditions, the Nusselt number, Nu, reaches an asymptotic value, Nu_∞ (see Eq. (6.56) and Table 6.3):

$$\mathrm{Nu} = \frac{\alpha \cdot d_t}{\lambda}; \ \mathrm{Nu}_\infty = 3.66 \text{ (circular tube)} \tag{6.36}$$

where λ is the heat conductivity of the reaction mixture and α is the heat-transfer coefficient.

To simplify the analysis, we will neglect any work done on the reacting fluid and assume constant heat capacity and reaction enthalpy. If only a single reaction takes place in the reactor, the steady-state temperature profile of the fluid can be calculated by considering the energy balance (Eq. (6.37)) simultaneously with the material balance (Eq. (6.38)):

$$\frac{dT}{dV} = \frac{dT}{S \cdot dz} = \frac{(-R_1)(-\Delta H_r)}{\dot{m} \cdot \bar{c}_p} - \frac{\alpha \cdot a \cdot (T - T_w)}{\dot{m} \cdot \bar{c}_p} \tag{6.37}$$

$$\frac{dX}{dV} = \frac{dX}{S \cdot dz} = \frac{-R_1}{\dot{n}_{1,0}} \tag{6.38}$$

For constant tube diameter, d_t, the specific heat exchange surface is given by $a = \frac{\Delta A}{\Delta V} = \frac{4}{d_t}$. With constant fluid density ρ, Eqs. (6.37) and (6.38) can be written as

$$\frac{dT}{dz} = \frac{(-R_1)(-\Delta H_r)}{w \cdot \rho \cdot \bar{c}_p} - \frac{\alpha \cdot 4 \cdot (T - T_w)}{d_t \cdot w \cdot \rho \cdot \bar{c}_p}$$

$$\frac{dT}{d\tau} = \frac{(-R_1)}{c_{1,0}} \Delta T_{ad} - \frac{\alpha \cdot 4 \cdot (T - T_w)}{d_t \cdot \rho \cdot \bar{c}_p} \tag{6.39}$$

With $\dot{n}_{1,0} = \dot{V}_0 c_{1,0}$ from Eq. (6.38) it follows:

$$\frac{dX}{d\tau} = \frac{-R_1}{c_{1,0}} \tag{6.40}$$

To determine the axial temperature profile in the tubular reactor, Eqs. (6.40) and (6.39) must be solved simultaneously by numerical integration.

Microreactors are often used for catalyst screening and for study the reaction kinetics. In the latter case, it is important to work under isothermal conditions. In an isothermal tubular reactor $dT/d\tau = 0$. This means that the heat removal from the reactor is equal to the heat produced by chemical reaction at any axial position:

$$\frac{\alpha \cdot 4 \cdot (T - T_w)}{d_t \cdot \rho \cdot \bar{c}_p \cdot \Delta T_{ad}} = \frac{\alpha \cdot 4}{\rho \cdot \bar{c}_o \cdot \Delta T_{ad}} \frac{\Delta T}{d_t} = -\frac{R_1}{c_{1,0}} \tag{6.41}$$

With the exception of a zero order reaction, the reaction rate varies along the reactor length and the condition (Eq. (6.41)) can only be satisfied by changing $\Delta T/d_t$ along the reactor, which in practice cannot be realized. But, the reactor can be designed with an acceptably small temperature gradient ΔT. Below will be shown an estimation of ΔT for a first-order reaction: $-R_1 = kc_1$. The rate constant k is related to the rate constant k_w corresponding to the wall temperature, T_w:

$$k_w = k_0 \exp\left(\frac{-E_a}{R \cdot T_w}\right)$$

$$k = k_w \left(\frac{k}{k_w}\right) = k_w \exp\left(\frac{E_a (T - T_w)}{R \cdot T \cdot T_w}\right) \cong k_w \exp\left(\frac{E_a \Delta T}{R \cdot T_w^2}\right) \tag{6.42}$$

The maximal tube diameter for an admissible small temperature gradient with $c_1 = c_{1,0}$ can be estimated as:

$$\frac{1}{d_t} = \frac{\rho \cdot c_p}{4 \cdot \alpha} \cdot \exp\left(\frac{E_a \Delta T}{R \cdot T_w^2}\right) \cdot \frac{\Delta T_{ad}}{\Delta T} \cdot k_0 \exp\left(\frac{E_a}{R \cdot T_w}\right) \tag{6.43}$$

Since for laminar flow in microchannels the heat-transfer coefficient depends on the channel diameter (see Eq. (6.36)) as

$$a = \frac{\mathrm{Nu}_\infty \cdot \lambda}{d_t} = \frac{3.66 \cdot \lambda}{d_t} \tag{6.44}$$

we obtain by combining Eqs. (6.43) and (6.44):

$$\frac{1}{d_t^2} = \frac{\rho \cdot c_p}{4 \cdot \mathrm{Nu}_\infty \cdot \lambda} \cdot \exp\left(\frac{E_a \Delta T}{R \cdot T_w^2}\right) \cdot \frac{\Delta T_{ad}}{\Delta T} \cdot k_0 \exp\left(\frac{E_a}{R \cdot T_w}\right) \tag{6.45}$$

Equation (6.45) allows estimating the diameter of microchannels to attain reactor operation under nearly isothermal conditions.

Let us consider an example: The catalytic partial oxidation of *o*-xylene should be studied in a multichannel microreactor. A temperature increase in the reactor should not be higher than $\Delta T = 3$ K. Which microchannel diameter is required for isothermal plug flow?

The following parameters can be estimated from literature data (see Table 6.2):

$$\frac{1}{d_t^2} = \frac{1.29 \times 1.05}{4 \times 3.66 \times 4 \times 10^{-5}} \cdot \exp\left(\frac{13\,630 \times 3}{600^2}\right) \cdot \frac{381}{3} \cdot 1.8 \times 10^{12} \exp\left(\frac{-13\,630}{600}\right)$$
$$= 8.1 \times 10^7\ \mathrm{m}^{-2} \qquad \Rightarrow d_t \cong 100\ \mu\mathrm{m}$$

From this estimation it follows that a microstructured reactor with a diameter of channels ~100 μm can be used for this extremely fast and exothermic reaction up to reaction temperatures of 600 K.

For given reaction kinetics the thermal behavior of the tubular reactor depends on three different parameters [8]:

The Arrhenius number:

$$\gamma = \frac{E_a}{R T_w} \tag{6.46}$$

The heat production potential:

Table 6.2 Estimated parameters for the partial oxidation of *o*-xylene [1].

Total pressure	$P = 1.013 \times 10^5$ Pa	Activation temperature	$E_a/R = 13\,630$ K
Partial pressure of *o*-xylene	$p_x = 2$ hPa	Frequency factor	$k_0 = 1.8 \times 10^{12}\ \mathrm{s}^{-1}$
Partial pressure of oxygen	$P_{O2} = 210$ hPa	Adiabatic temperature	$\Delta T_{ad} = 381$ K
Gas density	$\rho_g = 1.29$ kg/m^3	Reaction temperature	$T_w = 600$ K
Specific heat capacity of the reaction mixture	$c_p = 1.05$ kJ/kg	Maximum temperature rise	$\Delta T = 3$ K
Heat conductivity (gas)	$\lambda_g = 4 \times 10^{-5}$ kW/(m K)		

$$S' = \frac{\Delta T_{\rm ad}}{T_{\rm w}} \cdot \frac{E_{\rm a}}{RT_{\rm w}} \tag{6.47}$$

The ratio of the characteristic reaction time to the cooling time:

$$N = \frac{t_{\rm r}}{t_{\rm c}} = \frac{1}{k(T_{\rm w})c_{1.0}^{n-1}} \frac{a \cdot 4}{d_{\rm t} \cdot \rho \cdot c_{\rm p}} \tag{6.48}$$

The thermal behavior of tubular reactors will be considered below for an irreversible reaction of 1st order ($m=1$) by transforming Eq. (6.39) in a dimensionless form as proposed by Barkelew [8]. The characteristic reaction time is based on the inlet conditions. The inlet and wall temperatures are supposed to be identical ($T_0 = T_{\rm w}$):

$$t_{\rm r} = \frac{1}{k(T_{\rm w})} = \frac{1}{k_0 \exp\left(\frac{-E_{\rm a}}{R \cdot T_{\rm w}}\right)}; \ 1^{\rm st}\text{-order reaction} \tag{6.49}$$

$$\frac{{\rm d}X}{{\rm d}(\tau/t_{\rm r})} = \frac{{\rm d}X}{{\rm d}\tau'} = \exp\left(\frac{\Delta T'}{1 + \frac{\Delta T'}{\gamma}}\right)(1 - X) \cong \exp(\Delta T')(1 - X) \tag{6.50}$$

$$\frac{{\rm d}\Delta T'}{{\rm d}\tau'} = -N \cdot \Delta T' + S' \exp\left(\frac{\Delta T'}{1 + \frac{\Delta T'}{\gamma}}\right)(1 - X) \cong -N \cdot \Delta T' + S' \exp(\Delta T')(1 - X) \tag{6.51}$$

The dimensionless temperature is defined as

$$\Delta T' = \frac{T - T_{\rm w}}{T_{\rm w}} \frac{E_{\rm a}}{R \cdot T_{\rm w}} = \frac{T - T_{\rm w}}{T_{\rm w}} \cdot \gamma \tag{6.52}$$

The dimensionless space time τ' can be eliminated by dividing Eq. (6.51) by Eq. (6.50)

$$\frac{{\rm d}\Delta T'}{{\rm d}X} = S' - N\Delta T' \exp\left(\frac{-\Delta T'}{1 + \frac{\Delta T'}{\gamma}}\right) \cong S' - N\Delta T' \frac{\exp(-\Delta T')}{1 - X} \tag{6.53}$$

As it is seen from Fig. 6.6, the axial temperature profile is strongly influenced by the ratio between the characteristic reaction and cooling time, as shown for a given heat production potential, $S' = 15$, and Arrhenius number, $\gamma = 40$. At $N = t_{\rm r}/t_{\rm c} < 25$ the maximum temperature is very sensitive to small variations in N. The minimum ratio $t_{\rm r}/t_{\rm c} = N_{\rm min}$ for a safe reactor operation as a function of the heat-production potential can be estimated with the empirical relation (6.54) and is shown in Fig. 6.7 for different reaction orders [10].

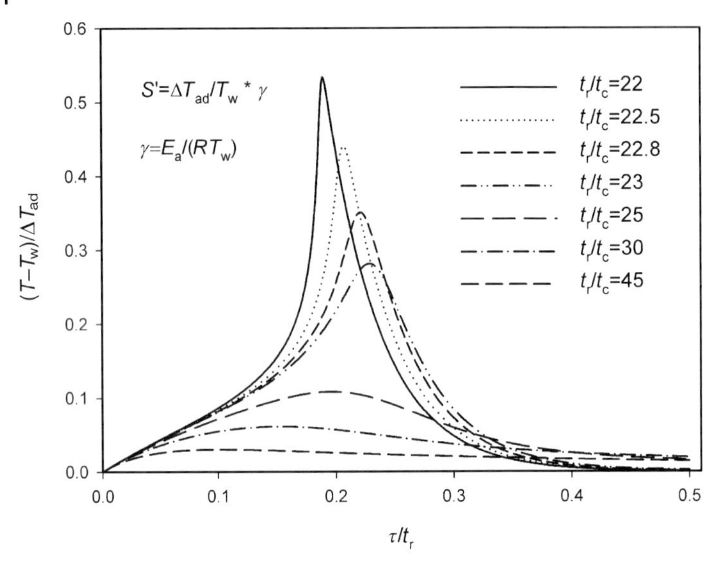

Fig. 6.6 Temperature profile for different ratios between characteristic reaction to cooling time. $S' = 15, \gamma = 40$.

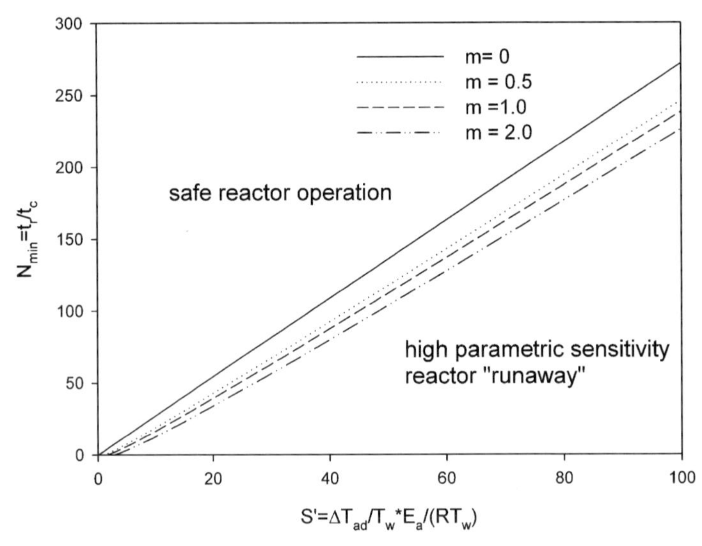

Fig. 6.7 Stability diagram for different reaction order m.

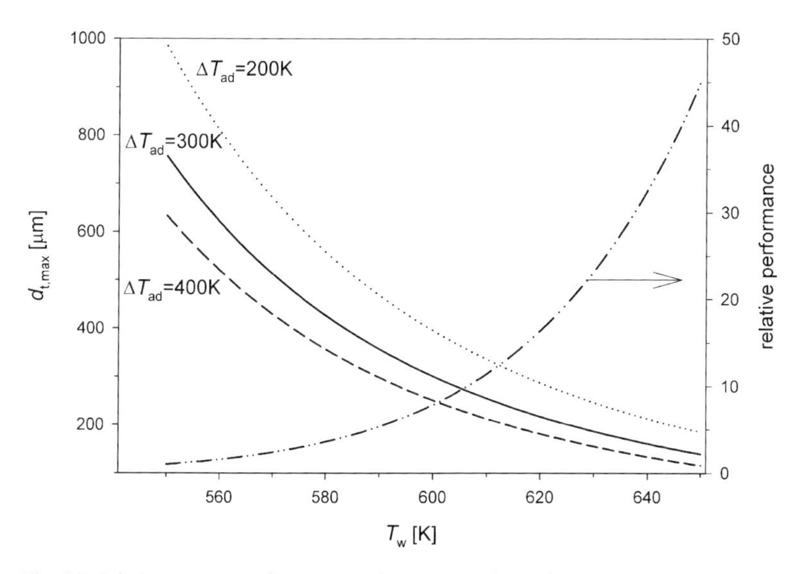

Fig. 6.8 Relative reactor performance and maximum channel diameter for safe operation (model parameters: Table 6.2).

$$N_{min} = \left(\frac{t_r}{t_c}\right)_{min} = 2.72 \cdot S' - B\sqrt{S'} \tag{6.54}$$

The parameter B is dependent on the reaction order, m, and is found to be $B=0$ ($m=0$), $B=2.60$ ($m=0.5$), $B=3.37$ ($m=1$), $B=4.57$ ($m=2$).

Respecting Eq. (6.54) allows to keep the maximum temperature (hot spot) independent of the adiabatic temperature at constant Arrhenius number. With the parameters given in Table 6.2 an Arrhenius number of $\gamma \cong 17$ and a heat production potential of $S' \cong 15$ can be estimated. Stable reactor operation can be assumed, if $N = t_r/t_c \geq 28$. The maximum acceptable channel diameter for safe operation is $d_t = 250$ µm. Under these conditions the hot-spot temperature will not exceed 30 K.

As the reaction rate increases exponentially with temperature, high wall temperatures are preferable for high reactor performances. This needs a reduction of the microchannel diameter with increasing wall temperature. In Fig. 6.8 the relative performance of the reactor $L_{P,T_w}/L_{P,550\,K}$ and the maximum channel diameter for safe operation are plotted as a function of the wall temperature for three different adiabatic temperatures. One can conclude from these data that for an effective and safe reactor operation a channel diameter of $\sim$200 µm is required when carrying out highly exothermic reactions with ΔT_{ad} up to 400 K and wall temperatures up to about 620 K.

6.4
Mass Transfer

As has been shown above, multichannel microstructured reactors are particularly suited for fast strongly exothermic or endothermic reactions, since they allow control of the RDT and nearly isothermal operation. For the reactions involving solid catalyst, which is mostly introduced into the reactor as a porous layer immobilized on the tube wall, the mass transfer has also to be considered. Prior to the reaction, the reacting molecules have to be transferred from the fluid bulk to the catalytic surface. Therefore, the influence of mass transfer on the overall kinetics has to be taken into account. Due to the small diameter of the channels, the flow regime is usually laminar with Reynolds numbers $\mathrm{Re} < 200$. The radial velocity profile in a single channel develops from the channel entrance up to the position where a complete Poiseuille profile has been established. The length of the entrance zone depends on the Re number and may be estimated from the following empirical relation [11, 12]:

$$L_e \leq 0.06 \cdot \mathrm{Re} \cdot d_t \tag{6.55}$$

Usually the entrance zone corresponds to less than 10% of the total tube length and can be neglected. Like the flow profile, the concentration profile develops within an entrance zone. Within this zone the mass transfer diminishes and finally reaches a constant value. In terms of Sherwood numbers, $\mathrm{Sh} = \beta d_t / D_m$, the following dependence has been proposed [13, 14]:

$$\mathrm{Sh} = \mathrm{Nu} = B\left(1 + 0.095\frac{d_t}{L}\mathrm{Re} \cdot \mathrm{Sc}\right)^{0.45} \tag{6.56}$$

The constant B in Eq. (6.56) corresponds to the asymptotic Sh number for constant wall concentration, which is identical to the asymptotic Nu number characterizing the heat transfer in laminar flow at constant wall temperature. The constant B depends on the geometry of the channel as summarized in Table 6.3.

Table 6.3 Mass- and heat-transfer characteristics for different channel geometries [13].

Geometry	Constant *B* in Eq. (6.56)
Circular	3.66
Ellipse; length : width = 2	3.74
Parallel plates	7.54
Rectangle; length : width = 4	4.44
Rectangle; length : width = 2	3.39
Square	2.98
Equilateral triangle	2.47
Sinusoidal	2.47
Hexagonal	3.66

If the entrance zone in the tube can be neglected, the mass transfer is constant and given by the constant B. For a circular tube we obtain:

$$\mathrm{Sh}_{\infty} = 3.66; \text{ for } L \geq 0.05\ \mathrm{Re} \cdot \mathrm{Sc} \cdot d_{\mathrm{t}} \tag{6.57}$$

If the mass transfer is accompanied by a chemical reaction at the catalyst surface placed on the reactor wall, the mass transfer depends on the reaction kinetics [15]. For a zero-order reaction the rate is independent of the concentration and the mass flux from the bulk to the wall is constant, whereas the reactant concentration at the catalytic wall varies along the reactor length. For this situation the asymptotic Sh number in circular tubes becomes $\mathrm{Sh}'_{\infty} = 4.36$. The same value is obtained for low reaction rates as compared to the rate of mass transfer. If the reaction rate is high (very fast reactions), the concentration at the reactor wall becomes zero within the whole reactor and the final value for Sh is $\mathrm{Sh}_{\infty} = 3.66$. As a consequence, the Sh number in a reacting system depends on the ratio of the reaction rate to the rate of mass transfer characterized by the 2nd Damköhler number:

$$\mathrm{DaII} = \frac{k_{\mathrm{S}}}{\beta} = \frac{k_{\mathrm{S}} \cdot d_{\mathrm{t}}}{D_{\mathrm{m}} \cdot 2}; \text{ for a first-order reaction, laminar flow} \tag{6.58}$$

Villermaux [15] proposed a simple relation to estimate the asymptotic Sh number as function of DaII:

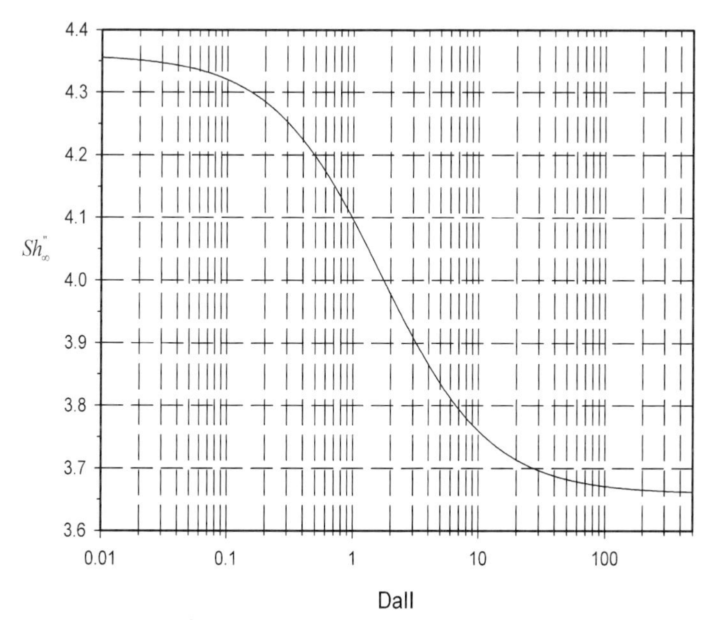

Fig. 6.9 Variation of the asymptotic Sherwood number with the second Damköhler number (1st-order reaction, circular channel).

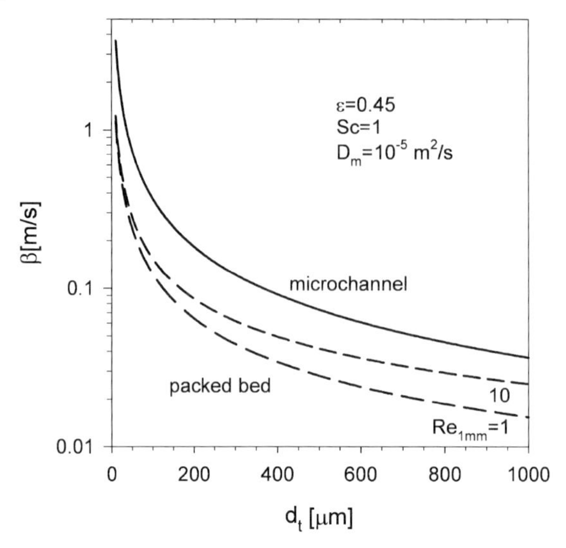

Fig. 6.10 Mass-transfer coefficient in microchannel and packed-bed reactors (DaII > 100).

$$\frac{1}{Sh''_\infty} = \frac{1}{Sh'_\infty} + \frac{DaII}{DaII + 1.979}\left(\frac{1}{Sh_\infty} - \frac{1}{Sh'_\infty}\right) \tag{6.59}$$

For the following discussion we will assume that the mass transfer is the rate-limiting step ($DaII \geq 100$) and the asymptotic Sh number of $Sh''_\infty = Sh_\infty = 3.66$ is attained (Fig. 6.9).

The mass transfer between the bulk and the particle surface in a packed-bed reactor depends on the Re-number and can be estimated with [16]

$$Sh_p = 2.0 + 1.8\, Re^{1/2} Sc^{1/3} \tag{6.60}$$

In Fig. 6.10 the mass-transfer coefficient in a circular tube as a function of the tube diameter is compared to the mass-transfer coefficient estimated for a packed bed at two different Reynolds numbers. The Re number is calculated for a particle diameter of $d_p = 1$ mm and is related to d_t as indicated in Eq. (6.35). As the residence time and specific surface areas are identical in the packed bed and in the microstructured reactor, the reactor performances are proportional to the mass-transfer coefficient. Clearly, the microstructured reactor demonstrates higher performances for reactions limited by mass transfer.

6.5 Pressure Drop

The pressure drop to be overcome during the passage of fluid through any reactor is an important characteristic related to the energy demand for the process optimization. Therefore, it will be considered herein for multichannel microreactors assuming noncompressible fluids. Gas properties will be used only up to ca. 600 K and at least atmospheric pressure ($p \geq 10^5$ Pa). Under these conditions continuum mechanics are valid [17]. In addition, fluid velocities lower than 10 m/s will be considered in channels with hydraulic diameters smaller than 1 mm. Therefore, fluid flow is laminar and compressibility effects can be neglected. The pressure drop in smooth tubes under laminar flow conditions is given by the following relation [18]:

$$\frac{\Delta p}{L_t} = 32 \frac{\eta \cdot w}{d_t^2} \tag{6.61}$$

The pressure drop through the fixed beds of hard spheres can be estimated by the Ergun equation, modified by Brauer on the basis of experimental results [19]:

$$\frac{\Delta p}{L_{bed}} = 160 \frac{(1-\varepsilon)^2}{\varepsilon^3} \frac{\eta \cdot w_0}{d_p^2} + 3.1 \frac{(1-\varepsilon)}{\varepsilon^3} \frac{\rho_f w_0^2}{d_p} \left(\frac{\eta(1-\varepsilon)}{\rho_f w_0 d_p} \right)^{0.1} \tag{6.62}$$

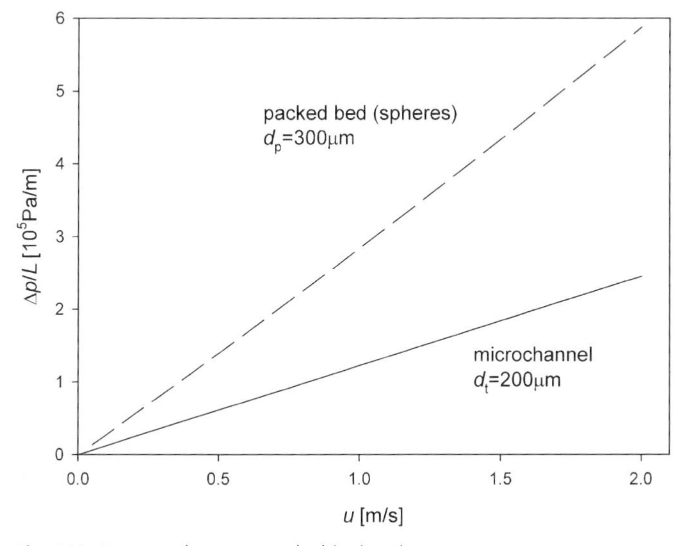

Fig. 6.11 Pressure drop in a packed bed and a microstructured multichannel reactor as function of the linear velocity of air (T=293 K; p=0.1 MPa).

With the equivalence conditions defined by Eq. (6.34) the pressure drops in both systems are compared in Fig. 6.11. As can be seen, the pressure drop in packed-bed reactors is ca. 2.2- to 2.5-fold higher compared to microstructured multichannel reactors.

6.6
Design Criteria for Microstructured Multichannel Reactors

The main characteristic of microstructured multichannel reactors is the small diameter of the channels ensuring short radial diffusion time. This leads to a narrow residence-time distribution, high heat and mass transfer. In addition, microstructured reactors possess a high surface to volume ratio of around 10 000 to 50 000 m^2/m^3 allowing efficient heat removal and high molar fluxes to avoid internal mass-transfer limitations for heterogeneous catalytic reactions. Therefore, microstructured reactors are suitable for embedded and portable devices. The small reactor dimensions facilitate the use of distributed production units at the place of consumption, thus avoiding transport and storage of dangerous materials. Based on the general relations presented so far, in the following paragraph criteria will be discussed for the rational design of microstructured reactors for achieving high specific reactor performance. We will restrict our discussion on simple irreversible reactions.

The specific reactor performance, $L_{\mathrm{P,V}}$, is defined as the production of the desired product per time and unit volume:

$$L_{\mathrm{P,V}} = \frac{\dot{n}_{\mathrm{product}}}{V_{\mathrm{R}}} = \frac{\dot{V}_0 c_{1,0} X}{V_{\mathrm{R}}} = \frac{c_{1,0} X}{\tau} \tag{6.63}$$

X is the conversion of the key reactant:

$$X = \frac{\dot{n}_{1,0} - \dot{n}_{1,\mathrm{s}}}{\dot{n}_{1,0}} \tag{6.64}$$

The conversion obtained in the reactor depends on the ratio of the characteristic time of operation t_{op} and the space time $\tau = V_{\mathrm{R}}/\dot{V}_0$ known as the first Damköhler number, DaI.

The characteristic time of operation depends on the limiting step of the process: it can be the chemical reaction, the mass transfer between fluid and a catalyst surface or the heat transfer for highly exothermic or endothermic reactions.

6.6.1 Homogeneous Reaction

If the kinetics of homogeneous reaction is determining the process performance, the characteristic time of operation is given by the characteristic reaction time: $t_{op} = t_r$.

For an irreversible reaction of mth order, the characteristic reaction time is defined by

$$t_r = \frac{k \cdot c_{1,0}^{m}}{c_{1,0}} = k \cdot c_{1,0}^{m-1} \tag{6.65}$$

and the Damköhler number becomes

$$\mathrm{DaI} = k \cdot c_{1,0}^{m-1} \cdot \tau \tag{6.66}$$

For a given inlet concentration and rate constant the necessary space time for a required conversion in a tubular reactor depends on the residence-time distribution. The highest reactor performance is obtained for plug-flow behavior: $\mathrm{Bo} \to \infty$. In practice, a Bodenstein number of $\mathrm{Bo} \geq 100$ is considered as sufficient to assume plug-flow behavior.

The Bo number is related to the Péclet number by Eq. (6.67). With the channel diameter d_t and its length L we obtain:

$$\mathrm{Bo} = \mathrm{Pe}_{ax} \frac{L}{d_t} \tag{6.67}$$

Under laminar flow conditions, the Péclet number depends on ReSc as shown in Fig. 6.4. If we assume that $L/d_t = 20$, a Pe number of $\mathrm{Pe}_{ax} \geq 5$ is required to assure plug-flow behavior. Therefore, $\mathrm{ReSc} = w \cdot d_t / D_m$ must be between 6 and 30, as can be seen in Fig. 6.4. Replacing the channel length by $L = 20\, d_t$ and $w = \mathrm{ReSc} \cdot D_m / d_t$ the channel diameter for plug-flow behavior can be estimated:

$$d_t = \sqrt{\frac{\mathrm{ReSc} \cdot \tau \cdot D_m}{20}}; \text{ with } 6 < \mathrm{ReSc} < 30 \tag{6.68}$$

As both linear velocity w and reactor length L are related to the channel diameter d_t, the pressure drop becomes independent of d_t and is only a function of the dynamic viscosity and the space time in the reactor:

$$\Delta p = \mathrm{const} \frac{\eta}{\tau} \tag{6.69}$$

6.6.2
Heterogeneous Catalytic Reactions Limited by Chemical Transformation

Heterogeneous reactions not influenced by transport phenomena exhibit a characteristic time proportional to the specific catalytic surface per reactor volume, $a = A/V_R$. Microstructured reactors are mostly designed as catalytic wall reactors and the specific catalytic surface is related to the channel diameter: $a = 4/d_t$. In this case the Da number becomes $\mathrm{DaI} = k_s \cdot \tau \cdot 4/d_t$. As a consequence, the specific reactor performance increases with decreasing channel diameter. For a required conversion and therefore, constant Damköhler number, the space time is proportional to the diameter of the channel:

$$\tau = \frac{\mathrm{DaI}}{4 \cdot k_s} d_t \tag{6.70}$$

But, changing the channel diameter will influence the Re number and as a consequence the RTD (Fig. 6.4). This must be taken into consideration for the reactor design.

6.6.3
Heterogeneous Catalytic Reactions Influenced by Mass Transfer

If the external mass transfer becomes rate limiting (DaII >100), the characteristic time of operation is given by the characteristic time of mass transfer in the channel: $t_{op} = t_D = 1/\beta \cdot a$. For fully developed velocity and concentration profiles the Sherwood number becomes constant and the mass-transfer coefficient is inversely proportional to the channel diameter: $\beta = \mathrm{Sh}_\infty \cdot D_m/d_t$. As the specific surface increases also with decreasing diameter, the characteristic time of mass transfer and the Damköhler number become:

$$t_D = \frac{d_t}{\mathrm{Sh}_\infty \cdot D_m} \frac{d_t}{4}; \; \mathrm{DaI} = \frac{\tau \cdot 4 \cdot \mathrm{Sh}_\infty \cdot D_m}{d_t^2} \tag{6.71}$$

For a required conversion the reactor performance increases with $1/d_t^2$. The highest specific reactor performance is obtained in reactors with thin and short channels, which could be obtained using catalytic grids or membranes. For this kind of reactor the Bodenstein number goes to zero ($\mathrm{Bo} \to 0$). The hydrodynamic behavior corresponds to a complete backmixed system. This has to be taken into account for the estimation of the reactor performance.

6.6.4
Heat Transfer and Reactor Safety

The rate of highly endothermic reactions may be limited by the rate of heat supply to the reactor. Heat evacuation can become crucial for exothermic reactions, when isothermal behavior is required or the risk of reactor runaway becomes important as discussed in Section 6.3. In all these situations the specific reactor performance is limited by the characteristic heating or cooling time:

$$t_c = \frac{\rho \cdot c_p}{\alpha \cdot a} = \frac{\rho \cdot c_p \cdot d_t}{\alpha \cdot 4} \tag{6.72}$$

For laminar flow with developed velocity and temperature profiles the Nusselt number reaches a final value, which corresponds to the asymptotic Sherwood number for mass transfer (see Section 6.4). As a consequence, the characteristic cooling time depends on the thermal conductivity of the fluid, λ, and the square of the channel diameter:

$$t_c = \frac{\rho \cdot c_p \cdot d_t^2}{\mathrm{Nu}_\infty \cdot \lambda \cdot 4} \tag{6.73}$$

As pointed out in Section 6.3, a maximal characteristic cooling time is required for safety reasons, avoiding reactor runaway ($t_c \leq t_r/N_{min}$). This maximum value depends on the characteristic reaction time, the adiabatic temperature increase and the sensitivity of the reaction rate to temperature variations (Eq. (6.54)). The estimated characteristic cooling times as function of the channel diameters are shown in Fig. 6.12 for water, gasoline and air. Physical properties were assumed for T=293 K and p=0.1 MPa.

The design criteria of a microchannel reactor for the safe operation of a fast highly exothermic reaction is demonstrated below for the liquid-phase oxidation of ethanol with hydrogen peroxide as a case study:

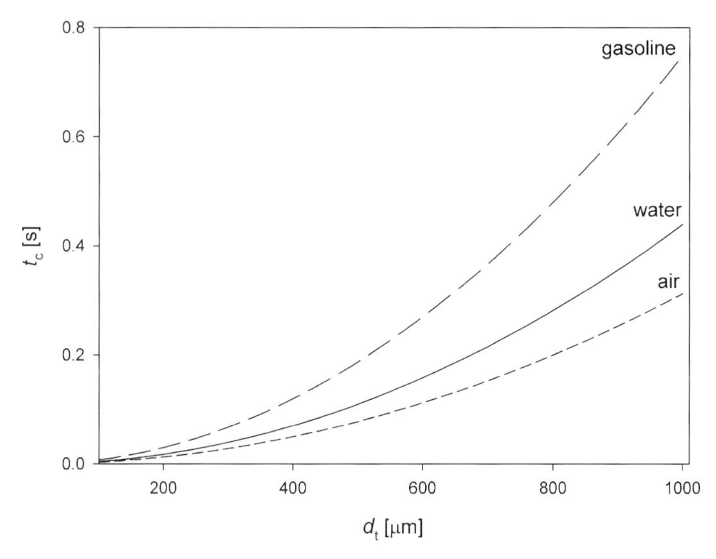

Fig. 6.12 Characteristic cooling time in a structured microchannel reactor for different fluids (T=293 K; p=0.1 MPa).

$$C_2H_4OH + 2H_2O_2 \xrightarrow{\text{cat}} CH_3COOH + 3H_2O \quad \Delta H_r = -690 \text{ kJ/mol} \tag{6.74}$$

The reaction is homogeneously catalyzed by ferric nitrate. The oxidation to acetic acid occurs in two steps with acetaldehyde as intermediate product. For a stoichiometric ratio of $C_2H_4OH/H_2O_2 = 2$ acetic acid is the only product observed.

In the literature, the ethanol oxidation served as a model reaction for theoretical and experimental studies on the stability and runaway behavior of chemical reactors [20, 21].

Kraut et al. [22] developed a multichannel microreactor to study the reaction at high reactant concentrations in the temperature range of 80–180 °C. The reactor consisted of 3 micromixers and 4 microreactor modules as shown in Fig. 6.13. The reactor modules were constructed from Hastelloy as cross-flow heat exchangers with asymmetric passages.

Each module (Fig. 6.14) had a length of 60 mm. The reaction passage consisted of 169 channels per module with a cross section of 150 μm×300 μm, and a length of 60 mm. The assembled 4 modules had a total reactor volume of $V_R = 3$ cm^3. The cooling fluid passed through 1960 channels with the same cross sections and a length of 18 mm.

The authors report results of the experiments performed using mass flows of 0.710 kg/h ethanol (98%), 3.025 kg/h hydrogen peroxide (35%), and 0.050 kg/h of an aqueous solution containing 1 kmol/m^3 $Fe(NO_3)_3$ and 1 kmol/m^3 acetic acid. Therefore, the space time is $\tau \cong 3$ s. Inlet temperatures of the thermofluid were varied between 70 and 115 °C.

A first-order kinetics with respect to hydrogen peroxide and the catalyst was found experimentally [20]:

$$-R_{\text{EtOH}} = k_0 \exp\left(\frac{-E_a}{RT}\right) \cdot c_{H_2O_2} c_{Fe(NO_3)_3} \tag{6.75}$$

with $k = 1.5 \times 10^{16}$ m^3 kmol^{-1} s^{-1}, $E_a/R = 12\,690$ K [23].

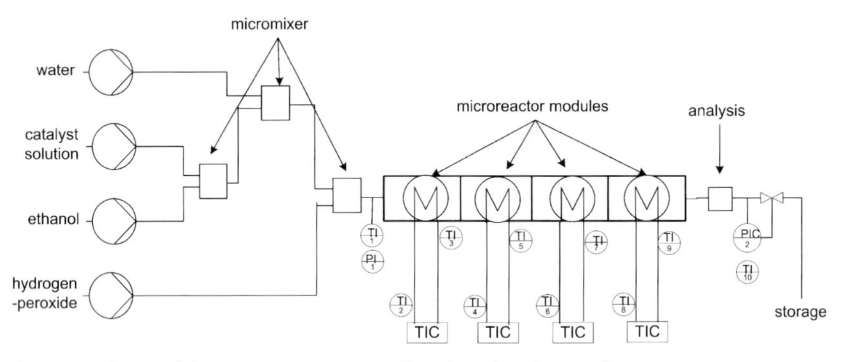

Fig. 6.13 Scheme of the microreactor system for ethanol oxidation after [22].

heat or mass exchanger. Int. J. Heat Mass Transfer, 1970. 13(1): pp. 145–146.

12 Hoebink, J. H. B. J. and G. B. Marin, *Modeling of Monolithic Reactors for Automotive Exhaust Gas Treatment*, in *Structured Catalysts and Reactors*, J. A. M. A. Cybulski, Editor. 1998, Marcel Dekker, Inc.: New York.

13 Cybulski, A. and J. A. Moulijn, *Monoliths in Heterogeneous Catalysis.* Catal. Rev. – Sci. Eng., 1994. 36(2): pp. 179–270.

14 Hayes, R. E. and S. T. Kolaczkowski, *Mass and heat transfer effects in catalytic monolith reactors.* Chem. Eng. Sci., 1994. 49(21): pp. 3587–3599.

15 Villermaux, J., *Diffusion dans un reacteur cylindrique.* Int. J. Heat Mass Transfer, 1971. 14(12): pp. 1963–1981.

16 Villermaux, J., *Génie de la réaction chimique.* 1993 Paris: TEC&DOC.

17 Commenge, J.-M., L. Falk, J.-P. Corriou, and M. Matlosz, *Optimal Design for Flow Uniformity in Microchannel Reactors.* AIChE Journal, 2002. 48(2): pp. 345–358.

18 *VDI-Wärmeatlas*, 9th, ed. Springer. 2002, Berlin, Heidelberg, New York: Verein Deutscher Ingenieure.

19 Brauer, M., *Grundlagen der Einphasen- und Mehrphasenströmungen.* 1971, Aarau: Sauerländer.

20 Hafke, C. 1972: PhD thesis, Universität Stuttgart.

21 Horak, J. and F. Jiracek, *61 Control of stirred tank reactors in open-loop unstable states.* Chem. Eng. Sci., 1980. 35(1/2): pp. 483–491.

22 Kraut, M., K. Schubert, and A. Nagel, *Oxidation of Ethanol by Hydrogen Peroxide in a Modular Microreactor System*, in *Proceedings of the 6th International Conference on Microreaction Technology (IMRET 6).* 2002, AIChE: New Orleans. pp. 100–105.

23 Kraut, M., *Literaturrecherche zur Oxidation von Ethanol in flüssiger Phase.* 1999, personal communication: FZ Karlsruhe.

7
Design Process and Project Management

Steffen Schirrmeister, Uhde GmbH, Dortmund, Germany,
Jürgen J. Brandner, Institute for Micro Process Engineering (IMVT), Forschungszentrum Karlsruhe, Germany,
Norbert Kockmann, Laboratory for Design of Microsystems, Department of Microsystem Engineering (IMTEK), University of Freiburg, Germany

Abstract
This chapter gives a short overview on the design process as a basic methodology in engineering combined with an appropriate project management. The design of chemical processes is described in general with the main emphasis on the application of microstructured equipment. The methodological design allows a general treatment of chemical processes and production plants and gives many hints for their layout and implementation. The general design process is sketched for the development of microstructured equipment. Emphasis is placed on the development of heat and mass transfer equipment with chemical reactions and their special requirements. Finally, the project management of industrial research and development projects embraces the entire process and equipment development.

Keywords
Chemical process design, equipment design, creativity techniques, research and development projects

Advanced Micro and Nanosystems Vol. 5. Micro Process Engineering. Edited by N. Kockmann

ISBN: 3-527-31246-3

The design process is a highly developed methodology in many engineering disciplines like civil, mechanical, or software engineering. For process engineers, appropriate textbooks [1, 2] and handbooks [3, 4] describe the design of processes and the dimensioning of plant equipment. Correlations for the layout of microstructured equipment are given in the previous chapters. The limits between the disciplines are moving and many tools can generally be used and applied. However, each engineer and scientist, who is engaged in this topic, has to gain his or her own experience, which can only be assisted and guided by this methodology.

The entire process from an idea or a concept to a product can be structured into four phases, see Fig. 7.1. Within the research phase the concept is developed in the frame of research and development projects. Often, new products or technologies are integrated by purchasing the necessary licenses or even entire companies. During the bid phase, a market study or feasibility study gives the financial frame of the future project. After winning the bid, the project implementation phase embraces the engineering, the procurement, the plant erection and commissioning, which is described in Section 7.1 for processes and plants, and in Section 7.2 for microfluidic equipment and devices. When the plant is operating, training, maintenance, and materials management is necessary as well as plant optimization and upgrading.

To organize these complex processes, project management and its tools are necessary for the engineers and scientists involved. Section 7.3 gives an insight into the project management of research and development projects. This entire chapter can only give a short introduction to the subject, more information can be found in the references and other special literature. An interested reader without prerequisites should read through the entire chapter to get an overview and then focus in his special interests due to the interconnected structure of design and project management.

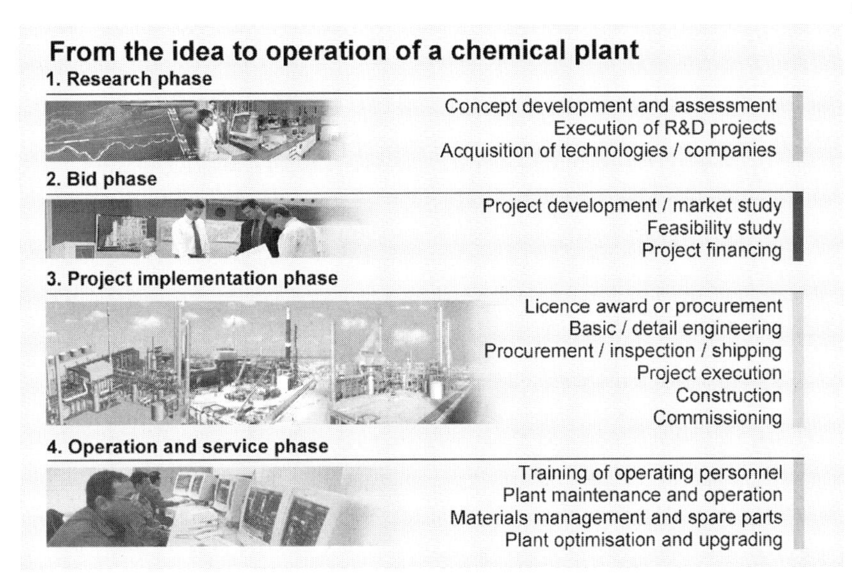

Fig. 7.1 Generation of a production facility from the idea to the product, courtesy of Uhde GmbH, Dortmund, 2005.

7.1 Design Procedure of Chemical Processes and Plants

Various levels of knowledge exist during the design and layout of technical processes [5, Chapter 1]. On the first and most general level, the basics are known as a rough, in general correct interpretation of the observed phenomena. On a second level, a detailed analysis, elaborated in the framework of experimental data or empirical correlations, offers a reliable solution for certain phenomena in reality. On the most detailed level of exact knowledge, a scientific theory shows the physical background of the investigated process and explains the correlated phenomena.

Mathematical modeling, which has long served for the calculation and prediction of events, was used in chemical engineering and process development with increasing industrialization in the 19th century. Owing to the lack of exact solutions, the complex balance equations have been simplified, and major parameters were combined into characteristic numbers. It could be shown that with keeping the characteristic numbers constant, objects with different size show the same behavior, see Chapter 7, Section 4. This so-called similarity theory serves even today as a fundamental principle in equipment development and design. It determines the scaling laws for the process mapping in miniplant and pilot plant.

7.1.1
Process Simulation, Scale-up, and Equal-up

Since the 1960s the fundamental balance equations could be solved with the help of the increasing computer performance. In the 1970s, process designers were relatively quickly able to calculate entire process chains with the introduction of unit operations, their mathematical modeling of the balance equations, and, in parallel, gathering information about the essential device parameters. The aim of process simulation is to deliver information for the understanding of the quantitative (and qualitative) properties of the process and system behavior. The process engineer determines design parameters for the reliable concept development of processes for industrial plants. The long practical experience of Uhde, GmbH, and other engineering companies has shown that the process simulation fulfils this requirement. The employed process models with model parameters are an accepted and approved tool in process design, which is actualized also with experimental data of larger production facilities.

Besides the application of mathematical and simulation tools, the appropriate, critical judgment and interpretation of the results are important for the reasonable and responsible work of an engineer. Experimental data for the evaluation are gained from model plants in the laboratory, miniplants and pilot plants, to industrial production plants modified for the special purpose. The conventional process development in the chemical industry (without microstructured equipment) starts from the laboratory scale (<1 m and <1 kg/h) and from miniplants (<3 m and <20 kg/h) over laboratory pilot plants (<10 m) and production pilot plants (<30 m and <2 t/h) to the final production plant in large scale. Zlokarnik [6] describes this scale-up procedure in detail.

Meanwhile, with the advanced scale-up knowledge, laboratory and production pilot plants are combined to save development time and investment costs. In the near future microstructured devices can help to yield laboratory data and information on the process and the plant [7], which are directly transferred to the production plant.

In conjunction with the integration of microstructured devices in laboratory and production plants, various implementation strategies have been developed to link the processes in single structures with the characterization of the entire device. The numbering-up process takes the results from processes in a single device and extrapolates them to the entire production unit by multiplying it by the number of devices. Parameters like pressure, temperature and concentration are kept constant. This simple approach exhibits various disadvantages like the inhomogeneous flow distribution or the parallel numbering-up of the measurement and control systems. Alternatively, the equal-up strategy was developed to set up a chemical process from the laboratory scale into the industrial production scale, see Fig. 7.2. With the equal-up method [8], the application of microstructured equipment in production plants can also simplify the scale-up process to more operational capacity using various fabrication techniques. The en-

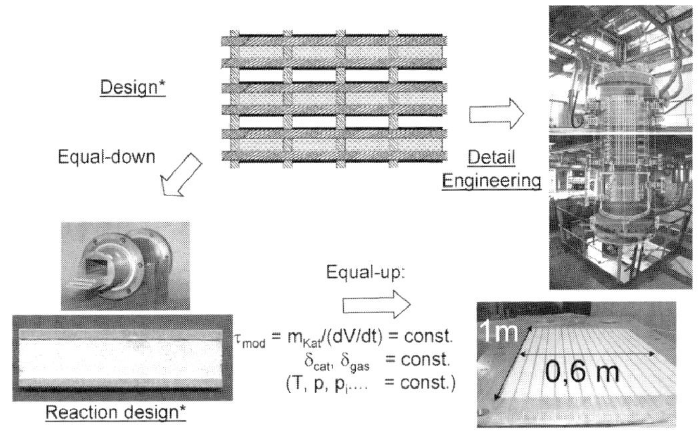

Fig. 7.2 Equipment design of a catalytic microchannel reactor with the equal-up and equal-down method acc. [9, 10]. The modified residence time, the catalyst thickness, the channel height, and the process conditions are kept constant. Courtesy of Degussa AG, Hanau and of Uhde GmbH, Dortmund.

tire equal-up strategy is described by Becker et al. [9] in more detail and embraces the following steps.

Starting from the process or product to be realized, the main effects and parameters are identified where a miniaturization could be beneficial. These key parameters, like enhanced transfer properties or integration possibilities, are transferred to the laboratory device in an equal-down step. The planned, future design of an industrial reactor also influences the design of the laboratory reactor. Process and reaction engineering with the constraints of economical design and fabrication guide the reactor development. Emig and Klemm [10] give a sophisticated description of the dimensioning of microstructured reactors concerning the fluid dynamics, the heat and mass transfer as well as mixing and reaction kinetics. The laboratory design is equal to the industrial design with respect to the governing micro effects, which implies similarity with respect to key geometries, fluid dynamics, mixing, reaction engineering, and heat management. It has clearly to be determined for the laboratory reactor, which dimensions cause the micro effects, which effects have to be transferred, and what is the shape and structure of the active area.

With the laboratory results, the detail design of the pilot reactor on the industrial scale is done in the next step, the equal-up step. The key parameters of the large device, like channel width, catalyst wall thickness and modified residence time, are equal to the laboratory equipment, but the channel length and number or the fabrication techniques may differ. For example, the microchannels have the same cross section or just only the same height, but differ in length, see Fig. 7.2.

With this procedure, comparable process conditions can be realized in the laboratory device to rapidly optimize many parameters and conditions as well as in the production device with high flow rates and controllable safety requirements. The first experimental results from the industrial plant give a reasonable reproducibility compared to the laboratory plant [9]. The scale-up design strategy and the consequent implementation of microstructures readily allow the enhancement of the throughput of devices from laboratory to production scale without the risks and costs of the conventional scale-up procedure.

Additionally, the modularization of microstructured equipment and the integration of measurement and automation devices open up new possibilities for process development [11]. A modular microreactor system with conventional pipe elements is presented by Henke and Winterbauer [12] that allows also the fast sampling of experimental data and transfer to industrial production scale for the processing of hazardous materials.

7.1.2
Method of Process and Plant Design

The process and plant design embraces engineering activities from the pilot study over basic and detail engineering to equipment design, fabrication, assembling and plant start-up, see Figs. 7.1 and 7.3. This process is organized with the help of professional project management tools, see Section 7.3.

The process design starts with the pilot study, which may have different origins from internal ideas and initiatives, customer requests, market research, or the reengineering of an existing plant or product. The pilot study shows the feasibility of the project, identifies some solution paths, and documents the specification in a measurable or testable description. It embraces the analysis of available, experimental data, the determination and analysis of material properties, and the process-simulation software like ASPEN [13, 14], MATLAB [15], or POLYMATH [16, 17]. The pilot study gives the starting point of the basic engineering in the main study, which develops a general functional process structure and generates the process flow sheet with main functional information, see also Fig. 7.4.

Starting the basic engineering involves the research and development group and the technical department in an engineering company. The following documents are prepared and elaborated during the basic engineering.

- Process flow diagram (PFD) with main control circuits;
- Mass and energy balances;
- Process description and specification, start-up and shut-down procedure;
- Equipment list with major devices;
- Dimensioning of major equipment: for example for heat exchangers: area, number of passages, channel cross section, channel length, channel distance and pitch; or separation columns with type and geometry of the inserts, special requirements for operation, maintenance and handling;
- Media list with hazard category;

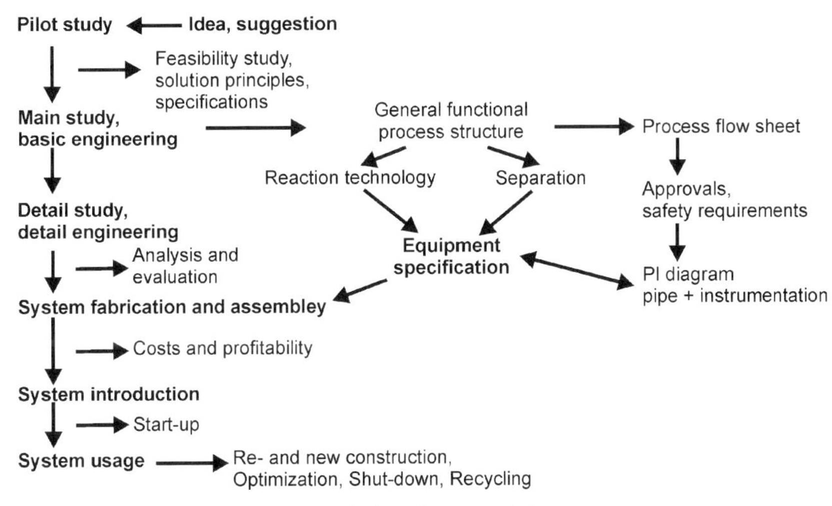

Fig. 7.3 Main steps of the process and plant design, including the basic steps of equipment design with engineering documentation (see Fig. 7.7), and of the plant construction and entire plant liftime cycle, acc. to [1].

- Choice of materials;
- Process performance and consumption values for resources and utilities, emissions of various kinds (chemicals, waste, noise, visual impression);
- Measurement and control list, valves, control circuits, safety requirements;
- Auxiliary systems, like tank farm, drains, vents, steam system, cooling water.

The process flow diagram shows the important currents of mass, energy, and signals; it is used for the specification of the process equipment and the control scheme. With the design of the process equipment and the preparation of the plant control system, the pipe and instrumentation diagram (PI diagram) is developed as the major information source for the process. During the detail engineering a permanent evaluation process adjusts the elements together and results in the design documentation of the entire plant. On this basis, the equipment is designed, fabricated, and assembled. At this stage of the detail engineering, the investment costs are mostly determined and the profitability of the plant may be fixed. With the first filling, priming and start-up, the evaluation of the specification requirements starts. The system is introduced to the operators and is used for the first production runs. Further succeeding optimizations or adjustments may be necessary.

The information levels and project steps during basic and detail engineering are illustrated in Fig. 7.4. On the functional level, the specification of the plant elements and the equipment functions are defined. The functional structure is displayed in the general flow sheet of the process, which looks like a black-box diagram. The contents of the individual process steps in the black boxes are

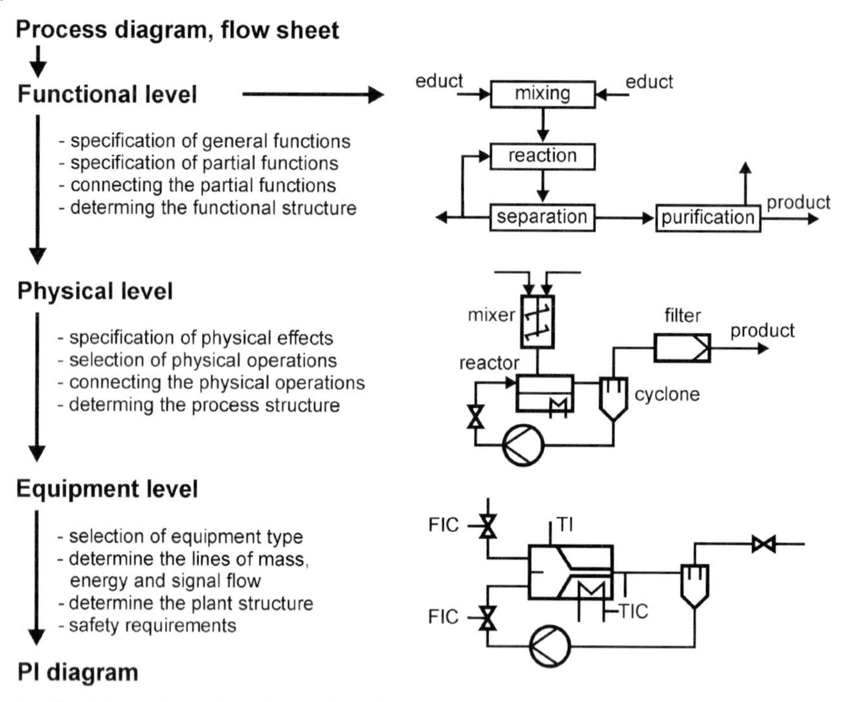

Fig. 7.4 Information and design levels in the process-design procedure, adopted from Blass [1]. More information on the process engineering symbols can be found in DIN ISO 10628.

specified on the physical level, where the process structure is defined with all the integrated unit operations. The information is displayed in the process diagram, which contains the main currents of mass, energy and signals (main control circuits). On the equipment level, the process is defined in detail and given in the PI diagram. Here, much information is given for the plant setup and the further equipment specification, design, and fabrication.

Parallel to the actual equipment level, the arrangement possibilities of the devices are important for the plant setup and the efficient plant performance. Figure 7.5 shows main types of frequently used basic combination possibilities. The serial switching of elements allows a performance enhancement, for example, a cascading of an effect for a better component separation. The parallel switching enables the disconnection of an element for partial load, for cleaning purposes, the parallel processing of a fluid stream, or the redundancy or safety enhancement of the entire process.

A bypass of a process unit serves for control purposes of the product or allows cleaning and service, or maintenance work at the bypassed equipment. A recirculation of a product reuses untreated material in a reactor and enhances the product quality and purity. Also, fluidized catalysts and solvents are reused within a reactor, an extraction, or absorption process. The recirculation of com-

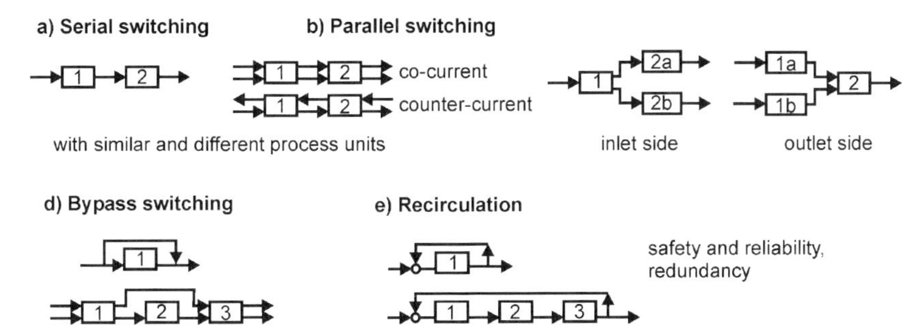

Fig. 7.5 Switching possibilities of process equipment according to [1].

ponents increases the process complexity, especially due to the possible and undesired enrichment of trace elements. Therefore, purging and filtering at appropriate locations should remove the enriched components from the circuits and equipment.

The arrangement of the different plant elements and devices requires connecting and guiding elements for the mass, energy and signal flow. In micro process engineering one has to distinguish between chip-to-chip and equipment-to-equipment connections, which is also combined with the internal and external numbering-up strategy. The guiding elements for fluids are various ducts like channels, pipes, tubes, or hoses, connected with flanges, screwed joints or other fittings. For electrical signals the guiding elements are electrical lines, wires, or RF signals by air. Active switches and valves are necessary to control the process. In conventional process technology, there exist many devices for control and automation, see for example in [1, Chapter 8; 4, Chapter 8]. The integration of various micro process equipment and measurement devices is described in Chapter 9 and further by Löbbecke et al. [11].

7.1.3 Methodological Process and Equipment Design

The general design and layout of chemical processes and microstructured equipment should follow methodological procedures to lead to successful results. The presented methodology is adopted from the mechanical engineering method of Claussen and Rodenacker [18] and accomplished by some methods of Koller [19], coming from the design of conventional process equipment. The given rules partly sound self-evident, but they are often ignored due to the diversity of the upcoming tasks of a new project. Therefore, a clear and structured working procedure and schedule are very helpful during the treatment and design of complex systems. These rules should not be treated as strict laws, but more as a rule of thumb or in a heuristic matter. They are more like a checklist and guideline to help the designer to get through the jungle of information and tasks.

1st Step: Specification At the beginning, a clear and precise specification of the work scope is essential. The effort put into this document is often hard to justify, but experienced project managers and designers know the value of a good specification. The following questions and issues should be answered or known in a good specification.

- Which products should be handled? What are the properties (value!)? Process data and states?
- What are the efforts for fabricating, assembling, and operating the device? Geometrical data, materials and properties, interfaces; Fabrication process; Operating environment.
- Detailed specification of the properties: Performance, purity, flow rates, guarantee values; Norms and regulations, safety and reliability, redundancy; Operational behavior; Amount, quality, costs of the media; Fluctuations and partial load.
- Other requirements?

According to Fig. 7.4, the designer starts with a black-box arrangement with connections of mass, energy, and signal flow, showing the amount and quality at the inlet and outlet as well as the property change of the product. The results are the specifications for the process and the equipment and devices involved.

2nd Step: Function The work scope is divided into subtasks, which can be fulfilled by an individual element. The following issues should be treated:

- Design of flow sheets for mass, energy and signal flow: transforming, augmenting, translating, measuring, controlling, switching.
- Which general logical demands are to translate into logical active correlations for operating the plant or device? Safety, operation, practical value, economy, reliability, interference with the ambient.
- Which functional elements are necessary? A matrix with the fundamental functions needed is helpful for an overview and a good starting point.
- Which functional structures have to be arranged from the basic elements? Variational and combinatorial possibilities; Variation of the basic elements and functional elements.

The function of a device is to change the product or the product properties. An appropriate set up of the functional structure, of coupling and separating elements is developed with interconnections in many variations and combinations. The process flow diagram is the main result of this step.

3rd Step: Physics The search for partial solutions regarding the partial tasks starts from the functional structure.

- Which physical effects have to be considered for the realization of the functional elements? Type, main characteristics, important parameters of the physical effect.
- Which kind of energy; which species and components should be transformed.

An overview of the possible and available physical effects is given in the work of Schubert [20]. Additionally, the design methodology of Koller [19] is based on the appropriate usage of physical effects for transformation, linking and coupling, separation, guidance, or storage. One can distinguish between static effects like mechanical forces, hydraulic pressure, electromagnetic forces, mechanical friction, or capillary forces on the one hand and dynamic effects like centrifugal forces, hydraulic forces by a jet, or vibrations on the other hand. The energy storage has to be considered for the momentum of a moving element, the heat capacity, or a chemical transformation and reaction. Special effects occurring in microstructures are for example mechanical forces in capillary filling, electro-osmosis and wall charges, or anisotropic electrical resistance or heat conductivity in crystals. A systematic variation of effects and functions can lead to new configurations, which is assisted by creativity techniques and experience. Also important is the order of magnitude of various effects in comparison to each other, like length and time scales, momentum, force or energy scales. Dimensionless numbers and groups can represent the different scales and ratios to display the various regimes.

The physical processes often take place at surfaces or interfaces with certain transfer kinetics that are called active areas. The processes can be enforced by the proper design of the active area like heat transfer augmentation or reduced friction losses in channel flow. By variation and combination of the surface properties, they can be adjusted to the actual requirements. The energy, mass and signal processing is influenced by many disturbances, which can augment or damp the actual process. A short estimation of influence parameters helps to assess the correct behavior of the system. In complex systems, the elements may influence each other and show new, emerging features. Aughenbaugh and Paredis [21] propose a method to deal with these new emerging properties and the corresponding uncertainties and lack of knowledge. Designers must recognize and appropriately treat the uncertainties, implement them into their models and calculations, and communicate them properly. An engineer or a scientist has not only to respect technological or scientific aspects, but also economical and ecological, social and cultural aspects. Vandenburg [22] describes in detail how these various aspects can be included into the design and engineering work procedure.

4th Step: Design The elaboration of solutions to the partial tasks can be done in the following way:

- Definition of the active area with known physical processes at the surface or interface; Variation of the surface; Guiding and switching; Separating and combining the materials, components, energies, and signals.
- Definition of the direction of the active force with the requirements for the direction; Simple motions and directions, straight pipes, channels, and guides, round bends and simple connections between the elements.

The primary result of this step is the pipe and instrumentation diagram (PI diagram) that displays all of the process equipment, control devices, and information flows. Another result of this step is the shape and arrangement of the active area in the devices to enforce the physical or chemical processes. The fabrication method plays an important role for the shape and the employed material. A more detailed view on the equipment design is given in Section 7.2.

7.1.4
Tools in Engineering Design

Engineering design is a comprehensive discipline, which is based on experience, on working with creativity methods to find new, adequate solutions, and which has to be done within a given time and financial frame. Besides a good technical and organizational knowledge, which is systematically implemented in design methodology, creativity techniques, equipment and process evaluation, and safety thinking are fundamental tools in process engineering.

Creativity techniques The design process with main emphasis on the creativity and problem-solving process is displayed in Fig. 7.6, according [23]. The initiative of a project, of developing a new product, process, or device, may come from outside, from the market or the customer (market pull). On the other hand, internal sources like new fabrication capabilities, new materials and combinations (technology push), or the need to enhance an existing product or process gives a strong need or challenge, the spark to initiate a project. A good starting point is to visualize the object by existing CAD drawings, blackboard outlines, or manual sketches. The Mind Map method is a very helpful tool for visualization and organization [24]. Brainstorming, brain writing, or other improvising creativity techniques can yield new items or solution possibilities. A systematic problem-solving process is the key to the successful design of processes, equipment, and devices.

Brainstorming in groups is regarded as an effective method to generate new ideas. A psychological study [25] showed that the communication within a group, the listening to and the waiting for others, disturbs the creative thinking process of a single person and can hinder the generation of ideas. Single persons, who can think undisturbed, are much more creative. The individual writing down of ideas and subsequent communication and exchange of the ideas show much better results than just group work. Hence, a combination of single persons and group work will lead to maximum results. Other methods to find new ideas or to assist the design process are check lists [3, 18], heuristic rules [1], morphological charts [26], or the TRIZ method [27]. The last method consists of 40 general rules or laws from engineering experience, which have been set up from an extensive patent review by Altschuller. Examples of these 40 rules can be found in [28] for microsystem engineering and microelectronics and for process engineering in [29].

New and existing items are connected to fulfill the new functions. When collaborating, new functions can emerge that are very helpful to meet the challenge. In synthesizing various objects, new items are generated, which have to

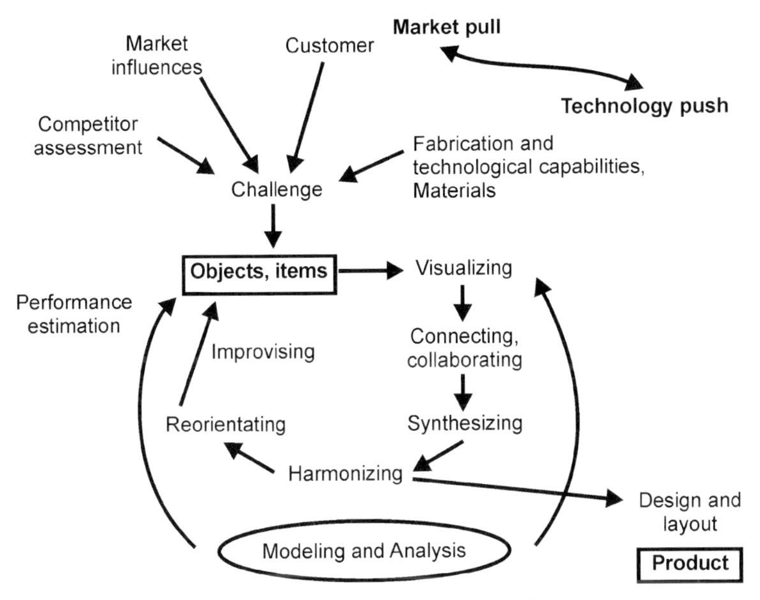

Fig. 7.6 The creativity process as one of the first steps in the design process, partially adopted from nbm [23].

be harmonized to fit together, to get a sound solution. In modeling and analyzing the new found items and systems, the performance and abilities are checked, which may already lead to a design and layout solution. If the required functions are not obtained, a reorientation gives new viewpoints to probably different solution possibilities. Improvising may bridge unconnected tasks and solutions to give further possibilities. Figure 7.6 makes the highly iterative process slightly clearer, but is only one description. In reality, the creativity process is very complex and individually different.

A systematic way for solving design tasks, which fits to the above described design methodology, is given in the following steps [18]:

1. Choose a constant attribute for all machines and equipment;
2. Fulfill the basic functions:
   2.1 arrange the mechanical elements,
   2.2 apply and integrate the physical effects;
3. Collect the physical effects;
4. Arrange the energy systems within the working elements;
5. Basic arrangement of equipment and devices;
6. Overview of exchange and transfer possibilities.

The yielded solutions have to be evaluated and rated against each other to find an optimum solution.

Characterizing and Evaluation of Equipment The efficiency of processes and the effectiveness of single equipment have to be treated regarding the relevant application and their special requirements. The performance can be measured with the mass flow, throughput, the transferred heat, the mixing quality or time, the selectivity or yield of a chemical reaction. The effort for a process may be measured in consumed energy or the pressure loss. The effort for a device comes from fabrication or assembly, or is expressed by the possible failure rate. Consequently, the equipment effectiveness is calculated as the ratio of the performance to the effort [30], relatively for the special application, like analysis, lab-on-a-chip, diagnostics, process screening or production of chemicals. The effectiveness of various solution possibilities are compared to find the best suitable solution.

Safety Requirements In process engineering, especially when dealing with toxic or hazardous chemicals, the topmost principle is: Safety First! This means a reliable performance of the functions and hazard minimization for human and environment. Safety engineering and design are wide, embracing, and emotionally loaded topics. A distinct safety consciousness and thinking is very important at the beginning of each activity. The process itself and eventually each device have to be operated under safe conditions, sometimes secured by safety means like temperature- or pressure-protection systems.

Protection measures have to be reliable, inevitable, and are not allowed to be bypassed. These safety measures and the proper and safe operation of a plant are discussed and documented in a special meeting during the detail engineering phase [3]. This meeting may follow a certain methodology, like the HAZOP-method (Hazard and Operability Studies) or the PAAG method (Prognose, Auffinden und Auswirkungen von Störungen, Gegenmaßnahmen, i.e. predicting, locating, and impacts of operational faults, counteractions).

7.2 Equipment Design

The detailed design of a chemical process device comprises the dimensioning of the physical or chemical process, the transfer processes within the device, the dimensioning of the active areas and the involved functional and structural elements. Together with the material choice and dimensioning according to the loads and stresses, with the selection of fabrication method, the interfaces, connections, and the housing are designed resulting in the shape and structure of the process device.

To reduce the number of variations from the problem-solving process, some criteria can be analyzed at first: Is the function performed with a simple structure? Do intrinsic effects achieve the physical process? May the active area be shaped by an easier fabrication process or may a different fabrication technique form a suitably active area? Whether the obtained solution is appropriate to the specified task will

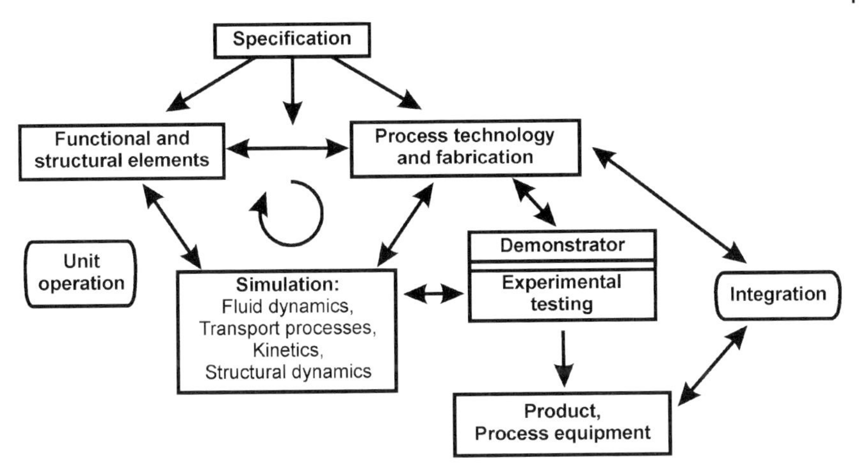

Fig. 7.7 The design process for microsystems with the design-iteration process between the functional and structural elements, the fabrication technology, and the simulation. The demonstrator is fabricated and experimentally tested. The narrow coupling of experiment and simulation enhances the design quality of succeeding devices. Over the integration with other elements or functions the process device or product is fabricated fulfilling the given specification.

be evaluated according the selection criteria of the quantity, quality, and costs. Here, the specification with countable and measurable targets is very helpful.

The equipment design in the detail study is illustrated in Fig. 7.7. The design of microstructured devices shows some characteristics that will be described in the following. Starting from the specified targets, the actual design process moves between the necessary prerequisites according to the functionality of the system, the available fabrication possibilities, the associated materials, and the simulation tools with different modeling sharpness. The entire process is highly interactive and strongly interdisciplinary.

Gerlach and Dötzel [31] list the functional elements according to their mechanical, thermal, fluidic, optic, electric, magnetic, electronic, chemical, and/or biological nature. The fabrication techniques can roughly be divided into surface and bulk methods and depend also on the appropriate materials determined by the mechanical, thermal and chemical requirements. A more detailed description is given in Chapters 10–12. The simulation starts with the calculation of the fluid dynamics and the transport processes employed. Eventually, the calculations of reaction kinetics or of the structural behavior are necessary.

7.2.1
Rules and Guidelines for the Equipment Design

An overview and selection of the main guidelines for the design of microstructured devices may be sufficient for this short introduction. Further discussions on microsystem design can be found in Kaspar [32] or Senturia [33]. The synthesis of microsystems is described by Ananthasuresh et al. [34] with different aspects of thermal problems and with emphasis on the fabrication and integration steps. From the system-design method [35], there are three levels of design rules known. The highest level comprises three fundamental design rules that are valid for each technical system: safe, simple, and well defined. On the next level, the design principles take into account constraints and boundary conditions that are very common for technical systems like mechanical strains, fabrication and materials, or cost aspects. On the third level, experience-based guidelines are formulated that will help designers to find solutions from other systems.

To assist the designer from long-time experience of successful devices, general design principles have been set up [35]. Dym and Little [26] have collected some major design principles, which they called: Design for ...

- fabrication and assembly: can this thing be made and with which means?
- materials, together with fabrication techniques,
- load and stress, together with material properties,
- expansion (thermal, mechanical, diffusion with time, etc.), together with material properties,
- affordability: how much does this thing cost? What is the benefit for the user, the added value of an upgraded plant?
- reliability: how long will this thing work? Maintainability, fouling and clogging, disposable devices,
- sustainability and lifetime cycle, see also [22].

Normally, regulations and norms of the local authorities have to be respected for the setup and operation of chemical equipment. More information can be found for example in [3, Chapter 3].

Design solutions have to be stable in the operation range, disturbances are not allowed to lead to uncontrolled operation. Instabilities may systematically be used for process enhancement, for example, flow instabilities for better mixing performance. A failure-tolerant system consists of simple and low-tolerance structures and minimizes the failure-influence variables. The main parameters of an important function should be as independent as possible from influences of possible failures. The compensation of the occurring disturbances needs buffer possibilities and components with high tolerance capabilities.

7.2.2
Design for Transfer Processes

In general, design rules for microstructured equipment follow the same laws as for conventional devices. Thus, the design process for microstructured equipment is similar to the design of conventional technology. Due to the much smaller dimensions, some special requirements have to be respected.

The design rules for the fabrication processes have to respect the dimensional precision ratio: A deviation of 50 μm in height is not much compared to a total

Table 7.1 Thermal properties and heat transfer.

Parameter	Dimensional proportionality
wall thickness	$\sim$ l [m]
surface area	$\sim$ l $\cdot$ l [m^2]
volume	$\sim$ l $\cdot$ l $\cdot$ l [m^3]
energy generation Q	$\sim$ l^3 [m^3]
heat dissipation H	$\sim$ surface area/wall thickness $\sim$ l [m]
H/Q	$\sim$ l^{-2} [m^{-2}]

structure height of 100 mm – the precision ratio here is 1:2000. A 2-μm deviation compared to a structural height of 40 μm means a precision ratio of 1:20. Similar examples can be given for all three dimensions. Hence, the precision ratio is dramatically increased with decreasing structure dimensions, which has to be taken into account for the design, for example the wall roughness in fluid dynamics. Other properties of microstructures can be considered similarly, also for transfer processes. The equations given in the following can all be found in the VDI Wärmeatlas [36] or in other books on fluid dynamics and heat transfer.

The ratio between the heat dissipation and the generated energy is inversely proportional to the product from lengths, which means: The smaller the dimensions, the better the heat transfer, see Table 7.1. This simple ratio makes highly exothermic or endothermic chemical reactions possible and explains the very short delay times arising at temperature changes applied to microstructure devices [37, 38].

The fluid flow and the pressure drop are characterized by the dimensionless Reynolds number, Re, which includes the convective and viscous forces:

$$\mathrm{Re} = \frac{w \cdot l \cdot \rho}{\eta} \tag{7.1}$$

with the mean velocity w, the characteristic length l, the fluid density ρ, and the dynamic viscosity η. The regime for laminar flow in straight channels is below Re=2300, as is well known from literature. Depending on the characteristic length, the flow velocity for liquids is in the range of some cm per second. For gases in laminar flow, the flow velocity may reach values of 100 m/s and more. The pressure drop in channel flow is described by the friction factor λ_R, the geometry, and the mean velocity w:

$$\Delta p = \lambda_R \cdot \frac{l}{d_h} \cdot \frac{\rho \cdot w^2}{2} \tag{7.2}$$

with the characteristic length l, the hydraulic diameter d_h, and the fluid density ρ. The friction factor λ_R is inversely proportional to the Re number in straight laminar flow, more details can be found in [36]. By decreasing the hydraulic diameter, the pressure drop will increase proportionally. For this reason, the microstructure sizes have to be chosen carefully.

The diffusion is mainly described by the mean free diffusion path length x

$$x^2 = 2 \cdot D \cdot t \tag{7.3}$$

with the diffusion coefficient D (about $10^{-5}\ \mathrm{m^2/s}$ for gases and $10^{-9}\ \mathrm{m^2/s}$ for aqueous solutions) and the diffusion time t. Assuming that x is in the range of the characteristic length, the diffusion time for gases lies in the milli- to microsecond range. Therefore, very efficient mixing is possible. The dimensions of the microstructures are normally ruled by three limiting factors: manufacturing, maximum pressure drop and heat transfer.

First, it should be possible to fabricate the desired dimensions by some means to generate the desired device (see Chapters 10–12). Fluid dynamics definitely influences the size and shape of microstructures and may be precalculated roughly by available correlations from the literature or, in a more detailed way, by CFD software. To reduce the pressure drop, the microstructures should be as large as possible, to increase the heat transfer, they should be as small as possible. The reduction of size will generally lead to higher flow velocities, which may result in a change of the flow regime and an even higher pressure drop.

The contradicting settings of reducing the pressure drop and increasing the heat transfer capabilities may be overcome by parallelization. Due to the need for a flow distribution as homogeneous as possible (and other reasons, see Chapter 10), a large number of microstructured devices in parallel is not a possible solution, but the so-called internal numbering-up or equal-up can be solutions. Here, a large number of microstructures are assembled in parallel inside a reasonable small number of single devices, leading to a suitable pressure drop combined with very good heat transfer properties [39]. The problem of equal fluid distribution to all microstructures can be solved by CFD simulations or experimentally [40, 41], see also Chapter 8. In the following, some general devices will be described in their design process very briefly. More details can be found in the literature.

7.2.3
Micro Heat Exchangers

The design of microstructured heat exchangers is based on well-known equations [36]. A precalculation of the thermal power to be transferred and the maximum allowable pressure drop will set the limits. The fabrication technique and the corresponding material is another limiting point in the design process. This gives at least a route to the microstructure number and dimensions. The temperature distribution is given by the design and the flow arrangement (cross-flow, counter-current or co-current flow) [39].

The design of microstructured heat exchangers has to be appropriate to allow a homogeneously distributed flow of the fluids. The design of heat exchangers is critical for phase-changing processes like evaporation or condensation [42, 43], see also Chapter 4. These processes are not fully understood in microstructures, so no general design rules can be given. Hence, trial and error is the easiest way to find a good design. The thermal insulation and separation of devices is critical due to the small length and heat capacities of the fluids and devices. For longer connecting tubes or residence time modules, a heat bath is recommended to guarantee a constant temperature, see [11].

7.2.4 Micromixers

Mixing is one of the best-investigated topics in micro process engineering. Nevertheless, a characterization of the mixing quality is not trivial and depends on the flow regime, the employed media, and the total mass flow applied to the system. Mixing principles normally used for micromixers are mixing by diffusion, by kinetic energy or by turbulence [44, 45]. For most gas–gas or liquid–liquid mixing, diffusive mixer processes are suitable. Mixing of immiscible fluids may take some more care, see Chapter 5. In general, mixing structures known from the macroscale can also be used in microstructure devices, as long as they are producible. A lot of literature and examples on this topic can be found in [46].

7.2.5 Chemical Reactors with Microstructures

A countless number of microstructure reactors suitable for various chemical reactions is described in the literature, made out of silicon, glass, ceramics, or metals. Here, the design process is more complex than for microstructured devices for heat exchange or mixing. According to the fabrication demands and the achieved heat transfer, the mixing of reactants and the homogeneous distribution is interesting for a good reaction yield. Also avoiding hot or cold spots are side aspects of the design for microstructure reactors. Moreover, catalyst support layers have eventually to be integrated into the microstructure system, which demands an appropriate design. In contrast to most chemical (batch) processes in conventional technology, microstructure devices normally run in continuous flow mode, which can be a tremendous advantage. Continuous operation needs are careful process and equipment design and a highly sophisticated process control system. More details can be found in various publications, for example in [39, 44, 46] as well as in Chapter 6.

7.2.6
General Design Issues for Microfluidic Devices

Aside from the chemical process and the device itself, some other substantial side parameters have to be taken in account. Depending on the fluid and the material of the device it might be mandatory to insulate the device thermally and/or electrically. In the extreme case, the complete device has to be decoupled from the rest of the process equipment to achieve its correct performance. In particular for the thermal behavior, the choice of material is essential [47]. If, for example, a small volume flow of gases has to be heated to a defined temperature, metal devices have to be nearly perfectly insulated. Electrical insulation might be necessary due to the use of explosive reaction mixtures or disturbances of integrated sensors.

Most fluids have to be driven through the microstructure arrangements by external pumping. The selection of the appropriate pump is obligatory: It should be stable against the fluid, the flow range has to be appropriate, and the pumping should be as pulsation free as possible. In the interest of cost effectiveness, it might be a good idea to look for a conventional pump offered "off the shelf" instead of having specially designed micropumps. The mass flow should be controllable and measurable, and the pumping pressure should be high enough to make sure that the desired flow is obtained. Moreover, to clean the fluid passages of larger particles and therefore prevent blocking of the microstructures, an adequate filter has to be applied either before or at least after the pump. Other methods of fluid transport like electrokinetic flow or electro-osmosis may also be worth thinking about, but will produce special demands on the fabrication, on the material, and the design.

Measurement and control of the process is one of the main tasks when applying microstructure devices to reaction engineering. Details on this topic can be found in Chapter 9, so only some general remarks will be given here. In principle, there are two possibilities to obtain data from a process device: Integral and local data acquisition. Integral measurement will lead to, e.g., temperatures, pressure and mass flow data at the inlet and outlet of the entire device. In most cases, this method is sufficient to control a process and leads to a good device performance. In some cases, local data acquisition is compulsory. Running a chemical reaction inside a microstructure device without knowing at least the rough process kinetics is always risky. Many undesired things can happen, starting with simple local overheating to locally increased corrosion or unexpected precipitation and channel blocking. Therefore, the integration of sensors of different types may be necessary. Unfortunately, this integration is not trivial due to the dimensions of the devices. In silicon devices numerous sensors are certainly possible [48], while the integration of a number of different sensors in metallic devices is a challenge. If data from inside the process are available, the design can be trimmed to fit the process demands and avoid things like "hot spots".

In the interest of cost reduction, the sensor interfaces should fit to standard measurement and control systems, which are normally applicable to microstructure devices. It has to be checked whether they may be too slow to follow the rapid process parameter changes, see [37]. A modular sensor and controller concept would be helpful, where only the sensors are applied that are really needed for the process. Other sensors may be adapted to the actual system where necessary. The first design suggestions for sensor modules have been made within the MicroChemTec program [49].

While much work was done for the appropriate design of microstructured equipment with regard to measurement and control, less attention has to be paid to the design with regard to the fluid properties. Almost all fluids show the same behavior in the microscale as in the macroscale. Three points have to be dealt with: capillary forces, pressure drop, and fouling.

Capillary forces should normally not be a problem, unless the fluid driving power is low and the microstructures are very small. By a simple increase of the dimensions, this problem may be solved. Increasing the dimensions or equal-up the device structures may also solve pressure-drop problems. Based on the equations given before, pressure-drop precalculations can easily be performed.

Fouling is a critical point for microstructure design and is not well understood [50]. By avoidance of dead volumes, especially narrow bends and holes, where a fluid is not moving at sufficiently high speed, fouling by sedimentation of particles may be suppressed. Beside this, the use of correctly sized filters upstream of the device may further improve the performance. Surface modifications with protective layers [51] or integration of mechanic energy (for example ultrasound [52] or pulsed flow) may help against fouling occurring by chemical processes. Avoidance of cold spots or hot spots is also mandatory for fouling mitigation.

While the flow behavior for most substances is the same in the macroscale and the microscale, there is definitely a difference in the corrosion behavior. A device material suitable for macroscale devices might not be applicable in the microscale. Although the corrosivity of the fluids has not changed, the corrosion rates might be too high for microstructure devices. Great efforts have to be made to prevent corrosion damage in microstructure devices. Some more details can be found in Chapter 10.

7.3 Project Management in Process and Plant Design

Project management is a general engineering discipline with many facets, see for example Seibert [53], Dym and Little [26], or Rinza [54]. This section describes the management and organization of the plant development and equipment design with microstructures, which still includes some uncertainties to be handled. Hence, the management of research and development projects (R&D projects) comes closest to this task.

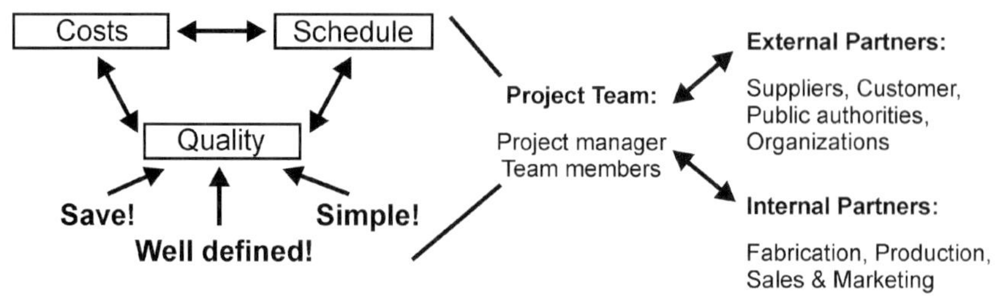

Fig. 7.8 Project management and constraints for the design and development of technical goods, for example chemical processes, plants, or equipment.

7.3.1
General Introduction

A project is a time-limited procedure where a project team works on a specified scope under fixed costs and given resources, see Fig. 7.8. A more precise definition of projects is given by Seibert [53]: Projects are schematic actions with defined start and closure that are characterized by the time limitation, uniqueness, complexity, and novelty, and which can cause organizational changes in an enterprise due to its interdisciplinary and cross-sectional nature.

The German Standard (DIN) additionally gives a definition of project management:

Project management is the entirety of leading and guiding tasks, organization and means for the treatment of a project (DIN 69901).

The development of a component or a complete system is a chronological procedure from the idea through the planning and definition of a technical specification, the design to the production and sales. The complex procedure is managed by an interdisciplinary project team with internal and external partners, see Fig. 7.8. The engineering task is segmented in independent compartments for the project team members. At the beginning of the project, measurable control values and milestones are set for the design and development, for the fabrication, assembly, and testing. Milestones are fixed for continuous target dates, on which these values are requested, checks with the target values and control activities may be arranged. The main task of the project manager is to control the milestones and track the project progress. He is also responsible for the cost structure and the budget of the project.

7.3.2
Project Management for R&D Projects

In micro process engineering the plant and equipment design consists of arranging many parts and elements whose behavior and properties are often unknown or where the interaction between the devices and elements can only be estimated. Hence, it is advantageous to organize a project dealing with the design and arrangement of microstructured equipment as an R&D project with many degrees of freedom.

An R&D project is a special form of a project, the R&D project management is a part of project management. The difference between an R&D project and an engineering project can be seen in the main cost structure and the procedure. The procedure of an R&D project is not generally defined. Typical are a short-term detailing of the engineering work with close monitoring and following more detailed planning.

Fundamental elements for a successful project management are the clear structure of a project coming from the project planning phase and the target/actual value comparison during the project progress in the form of an appropriate project control.

Before starting a development project the goals or targets have to be clearly defined and widely communicated. This is done on the base of a market study and analysis, where the product or the process to fabricate the product is placed or established. The first economical estimation shows the possibilities to successfully bring the product to the market. The elaborated market prognosis gives the reasons to start a new development. On the basis of an economic calculation, for example a net present value calculation (NPV), it has to be determined whether the start of the new development is economical.

On the basis of the development goals, the project objective specification is elaborated. The specification embraces potential targets to be reached, formulates the starting point for the project team for further planning and for the evaluation of the different methods and procedures for its realization. The specification is also the basis for the definition of milestones and for the evaluation for the project progress.

The difference between the actual state-of-the-art and the project targets determines finally the project contents. The quantitative estimation of the development effort, which is necessary for the efficient project management, belongs to the most difficult tasks in the planning phase. Here, appropriate estimation procedures that are usual for different industrial sectors [53, 55], can assist the planning. A measure of uncertainty is typical for development projects.

Depending on the complexity and the project scope one has to select the organization form of the project. Typical forms are, according to [56], the contract model and the cooperation model. The mutual exchange of services is typical for the contract model. The cooperation model is chosen in the case of a succeeding cooperation for a common purpose. The selection of the organization form regulates the elaboration of the cooperation contract.

In the further structuring the project is divided into different project phases and segmented into work packages. The number of project phases depends on the entire scope and on the partial targets. Examples for project phases are the
- laboratory development,
- development of specific devices,
- installation of pilot plants or miniplants.

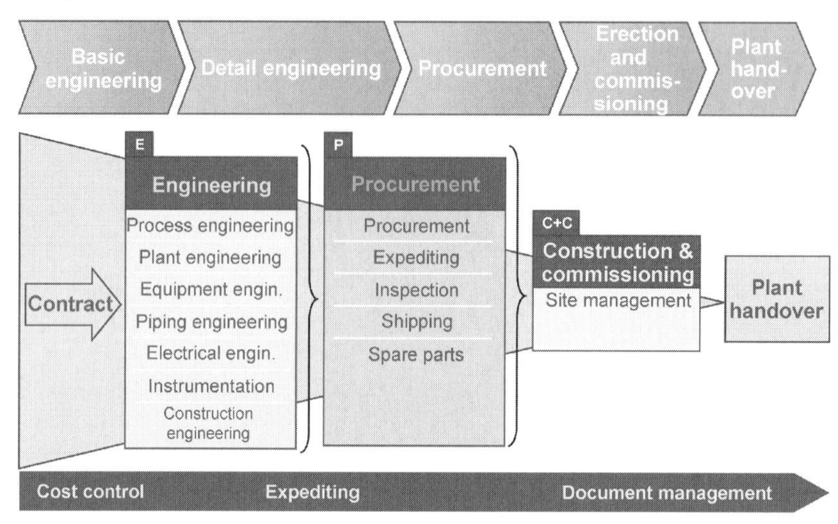

Fig. 7.9 Project execution phases and engineering tasks, courtesy of Uhde GmbH, Dortmund, 2005.

The project phases during the design and erection of a chemical production plant consists of enclosed packages which are described in literature in detail [57].

7.3.3
Project Execution Phases for Chemical Plants

The various engineering tasks and work packages during the engineering and plant setup are shown in Fig. 7.9. Starting with the contract or the research grant, the basic and detail engineering embraces the groups of process engineers, plant engineers, equipment engineers for the various devices and machines, piping, electrical engineering, and instrumentation, measurement and control. For larger plant with civil engineering tasks, the construction department is included in the project.

The procurement of devices and services is integrated in or parallel to the engineering phase, as well as the first engineering activities in basic engineering are parallel to and assist the contract phase. The parallel and interconnected project management is called simultaneous engineering and should serve for a fast and optimized project execution.

The procurement of equipment and services starts during the engineering phase and consists mainly of the ordering, inspection, and quality control as well as the expediting and shipping of the parts and devices. Sometimes the procurement of spare parts is included and has to be managed, especially for oversea projects.

Onsite construction has to be managed during the erection and commissioning phase. The legal regulations and norms have to be fulfilled by the equipment, which is checked during commissioning and startup by the authorities.

During the startup of the plant process engineers are often involved as well as special startup engineers. The plant operators are involved in the startup to train and to assist the commissioning. The plant documentation is finished and handed over to the operators or operating department. With the plant hand over, the operational responsibility shifts to the operators and to the operation organization.

The engineering process is assisted by many software tools during the various project phases, which is summarized under the idea of computer-aided process engineering (CAPE). An overview on the various software packages in process engineering and plant design is given in Fig. 7.10. The process simulation is assisted by programs like Aspen [13] or ChemCAD, the equipment layout calculations are performed by special software packages like HTRI for heat exchangers. It has to be checked whether these programs are suitable for the design and layout of microstructured equipment.

The mechanical design of equipment like pressure vessels, heat exchangers or micro reactors is assisted by programs as well as the piping work and material selection and management. Standard CAD programs (computer-aided design) include packages for the layout of process flow sheets or PI diagrams. Often engineering companies have their own software tools and tool boxes, where also the data transfer and communication is optimized.

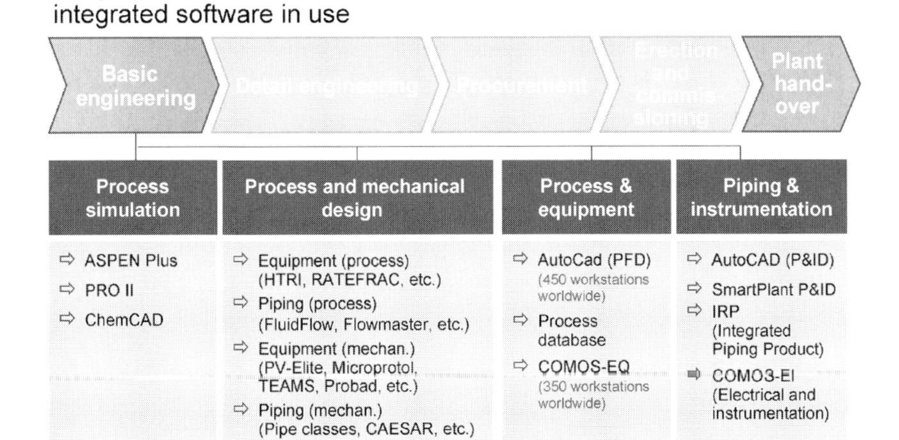

Fig. 7.10 Employed software in the basic engineering, which are decisive for the detail engineering, courtesy of Uhde GmbH, Dortmund, 2005.

The project phases typically contain also milestones for the most important working packages that form the decision criteria for the step into the next project phase. Within the project phase, the time schedule determines the scheduled proceeding of the single working steps.

7.3.4 Project Controlling for R&D Projects

Besides a technically qualified team for the project execution, which provides all necessary competences for the project, the project control forms a crucial base for the success of the project. The project control has the function to compare the project progress concerning contents, quality, time, and costs with the specified targets. Appropriate tools for this task are bar charts, schedules or the network planning technique. Some commercial software tools are available for assisting these controlling tasks. For the project manager, the controlling cycles and check points are very important to guide the project. Table 7.2 gives recommended cycle times depending on the project duration. These times may change during the project. At the beginning and near to the end, the control cycles are shorter, within a project the times should be flexibly handled.

Especially helpful for R&D projects is the detailed elaboration of the project plan for the required control period. The following target/actual value comparison is precise and measurable. The reasons for time delays can instantaneously be registered and appropriate measures can be determined and realized.

Besides the time schedule controlling, cost control plays an important role in projects. This gets more important for R&D projects, because their cost planning is based on estimations. Efficient cost management requires a prompt monitoring of all costs and information provided. Planning and estimation methods allow a prognosis of the costs evolution to the project end. In addition to the cost calculation of planned work packages, unforeseeable tasks should also be included with the help of adequate prognosis methods.

The schedule and costs situation should be prepared for the project management in a way to rapidly identify deviations. Besides the classical tools for project control, colored marking is useful for visualization: green for regular project status, yellow for deviations, and red for critical scheduling delay or cost overshoot. With this, the actual project status can efficiently be summarized from a

Table 7.2 Scheduling the check points in a project.

Duration of a project	Recommended controlling cycle
up to 3 months	weekly
3 to 6 months	every 2 weeks
7 to 60 months	monthly
more than 60 months	2 to 3 months

single step in a work package to a partial target. These summaries are supplemented by measures to hold the schedule and costs in the future.

To conclude, in project management it is considered that the correct documentation of the project progress and the project completion are very important. This includes the correct filing of the correspondence with the project partners. In larger companies this is managed by a quality management system that embraces guidelines for the project planning, for the procedure including the document filing and the completion.

7.3.5
Cost Structure and Development

The cost structure regarding the design of microsystems [32] shows special attributes, which are different from conventional technology. The material costs are low; the costs of the fabrication facilities are high. The mass production of microsystems is cost efficient due to the effective replication techniques (molding, casting, ...) and parallel processing of a multiplicity of structures (lithography, etching, ...). The prototype fabrication of microsystems is very time and cost intensive. Hence the possibility of an efficient microscale rapid prototyping (laser production, ...) is very important. The proper simulation of processes, structures, and systems is very important to reduce the number of prototypes. The design and development bear a high responsibility for the costs. Typically, the design consumes 10 to 20% of the development budget, but is responsible for 70 to 80% of the entire product costs (in process engineering) see Fig. 7.11. The decisions at the project start phase are made on the basis of little information but with high cost relevance.

At the beginning of a project, the degree of freedom is relatively high, but on a sparse knowledge base. At the end of the project, a lot of knowledge was generated, but a system change to correct errors is very expensive. In micro process engineering many projects are governed by research demands, where the engi-

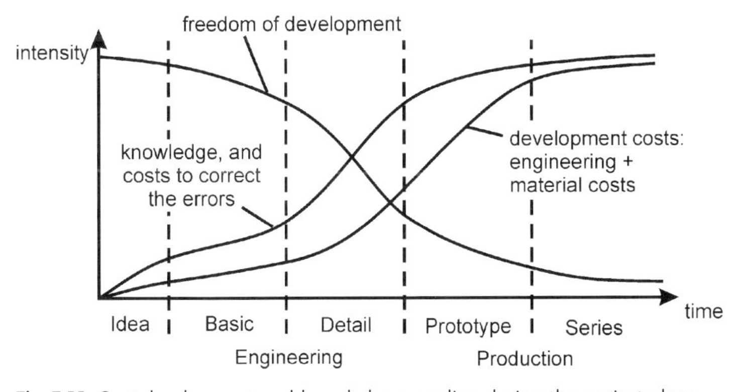

Fig. 7.11 Cost development and knowledge sampling during the project phase.

neering part is dominant. Therefore, a good part of the budget should be invested in the pre- and basic engineering with extensive simulations, design studies, and the evaluation of different variants. This saves money in the cost-intensive fabrication and testing phase.

7.4 Conclusions and Outlook

A well-designed chemical process and appropriate process equipment together with a comprehensive project management are essential for the successful and cost-effective application of not only microstructured equipment. Therefore, the tools and information given in this chapter are collected from various engineering disciplines and are useful for many engineering application. Problem-solving strategies, design rules, and creativity techniques are given together with heuristic rules and our own experience for the design of microstructured devices and their usage in chemical process engineering. This information should serve as guidance, should not delimit but enhance the creative process, which is absolutely necessary for the development of new, complex systems and processes. Sophisticated project management is an essential tool for the successful implementation of technical and chemical plants, here for microstructured equipment with the emphasis on research and development projects.

With the help of microstructured plants it is intended in the near future to yield laboratory data and information on the process and the plant [7], which can directly be transferred to the production plant, see Fig. 7.12. The established

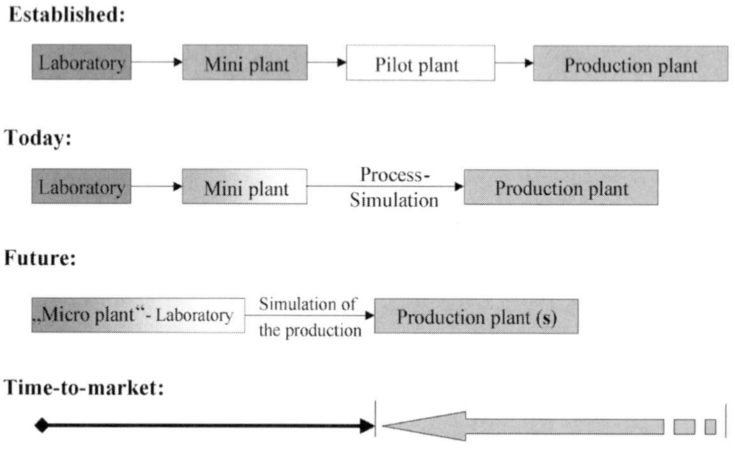

Fig. 7.12 Steps of the process development and future possibilities with microstructured equipment to a shorter time-to-market of processes and products, courtesy of Uhde GmbH, Dortmund, 2005.

scale-up procedure with advanced simulation tools combines miniplant with the pilot plant and abridges one or two development phases.

With the equal-up strategy, the sophisticated exploitation of diverse effects and numbering-up of microstructured elements, even the miniplant investigations can be abridged. The simulation of process parameters and equipment geometry are similar for laboratory micro plants with microstructured equipment and production facilities. This will lead to an easy data transfer and scale-up procedure, to a more cost-efficient process development, and to a shorter time-to-market period. But, the guiding line for the application of microstructured devices in production plants should be as much microstructures as necessary and beneficial to intensify the predominant processes. With a smart integration microstructured devices are playing an essential role in process intensification.

References

1 E. Blass, *Entwicklung verfahrenstechnischer Prozesse – Methoden, Zielsuche, Lösungssuche, Lösungsauswahl*, Springer, Berlin, **1997**.

2 W. D. Seider, J. D. Seader, D. R. Lewin, *Product and Process Design Principles: Synthesis*, Wiley, New York, **2004**.

3 G. Bernecker, *Planung und Bau verfahrenstechnischer Anlagen*, VDI-Verlag, Düsseldorf, **1980**.

4 R. H. Perry, D. W. Green, *Perry's Chemical Engineers' Handbook*, McGraw Hill 7th Int. edn., **1998**.

5 W. R. A. Vauck, H. A. Müller, *Grundoperationen chemischer Verfahrenstechnik*, 11. Aufl., Deutscher Verlag für Grundstoffindustrie, Stuttgart, **2000**.

6 M. Zlokarnik, *Scale-up in Process Engineering*, Wiley-VCH, **2002**.

7 T. Bayer, M. Kinzl, Werkzeug für Verfahrensentwickler. Mikroverfahrenstechnik – Beispiele aus der industriellen Praxis, *Verfahrenstechnik*, **2004**, 38, 7–8.

8 E. Klemm, M. Rudek, G. Markowz, R. Schütte, *Mikroverfahrenstechnik*, in: Winnacker-Küchler: Chemische Technik – Prozesse und Produkte, Vol. 2, Chapter 8, Wiley-VCH, Weinheim, **2004**.

9 F. Becker, J. Albrecht, G. Markowz, R. Schütte, S. Schirrmeister, K. J. Caspary, H. Döring, T. Schwarz, E. Dietzsch, E. Klemm, T. Kruppa, F. Schüth, DEMiS: Results from the development and operation of a pilot-scale microreactor on the basis of laboratory measurements, *IMRET 8*, **2005**, TK131 f.

10 G. Emig, E. Klemm, *Technische Chemie, Einführung in die chemische Reaktionstechnik*, Chapter 16, Springer, Berlin, **2005**.

11 S. Löbbecke, W. Ferstl, S. Panic, T. Türcke, Concepts for modularization and automation of microreaction technology, *Chem. Eng. Tech.*, **2005**, 28, 484–493.

12 L. Henke, H. Winterbauer, Modularer Mikroreaktor zur Nitrierung mit Mischsäure, *Chem. Ing. Tech.* **2004**, 76, 1783–1790.

13 W. D. Seider, J. D. Seader, D. R. Lewin, *Product and Process Design Principles: Synthesis, Analysis and Evaluation*, Wiley, New York, **2004**.

14 D. R. Lewin, W. D. Seider, J. D. Seader, E. Dassau, J. Golbert, D. N. Goldberg, M. J. Fucci, R. B. Nathanson, *Using Process Simulators in Chemical Engineering (CD-ROM)*, Wiley, New York, **2003**.

15 A. Löwe, *Chemische Reaktionstechnik – mit MATLAB und SIMULINK*, Wiley-VCH, Weinheim, **2001**.

16 J. Hagen, *Chemiereaktoren – Auslegung und Simulation*, Wiley-VCH, Weinheim, **2004**.

17 D. M. Himmelblau, J. B. Riggs, *Basic Principles and Calculation in Chemical Engineering*, Prentice Hall, Upper Saddle River, **2004**.

18 U. Claussen, W.G. Rodenacker, *Maschinensystematik und Konstruktionsmethodik – Grundlagen und Entwicklung moderner Methoden*, Springer, Berlin, **1998**.

19 R. Koller, *Konstruktionslehre für den Maschinenbau*, Springer, Berlin, **1985**.

20 J. Schubert, *Physikalische Effekte – Anwendungen, Beschreibungen, Tabellen*, Physik-Verlag, Weinheim, **1982**.

21 J.M. Aughenbaugh, C.J.J. Paredis, *The role and limitations of modeling and simulation in systems design*, ASME meeting, Anaheim, IMECE2004-59813, **2004**.

22 W.H. Vandenburg, *The Labyrinth of Technology*, University Press of Toronto, **2000**.

23 nbm, Creativity techniques for engineers, URL: *www.eweek.org/site/nbm/*, **2004**.

24 T. Buzan, *Mind maps at work*, HarperCollins, **2004**.

25 A. Otto, Kreativität, *Psychologie heute* 32, 3, p. 8, **2005**.

26 C.L. Dym, P. Little, *Engineering Design – A project-based introduction*, John Wiley, New York, **1999**.

27 M.A. Orloff, Inventive Thinking through TRIZ – A Practical Guide, Springer, Berlin, **2003**.

28 G. Retseptor, 40 Inventive Principles in Microelectronics, TRIZ Journal 8 **2002**, www.triz-journal.com/archives/2002/08/b/index.htm.

29 J. Hipple, 40 Inventive Principles with Examples for Chemical Engineering, TRIZ Journal 6 **2005**, www.triz-journal.-com/archives/2005/06/06.pdf.

30 N. Kockmann, T. Kiefer, M. Engler, P. Woias, Convective Mixing and Chemical Reactions in Micro Channels, *SensAct B*, acc. paper, **2005**.

31 G. Gerlach, W. Dötzel, *Grundlagen der Mikrosystemtechnik*, Hanser, München, **1997**.

32 M. Kaspar, *Mikrosystementwurf, Entwurf und Simulation von Mikrosystemen*, Springer, Berlin, **2000**.

33 S.D. Senturia, *Microsystem Design*, Kluwer, Boston, **2000**.

34 G.K. Ananthasuresh (Ed.), *Optimal Synthesis Methods for MEMS*, Kluwer, Boston, **2003**.

35 G. Pahl, W. Beitz, *Konstruktionslehre – Methoden und Anwendungen*, Springer, Berlin, **2004**.

36 *VDI Wärmeatlas, Berechnungsblätter für den Wärmeübergang*, VDI-Verlag, **2000**.

37 J.J. Brandner, *Entwicklung von Mikrostrukturreaktoren zum thermisch instationären Betrieb chemischer Reaktionen*. Forschungszentrum Karlsruhe, Wissenschaftliche Berichte FZKA 6891, Diss. Universität Karlsruhe (TH), **2003**.

38 J.J. Brandner, G. Emig, M.A. Liauw, K. Schubert, Fast temperature cycling in microstructure devices, *Chem. Eng. J.* **2004**, 101, 217–224.

39 K. Schubert, J.J. Brandner, M. Fichtner, G. Linder, U. Schygulla, A. Wenka, Microstructure devices for applications in thermal and chemical process engineering, *Microscale Thermophys. Eng.*, **2001**, 5, 17–39.

40 J.M. Commenge, L. Falk, P. Corriou, M. Matlosz, Optimal Design for Flow Uniformity in Microchannel Reactors, *IMRET 4*, **2000**, 23–30.

41 S. Walter, G. Frischmann, R. Broucek, M. Bergfeld, M.A. Liauw, Fluiddynamische Aspekte in Mikrostrukturreaktoren, *Chem. Ing. Tech.*, **1999**, 71, 447–455.

42 T. Henning, J.J. Brandner, K. Schubert, Characterisation of electrically powered micro heat exchangers. *Chem. Eng. J.* **2004**, 101, 339–345.

43 T. Henning, J.J. Brandner, K. Schubert, High speed imaging of flow in microchannel array water evaporators, *Microfluidics Nanofluidics*, **2005**, 1, 128–136.

44 V. Hessel, H. Löwe, Mikroverfahrenstechnik: Komponenten – Anlagenkonzeptionen – Anwenderakzeptanz, *Chemie Ingenieur Technik*, **2002**, 74, 17–30, 185–207, 381–400.

45 N. Kockmann, M. Engler, T. Kiefer, P. Woias, Silicon Microstructures for High Throughput Mixing Devices, *Proc. of 3rd Int. Conf. on Micro and Minichannels ICMM2005*, June 13–15, **2005**, Toronto, Canada.

46 V. Hessel, S. Hardt, H. Löwe, *Chemical Micro Process Engineering*, Wiley-VCH, Weinheim, **2004**.

47 T. Stief, O. U. Langer, K. Schubert, Numerical investigations on optimal heat conductivity in micro heat exchangers, *IMRET 4*, **2000**, 314–321.

48 M. Madou, *Fundamentals of Microfabrication*, CRC Press, London, **1997**.

49 A. Müller, V. Cominos, V. Hessel, B. Horn, J. Schürer, A. Ziogas, K. Jähnisch, V. Hillmann, V. Großer, K. A. Jam, A. Bazzanella, G. Rinke, M. Kraut, Fluidic bus system for chemical process engineering in the laboratory and for small-scale production, *Chem. Eng. J.* **2005**, 107, 205–214.

50 N. Kockmann, M. Engler, P. Woias, Fouling processes in microstructured devices, Proc. of ECI Conference on Heat Exchanger Fouling and Cleaning – Challenges and Opportunities, Kloster Irsee, Germany, June 5–10, **2005**.

51 M. Fichtner, W. Benzinger, K. Hass-Santo, R. Wunsch, K. Schubert, Functional coatings for microstructure reactors and heat exchangers, *IMRET 3*, **2000**, 90–101.

52 W. Benzinger, M. Jäger, U. Schygulla, K. Schubert, Anti Fouling Investigations with Ultrasound in a Microstructured Heat Exchanger, Proc. of ECI Conference on Heat Exchanger Fouling and Cleaning – Challenges and Opportunities, Kloster Irsee, Germany, June 5–10, **2005**.

53 S. Seibert, *Technisches Management*, Teubner, Kap. 10, Stuttgart, **1998**.

54 P. Rinza, *Projektmanagement, Planung, Überwachung und Steuerung von technischen und nichttechnischen Vorhaben*, Springer, Berlin, **1998**.

55 M. Burghardt, *Projektmanagement*, Publics-MCD-Verl., Berlin, **1997**.

56 Projektkooperation beim Internationalen Vertrieb von Maschinen und Anlagen. VDI Gesellschaft Entwicklung Konstruktion Vertrieb.

57 Auftragsabwicklung im Maschinen- und Anlagenbau. VDI Gesellschaft Entwicklung Konstruktion Vertrieb.

8
Simulation and Analytical Modeling for Microreactor Design

Osamu Tonomura, Department of Chemical Engineering, Kyoto University, Katsura Campus, Nishikyou-ku, Kyoto, Japan

Abstract
Microchemical process technologies offer much improved controllability of reaction conditions such as residence time, temperature and concentration distributions. Since the performance of a microreactor depends largely on the shape and size of microchannels and the longitudinal heat conduction inside walls, the design problem of microreactors is regarded as that of distributed-parameter systems. In this chapter, the several design methods based on computational fluid dynamics (CFD) and/or simplified thermofluid models are described to optimally design microreactors under the various constraints.

Keywords
Optimal design, shape design, computational fluid dynamics, compartment modeling

Advanced Micro and Nanosystems Vol. 5. Micro Process Engineering. Edited by N. Kockmann

ISBN: 3-527-31246-3

8.1 Introduction

Microchemical plants potentially enable us to handle materials that cannot be produced in conventional chemical plants and to improve the present production efficiency drastically. Extensive studies have been conducted on microchemical process technologies (Abrahamse et al. [1]; Ehrfeld et al. [2, 3]; Fletcher et al. [4]; Hessel et al. [5]; Jensen [6]; Stone and Kim [7]; Zech and Hönicke [8]). In particular, chemical reactions, mixing, separation, and transport phenomena in microspace have been energetically analyzed, and many types of microdevices have been proposed for each microunit operation. However, few microchemical plants have been used for real production, and engineers have less experience in designing microchemical plants. In fact, most microdevices are designed and fabricated by trial and error – there is no systematic procedure for developing microdevices. Unless this situation is improved, the design period cannot be shortened. To achieve further breakthrough in microchemical plants' technologies, proper modeling and simulation methods are indispensable. Much research has already been carried out in the field of design optimization and numerous methods. In the field of MEMS (microelectromechanical systems), many efficient simulators have been developed to predict device performance in the last few years. A lumped-parameter system is usually used in MEMS simulators to describe flow behavior, because a model should be simple in order to enhance design efficiency. But, when microchemical plants are designed, microdevices need to be treated as distributed-parameter systems in order to analyze rigorously the characteristics of flow and heat transfer in them. Thus, MEMS simulators are inadequate for designing microchemical plants. In this chapter, the several design methods based on computational fluid dynamics (CFD) and/or simplified thermofluid models are described to optimally design microreactors under the various constraints.

8.2 Design Using CFD Simulations

CFD is concerned with obtaining numerical solution to fluid-flow problems by using high-speed and large-memory computers. The differential equations governing the fluid flow are transformed into a set of algebraic equations, which can be solved with the aid of a digital computer. The well-known discretization methods used in CFD are finite difference method (FDM), finite volume method (FVM), finite element method (FEM), and boundary element method (BEM). CFD not only predicts fluid flow behavior, but also the transfer of heat and mass, phase change, and chemical reaction. CFD analysis often shows you the phenomena happening within a system or device that would not otherwise be visible through any other means. In design and development, CFD is now considered to be standard numerical tool, widely used within industry, and it is an essential tool for shortening the design and development cycles.

Over the last few years, a variety of microdevices such as T- and Y-type flow mixers, multilaminating mixers, and split-and-recombine mixers have been reported. These studies on mixing were motivated by the evolving interest in the miniaturization of conventional macroscale processes into microscale ones. The cross-sectional dimensions of miniaturized systems are of the order of micrometers to millimeters. The flow in a microchannel is usually laminar and thus mixing is dominated by molecular diffusion. To shorten diffusion or mixing time, the interfacial area between the fluids should be increased and the diffusion length should be decreased. CFD is a powerful tool for investigating the effect of design parameters of microdevices on the mixing performance. Aubin et al. [9] investigated the effect of various geometrical parameters of a grooved staggered herringbone micromixer on the mixing performance by using CFD. The grooved staggered herringbone micromixer is shown in Fig. 8.1 and creates a transverse velocity component in the flow field. The direction of asymmetry of the herringbones switches with respect to the centerline of the channel every half cycle. The mixer consists of a rectangular channel (w=200 μm, h=77 μm, L=0.01 m) with grooves of depth, d_g, and width, W_g. Table 8.1 presents the different cases investigated and the values of the geometrical parameters. The simulation results show that the deep grooves improve the mixing quality. The number of grooves per mixing cycle does not affect the mixing quality. If the groove width is very small, the flow in the microchannel is hardly affected. For wide grooves, the mixing quality is not necessarily improved, and dead zones in

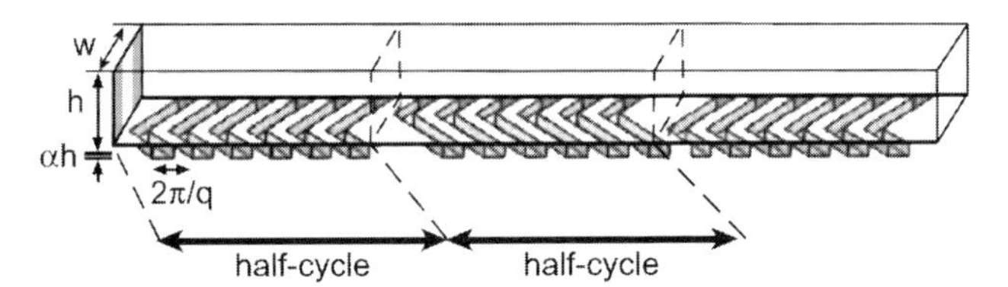

Fig. 8.1 The staggered herringbone micromixer (Stroock et al. [10]).

Table 8.1 Geometrical parameters of the different micromixers (Aubin et al. [9]).

Case	Groove depth, d_g (h)	Groove width, W_g (μm)	Number of grooves per cycle
Reference (Srock et al., 2002 a)	0.23	50	20
1 (d_g=0.30 h)	0.30	50	20
2 (d_g=0.35 h)	0.35	50	20
3 (W_g=25 μm)	0.23	25	20
4 (W_g=75 μm)	0.23	75	20
5 (N_g=10)	0.23	50	10
6 (N_g=30)	0.23	50	30

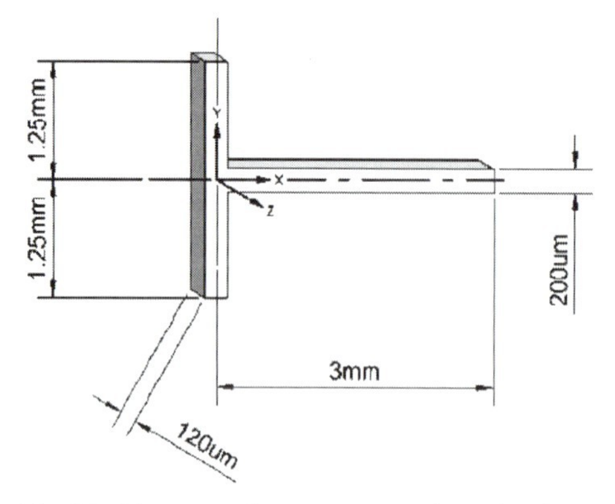

Fig. 8.2 Schematic of a T-type microchannel (Glasgow et al. [11]).

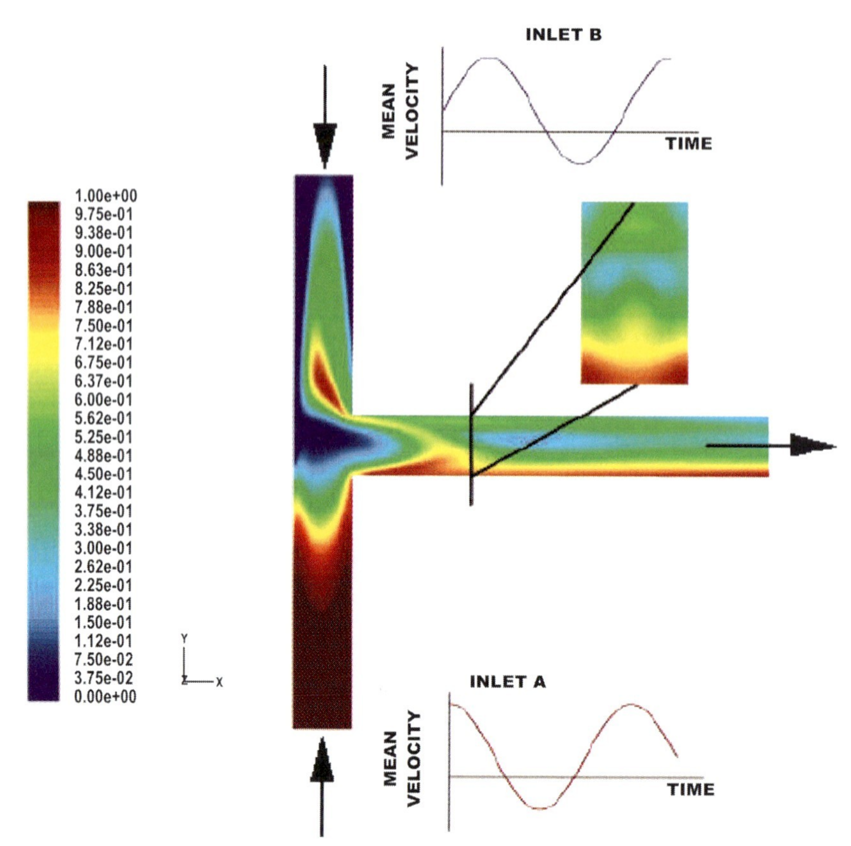

Fig. 8.3 Contour plots of concentration of the liquid. The flow rate at inlet A leads to flow rate at inlet B by ¼ cycle (Glasgow et al. [11]).

the microchannel are created. In order to improve the mixing quality, one is advised to use a mixer with deep grooves, having a depth of 30–40% of the channel height.

The mixing performance is also improved by introducing pulsing. Glasgow et al. [11] investigated the pulsed flow mixing in a microchannel with the simplest confluence geometry by using CFD simulation. A schematic drawing of it is shown in Fig. 8.2. The three channel branches are each 200 μm wide by 120 μm deep, a size range appropriate for many applications. CFD simulations are executed to examine the effect of varying several nondimensional parameters on the mixing performance. These parameters include the following: the Reynolds, Strouhal, Stokes, Schmidt, and Péclet numbers, the ratio of the pulse volume to the intersection volume (PVR), and the pulse shape. The Strouhal number means the ratio of flow characteristic time scale to the pulsing time period. Figure 8.3 illustrates the contour plots of concentration of the liquid at half the depth of the channel in a case study. Glasgow et al. [11] conclude that the higher Strouhal number and PVR lead to better mixing. However, the contact time of the two fluids and the shape of the pulse's waveform do not have a strong influence on the degree of mixing.

8.3 Design Using Simplified Models

In a conventional design problem, unit operations are modeled as lumped-parameter systems under conditions such as perfect mixing, plug flow, and overall coefficient of heat transfer. Recent advances in CFD enable us to know both flow and temperature distributions in a microdevice precisely without conducting any experiments. However, it is not practical to apply CFD simulations directly to the optimal design problem of microreactors, since they require too much computational time and effort. From this viewpoint, a new systematic design approach based on a simplified thermofluid model is introduced in this section.

8.3.1 Necessity for Shape Design of Microdevices

Microreactors enhance the heat and mass transfer and realize the precise control of residence time. These characteristics enable us to handle highly exothermic/endothermic and rapid chemical reactions. The use of microreactors for industrial-scale production requires a large number of parallel microchannels, because each microchannel provides only a small number of products. At first glance, the repetition of a microchannel may appear to be simple, but there are issues that have not been addressed in the design and operation of conventional chemical plants. The inadequate shape design gives rise to poor uniformity in the temperature and residence-time distributions among parallel microchannels,

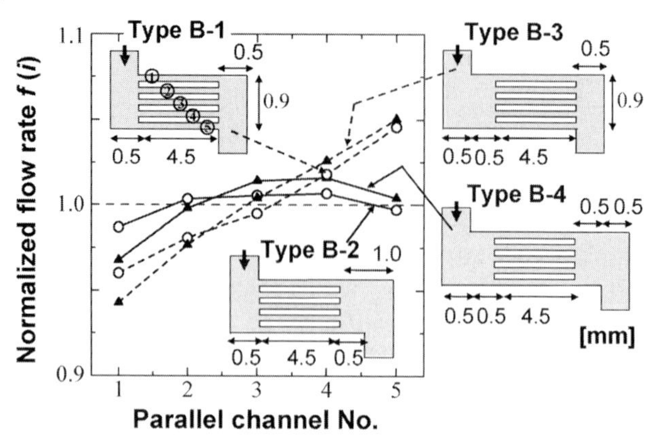

Fig. 8.4 Effect of device shape on flow distribution.

which will make product quality worse. To show how the device shape affects the flow pattern in the device, three-dimensional flow simulations are executed by using commercial CFD software (Fluent®). Figure 8.4 shows the normalized mass flow rate at each parallel channel (Tonomura et al. [12]). From the viewpoint of flow uniformity, Type B-2 is better than Type B-1. The sizes of inlet manifold in Types B-3 and B-4 are larger than those in Types B-1 and B-2. The flow uniformity in parallel channels is not markedly affected by making the inlet manifold larger. These results fully demonstrate the importance of shape design for microdevices. Ehrfeld et al. [3] pointed out the importance of shape design through a similar result.

8.3.2
Design Problem of Plate-fin Microreactors

In this section, the optimal design problem of a plate-fin microreactor is investigated. It is typically composed of three sections: inlet manifold section for flow distribution, parallel microchannels' section for reaction, and outlet manifold section for mixing. An optimal design problem of the plate-fin microreactor is formulated as shown in Table 8.2. It is not practical to apply CFD simulation directly to the optimal design problem due to the large number of degrees of freedom. In the following sections, a new systematic design approach is explained.

8.3.3
Thermofluid Design Approach

The proposed thermofluid design approach for a plate-fin microreactor mainly consists of two stages (Tonomura et al. [13]; Noda et al. [14]). The flowchart is shown in Fig. 8.5. In the first stage, the optimal microchannel shape is derived by using a thermal-compartment model: the "thermal design stage." In the second stage, the optimal manifold shape is determined by using a pressure-drop

Table 8.2 Optimal design problem of a plate-fin microreactor.

Objective function	Maximization of product yield
	Maximization of production rate
Optimization variables	Microchannel shape and size
	Manifold shape and size
	Number of microchannels
	Feed flow rate and Inlet temperature
	Coolant temperature
Constraints	Mass balance for a given species
	Energy balance for fluid and wall
	Reaction temperature conditions
	Minimum product yield
	Total flow rate
	Residence-time distribution among microchannels
	Maximum pressure drop (pump performance)
	Upper and lower limits of reactor dimensions

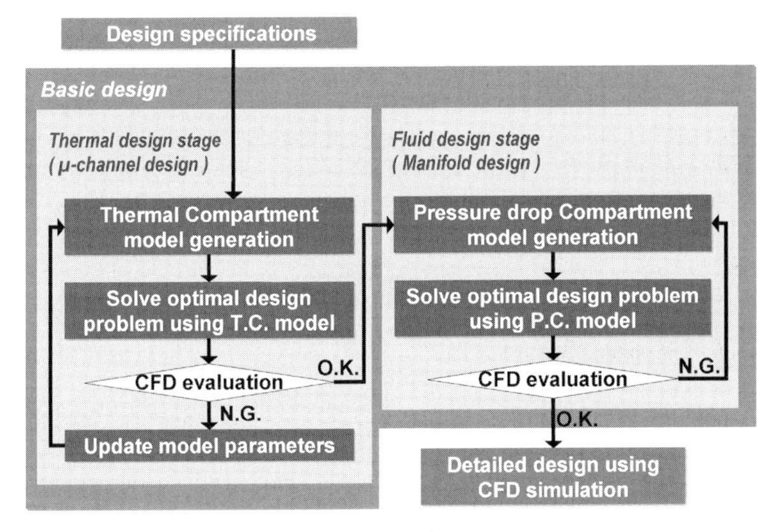

Fig. 8.5 Flowchart of proposed design approach.

compartment model: the "fluid design stage." After these two design stages, the obtained shape of the microreactor is validated by CFD simulation. The proposed approach will accelerate the microreactor design.

8.3.3.1 Thermal Design Stage

In the thermal design stage, the parallel channels' section in a microreactor are divided into an adequate number of thermal compartments to describe heat transfer between fluids and walls. Each compartment is regarded as a lumped-

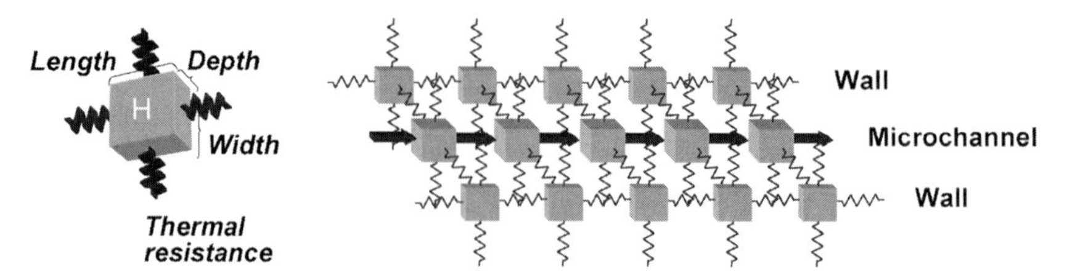

Fig. 8.6 Thermal compartment (left) and thermal-compartment model (right).

parameter system and has information on its shape and size, physical properties, and heat-transfer coefficients between connected compartments. Figure 8.6 represents a thermal-compartment model for one microchannel and the walls that surround it.

In a thermal-compartment model, the longitudinal heat conduction inside the walls is considered, which is neglected in the conventional models. A microchannel and its walls can be representative of the parallel microchannels' section on the assumption that flow distribution is uniform. Energy balances of a microchannel compartment i and a wall compartment i are given by Eqs. (8.1) and (8.2):

$$C_{\mathrm{p}}(F^{i}_{\mathrm{in}}T^{i}_{\mathrm{in}} - F^{i}_{\mathrm{out}}T^{i}_{\mathrm{out}}) - Q^{ii} + \Delta Hr^{i} = 0 \tag{8.1}$$

$$k_{\mathrm{w}}A^{i}_{\mathrm{w}}(T^{i-1}_{\mathrm{w}} + T^{i+1}_{\mathrm{w}} - 2T^{i}_{\mathrm{w}})/\Delta x + Q^{ii} - Q^{ii}_{\mathrm{w}} = 0 \tag{8.2}$$

where F^i, T^i, C_{p} and ΔHr^i are flow rate, temperature, specific heat and reaction heat generated in the microchannel compartment i. Q^{ii} is heat-transfer rate from the microchannel compartment i to the wall compartment i. T^i_{w}, A^i_{w}, k_{w} and Δx are temperature, cross-sectional area, thermal conductivity and compartment length in the wall compartment i. Q^{ii}_{w} is heat-transfer rate from the wall compartment i to the surrounding. The total thermal balance of the microreactor is solved under given boundary conditions.

The microchannel width is optimized as a function of the longitudinal position by taking into account the constraints related to thermal balance, pressure drop, reaction yield, temperature, and reactor dimensions. By changing the channel width, that is, by controlling residence time of fluid at each longitudinal position, a uniform temperature distribution with no hot spot is realized even for a rapid exothermic reaction. The thermal design result is evaluated by CFD simulation. If the derived shape of the microchannel does not satisfy the given design specifications, model parameters, such as heat-transfer coefficients, are adjusted to the value obtained by CFD simulation, and the thermal design is repeated.

8.3.3.2 Fluid Design Stage

The detailed fluid dynamics of a microreactor is expressed by Navier–Stokes equations. However, it is difficult to solve those equations under the various boundary conditions. The pressure distribution usually determines the flow pattern at low Reynolds number. Therefore, the fluid-design problem is regarded as a problem to find an appropriate pressure distribution over a microreactor.

In the fluid design stage, a pressure-drop-compartment model is used to estimate the pressure distribution over the microreactor including the inlet and outlet manifolds. Four types of pressure-drop compartments with three design parameters (length, depth, and width) are introduced to describe the fluid dynamics in the microreactor: 1) a distributor dividing one stream into two streams, 2) a junction joining two streams into one, 3) a microchannel, and 4) a reaction-microchannel. A reaction-microchannel differs from a microchannel in having reaction kinetic models. In this investigation, each compartment is assumed to be a rectangular duct. For laminar flow in channels of rectangular cross section, the velocity profile can be determined analytically. The pressure drop (ΔP) is calculated by Eqs. (8.3) and (8.4):

$$\Delta P = \lambda \left(\frac{Z}{4d_\mathrm{h}} \right) \left(\frac{\rho \bar{u}^2}{2} \right) \tag{8.3}$$

$$\lambda = \frac{64}{\mathrm{Re}} \cdot \frac{3/2}{(1+\varepsilon)^2 \left[1 - \frac{48}{\pi^5} \varepsilon \sum_{k=1,3,5,\ldots}^{\infty} \frac{4}{k^5} \tanh\left(\frac{k\pi}{2\varepsilon} \right) \right]} \tag{8.4}$$

Here, λ is the friction factor, Z the duct length, m the hydraulic diameter, ρ the density of fluid, $\bar{u}$ the average flow velocity, ε the aspect ratio of cross section, and Re the Reynolds number.

The pressure distribution over a microreactor is modeled by combining compartments, which have design parameters. Figure 8.7 shows an example of a combination of pressure-drop compartments. While being distributed into each reaction-microchannel, the input stream passes through distributor compartments. The partitioned fluid is put together and becomes the output stream, while passing through junction compartments. Pressure-balance equations are formulated for the combination of pressure-drop compartments. Equation (8.5) represents the pressure balance among four compartments within the light gray region in Fig. 8.7,

$$\Delta P_\mathrm{C}^{n-2} + \Delta P_\mathrm{O}^{3} = \Delta P_\mathrm{I}^{n-1} + \Delta P_\mathrm{C}^{n-1} \tag{8.5}$$

By repeating the derivation of pressure balance among other sets of compartments, algebraic equations that determine the relation between design parameters and pressure distribution are obtained.

Through the fluid design, the shapes of the inlet and outlet manifolds and the number of parallel microchannels are determined so as to optimize a given

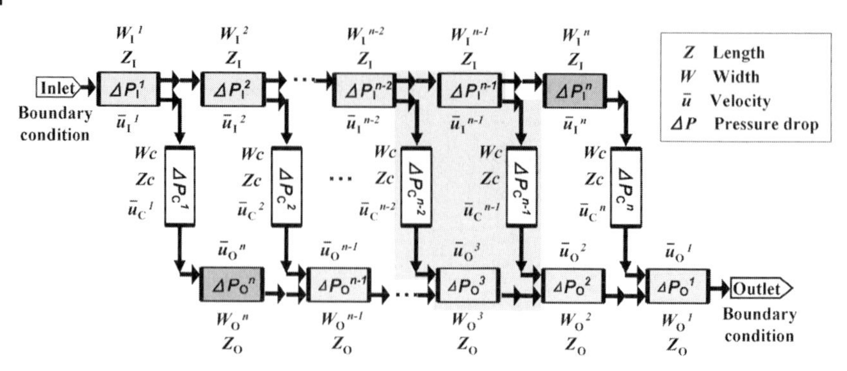

Fig. 8.7 Pressure-drop compartment model.

performance index such as the minimization of the total volume in the microreactor. Many constraints related to the pressure and mass balances, flow uniformity, pressure drop, total flow rate and reactor dimensions are considered in this design stage. The fluid-design result is evaluated by CFD simulation. The compartment model is regenerated, and fluid design is repeated until the derived shape of manifold does not satisfy the design specifications.

8.3.3.3 **Detailed Design Stage**

In the above-mentioned approach, the thermofluid design of the plate-fin microreactor is divided into two successive optimization problems. Therefore, the final shape of the microreactor, which is obtained by combining the thermal design and fluid-design results, needs to be validated rigorously by CFD simulations.

8.3.4
Case Study: Design Conditions and Formulation

The thermofluid design approach is applied to a microreactor design problem. The objective of the design is to realize uniform temperature and residence-time distributions. An exothermic first-order reaction $A \rightarrow B$ is taken up in this case study.

In the thermal-design problem, it is assumed that a perfect flow distribution among parallel microchannels can be achieved, which means that all microchannels have the same temperature profiles axially. Therefore, symmetric boundary conditions are introduced into the thermal-design problem of the plate-fin microreactor. These conditions can reduce the design region from the whole channel part to the one microchannel with a surrounding wall. In the case study, the basic design region is defined as a rectangular glass. To calculate the temperature distribution, the basic design region is divided into many thermal compartments representing the microchannel and the wall. The objective function of the thermal-design problem is minimizing the integrated temperature error along the microchannel. The integrated temperature error IE is calculated by Eq. (8.6) where x means the axial position in the microchannel and T_{set} is a desired reaction temperature.

$$\mathrm{IE} = \int (T(x) - T_{set})\mathrm{d}x \tag{8.6}$$

For a conventional macroreactor, it is not common to optimize tube diameter as a function of the tube position. However, a microchannel with varying width is not so difficult to fabricate by the techniques developed in the field of microelectromechanical systems (MEMS).

In a fluid-design problem, the microreactor including the inlet and outlet manifolds is compartmentalized as shown in Fig. 8.7. In this case study, it is assumed that the pressure-compartment configuration has point symmetry and that the manifold shape is a trapezium. In addition, the first distributor compartment width (i.e. reactor inlet width), channel depth, and outlet pressure are fixed at constant value. The nth distributor compartment width and the number of microchannels are optimized under the constraints related to the pressure and mass balances, the flow uniformity, pressure drop, and reactor dimensions.

8.3.5 Case Study: Design Results and Discussion

Figure 8.8 illustrates the thermal design result. The solid line in the upper graph shows the optimal channel width along the microchannel, and the dashed line shows the channel width, which is optimized as a constant value. The optimal channel width increases gradually along the channel except for the inlet zone. The inlet width is slightly wider to preheat the feed stream. The width of the middle part becomes narrower due to a faster reaction rate. The width of the last part

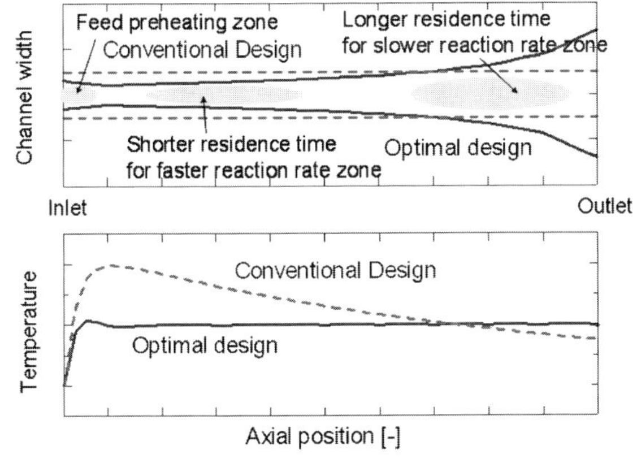

Fig. 8.8 Optimal channel width (upper) and temperature profiles (lower).

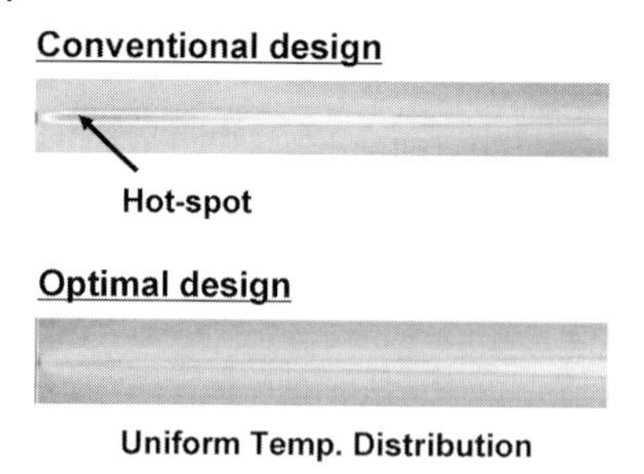

Fig. 8.9 Temperature contour plots by CFD.

becomes greater due to a slower reaction rate. The bottom figure illustrates that the optimal microchannel shape is effective to keep the fluid temperature distribution uniform along the microchannel. On the other hand, the conventional straight channel has a hot spot near the inlet even if it is optimized.

To assess the validity of the thermal-compartment model, the design results obtained by using the thermal-compartment model are compared with CFD simulation results. In this work, Fluent® code is used to calculate three-dimensional temperature distribution in the microchannel. The finite-volume method is used in Fluent® to solve the conservation equations for mass, momentum, and energy. The equations are solved by using the SIMPLE algorithm. Figure 8.9 shows the contour plots of the temperature distribution in the microchannel including walls, which are obtained through CFD simulation. In the conventional channel, there is a red area near the inlet, which means the existence of a hot spot. On the other hand, the temperature distribution in the optimized channel is more uniform than that in the conventional channel.

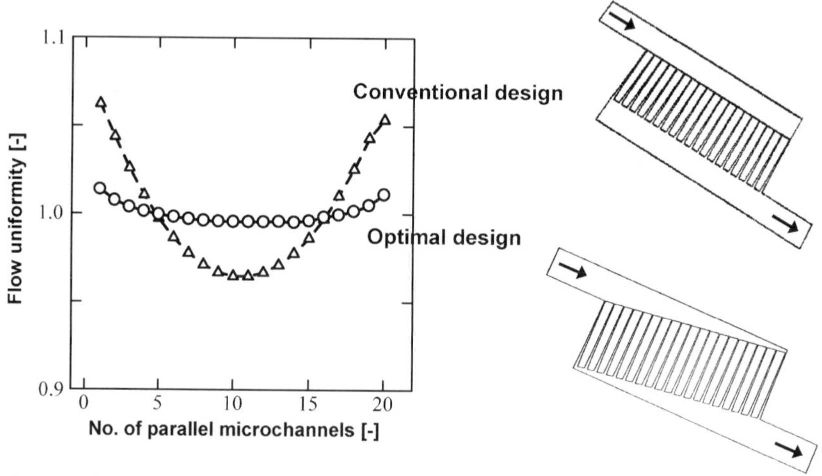

Fig. 8.10 Fluid-design results.

Finally, the fluid-design result is discussed. Figure 8.10 shows a schematic drawing of the final shape of a plate-fin microreactor and the normalized flow distribution among parallel microchannels. The circle points correspond to the optimal design, and the triangle points represent the conventional design. The optimal manifold shape can realize flow equipartition in all parallel channels and avoid deterioration in the reactor performance.

8.4 Summary

In this chapter, the simulation and analytical modeling for microreactor design are described to qualify and improve the design and reduce time to market. First, the CFD-based analysis and design of the grooved staggered herringbone micromixer and T-shape microchannel with pulsed flow mixing are described. CFD is the effective method to calculate the complex flow in the mixer. However, it requires considerable computational time and high skills in generating models and meshes. Therefore, the thermofluid design approach based on the simplified models is described in the latter half of this chapter. This approach is applied to the optimal design problem of a microreactor with uniform temperature and residence-time distributions. In most of the conventional design approaches, the cross-sectional area of each microchannel is assumed to be constant, and the channel design under that condition sometimes causes an unnecessary increase of pressure drop. In the design strategy shown in Section 8.3, the fluid temperature along the microchannel is equalized by changing channel width, that is, the fluid residence time is controlled. In addition, the optimally designed manifold shape ensures the same residence time in all parallel microchannels and avoids deterioration in the reactor performance. The thermofluid compartment model is a very simple but powerful tool to shorten the computational time of microreactor design.

References

1 Abrahamse, A. J., van der Padt, A., Boom, R. M., and de Heij, W. B. C. (2001) Process fundamental of membrane emulsification: simulation with CFD. AIChE J. 47 (6), 1285–1291.

2 Ehrfeld, W., Hessel, V., and Löwe, H. (eds) (2000 a) Microreactors: New technology for modern chemistry. Wiley-VCH, Weinheim.

3 Ehrfeld, W., Hessel, V., and Löwe, H. (2000 b) Extending the knowledge base in microfabrication towards chemical engineering and fluid dynamic simulation. In: W. Ehrfeld, U. Eul, and R. S. Wegeng (eds) 4th International Conference on Microreaction Technology (IMRET), 5–9 March 2000, Atlanta, 3–20.

4 Fletcher, P. D. I., Haswell, S. J., Pombo-Villar, E., Warrington, B. H., Watts, P., Wong, S. Y. F., and Zhang, X. (2002) Micro reactors: principles and applications in organic synthesis. Tetrahedron 58, 4735–4757.

5 Hessel, V., Hardt, S., and Löwe, H. (eds) (2003) Chemical micro process engineering. Wiley-VCH.

6 Jensen, K.F. (2001) Microreaction engineering – Is small better? Chem. Eng. Sci. 56, 293–303.

7 Stone, H.A. and Kim, S. (2001) Microfluidics: Basic issues, applications, and challenges. AIChE J. 47 (6), 1250–1254.

8 Zech, T. and Hönicke, D. (1998) Microreaction technology: Potentials and technical feasibility. Erdöl Erdgas Kohle 114, 578–581.

9 Aubin, J., Fletcher D.F., Wuereb C. (2005) Design of micromixers using CFD modeling. Cheme. Eng. Sci. 60, 2503–2516.

10 Stroock, A.D., Dertinger, S.K.W., Ajdari, A., Mezic, I., Stone, H.A., Whitesides, G.M. (2002) Chaotic mixer for microchannels. Science 295, 647–651.

11 Glasgow, I., Lieber, S., Aubry, N. (2004) Parameters influencing pulsed flow mixing in microchannles. Anal. Chem. 76, 4825–4832.

12 Tonomura, O., Shotaro, T., Noda, M., Kano, M., Hasebe, S., and Hashimoto, I. (2004a) CFD-based optimal design of manifold in plate-fin microdevices. Chem. Eng. J. 101, 397–402.

13 Tonomura, O., M. Noda, M. Kano, and S. Hasebe (2004b) Optimal design approach for microreactors with uniform residence-time distribution. In: the 10th Asian Pacific Confederation of Chemical Engineering (APCChE), 4B-07.

14 Noda, M., O.Tonomura, M. Kano, and S. Hasebe (2004) Systematic approach for thermal-fluid design of microreactors. In: the 10th Asian Pacific Confederation of Chemical Engineering (APCChE), 4B-08.

9
Integration of Sensors and Process Analytical Techniques

Stefan Löbbecke, Fraunhofer Institut für Chemische Technologie ICT, Pfinztal, Germany

Abstract
Microreaction technology is nowadays widely used as a lab tool for testing, screening and development of chemical processes. Moreover, it is becoming increasingly accepted as a suitable technology for industrial production purposes. The latter requires functional concepts for process analytical techniques that allow a precise control and regulation of microchemical production processes. Precise analytical data are also required for mechanistic, kinetic and calorimetric studies. Therefore, suitable sensors and analytical tools have to be integrated into continuous microreaction processes without disturbing and degrading the performance of the process analytical techniques or even that of the entire microprocess. Selected examples are briefly discussed in this chapter that show promising developments for the analytical and sensoric monitoring of microreaction processes.

Keywords
On spectroscopy, microstructured sensors, sampling, calorimetry, process monitoring

Advanced Micro and Nanosystems Vol. 5. Micro Process Engineering. Edited by N. Kockmann

ISBN: 3-527-31246-3

9.1 Introduction

Microreaction technology (MRT) is today one of the most exciting innovations in chemical and pharmaceutical synthesis, chemical processing and process technology. It is at the threshold of widespread application in industry and research.

Worldwide research over the last few years has clearly demonstrated the impressive application potential of MRT – in particular because of the intensification of heat and mass transport seen in microfluidic devices. Users in chemistry and chemical engineering find the most convincing arguments in the concrete technical advantages of microreactors: process intensification in terms of higher product yield, selectivity and purity, improved safety and access to new products and processes – to name just a few. In the meantime, a remarkable number of successful applications of MRT have been described in the literature. Recently published reviews [1–4] and a very comprehensive collection of MRT applications [5] are recommended to interested readers.

Nowadays, microreaction technology has successfully lost its mark of a mere academic plaything and is now broadly accepted as a tool for process screening and optimization in the R&D labs of chemical companies and research institutions. An indication of a considerable progress in this direction is the appearance of new companies that provide products for MRT applications such as microfluidic devices made of different materials and even complex microreaction systems, for example based on toolkit concepts [6].

However, despite these promising trends the breakthrough to wide application of microreaction technology in *production processes* has not yet occurred. At least in the literature only a few examples can be found that report on chemical production processes on the basis of microfluidic components, for example [7–9]. To establish microreaction technology in industrial production processes a huge number of technical, economic and regulative conditions have to be fulfilled. Technical requirements, for example, are:

- to accomplish sufficient throughput and production capacity accordingly (for example, by "numbering-up" or "equaling-up" strategies, see Chapter 1)
- to ensure long-term stability, robustness and safety of microreaction production processes
- and to enable an active regulation and control of the microreaction process by suitable process measuring and control techniques (process control engineering).

The latter is essentially dependent on the acquisition and measurement of relevant process data. Hence it follows that in the context of growing industrial applications of microreaction processes suitable sensoric and analytical techniques are coercively required that can be integrated into continuously operated microfluidic processes. Surprisingly, for a long time no great importance has been ascribed to adequate sensors and analytical interfaces for the monitoring of microchemical processes.

In the following some general aspects concerning the integration of process analytical techniques into microchemical processes will be discussed and exemplary applications will be given.

9.2
Process Analytical Techniques for Microchemical Processes: Characteristics and Requirements

As described above, the interest in process analytical techniques (PAT) for the monitoring and control of microchemical processes has significantly increased as production purposes moved more and more into the center of industrial needs. Moreover, process analytical techniques are currently very intensively discussed to help the industry to move away from empirical, and towards science-based standards for control of chemical and pharmaceutical production processes. The PAT initiative of the US Food and Drug Administration (FDA) illustrates the increasing regulative, political and eventually economic impact on the chemical and pharmaceutical industry [10]. The FDA initiative is intended to foster improvements in manufacturing efficiency and product quality by encouraging the application of advanced analytical technologies.

It is important to note in this context that the term "analytical" in PAT is viewed broadly to include chemical, physical, microbiological, mathematical, and risk analysis conducted in an integrated manner. Therefore, PAT is defined as a system for designing, analyzing, and controlling manufacturing through timely measurements (i.e. during processing) of critical quality and performance attributes of raw and in-process materials and processes with the goal of ensuring final-product quality [11]. There are many current and new tools available in the PAT framework that can be categorized as:

- modern process analyzers or process analytical chemistry tools
- process and endpoint monitoring and control tools
- multivariate data acquisition and analysis tools
- continuous improvement and knowledge-management tools.

An appropriate combination of some, or all, of these tools may be applicable to a single unit operation, or to an entire production process and its quality assurance.

At the beginning of microreaction technology most R&D activities were focused on the development of suitable microfluidic structures and components with certain capabilities (mixing performance, heat transfer, throughput ...). The aim was to provide evidence that several chemical and process technological benefits can be achieved by running continuous microfluidic processes instead of macroscopic batch processes. At that time, product compositions obtained at the end of a microreaction process were typically investigated by off-line analytical techniques (e.g. sampling and subsequent chromatographic separation by HPLC or gas chromatography) and compared with data known from macroscopic (batch) processes. Usually, the reaction mixture of a microreaction process was collected for a while in a vessel and afterwards analyzed – a common procedure in chemical laboratories.

However, this procedure disregards the fact that the analytical data obtained do not only describe the performance of the microreactor solely but also that of its macroscopic periphery that directly influences the chemical process under investigation. One has to consider that the hold-up of the fluidic interconnections (e.g. tubings, capillaries) between the microreaction process and the sampling vessel, and the hold-up of the sampling vessel itself exceed that of the microstructured components by orders of magnitude. The same applies to the residence time that is provided by both the microreaction process and its macroscopic periphery. In other words: the chemical and technological benefits of microreactors in terms of precise thermal conditions and residence times can be significantly adulterated by macroscopic downstream process units (such as fluidic interconnections, analytical tools, sampling and collecting vessels, etc.). For example, unwanted thermally induced side or decomposition reactions that can be successfully suppressed in microreactors due to precise isothermal processing are shifted towards subsequent macroscopic process units in which the conditions of the microchemical process can no longer be maintained.

Hence, the integration of process analytical techniques into a microchemical process in its immediate vicinity is of significant importance regardless whether sampling and subsequent offline analysis, online analysis or process sensors are considered. One has to make sure that the chosen conditions for the microchemical process (in particular temperature and residence time) are not distorted within the process segment that is covered and depicted by the process analytical tool. In most cases this can not only be achieved by a spatially closed arrangement of the microchemical process and the required analytical tools but also, for example, by an active temperature control of the analytical components and sensors. Moreover, the physical dimensions of the sensors and analytical components that are integrated into a microchemical process are also of significant importance. Many already realized concepts for the monitoring of microchemical processes have failed in accurate data acquisition since macroscopic sensors have been used that have recorded data apart from the actual site of reaction (the microchannel). In addition, the large hold-ups and dead volumes as well as the huge thermal mass of macroscopic sensors worsen this situation.

Nowadays the prevailing opinion is that only the development of microstructured or at least miniaturized sensors will meet the technical requirements that will allow a precise monitoring and control of microchemical processes as they are required for industrial production purposes. So far, existing provisional arrangements such as the integration of macroscopic process measuring and control equipment should be consequently replaced by new miniaturized concepts in the medium term. Respective R&D activities have just started that comprise not only the development of appropriate sensors but also of fluidic and electrical interfaces to allow a functional integration into microchemical processes.

Similar considerations apply to the integration of spectroscopic analysis to allow a more substance specific monitoring of microchemical processes, for example for monitoring of conversion, yield, byproduct formation, etc. In lab-scale applications of microreaction processes (testing of chemical reactions, screening of process parameters) the integration of UV/Vis, infrared or Raman spectrometers has been realized so far quite rarely. As optical interfaces to the microfluidic process, usually commercially available optical flow-through cells were used and were fluidically connected to the microfluidic components. In general, such optical interfaces have to meet the same demands with respect to hold-up, dead volume, residence time, and temperature control as it was discussed above for the integration of sensors.

When microchemical processes are used for production purposes the integration of high-value spectrometers will in most cases not be accepted by industrial users, mainly for economic reasons but also because of a quite complex interaction with the measurement and control algorithms of the process-control system. Instead of that, industrial users are highly interested in the development of miniaturized spectroscopic process analytical tools for microchemical processes that are inexpensive, easy to handle and thus easy to integrate into a process-control system. The latter should provide a controlled interaction between miniaturized PAT components and process actuators (such as pumps, valves, thermostats ...) to enable automated and thus stable processing under industrial conditions.

To sum up, one may say that process analytical techniques for industrial applications of microchemical processes are currently in the early stages of development. Nevertheless, many R&D projects have been carried out that demonstrate promising concepts and strategies for the integration of PAT components into microreaction processes. Some of them will be described as selected examples in the following chapters. It should be pointed out in this context that the development of suitable PAT concepts for microchemical production processes is clearly differentiated from the development of microfluidic sensors and analytical devices that are often described by the term "micro total analysis systems μTAS" and "lab on a chip" [12]. Such devices are usually independent sensor and analysis systems that are mainly used in the areas of medical, bio and environmental analysis. Their microfluidic structures are usually of significantly smaller size than is required in microchemical (production) processes. Moreover, these lab tools do not usually fulfil the requirements of microchemical processes in terms of throughput (mass or volume flow), pressure and temperature resistance, chemical inertness and functionality (mixing efficiency, residence time distribution, heat-transfer characteristics ...).

9.3
Integration of Spectroscopic Analysis into Microchemical Processes

The integration of spectroscopic analysis into microchemical processes has aroused increasing interest over the last years. Most of the so far conducted R&D work has focused on the analysis of chemical processes in the liquid and liquid/liquid regime. Usually optical flow-through cells were fluidically connected to the outlet of microfluidic process units (such as micromixers or microreactors) by capillaries or tubings. Depending on the investigated spectral range the small-sized optical cells were either connected to the spectrometer via fiber optics or directly placed into its sample compartment (free optical path). Most of the optical cells used are commercially available products such as flow-through cuvettes for UV/Vis and near-infrared spectroscopy, special flow-through cells for Raman spectroscopy or optical flow-through cells for transmittance and reflectance (e.g. attenuated total reflection, ATR) infrared spectroscopy.

The main objective of spectroscopic measurements in microchemical processes has been so far the qualitative and semiquantitative monitoring of individual analytes in the reaction mixture. Depending on the frequency of spectroscopic measurements time-resolved information about the progress of a chemical reaction and the completeness of conversion can be obtained. In particular for screening applications of microreactors spectroscopic online analysis is a very helpful tool to provide qualitative information about the suitability of certain reactants and process conditions (e.g. temperature, residence time, and stoichiometry). In the case of a sufficient time resolution even kinetic information can be obtained online. A recently published example is the application of fiber optical UV/Vis spectroscopy for the monitoring of high-pressure reactions up to 600 bar in a capillary microreaction process with an integrated 3-μL optical cell [13]. The reaction rate constants and activation volumes of a Diels–Alder reaction and a nucleophilic aromatic substitution were spectroscopically determined under systematic variation of the process pressure.

Manz and coworkers [14] report on the time-resolved FTIR spectroscopic monitoring of microchemical processes. A commercial IR transmission cell (d=25 μm) was connected to the outlet of an interdigital micromixer to investigate chemically induced changes of protein conformations (here: ubiquitin) in almost real time. The micromixer enables a very fast and intensive mixing (within 0.5–1.4 s) that is monitored by rapid scan measurements of infrared spectra (one spectrum each 65 ms) to provide kinetic data.

Even miniaturized measuring cells for nuclear magnetic resonance (NMR) spectroscopy are nowadays available that makes their direct integration into chemical reaction processes possible. This makes NMR spectroscopy highly attractive as a further spectroscopic technique for PAT applications since it provides very molecule-specific information. In recent years NMR spectroscopic process analysis has made remarkable progress and its application to microfluidic processes became possible, as described by Maiwald et al. [15]. For example, a time-resolved NMR spectroscopic analysis in connection with a micromixer is described by Kakuta et al. [16]. They used a microcoil NMR probe that allowed them to carry out NMR measurements of μL samples without losing significant sensitivity. The setup was also used to study solvent-induced changes in protein conformations (see above) under systematic variation of volume flow and residence time. A time resolution in the range of seconds provided a sufficient accuracy for kinetic evaluations.

For certain chemical reactions it is not sufficient to conduct the spectroscopic analysis subsequent to the microreaction process. It is rather required that an analyte is directly monitored in the immediate vicinity of its formation or conversion – in other words: within the microchannels. Different reasons may be responsible for such a requirement:

- fast reaction kinetics
- detection and analysis of short-lived species or instable intermediates
- additional residence time caused by fluidic interconnections to the optical cell must be avoided (see discussion in the previous chapter)
- uncontrolled thermal conditions in the fluidic interconnections to the optical cell must be avoided (see discussion in the previous chapter).

For example, even a capillary of a few centimeters length that connects a microreactor with a subsequent optical cell may significantly adulterate strong exothermic reactions or reactions that depends highly on accurate residence times in such a way that an uncontrolled and unwanted reaction progress takes place until the analysis is carried out.

To enable spectroscopic analysis in individual microchannels a sufficient spatial resolution of the spectrometer is required. For microreactors made of glass (or other transparent material) spatially resolved UV/Vis and NIR spectroscopy can be realized by fixing fiber optics (e.g. with commercial SMA connectors) at defined positions of a microchannel, for example to carry out transmission spectroscopy. Individual glass fibers can also be integrated into the assembly of microfluidic components, for example by laying them in cavities or hollow microchannels close to the reaction channels [17]. Since nowadays many miniaturized, low-cost fiber optic UV/Vis spectrometers are commercially available and can be easily operated via conventional PCs a quite simple and economic way for realizing spatially resolved UV/Vis process analysis is achievable.

To obtain more molecule- and structure-specific information about analytes in microchannels spatially resolved infrared spectroscopy turned out to be a more appropriate method [18–20]. However, an indisputable precondition is that the material in which the microchannels are incorporated is transparent in the infrared spectral range. In particular, microfluidic components made of silicon are especially suited to enable spatially resolved infrared spectroscopic process analysis due to the transparency of silicon in wide regions of the infrared spectral range. Since silicon provides other interesting features such as chemical inertness and good heat conductivity and can be easily microstructured by standard procedures its use in microchemical processes is of general interest.

To increase the sensitivity of IR spectroscopic process analysis in silicon microreactors a reflectance layer (e.g. gold layer) can be deposited on the outer side of the microreactor. In contrast to conventional transmittance spectroscopy the infrared beam is guided again through the microchannel after being reflected at the gold layer [21].

One practical approach to realize spatially resolved IR analysis is the use of FTIR microscopes (spatial resolution $\geq 10\ \mu m$) that allow a successive addressing of different areas of a microchannel to analyze and distinguish individual components of the reaction mixture at different positions in the microreactor [19, 20]. By focusing the IR beam consecutively on different predefined areas of a microchannel series of infrared spectra can be measured (either transmittance or reflectance spectra). Hence, the progress and kinetics of a reaction can be spectroscopically monitored along the flow direction and information about the completeness of conversion in a certain area of the microreactor are obtained [20]. The latter will allow an immediate regulation of the residence time via the volume or mass flow of the reactants. Moreover, the mechanism of a reaction can be revealed, for example by identifying intermediates in the reaction mixture.

Figure 9.1 shows, as an example, the use of FTIR microscopy for the monitoring of fast and strong exothermic nitrations of dialkyl-substituted ureas carried out in a silicon microreactor. The measured infrared spectra show clearly the progress of the reaction along the microchannel and the identification of a new – at that time unknown – intermediate (here: nitroso urea) [20]. By analyzing parallel microchannels in more complex microreactor geometries the conversion in different channels can be directly compared. This procedure gives evidence on the uniformity of the flow distribution in a microfluidic component consisting of parallelized microchannel arrays.

When visualization of fluid dynamics and flow distribution is the predominant reason for process monitoring even a two-dimensional mapping of the entire microfluidic structure is possible. For example, thermographic systems (infrared cameras) allow the detection of heat that is released in the microchannels during chemical reaction (Fig. 9.2). Since all microchannels are monitored simultaneously a real-time visualization of all individual flows in the microfluidic component is obtained that allows a spatial resolved detection of hot spots, blockages, fouling, etc. Moreover, the observation of fluid dynamics helps to evaluate and improve the three-dimensional microfluidic design of microreactors and micromixers [20, 22, 23].

A two-dimensional monitoring of flow distributions in microfluidic components made of transparent materials (e.g. glass) can also be easily realized by using optical or fluorescence microscopy. However, in most cases no real synthesis processes are monitored but rather reference processes with colored or fluorescent species (tracers) are used for visualization purposes only.

Apart from the detection and monitoring of single analytes an increasing interest in the simultaneous analysis of multicomponent systems can be observed for both production and screening applications of microchemical processes. To

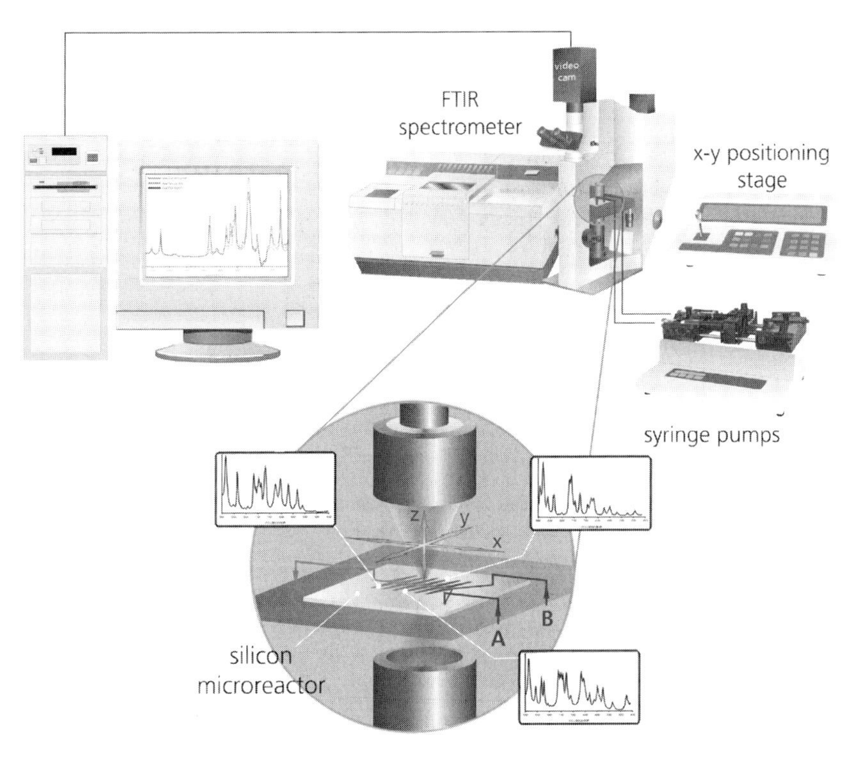

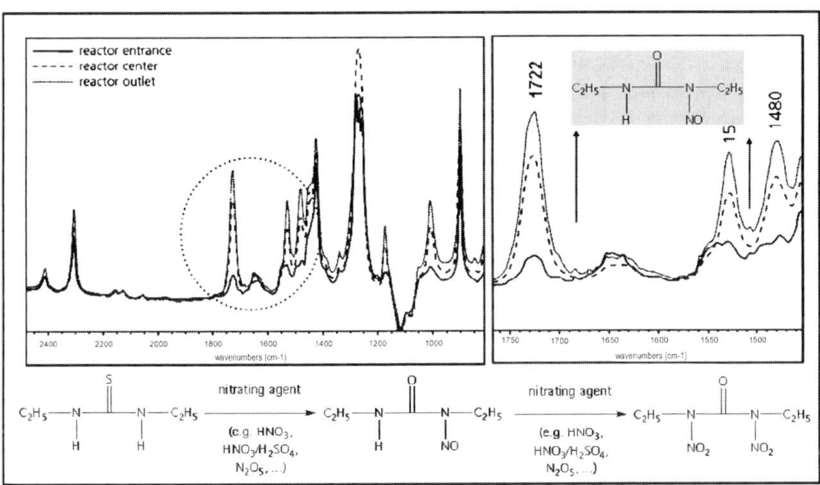

Fig. 9.1 Application of FTIR microscopy for the spatially resolved process monitoring in silicon microreactors: working principle (above) and exemplary monitoring of the reaction progress along one microchannel during the nitration of a dialkyl substituted urea: identification of a nitroso intermediate (below) [20].

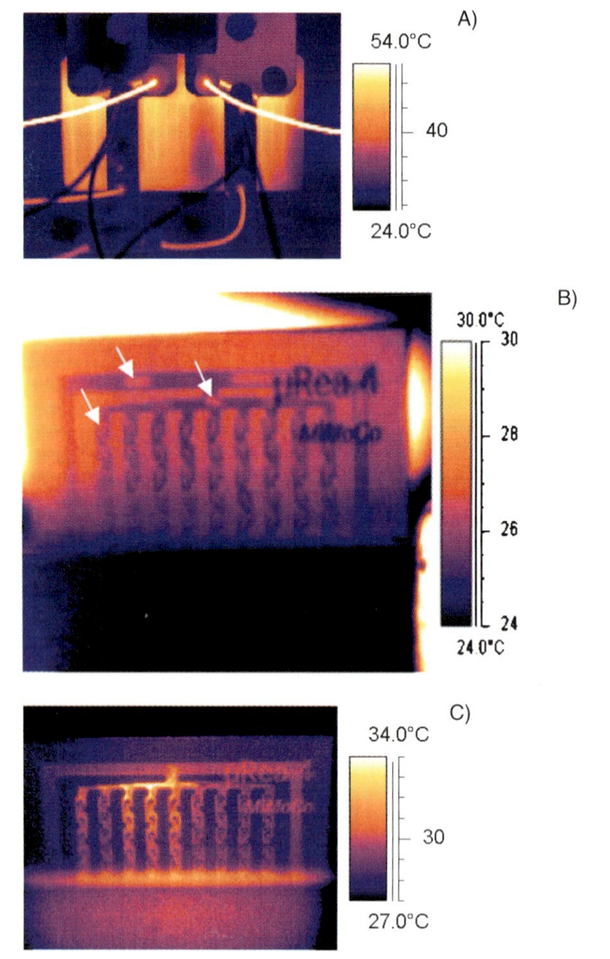

Fig. 9.2 Application of infrared thermography for the two-dimensional monitoring of microreaction processes (A: visualization of flow distribution in parallel microchannels within a microreactor made of glass, B: monitoring of gas bubbles in a silicon microreactor formed during reaction, C: hot-spot monitoring in a silicon microreactor during strong exothermic reaction) [20, 22].

realize a quantitative online analysis of different species in the reaction mixture (reactants, target product, unwanted byproducts ...) extensive calibration procedures (e.g. chemometric calibration methods) and multivariate data-analysis techniques are required as in macroscopic processes.

One example of an online multicomponent analysis that was integrated into a microchemical process is described by Ferstl and coworkers [24]. Online Raman spectroscopy was used in combination with a specific quench and sampling device for chemometric calibration purposes (Fig. 9.3). The sampling device was directly connected to the Raman optical flow-through cell that enables a well-defined quench of an aliquot of the reaction mixture at a certain time in the reaction progress. The sampling device was designed in such a way that the fluid dynamics of the microchemical process are not disturbed during operation. Once a quenched sample is taken it is immediately transferred to an offline chromatographic system (HPLC or GC). Since the Raman spectrum of the reaction mixture is recorded immediately before sampling, the spectroscopic and chromatographic data represent the same composition of the reaction mixture at a certain point of the process ("snapshot"). Hence, all data from the chromatographic analysis are used to generate a chemometric model for the calibration of the Raman spectrometer. Once the Raman spectrometer is calibrated for a certain chemical process a qualitative and quantitative multicomponent analysis of the running microreaction process is possible in realtime.

A successful chemometric calibration of the spectrometer depends highly on the completeness of the quench since otherwise uncontrolled subsequent reactions will occur during sample transfer and chromatographic analysis that will significantly alter the composition of the process snapshot [20]. For this reason a flexible quench module can be integrated into the microchemical process that allows both quenching of a homogeneous (liquid phase) and heterogeneous reaction mixture (two liquid phases) dependent on the actual chemical reaction process that is considered (Fig. 9.3). For homogeneous reaction mixtures quenching and sampling is realized by a valve that turns for a short time from the production line to a microliter sampling loop that is flushed with a certain volume of a quenching agent. If the reaction mixture is heterogeneous after quenching (i.e. nonmiscible fluids) the sampling loop is substituted by an additional micromixer to achieve an instantaneous and intensive mixing of the sam-

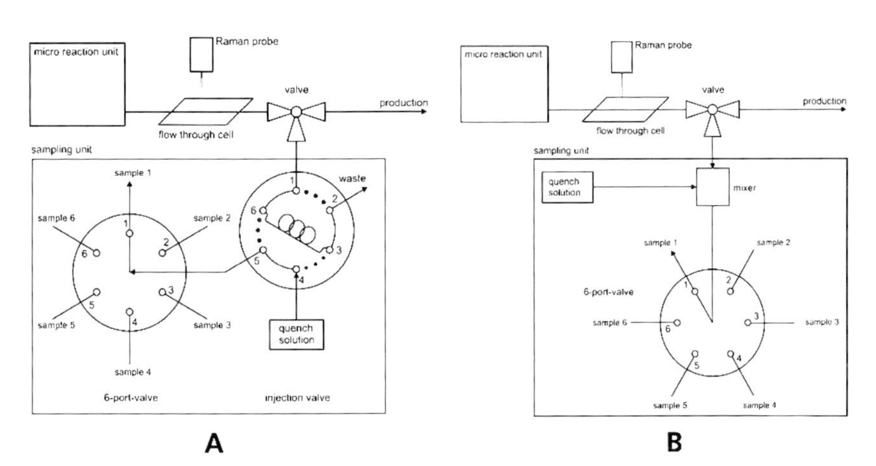

Fig. 9.3 Scheme of sampling and quenching devices for homogeneous (A) and heterogeneous (B) reaction mixtures. The devices are used for the chemometric calibration of a Raman spectroscopic online analysis integrated into an automated microreaction system [24].

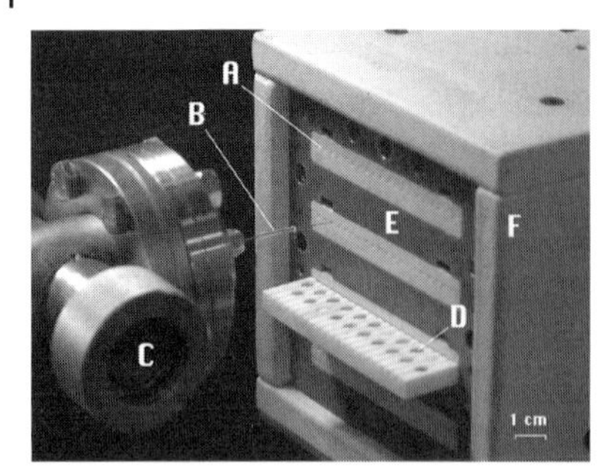

Fig. 9.4 Mass spectrometric gas analysis at the outlet of an array of microchannels that are part of a catalyst-screening system (A: microreactor array; B: sampling capillary; C: mass spectrometer; D: catalyst pellets; E: heating block;
F: insulation) [28].

ple and the quenching agent. Quenched samples are directly transferred via 1/6- or 1/16-port valves to vials for subsequent chromatographic analysis [24]. Once chemometrically calibrated the Raman spectroscopic process analysis was successfully used for the simultaneous quantitative monitoring of up to six individual analytes of different microchemical nitration processes in realtime [24].

In addition to the above-described examples of spectroscopic PAT for microchemical processes in the liquid regime also much work was done concerning the connection of analytical methods to microchemical gas-phase processes. Apart from gas-chromatographic process analyzers (which are also the subject of extensive miniaturization and microfabrication [25, 26]) the connection of mass spectrometers to microchemical gas-phase processes turned out to be a suitable process analytical method.

Catalytic gas-phase reactions are among the import applications of microreaction technology. Heterogeneous catalysts can be immobilized on the huge internal surfaces of microreactors and used with high efficiency under precise process conditions (temperature, pressure, residence time ...) [27]. Therefore, microreaction processes are also ideally suited for catalyst-screening purposes. A certain parallelization of microchannels that are individually charged with different catalyst materials enables high throughput screening (HTS) applications that require appropriately fast and specific analytical techniques such as mass spectrometry. For example, Senkan et al. [28] as well as Claus et al. [29] have successfully demonstrated the connection of a mass spectrometer (MS) to a microstructured catalyst-screening system. The sampling capillary of the MS moves gradually to the outlet of each microchannel that is charged with an individual catalyst system and analyzes the composition of the reaction-gas mixture (Fig. 9.4). From the gathered MS data conclusions can be drawn concerning the performance, stability and range of operation of catalysts (for example with respect to temperature, pressure, concentration) for specific gas-phase processes.

9.4 Microstructured Sensors for Microchemical Processes

As already described above, the development of sensors that are specially designed for microchemical production processes has just begun. Nowadays usually commercially available macroscopic flow-through sensors are fluidically connected to microstructured components to gather relevant process parameters such as temperature, pressure, and mass flow (resp. volume flow). Sometimes additional process data such as pH value, density or electric conductivity are measured. To overcome the difficulties arising from the huge hold-up, dead volume and thermal mass macroscopic sensors have to be replaced by suitable microstructured flow-through sensors with dimensions that are compatible with that of the microfluidic process.

For example, promising work towards this direction was conducted by Siemens and partners who have developed the first prototypes of microstructured flow-through sensors made of silicon that were specially designed for microchemical production processes. Both a combined pressure/temperature sensor and a combined mass flow/density sensor meet the industrial demands for chemical stability and robustness, small hold-up and operability in a broad mass-flow range. Originally the sensors were developed as part of an automated microreaction system for nitration processes at mass flows up to 50 g/min. The sensors can be integrated at different positions of the microchemical process and are electrically connected via sensor-specific ASICs with a process-control system [30].

The microstructured silicon pressure sensor (that can be equipped with an additional temperature sensor) was designed for flow-through applications in the pressure range of 0–10 bar. The corrosion-resistant sensor can be operated at temperatures up to 200 °C in direct contact with the reaction mixture and has a dead volume of <0.5 μL. It is thus ideally suited for microchemical production processes. Figure 9.5 shows the repeatable linear response of the sensor over the entire measuring range at different temperatures.

The microstructured mass-flow sensor that is based on the Coriolis principle consists of two silicon loop-like shaped tubes that are electrostatically excited by

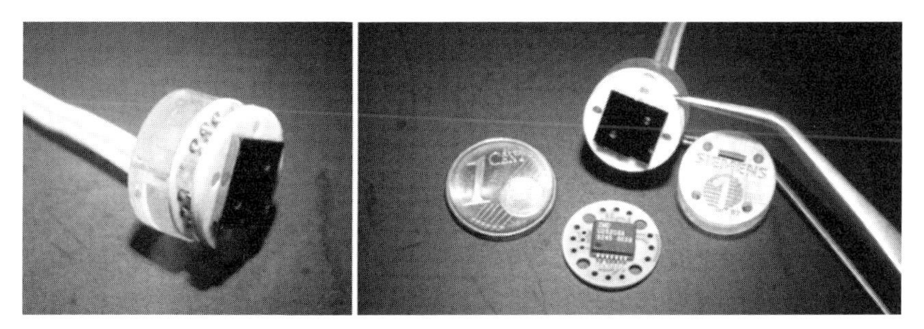

Fig. 9.5 Microstructured pressure/temperature sensor made of silicon for flow-through applications [30].

precisely positioned electrodes to achieve oscillation while the fluids are flowing through the tubes (Fig. 9.6). The design of the sensor is based on a concept first described by Enoksson et al. [31] and was further improved. When the fluid flows through the silicon tubes of this micromechanic resonator the resonance frequency changes as a function of the fluid density. Hence, the sensor can also be used for density measurements. Very high accuracies of the sensor ($<10^{-4}$ g/cm^3) makes it possible to measure densities of both liquids and gases in flow-through operations. Due to the loop-like design of the silicon tubes the Coriolis force causes an additional tilting of the structure within certain amplitudes [31]. Since the mass flow is directly proportional to the ratio of the excitation amplitudes it can be determined just by measuring these amplitudes, for example optically.

Concepts for the calorimetric monitoring of microchemical processes are described rarely in literature. Köhler and coworkers [32], for example, have devel-

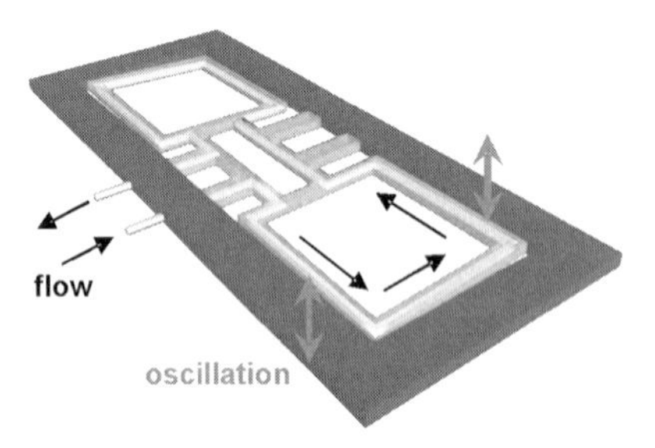

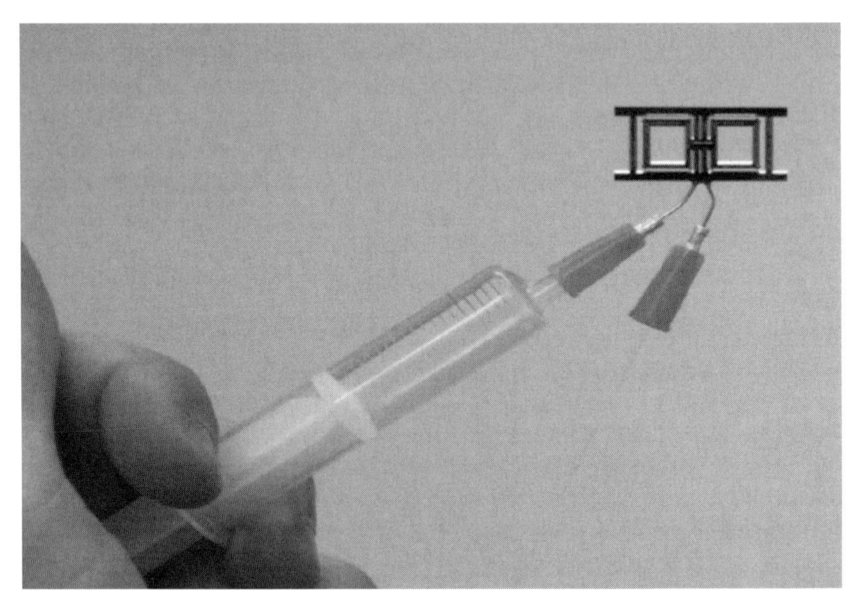

Fig. 9.6 Microstructured mass flow/density sensor made of silicon for flow-through applications [30].

oped a microfluidic chip for the determination of concentrations in liquid reaction processes on the basis of a thermoelectric measurement principle using thin-film thermopiles. Antes et al. [33] have recently developed a new kind of microliter flow-through calorimeter that allows a fast and quantitative measurement of reaction heats. The μL-calorimeter consists of a flat microreactor made of glass or silicon that is embedded between thermoelectric modules as sensors (two Seebeck elements and one Peltier element). This sandwich setup is placed into a thermostated heat reservoir to control the reaction temperature and allow isothermal processing (Fig. 9.7).

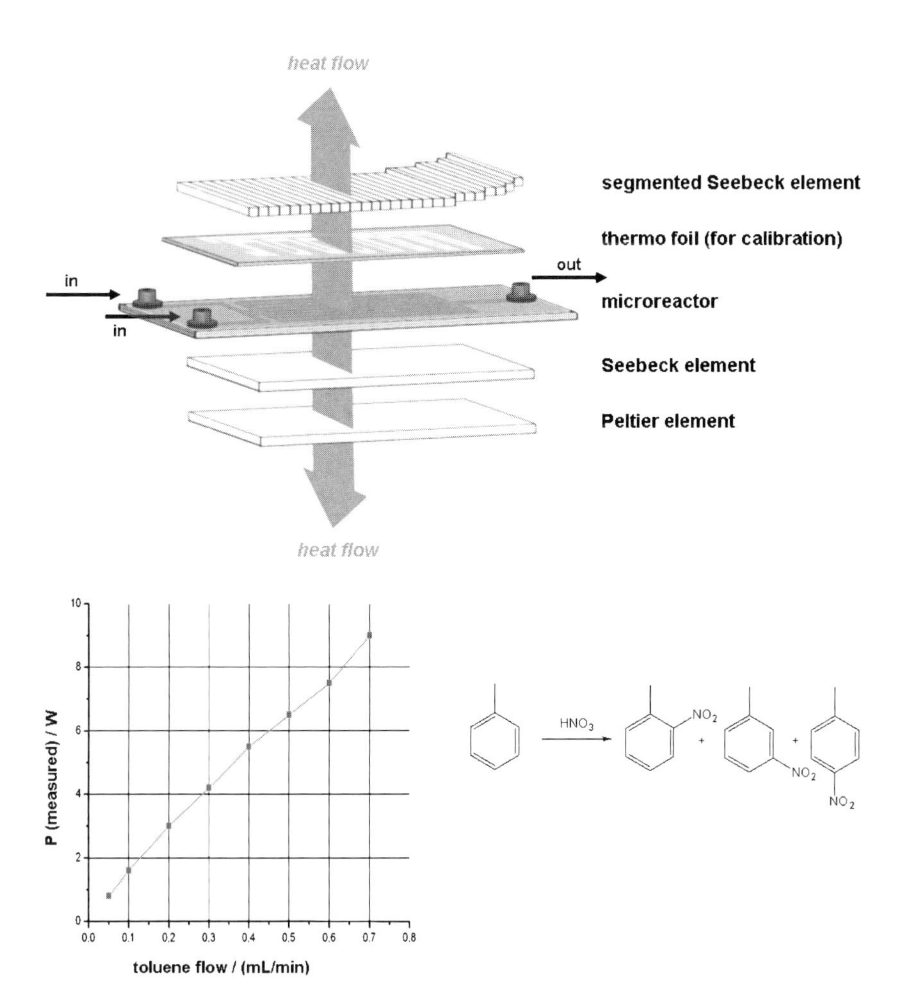

Fig. 9.7 Scheme of a μL flow-through calorimeter based on a sandwich setup of a microreactor embedded between thermoelectric modules and its application for the calorimetric monitoring of strong exothermic reactions (here: nitration of toluene at 50 °C). The heat of reaction can be determined from the slope of the measured heat flow (here: 110 kJ/mol) [33].

The calorimetric measurement is performed by recording continuously the heat flow that is released by the chemical reaction in the microreactor and is directed through the Seebeck element towards the Peltier element. The latter is regulated by a PID controller to generate a constant temperature difference between the microreactor and the opposite Seebeck element. By integrating the voltage signals of the Seebeck element over the time the released reaction heat can be calculated quantitatively. The huge surface-to-volume ratio of the microchannels ensures an instantaneous transfer of the reaction heat towards the thermoelectric modules. Hence, the μL-calorimeter has a very small time constant of about 3 s that is smaller by a factor of 20 than that of conventional (macroscopic) reaction calorimeters. Therefore, this device is ideally suited for measuring fast and highly exothermic reactions at isothermal conditions. Moreover, it can be deliberately used as a monitoring tool for process control and safety analyses. The performance and the accuracy of this new device have been demonstrated for different strong exothermic reactions in the liquid and liquid/liquid regime. One example is shown in Fig. 9.7.

9.5 Conclusions

In the last decade microreaction technology has provided promising developments that offer significant benefits for chemical and pharmaceutical processes. In addition to its application as a tool for testing, screening and development of chemical processes microreaction technology is increasingly accepted and foreseen as a technology for industrial production purposes. The latter requires functional concepts for process analytical techniques that allow a precise and stable monitoring of microchemical processes. Precise analytical data are indispensably required for process optimization, mechanistic, kinetic and calorimetric studies as well as for the control and regulation of microchemical production processes.

In most cases the integration of macroscopic sensors or analytical modules is not sufficient to monitor precisely the current status of a microchemical process. Huge hold-ups and dead volumes or uncontrolled thermal conditions will significantly alter the analytical measurement or even the entire microprocess. Hence, miniaturized and partly microstructured sensors and analytical tools are required that meet the demands for a reliable process analysis of microchemical processes. Promising developments towards this direction can be observed. The selected examples that are briefly discussed in this chapter might inspire interested readers to look more in detail into new developments and concepts for microprocess analytical techniques.

References

1 K. Jähnisch, V. Hessel, H. Löwe, M. Baerns, *Chemistry in Microstructured Reactors*, Angew. Chem. Int. Ed. **2004**, *43*, 406

2 H. Pennemann, P. Watts, S.J. Haswell, V. Hessel, H. Löwe, *Benchmarking of Microreactor Applications*, Organic Process Research & Development **2004**, *8*, 422

3 V. Hessel, P.Löb, H. Löwe, *Development of Microstructured Reactors to Enable Organic Synthesis Rather than Subduing Chemistry*, Current Organic Chemistry **2005**, *9*, 765

4 K.F. Jensen, *Microreaction engineering – is small better?*, Chem. Eng. Sci. **2001**, *56*, 293

5 V. Hessel, S. Hardt, H. Löwe, *Chemical Micro Process Engineering: Fundamentals, Modelling and Reactions*, Wiley-VCH, Weinheim **2004**

6 see for example: www.microchemtec.de, www.microreaction-technology.info, www.ehrfeld.com

7 G. Markowz, S. Schirrmeister, J. Albrecht, F. Becker, R. Schütte, K.J. Caspary, E. Klemm, *Microstructured Reactors for Heterogeneously Catalyzed Gas-Phase Reactions on an Industrial Scale*, Chem. Eng. Technol. **2005**, *28*, 459

8 T. Schwalbe, A. Kursawe, J. Sommer, *Application Report on Operating Cellular Process Chemistry Plants in Fine Chemical and Contract Manufacturing Industries*, Chem. Eng. Technol. **2005**, *28*, 408

9 V. Hessel, H. Löwe, A. Müller, G. Kolb, *Chemical Micro Process Engineering: Processing and Plants*, Wiley-VCH, Weinheim **2005**

10 http://www.fda.gov/cder/guidance/6419fnl.pdf

11 http://www.fda.gov/cder/OPS/PAT.htm

12 E. Oosterbroeck, A. van den Berg (Eds.), *Lab-on-a-Chip: Miniaturized Systems for (Bio)Chemical Analysis and Synthesis*, Elsevier Science, London **2003**

13 F. Benito-Lopez, W. Verboom, M. Kakuta, J.G.E. Gardeniers, R.J.M. Egberink, E.R. Oosterbroek, A. van den Berg, D.N. Reinhoudt, *Optical fiber-based on-line UV/Vis spectroscopic monitoring of chemical reaction kinetics under high pressure in a capillary microreactor*, Chem. Commun. **2005**, *22*, 2857

14 M. Kakuta, P. Hinsmann, A. Manz, B. Lendl, *Time-resolved Fourier transform infrared spectrometry using a microfabricated continuous flow mixer: application to protein conformation study using the example of ubiquitin*, Lab Chip **2003**, *3*, 82

15 M. Maiwald, H. Fischer, Y.-K. Kim, K. Albert, H. Hasse, *Quantitative High-Resolution On-line NMR Spectroscopy in Reaction and Process Monitoring*, J. Magn. Reson. **2004**, *166*, 135

16 M. Kakuta, D.A. Jayawickrama, A.M. Wolters, A. Manz, J.V. Sweedler, *Micromixer-Based Time-Resolved NMR: Applications to Ubiquitin Protein Conformation*, Anal. Chem. **2003**, *75*, 956

17 S. Bargiel, A. Górecka-Drzazga, J. Dziuban, P. Prokaryn, M. Chudy, A. Dybko, Z. Brzózka, *Nanoliter detectors for flow systems*, Sens. Actuators A: Phys. **2004**, *115*, 245

18 R.J. Jackman, T.M. Floyd, M.A. Schmidt, K.F. Jensen, *Development of Methods for On-Line Chemical Detection with Liquid-Phase Microchemical Reactors Using Conventional and Unconventional Techniques*, in Proc. of 4th Int. Symp. on Micro Total Analysis Systems (microTAS), Enschede, The Netherlands **2000**, 155

19 A.E. Guber, W. Bier, K. Schubert, *IR spectroscopic studies of a chemical reaction in various micromixer designs*, in Proc. of 2nd Int. Conf. on Microreaction Technology (IMRET 2), New Orleans, USA **1998**, 284

20 J. Antes, D. Boskovic, H. Krause, S. Löbbecke, N. Lutz, T. Türcke, W. Schweikert, *Analysis and Improvement of Strong Exothermic Nitrations in Microreactors*, Trans IChemE **2003**, *81* Part A, 760

21 R. Keoschkerjan, M. Richter, D. Boskovic, F. Schnürer, S. Löbbecke, *Novel Multifunctional Microreaction Unit for Chemical Engineering*, Chem. Eng. J. **2004**, *101*, 469

22 E. Marioth, S. Löbbecke, M. Scholz, F. Schnürer, T. Türke, J. Antes, H.H. Krause, N. Lutz, *Investigation of Micro-*

fluidics and Heat Transferability Inside a Microreactor Array Made of Glass, in Proc. of 5th Int. Conf. on Microreaction Technology (IMRET 5), Strasbourg, France **2001**, 262

23 C. Wille, W. Ehrfeld, V. Haverkamp, T. Herweck, V. Hessel, H. Löwe, N. Lutz, K.-P. Möllmann, F. Pinno, *Dynamic Monitoring of Fluid Equipartition and Heat Release in a Falling Film Microreactor Using Real-Time Thermography*, MICRO.tec 2000 VDE World Microtechnologies Congress, Hannover **2000**, 349

24 S. Löbbecke, W. Ferstl, S. Panic, T. Türcke, *Concepts for Modularization and Automation of Microreaction Technology*, Chem. Eng. Technol. **2005**, *28*, 484

25 F. Müller, *Micro Electromechanical Systems Applied to Gas Chromatographs*, atp **2004**, *1*, 3

26 J. Dziuban, J. Mroz, M. Szczygielska, M. Malachowski, A. Gorecka-Drzazga, R. Walczak, W. Bula, D. Zalewski, L. Nieradko, J. Lysko, *Portable Gas Chromatograph with Integrated Components*, Sens. Actuators A: Phys. **2004**, *115*, 318

27 G. Kolb, V. Hessel, *Micro-structured reactors for gas-phase reactions: A review*, Chem. Eng. J. **2004**, *98*, 1

28 S. Senkan, K. Krantz, S. Ozturk, V. Zengin, I. Onal, *High-Throughput Testing of Heterogeneous Catalyst Libraries Using Array Microreactors and Mass Spectrometry*, Angew. Chem. Int. Ed. **1999**, *38*, 2794

29 P. Claus, D. Hönicke, T. Zech, *Miniaturization of Screening Devices for the Combinatorial Development of Heterogeneous Catalysts*, Catalysis Today **2001**, *67*, 319

30 W. Ferstl, S. Löbbecke, J. Antes, H. Krause, M. Grund, M. Häberl, H. Muntermann, D. Schmalz, J. Hassel, A. Lohf, A. Steckenborn, T. Bayer, M. Kinzl, I. Leipprand, *Development of an Automated Microreaction System with Integrated Sensorics for Process Screening and Production*, Chem. Eng. J. **2004**, *101*, 431

31 E. Enoksson, G. Stemme, E. Stemme, *A Silicon Resonant Sensor Structure for Coriolis Mass-flow Measurements*, J. Microelectromech. Sys. **1997**, *6*, 119

32 S. Beißner, T. Elbel, J.M. Köhler, M. Zieren, *Thermoelectrical Measurement System for Chemical Instrumentation*, in Proc. of 3rd Int. Conf. on Microreaction Technology (IMRET 3), Frankfurt a.M., Germany **1999**, 597

33 J. Antes, D. Schifferdecker, H. Krause, S. Löbbecke, *A New Concept for the Measurement of Strong Exothermicities in Microreactors*, in Proc. of 8th Int. Conf. on Microreaction Technology (IMRET 8), Atlanta, USA **2005**, 134b

10
Microfabrication in Metals and Polymers

Jürgen J. Brandner, Thomas Gietzelt, Torsten Henning, and Manfred Kraut,
Institute for Micro Process Engineering (IMVT), Forschungszentrum Karlsruhe, Germany
Holger Moritz, Institute for Microstructure Technology (IMT), Forschungszentrum Karlsruhe, Germany
Wilhelm Pfleging, Institute for Materials Science I (IMFI), Forschungszentrum Karlsruhe, Germany

Abstract
In this chapter, the fabrication of microstructure components and devices out of metals and polymers will be described briefly. Influences of the material choice and the basic process parameters with regard to the manufacturing of microstructure components are discussed as are bonding, sealing and packaging of devices. Examples for the choice of material are given. Manufacturing processes for metal microstructure components and microstructures made of certain polymers are described briefly with a clear focus on some special microstructuring methods well established in microstructure technology. Examples of those techniques are given. In the bonding section, the most common bonding and sealing techniques are briefly described. The main focus here is on techniques that allow a prototype manufacturing as well as a small-scale or large-scale series production of devices suitable for reaction or process engineering. Special regard is given to the different methods to obtain devices with mass-flow ranges suitable for industrial production of, e.g., chemicals in microstructure devices. Quality assurance and the long-term stability are also taken into account. Finally, a short view to series production of microstructure components is given. This chapter does not give a complete overview of the large field of generating microstructure devices, but may show some opportunities and, with the references placed at the end of the chapter, some good hints to find more detailed descriptions for the case under consideration.

Keywords
Microstructuring, metal, polymer, bonding, sealing, equaling

Advanced Micro and Nanosystems Vol. 5. Micro Process Engineering. Edited by N. Kockmann

ISBN: 3-527-31246-3

Metals and polymers are powerful materials for microstructure devices for microreaction engineering. Thus, development of technologies for the manufacturing of devices out of materials taken from one of these classes have achieved increasing interest within the microstructure technology community as well as in industrial applications.

Most manufacturing technologies have their roots either in silicon device production (see Chapter 11), precision machining or polymer processing. The processes well known from those techniques have been tried out for microstructure dimensions, adapted and improved to reach the desired precision and surface quality. Rarely was it possible to use the same manufacturing process for macroscale and microscale devices to obtain sufficient results. In most cases, more or less strong changes within the design of the device, the methodology of the process and the manufacturing process itself was necessary to provide the accuracy and quality needed for microstructure devices suitable for process or reaction engineering.

The manufacturing process is driven by technological and application-defined limitations that have to be considered not only before starting with manufacturing but also along the complete manufacturing pathway. Unlike the general design procedure (see Chapter 7), no straight-on or somehow branched manufacturing roadmap can be given. The manufacturing process is more or less a continuous improvement by using a "manufacturing circle". Several tasks have to be considered in addition to the manufacturing process, and results from steps of the process may lead directly to a completely different manufacturing technology from that originally thought to be the right. In Fig. 10.1, a rough scheme of a design and manufacturing consideration process is shown.

In this chapter, an overview of the way to manufacture microstructure devices will be given. Some manufacturing methods suitable for microstructure devices will be briefly described or at least referenced. We will concentrate on technologies for metals and polymers, though the manufacturing processes for silicon are described in Chapter 11, and those for ceramics are given in Chapter 12.

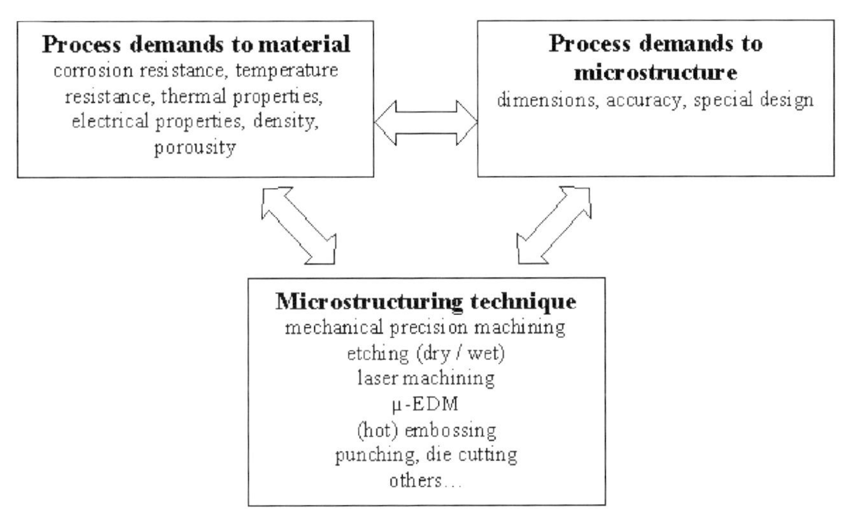

Fig. 10.1 Scheme of the "manufacturing circle". Driven by process and material properties, the manufacturing technique may change several times unless the desired microstructure device can be generated.

Moreover, most descriptions will not go deeply into details. Thus, a detailed description of most manufacturing processes can be found in [1].

10.1
Influences of Applications to Material Choice

Several factors influence the choice of materials for reaction engineering devices. The most important are temperature resistance, corrosion resistance and thermal properties; less important are the electrical properties.

For nonelectrochemical applications, electrical properties are not that critical to microreaction engineering. Mostly, it simply does not matter whether the material provides a high electrical conductivity or not. Only for processes driven by, i.e., voltage differences applied from outside, the electrical properties of the microstructure device material are of great interest. In these cases, nonmetallic materials like glass or ceramics should be applied (see Chapter 12), or at least surface modifications like insulating layers should be used to cover the active surface of the device.

For many processes, the demands on the device in terms of thermal properties depend on the type of fluids used (e.g. gases, liquids or mixtures of both) and are not difficult to estimate. The amount of heat needed by endothermic or generated by exothermic chemical reactions can be precalculated easily. Thermal properties of the material of choice are well known, and heat transfer through microstructures is not very difficult to calculate. In fact, for most applications either standard equations or CFD can be applied to calculate at least the integral temperature behavior of microstructure devices. For a more detailed view, local measurements or highly resolved CFD calculations are necessary. In those cases, the thermal properties of the material may have a large influence. For the manufacturing process of the device, the thermal properties of the material have no influence.

The boundary conditions of the process are often given by temperature, pressure and the fluids used. Evidently the construction materials should sustain these. There are norms for temperature and pressure devices such as the European Pressure Equipment Directive (PED) 97/23/EG [2]. But also national organizations have published directives, such as the AD-Merkblätter [3] in Germany. The directives given in the references define their validity in terms of pressure times volume. They give another boundary condition stating a volume below which these rules no longer apply. Microstructure devices are often exempted from these rules, since the active volume (the hold-up, respectively) is small. However, if technically relevant throughputs are to be obtained the active volume is of an order that the rules apply.

Another point is, of course, the volume needed for fitting microstructure devices to the real world. Here we also have comparatively large volumes. So if one is to set up a unit the rules have to be carefully checked. Here the limits of the microstructure have to be assessed separately. Normally the testing organizations, e.g. the TÜV in Germany, are only concerned with the 'outer pressure-bearing shell', so that the actual microstructure is not relevant, as long as it is included in a housing resisting a pressure that exceeds the process operating pressure by a certain amount (defined by the directives). However, the pressure resistance is still of concern in cases where two fluids of different pressure are processed inside the device. This often is the case for heat exchangers. A problem occurring for this case is that the applicability of standard formulas for the calculation of relevant measures, like the bending moment, is not yet cleared up completely. Consequently, experimental tests have to be performed on a case-by-case basis.

Another extremely important factor for the choice of materials for devices to apply in reaction engineering is the question of corrosion. This topic needs special attention. Standard literature on corrosion, e.g. the corrosion handbook by Dechema [4] takes a corrosion rate of 0.1 mm y^{-1} as resistant and a corrosion rate of 1 mm y^{-1} as fairly resistant (see Table 10.1). These values are chosen with respect to standard, macroscopic reaction vessels, tubes and fittings, which normally provide wall thicknesses in the range of several mm or higher. When talking about microstructure devices, we often have wall thickness in the order of 0.1 mm (=100 μm) or even below. Consequently, the corrosion-rate limitations given in the cited literature can not be applied to microstructure devices. Sadly though corrosion has not been a topic in most publications dealing with microstructure devices, but it has normally been assumed that the material of choice is completely resistant against corrosion by the given reaction system. So a material deemed resistant according to the corrosion handbook can prove unsuitable for the process running in the microstructure device.

Presently, no common design rules or regulations from any federal or private organization dealing with corrosion rates for microstructure devices exist. Thus, for each process that should be handled with microstructure devices, the corrosivity of the fluids and the resistivity of the material have to be checked carefully to prevent damage and losses within extremely short times. But measuring extremely small corrosion rates (e.g. gravimetric measurements) is time consuming and prone to errors, due to the small losses in weight.

To assess the suitability of a construction material for a given process, careful experiments under the expected conditions have to be undertaken. The choice

Table 10.1 Definition of suitability of materials according to Corrosion Handbook [4].

Behavior	Maximum material consumption rate per unit area [g(m² d)$^{-1}$]	Corrosion rate [mm y^{-1}]
Resistant	2.4	0.1
Fairly resistant	24	1.0
Hardly usable	72	2.5–3.0
Unsuitable	>72	>3.0

of material in relation to corrosion should, to a first approximation, be limited to ones that are expected to be completely resistant. However, if the use of the device is limited to comparatively short times, e.g. in a production or measurement campaign, the above-mentioned limitation need not be strict. One should be sure that the complete destruction of the functionality is not to be feared during the time of the campaign.

10.2 Microfabrication Techniques for Metals

For metallic microstructure devices suitable for reaction engineering, different process stages have to be considered in advance of the manufacturing.

As a first stage, the microstructure has to be manufactured in the appropriate way. All considerations described in Section 10.1 have to be made, and an appropriate microstructuring method has to be found. A rough scheme of this is shown in Fig. 10.2.

Metals may be microstructured with different techniques, reaching from "simple" mechanical micromachining (turning, milling, drilling), μ-EDM and laser ablation to dry and wet chemical etching. Here, we will focus on mechanical micromachining, laser machining and wet chemical etching. In addition to these, a brief description of a nonabrasive method, selective laser melting, will be given. Depending on the microstructuring technique used, the accuracy, the surface quality, the cross section and the design of the microstructures change. A more precise overview is given in [1, 5–9].

10.2.1 Mechanical Micromachining

Mechanical micromachining is the ablation of metallic material from the surface. Special tools are needed, either natural diamond tools for cutting, e.g., in brass, or hard metal tools for, e.g., stainless steel. Mechanical micromachining is a very flexible and common method for manufacturing prototypes and small-scale series, e.g. for microfluidic devices or tools for polymer-replication techniques (see polymer microfabrication). More information may be found in [1, 5, 10].

The dimensions of the tools used are below 1 mm, preferably below 0.5 mm. In Figs. 10.3 and 10.4, examples of such tools are shown.

For small dimensions new effects appear that are not known in the macroscopic world. As an example, the radii of cutting edges and the wear are nearly constant, regardless of the tool diameter. Coatings to decrease wear are often worse. In consequence, tool deflections occur requiring special machining strategies to meet the tolerance requirements [10, 11].

Another topic is the surface roughness: Since tight tolerances raise the machining effort exponentially the suitable surface roughness should be about one order of magnitude larger than the achievable surface roughness. Nevertheless,

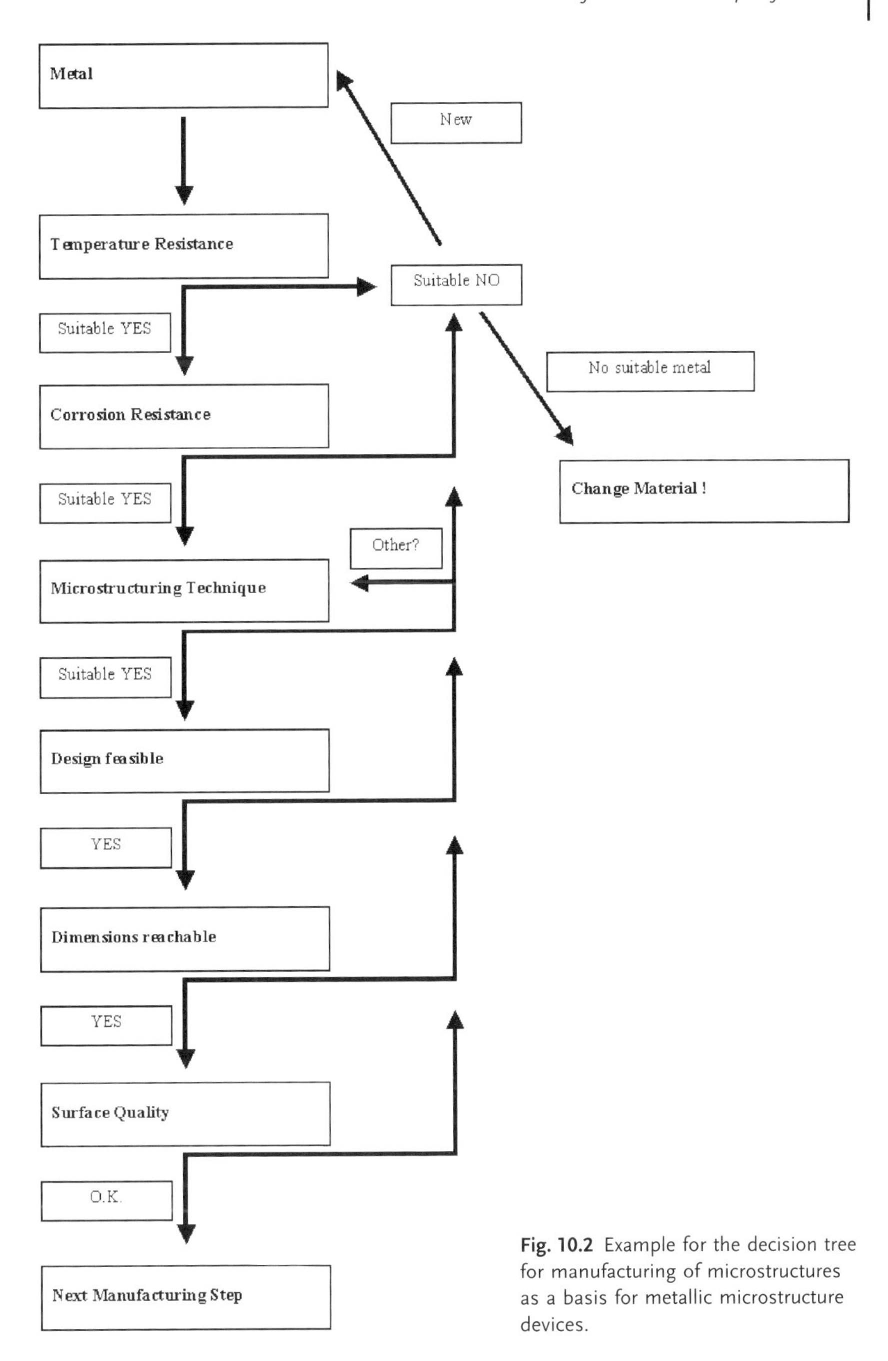

Fig. 10.2 Example for the decision tree for manufacturing of microstructures as a basis for metallic microstructure devices.

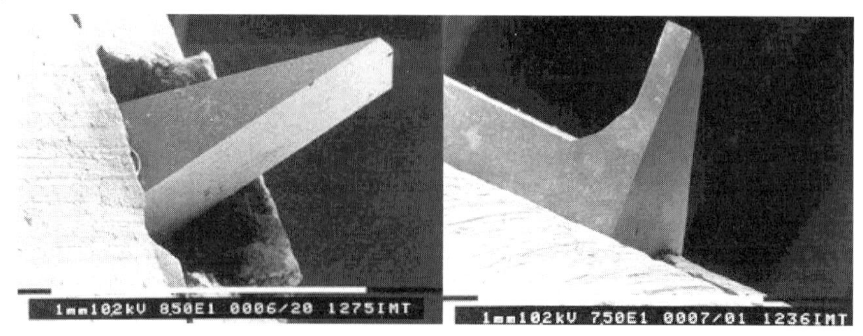

Fig. 10.3 Natural diamond tools. Left: to machine triangularly shaped microstructures, right: tool for precision micromachining of metal foils.

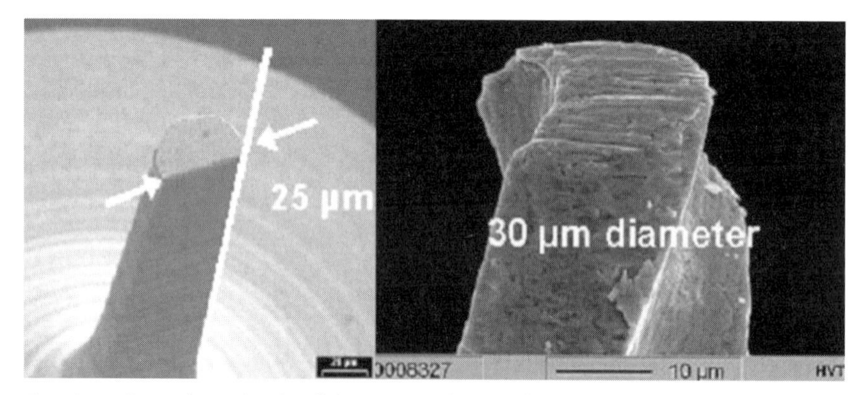

Fig. 10.4 Left: Hard metal end mill for micromachining of metals. The diameter of the cutting edge is 25 µm. Right: SEM of a hard metal microdrill providing a diameter of 30 µm.

extremely high surface qualities with a mean roughness in the range of several 10 nm can be reached, depending on the metal or alloy used. In Figs. 10.5, 10.6 and 10.7, SEM photos of mechanically machined linear microchannel structures in different materials are shown. The techniques to measure these surface qualities and the accuracy of the microstructures have to be sophisticated so that the compliance of tolerances can be proved by independent operators.

The range of materials available for micromachining includes polymers, ferrous and nonferrous alloys as well as some types of ceramic materials. For ferrous alloys (and ceramic materials), tools made of ultrafine hard metals are the best choice. Here, the grain size is about 0.4 µm. However, since the cutting edge consists of hard WC grains embedded in soft cobinder matrixes, burr formation at the microstructures is often a problem. If the burr is regular to a certain degree, electropolishing can be applied for removal.

Fig. 10.5 Left: SEM of linear microchannels in stainless steel, generated by mechanical micromachining. The channels are about 200 µm wide and 70 µm deep. Right: SEM detail of the photo left.

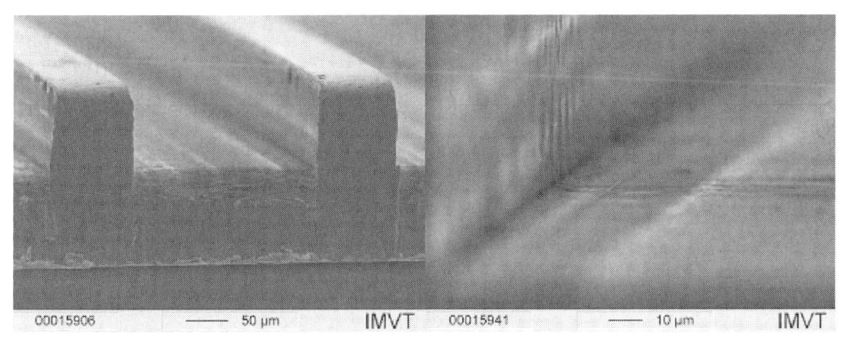

Fig. 10.6 Left: SEM photo of linear microchannels, generated by mechanical micromachining in copper. Right: SEM detail of the left photo. Some irregularities at the bottom and the sidewall of the copper microchannel can be seen. These originate from the copper material and the tool.

Recently, single cutting edge tools down to 37 µm diameter have been realized. Two-flute tools down to 50 µm diameter are commercially available today. The aspect ratio of the microstructures reachable with mechanical micromachining depends strongly on the tool diameter because of machining forces and tool deflection. For a 50-µm tool the aspect ratio is only 1:1. 100-µm tools with aspect ratio of 1:4 are available, and for 200 µm diameter, an aspect ratio of 1:10 is possible. Further diameter miniaturization comes to a limit since the cross section decreases by a power of 2, and for composite materials a certain number of grains in the cross section is necessary for homogeneous material properties. Beside this, the danger of flaws in the material causing short tool life and finally rupture increases exponentially with decreasing tool diameter [12].

For nonferrous alloys (and polymers), tools made of monocrystalline diamond are used (see Figs. 10.3 and 10.4). A very smooth cutting edge guarantees per-

Fig. 10.7 SEM photo of linear microchannels, generated by mechanical micromachining in OF copper. It is clearly seen that the surface quality of the microchannel bottom as well as the sidewalls is extremely high. The shadows inside of the channels remain from cleaning.

fect edges of the microstructures. However, since diamond is a brittle material, the reachable aspect ratio is lower than for hard metal tools, and the smallest diameter is 100 μm.

Consequently, the possible aspect ratio of micromachined microstructures depends on the design (positive or negative), the feature size and, due to cutting force and tool deflection, on the material to be machined. Further information can be obtained from the literature.

10.2.2
Laser Micromachining

Beside mechanical micromachining, laser machining is possible with most metallic materials. Laser micromachining of metals includes the following process technologies: laser cutting, laser drilling and laser patterning [13].

Laser cutting of metal foils with thicknesses between 50 μm up to 2 mm is well established with a cutting width of about 20–100 μm. This process is used routinely for the manufacturing of stents made of steel or nitinol (NiTi) as well as for the fabrication of small SMA (shape memory alloy) actuator parts [14] and small flexible instruments for minimal invasive surgery [15]. Figure 10.8 shows some examples.

Laser drilling in microsystem technology can be performed with different kinds of process strategies. In most cases, drilling can be performed via laser trepanning or laser percussion (see Fig. 10.9). In percussion mode, the laser beam is kept stationary while in trepanning mode the laser beam is rotated. For

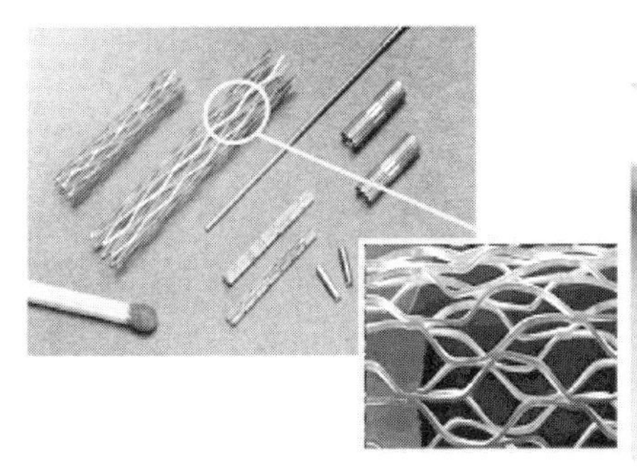

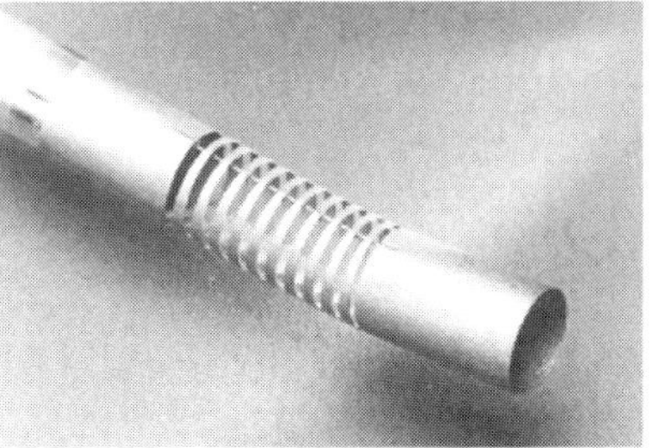

Fig. 10.8 Components fabricated by laser cutting; left: stents (overview and detail view) made of NiTi; right: flexible distal tip made of NiTi for a steerable endoscopic camera system.

holes larger than 60 μm in diameter, also laser cutting can be performed if the material thickness is smaller than 2 mm. The main advantage of laser cutting compared to the laser drilling via percussion is that steep sidewalls can be established. Nevertheless, percussion is a fast method, and is therefore good for drilling hundreds to thousands of holes. The smallest attainable diameter of laser-generated holes is in the μm-range.

The laser-assisted micropatterning of steel alloys such as hardened steel is of special interest in microsystem technology. The main challenges are to realize defect-free and smooth surfaces during laser processing. Two different kinds of laser-ablation technique are suitable: ablation via sublimation and laser microcaving (LMC). Ablation via laser cutting with two laser beams or ablation via melt ejection is not precise enough for the desired applications (see Figs. 10.10 and 10.11). The surfaces will be very rough (R_z >>10 μm), and therefore the accuracy of the generated shapes and holes is only suitable for mm-scaled applications. Furthermore, debris formations and contaminations are too large.

A significant decrease of debris was realized via laser-assisted patterning of hard materials such as cemented carbide with frequency-tripled Q-switch Nd:YAG (wavelength 355 nm) [16] and Q-switch Nd:YAG (wavelength 1064 nm) [17] lasers. Nevertheless, a significant surface roughness and pore formation at the sidewalls is still observed. Therefore it was decided to develop the process of laser microcaving (LMC) [18–21]. LMC can be described as a laser-induced oxidation. A simple description of the mechanisms during LMC is the following: LMC is performed with laser powers of P_L=2–7 W in cw (continuous wave) mode. The laser intensity profile is a gaussian one with a focal diameter of about 10–20 μm. The steel substrate is locally heated by laser radiation. The laser energy causes a temperature rise above the melting temperature. In combination with oxygen as the processing gas, a laser-induced oxidation of the melt occurs. Under special conditions, the mechanical tension inside the oxide layer reaches a critical value, and the oxide layer lifts off from the bulk material. LMC is performed with a solid-state laser-radiation source (Nd:YAG, wavelength

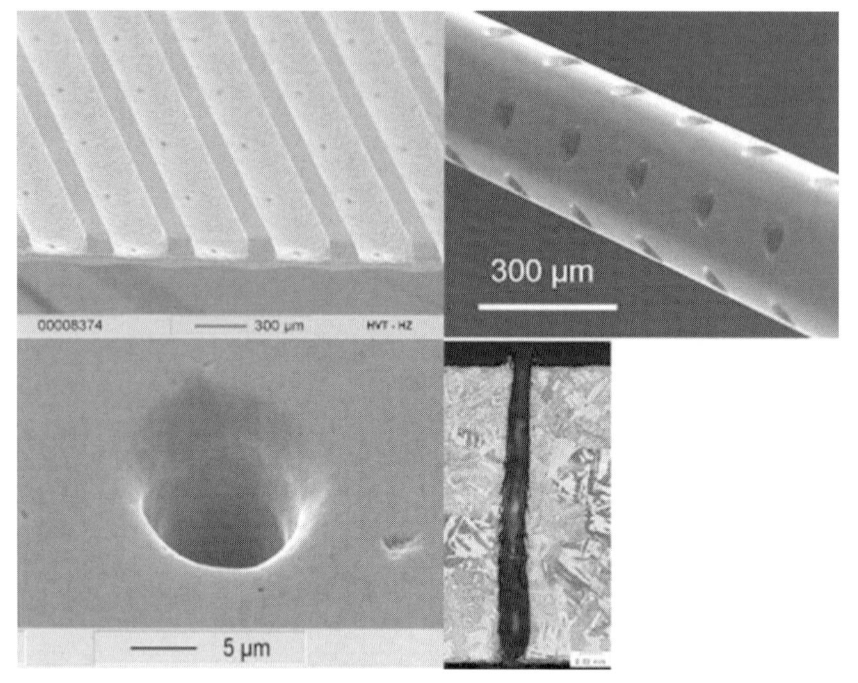

Fig. 10.9 SEM images of small via holes in stainless steel obtained by laser-percussion drilling (Nd:YAG); top: holes with a diameter of 20 µm and 50 µm in a sheet (micro reactor) and in a tube (medical needle); bottom: blind hole with a diameter of 11 µm and cross section of a hole with a diameter of 20 µm and an aspect ratio of 12.

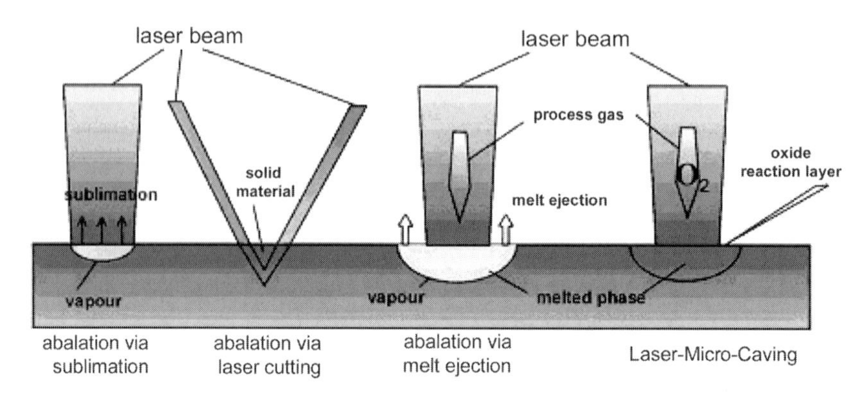

Fig. 10.10 Different experimental setups for laser patterning of metal surfaces.

$\lambda = 1064$ nm). The laser beam is focused onto the sample surface by a lens and can be scanned over the sample surface via deflection mirrors up to speeds of 2000 mm s^{-1}. Large areas of up to 110×110 mm^2 can be treated. The patterning of metallic surface via LMC enables a "clean" patterning process with only a

Fig. 10.11 SEM of a stainless steel foil with laser-patterned linear microchannels. The laser parameters have not been set correctly, so debris deposition is high, and the surface roughness is poor.

small amount of debris and melt. Surface qualities with a roughness of about $R_a = 100$ nm can be realized. Ablation rates are in the range of 10^5–10^8 $\mu m^3\ s^{-1}$. Microstructures with aspect ratios up to 10 were achieved. Fundamental aspects of LMC with respect to rapid tooling were investigated in detail [22]. In the actual research and development three-dimensional geometries are created in a 3D-CAD tool. In dependence on the ablation rate per laser scan, the geometry is subdivided into layers with a thickness of about 1–20 µm (CAM module). Each of the layers can receive an individual filling strategy. One important strategy is the change of scanning direction after each layer. The value of the angle between two scanning directions can be changed as desired. The change of scanning direction should avoid an accumulation of scanning patterns. Furthermore, the steepness of the sidewalls as well as their surface quality can be optimized by using appropriate scanning directions. An additional process strategy is the choice of the appropriate scan offset which is in the range between 1 µm and 15 µm for optical focus diameters of about 10–20 µm. This parameter determines the average surface roughness and ablation depth per layer. The best average surface roughness without any postprocessing is of about $R_a = 100$ nm while the ablation rate is in the range of 1–20 µm/layer.

10.2.3 Wet Chemical Etching

Wet chemical etching might be the most flexible process for microstructuring of metallic materials. Almost all metals are etchable with certain reagent mixtures (see [22, 23]), and the etching process itself is well defined, quite well

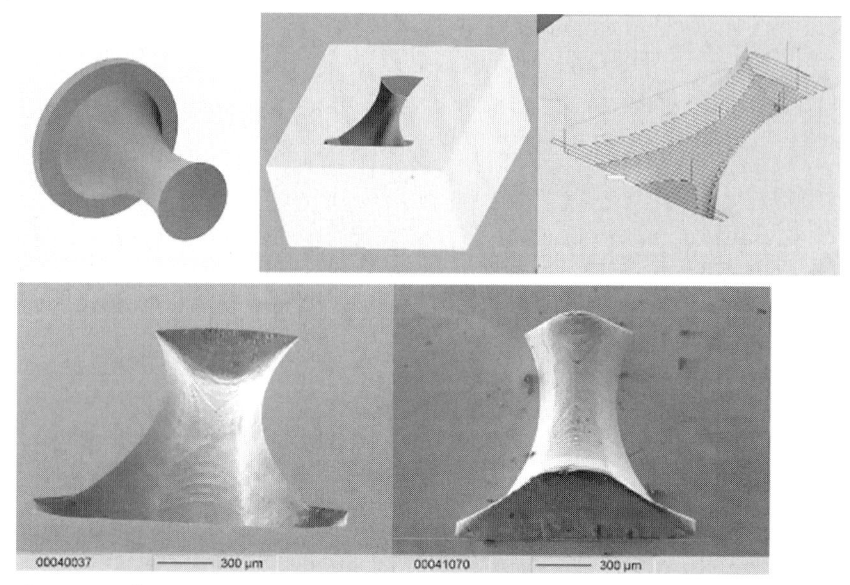

Fig. 10.12 Different stages of prototype fabrication (from top left to bottom right): initial design of a three-dimensional part (CAD-module); individual filling strategy for each of the 40 layers (CAM-module); laser patterned steel substrate (LMC) and demolded part made of PMMA.

understood, controllable and very cheap. Moreover, in principle it is possible to generate any microstructure design, like bent channels, freestanding columns or meander structures [24–29]. Examples are shown in Fig. 10.13. Most designs are also possible to obtain by mechanical micromachining, but the costs are much higher. Due to the low prices of etching microstructures, almost all microstructure mass products in common industry (like aperture masks for TV and computer screens, SMD samples, ICs, etc.) are produced with etching technologies (not only wet but also dry etching). Nevertheless, there are some limitations to this process that makes it not suitable for all microstructuring tasks.

Wet chemical etching is based on a photolithographic process to generate an etching mask. This process is taken from silicon technology and is described in detail in Chapter 11 or [1, 5, 27–30]. Here, we should keep in mind that there is a resolution limitation by the optical processing and the mask material used.

In most cases, wet chemical etching means isotropic etching. Thus, the microstructure dimensions reached with this masking technology is defined by the width of the mask opening plus twice the etching depth. Due to the isotropic etching, the edges of the cross section of the structures are more or less rounded, in the case of relatively small structures the cross section becomes semielliptic or semicircular. Figure 10.14 shows an example of such a semielliptic structure etched in stainless steel.

Fig. 10.13 Left: SEM of sinusoidally shaped microchannels wet chemically etched in stainless steel. Right: SEM of a linear microcolumn structure wet chemically etched in stainless steel. The rounding of the edges at the bottom of the sidewall of the single microcolumns is clearly seen.

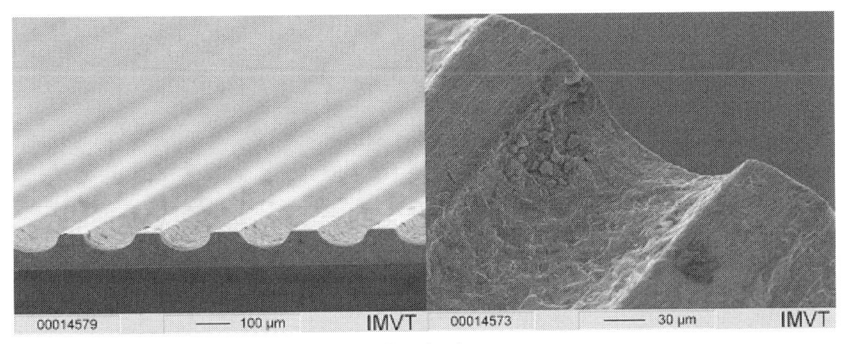

Fig. 10.14 Left: SEM of linear microchannels. Clearly seen is the semielliptical cross section of the etched microchannels. Right: SEM detail of a wet chemically etched microstructure in stainless steel. The surface roughness is in the range of several micrometers.

Another limitation is the more or less poor surface quality. Roughness in the range from about half a micrometer (obtained with copper) up to several micrometers for stainless steel are normal (see Fig. 10.14). For some applications (like deposition of secondary layers on the sidewalls of microstructures), this surface roughness may be an advantage. With regard to the fouling problems occurring in microstructure devices, they are definitely a disadvantage.

10.2.4 Selective Laser Melting

The techniques described so far are abrasive techniques, which means they remove material from the surfaces of the machined materials. Another pathway can be chosen by using selective laser melting (SLM). This technique is based on the microstereolithographic process (see Section 10.3.2) and belongs to the so-called rapid prototyping processes, in which a structure is generated on basis of CAD models only [1, 31–33].

Starting with a three-dimensional CAD model, the microstructure is built by melting metal powder particles locally together using a focused laser. With this method, a microstructure can be generated layer by layer with, in principle, any design. The surface quality is not very high, and it is not quite clear how leak-tight or porous the generated metal walls are. Nevertheless, this is one of the rare methods to create 3D microstructures without bonding technologies. The resolution and density of the generated microstructures is mainly influenced by the laser energy and the exposure strategy. For SLM, stainless steel powder with grain sizes of 10 μm to 30 μm was used so far. Other materials may be possible. Figure 10.15 shows some examples of the surface quality of a microstructure device generated by SLM.

At the CAD model in the center, four areas of interest are marked in red. Recent results presented here show high-quality welding beads 80 μm thick and 50 μm high. At some points, single powder grains are deposited, but the surface coverage with those grains is below 10% of the surface.

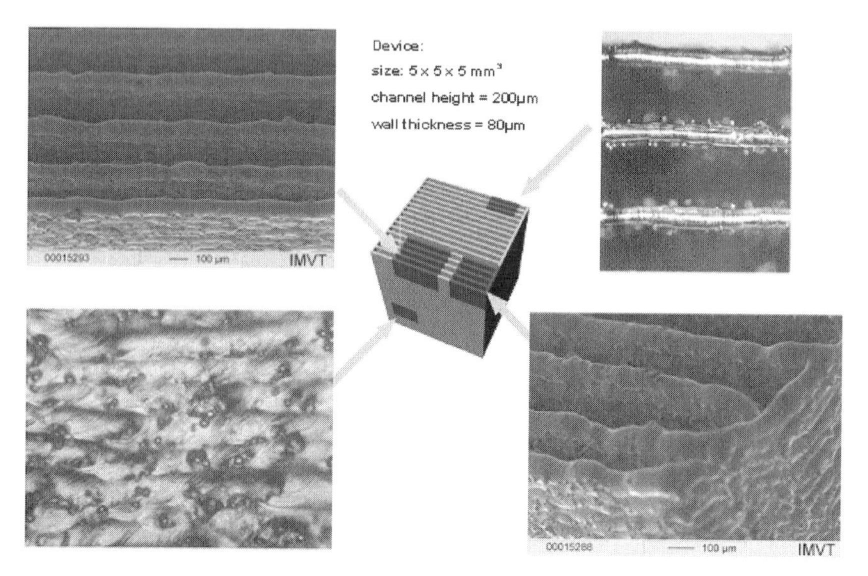

Fig. 10.15 Examples for SLM – microstructure devices made of stainless steel.

10.3 Microfabrication Techniques for Polymers

Although most devices in microreaction technology are made of metals, there is evidence for a growing role of polymers in the future. Polymers, often named 'plastics', provide some advantages that justify their use.

Compared to metals, polymers are lower in their thermal, chemical and mechanical resistance. In many cases, the opportunity of optical transparent components is a good reason to use polymers (or, in some cases, glass, see Chapter 11). Sometimes thermal and electrical isolating properties of polymers or other, more special options of polymeric materials are the reason to use them, but in most cases the option of a cheap mass production, namely by molding technologies like injection molding, is the unique selling point of polymers in technologies.

One example of the use of polymer specific material properties is given in Fig. 10.16, a fine perforated membrane, made of a deep-draw polycarbonate film. The pores are produced by irradiation with xenon ions followed by subsequent etching. This film is used as a filter to separate cell cultures from a nutrient solution. Neither metals nor ceramics are able to form a porous membrane in such an easy manner [34].

10.3.1 Materials Selection

While two of the three main classes of polymeric materials (duromers, elastomers and thermoplastics) are rarely used, most known applications are from the moldable thermoplastics. An example of the use of a duromer material is

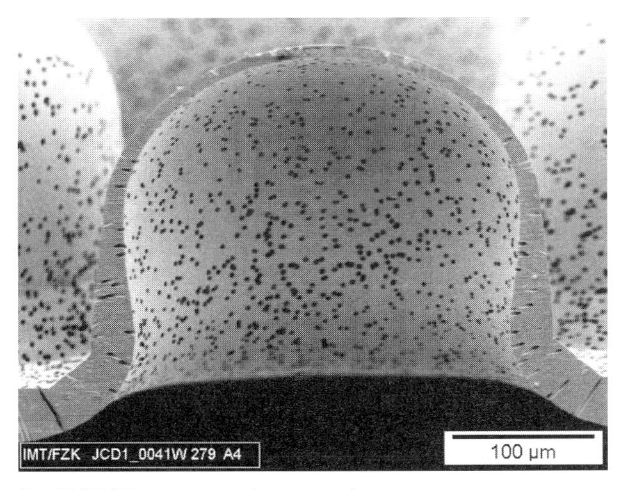

Fig. 10.16 Microporous polymer membrane.

given in Section 10.3.5, Microstereolithography. Elastomers are rarely used as sealing, for example as O-rings made of Viton® (see Section 10.4.4 Packaging and Sealing).

Thermoplastics are the usual plastics, built up from long organic molecule chains. The chemistry along this chain defines the type of the material, the length and the number of branches modifies the properties as well as additives. Well-known examples of these are particles or short fibers, used to enhance the stiffness and the temperature stability.

Nearly all thermoplastics are moldable for microfluidic devices. Here, the application parameters define the material, in particular the chemical resistance is crucial (see Section 10.1). The highest application temperature should be well below the glass transition temperature. The polymer with the highest temperature resistance is PEEK. It can be used up to 300°C. If any optical detection is needed, transparent amorphous polymers like PMMA, PC, COP or PSU will do.

More problems are given from the media influence. No polymer is stable against every liquid. Choosing a material different in chemical construction to the liquid is preferable. For example, COP is a class of amorphous polyolefines that are heavily damaged by nonpolar organic liquids like oils or benzene. They are pretty stable against nearly all aqueous solutions, except very oxidizing agents. PVA (Polyvinylalcohol) is very watery, very stable against all oils, but in water perfectly soluble.

10.3.2
Molding

Constructing parts by molding is simple in principle and has been done for hundreds of years, e.g. in bell construction. A material has to be melted, then filled into a form and resolidified. A product in the shape of the inner lumen of the form is created – that is it.

Polymers melt at temperatures below 350°C. This allows molding with metallic tools many times without damaging them. A form for a bell is only to be used once and is destroyed when the bell is demolded. So, the bell has at least the price of the mold. Using a mold for lots of products divides the mold price by the number of products, which means even a complex and costly mold allows very cheap pieces. This is especially true, if more functionalities are added in the mold and are transferred into the product in one step [30, 35].

Figure 10.17 gives an example of this principle. Here, optical as well as mounting and adjusting functions are integrated into one mold [36].

10.3.2.1 Generating a Mold

For all replication techniques, a precise microstructure mold is needed. All methods described in Section 10.2 are suitable to manufacture molds. Thus, since the price of the mold is not the determining factor, other structuring methods are also applicable, even combined methods to get more complex

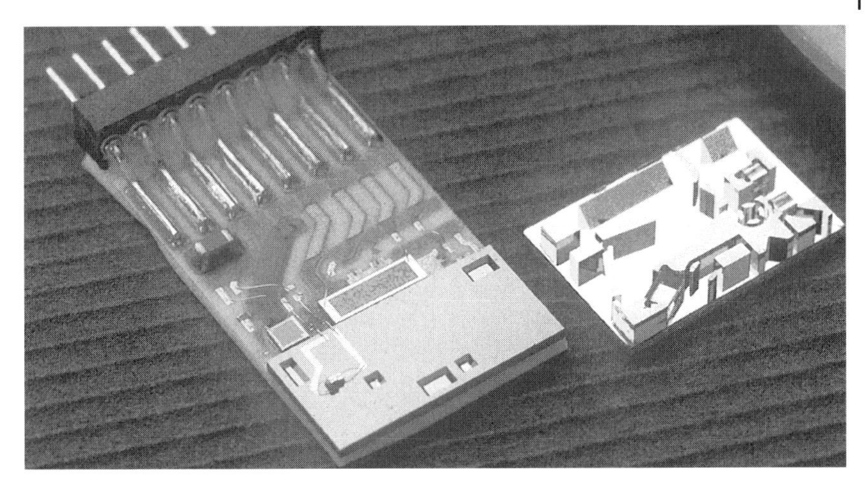

Fig. 10.17 Components of an optical distance sensor. The part on the right is made from polycarbonate by molding and contains different functions (mirrors, lenses, alignment, mounting structures).

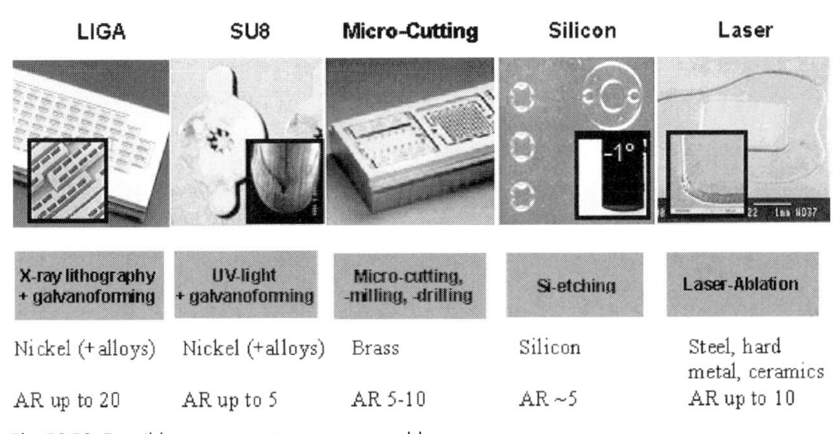

Fig. 10.18 Possible processes to generate molds.

structures. Figure 10.18 gives some examples, including the aspect ratio AR for the processes [1, 30].

10.3.2.2 **Injection Molding**

Injection molding is a standard process in the plastics industry. Injection-molding systems are produced in numbers and in a wide range of size by several companies. The process is well established and highly automated, producing most of daily-life plastic parts.

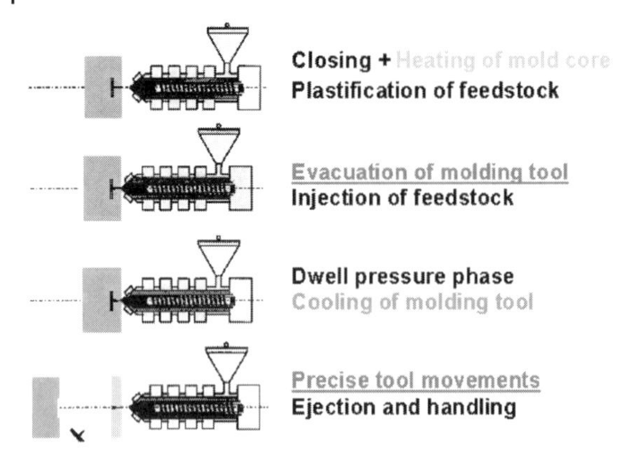

Fig. 10.19 Principle steps of the injection-molding process.

The sketches in Fig. 10.19 demonstrate the injection-molding process (black letters: standard process, colored letters: extra steps that are sometimes necessary for the production of microstructured parts) [37].

The cycle time of injection molding can be as low as one second (in rare cases even slightly below), but for complex parts it is larger. Especially, when a variothermal process is enforced (heating and cooling of the mold in each cycle) it may reach several minutes. In industrial applications the process is highly automated and can be used to produce devices in the order 10 million parts per year and more.

An example of an injection-molded microstructured fluidic device, a disposable device for medical applications, is shown in Fig. 10.20.

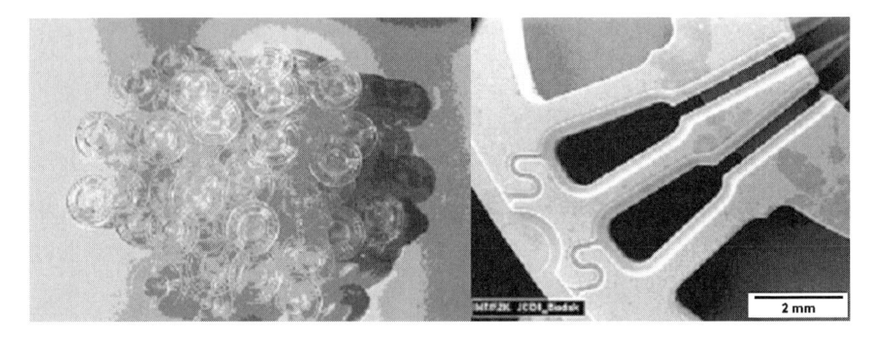

Fig. 10.20 Left: Biodisk, injection molded in PMMA, mold insert manufactured by precision milling. Right: SEM detail of the Biodisk. This photo demonstrates the diversity of functional structures that can be produced for the cost of one molding shot.

10.3.3 Hot Embossing

Injection molding not only needs a microstructured mold but additionally a complex tool that forms the other parts of the desired products. It contains structures for the runner system, cooling channels, ejector pins and other functionalities and is normally quite costly. For sample development, design variations or prototyping this is a major drawback.

Hot embossing is a reasonable alternative, which allows the efficient production of samples and even of small test series. Only a mold insert and a hot press is needed. The mold insert – mounted in a ground plate, if needed – is covered with a plate of the polymer and a counter plate. This sandwich is heated up and then pressed together. The polymer material is slightly plasticized and flows into the cavities. In Fig. 10.21, the principle of hot embossing is shown.

Automation is not easy, and the long cycle times of hot embossing makes it not a method of choice for usual mass production [38].

For laboratory use and some exceptional products, specialized presses for micro hot embossing are available by, e.g., JENOPTIK (HEX03).

10.3.4 Polymer Laser Micromachining

UV-laser-assisted micromachining of polymers is a powerful tool for a rapid manufacturing of complex three-dimensional microstructures in polymer surfaces. The major drawback for this process is costs: Laser machining is time in-

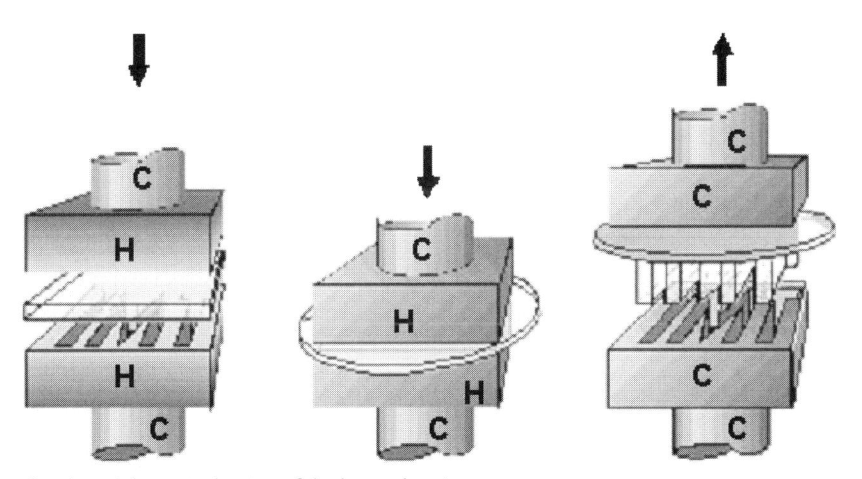

Fig. 10.21 Schematic drawing of the hot embossing process. H: hot, C: cold, transparent: polymer, lower plate: mold insert, upper plate: counter plate.

tensive, which makes it more or less a method for prototyping or small series production, but not for mass production of devices. Nevertheless, the process is very important due to the high quality of devices that can be manufactured.

Laser-machined polymer surfaces can have depths between 100 nm and 500 μm and aspect ratios up to 10. Minimum structure sizes are in the μm range. Typical application fields are in micro-optics, microfluidics, life science and rapid tooling [19, 39–44]. There are two types of UV-laser radiation sources: Excimer (wavelength: 157, 193 or 248 nm) and frequency-multiplied Nd:YAG (266 or 355 nm). In general, excimer lasers are used for large area patterning while Nd:YAG are used for small-volume ablation.

The UV-assisted ablation process is mainly influenced by the laser fluence ε. The etch rate as a function of ε for the polymers PSU, PEEK and PI is shown in Fig. 10.22. It is obvious for PEEK and PSU that the etch rate increases linearly with increasing ε for values between $1\ \mathrm{J\,cm^{-2}}$ and $5\ \mathrm{J\,cm^{-2}}$. PI reveals the smallest etch rate of all of the selected polymers. Furthermore, the etch rate of PI as function of ε reveals three regions (see Fig. 10.22) that can be referred to:

(0.4–$1.8\ \mathrm{J\,cm^{-2}}$) saturated excimer laser absorption with $R \propto \frac{1}{a}(e - e_0)$,

where ε_0 denotes a characteristic laser fluence usually not equal to the threshold laser fluence and a is called "effective absorption coefficient",

(1.8–$3.0\ \mathrm{J\,cm^{-2}}$) plume attenuation of the laser intensity that finally results in a
($> 3.0\ \mathrm{J\,cm^{-2}}$) saturated ablation rate.

For significantly smaller ε a further region referred to linear excimer laser absorption is expected that can be described by Beer's absorption law [39]. For $\varepsilon > 3\ \mathrm{J\,cm^{-2}}$, the ablation rate for PI tends to saturation. For applications it is suitable to work within this third region in order to guarantee a high accuracy in structure depth. Small variations in pulse energy lead to no significant change of the ablation rate. Furthermore, it is observed that with increasing ε the contours of sidewalls become more rectangular. The inclination angle of a

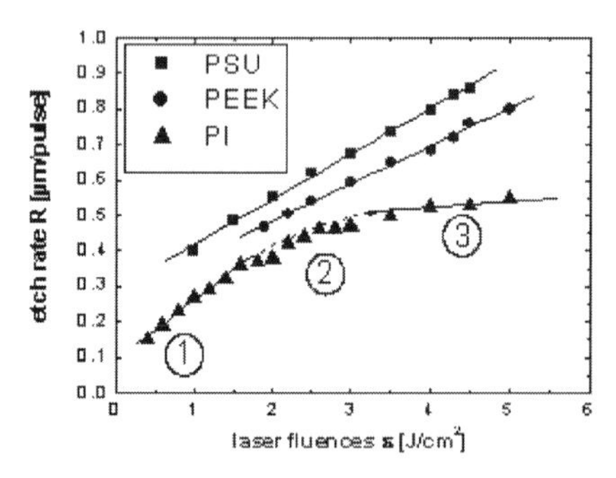

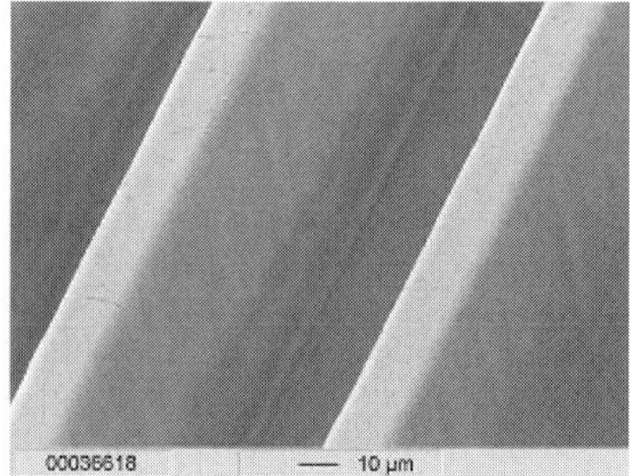

Fig. 10.22 Left: Etch rate for polysulfone (PSU), polyetheretherketone (PEEK) and polyimide (PI) as function of laser fluences (KrF-excimer, wavelength 248 nm, pulse width 25 ns, laser repetition rate 10 Hz, number of laser pulses $N = 100$); right: SEM of small trenches in PI obtained by excimer laser ablation.

sidewall in PSU decreases from 17° ($\varepsilon = 1$ J cm^{-2}) down to 6° ($\varepsilon = 5$ J cm^{-2}). The inclination angle for PEEK is somewhat smaller (7° for $\varepsilon = 2$ J cm^{-2} and 5° for $\varepsilon = 4$ J cm^{-2}). At very high ablation rates, a slight rounding is indicated at the bottom of laser-patterned walls.

With increasing ε a decrease of debris formation is observed. The size of material fragments as well as the surface roughness of the sidewalls decrease also significantly. In the case of an increasing laser pulse number, N, an enhanced deposition of debris is observed and the size of ejected material fragments increases. Rounding of edges on the ground of laser-patterned grooves occurs. This effect becomes more important with increasing N.

The numerical aperture NA of the objective has a great influence on the cross section of laser-patterned trenches. A rectangular cross section is achieved with numerical apertures under 0.25. The appropriate slope of the sidewalls depends on a suitable laser fluence as described above. High numerical apertures cause significant rounding on the patterned ground. In the case of NA > 0.4, undercuts of the sidewalls and rounded edges on top of the trenches can hardly be avoided. Steep sidewalls and well-defined contours of edges are required for the fabrication of polymeric mold inserts. For this purpose, objectives with NA = 0.10 are used.

The wet-chemical pretreatment of the polymer and the processing gas is an important aspect for improved surface quality: The cleaning of PSU and PEEK in the ultrasonic bath and the use of helium as process gas caused a drastic reduction of the debris formations. Thus, the surface roughness could be reduced to $R_a < 200$ nm. For PSU, the effect of wet-chemical pretreatment is very significant and may be caused by a penetration of organic solvent in the outer polymer surface [45].

The best attainable surface roughness after excimer laser patterning of the selected polymers is in the range of $R_a = 100$–200 nm for PEEK and PSU. For PI, the average surface roughness is significantly lower than 100 nm. The average roughness for PI is between 50 nm and 60 nm for high ε. This is an excellent condition for the laser-assisted fabrication of polymeric microstructures.

Excimer laser ablation can be performed with different mask techniques in order to meet the desired pattern: Direct optical imaging of complex structures or motorized, moveable masks are established. Furthermore, a motorized rotating mask can be used that allows changing of the complex mask structures with high precision and without stopping the ablation process. This mask technique is used for the patterning of large complex structures. Another technical option is the use of a motorized aperture mask: The CAD data can be directly transmitted into the polymeric surface. The aperture consists of four independent shields that can be positioned with μm accuracy. This mask technique is a precise processing technique for the fabrication of well-defined groove structures such as cross and T-cross designs (see Fig. 10.23) or grooves with a variation of cross section. This scanning technology ("laser direct writing") is flexible and needs no additional masks, but the processing time may be large compared to the direct optical imaging of a large mask geometry.

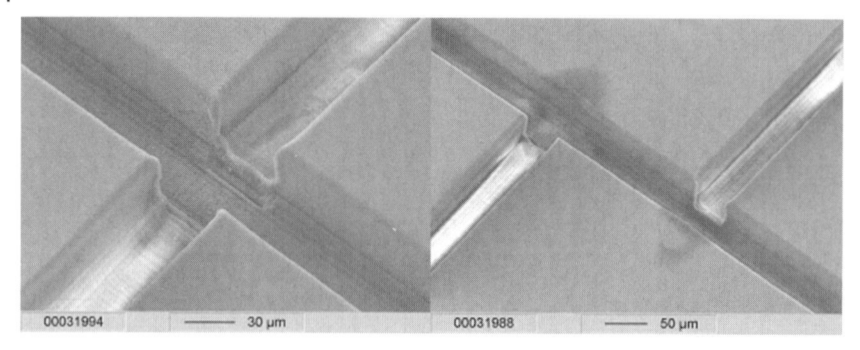

Fig. 10.23 Laser-patterned microchannels made of PSU with cross (left) and T-cross design (right).

The patterning of polymer chips is performed with the motorized aperture. The functionality of, e.g., a capillary electrophoresis chip-prototypes is successfully demonstrated. The top view of the chip with cross design (left part of Fig. 10.23) reveals that the edges of the grooves are in a rectangular shape. This is a very important fact, because a possible rounding of these edges may cause turbulences of the sample fluid during chip operation. The rounding of the edges in such polymeric components may be observed when conventional mechanical micromachining is used; because the micro-end-mills have a diameter of about 50 µm in general.

Recently, new types of excimer laser radiation sources were developed with increased laser pulse repetition rate up to 500 Hz or 1000 Hz compared to the "conventional" high-power excimer lasers with a repetition rate of about 50–100 Hz (energy up to 1000 mJ/pulse and a pulse width of 20–25 ns). These high "repetition-rate excimers" reveal a smaller laser pulse energy (up to 20 mJ) but also a significant smaller laser pulse width (6 ns). It is observed that the debris formation during patterning is significantly reduced for all polymers by the use of small laser pulse widths. A small pulse width leads to a significant reduction of thermal contributions during ablation. This is also indicated by the laser structuring of test structures such as very small walls with a width of about 3–5 µm and a height of about 100 µm (Fig. 10.24).

If the scanning process is too slow because of the low laser repetition rate (<500 Hz) it may be useful to establish UV-Nd:YAG laser radiation sources where laser repetition rates up to several kHz are possible. This could cause a reduction of processing time of more than one order of magnitude (see Fig. 10.25). High ε of >40 J cm^{-2} enable polymer ablation as well as ablation and drilling of glass, steel and ceramics. For PC and PI investigations concerning ablation rate and surface roughness were performed. The surface roughness for PC varies only between $R_a=140$–400 nm for all laser pulse numbers. PI reveals a lower etch rate but an increased surface roughness ($R_a=0.3$–1.3 µm). The surface roughness largely depends on the pulse overlap during laser scanning as well as on the optical focus diameter.

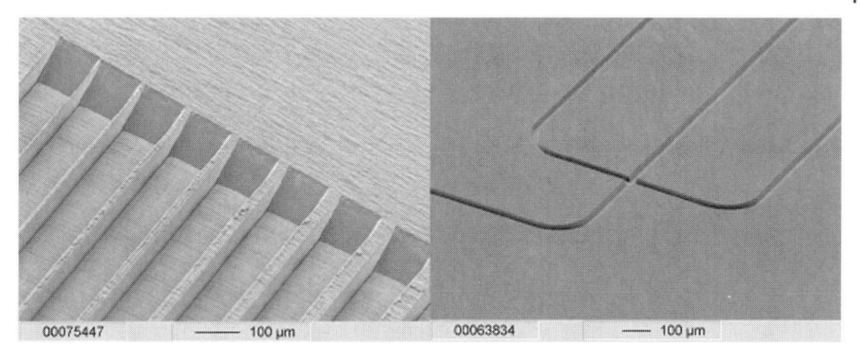

Fig. 10.24 SEM of microstructures in PI obtained by excimer laser ablation.

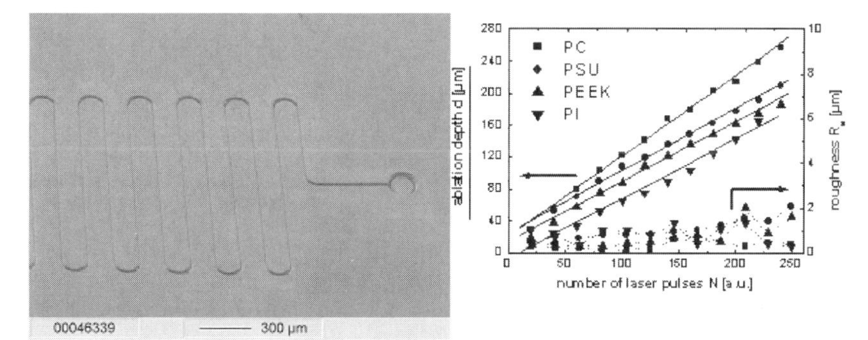

Fig. 10.25 Left: SEM of a channel structure in PI obtained by laser direct writing with high laser repetition rates; right: ablation depth and average surface roughness as function of laser pulse number (4ω Nd:YAG, wavelength 266 nm, pulse width 400 ps, laser repetition rate 1000 Hz, focal length 75 mm) for polycarbonate (PC), polyetheretherketone (PEEK), polysulfone (PSU) and polyimide (PI).

Finally, it has to be mentioned that for structure sizes in the range of 100 µm the direct patterning of polymers with CO_2-laser radiation may be appropriate. Nevertheless, for the state-of-the-art in CO_2-laser-assisted ablation, the inclination angle of the sidewalls of the generated channels are not well defined. A significant formation of a U- or V-shape is observed [46].

10.3.5 Microstereolithography

Microstereolithography is an additive process to generate 3D polymer microstructures. The technology used is similar to the previously described SLM technique (see Section 10.2.4). In Fig. 10.26 (left), the process is schematically shown.

The model is built within a liquid monomer substrate on a building platform that can be moved vertically, following the lines of a 3D-CAD model. The platform is lowered for the distance of a layer thickness. Liquid monomer floods

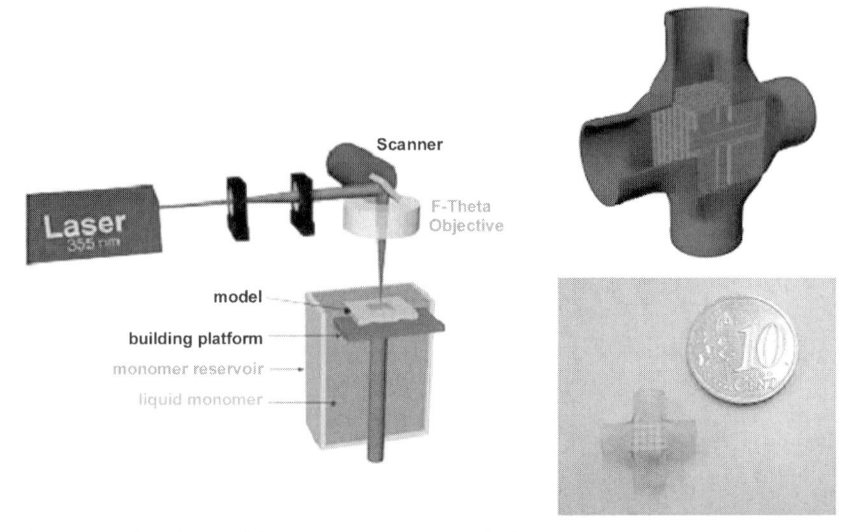

Fig. 10.26 Left: Scheme of the SL process (see text). Right: 3D-CAD model and by SL realized crossflow microstructure heat-exchanger device.

the building platform, and a focused laser exposure for the next layer starts. With the exposure, the plastic is polymerized, and the next layer is generated [31, 47–50]. With the SL process, a wall thickness of about 100 μm is feasible. Smaller designs are possible, the question here is the applicability of these to (industrial) processes [48–50]. In Fig. 10.26 (right), a CAD model of a crossflow microstructure heat exchanger and the device realized with SL is shown.

10.4 Assembling, Bonding and Packaging

The second stage in the microfabrication process is assembling, directly followed by the third stage, the bonding process, and the fourth, packaging and sealing the devices to make them connectable to conventional process-engineering equipment. Here, we will give only a short review about some demands, difficulties and techniques for the three stages for metal and polymer microstructure devices.

10.4.1 Assembling: General Remarks

While assembling of a number of device parts is not really a problem in the macroscale world, on the microscale this step is more delicate to handle. The main point is the adjustment and alignment accuracy of the parts, in addition to sealing, fixa-

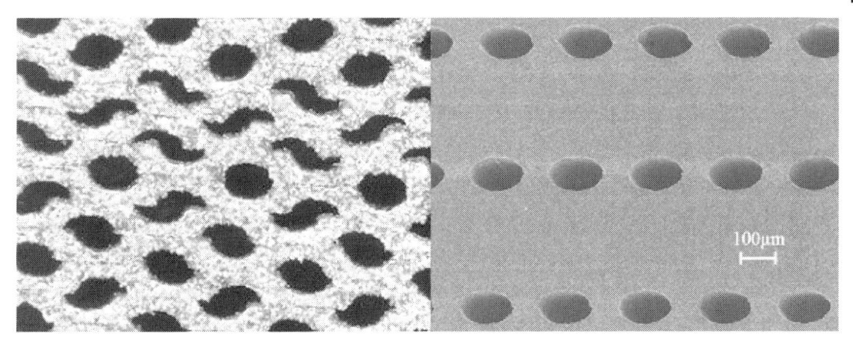

Fig. 10.27 Left: Misalignment of microchannels.
Right: correctly aligned microchannels forming full elliptical cross section.

tion and bonding technology problems. Depending on the surface quality and the bonding technology applied, alignment errors may reach similar dimension values to the microstructure themselves! An example is shown in Fig. 10.27. Here, a number of microstructure foils with wet chemically etched microchannels have been aligned to form fully elliptic channels. As can be seen in Fig. 10.27, misalignment will lead to nonregular channels and therefore may interfere with the bonding technique. A correct alignment will lead to only small deviations from the desired elliptical shape, and the distortion while bonding will be small.

Alignment techniques used to avoid errors can be simple mechanical methods (e.g. use of alignment pins), edge catches in a specially designed assembling device or optical methods like assembling under a microscope. Automation is possible by using laser diodes and photosensors. Most of these methods come from silicon processing technology, where precise alignment of multiple mask layers is needed to guarantee the functionality of the manufactured devices [1, 30].

A speciality of microstructure assembling is directly correlated to the microstructuring technology used. While in assembling macroscale devices, burr formation on the top or at the sidewalls of the device structure do not play a major role. In the microscale, burr formation generated by mechanical micromachining or laser machining may lead to significant problems with assembling of device parts as well as with the bonding. Thus, special attention has to be paid to burr microstructures or to avoid burr formation.

10.4.2
Bonding of Metals

In most cases, microstructure devices can not be manufactured from one part only but have to be assembled, eventually sealed and bonded together. Exceptions are microstructure devices made by SLM (see Section 10.2.4).

Metals can be connected in various ways, e.g. by welding, soldering, screwing, compression, bolting, etc. The application of most of these techniques in building macrodevices is straightforward and has been well known for decades. However, if we come to the connection of microstructured devices care has to be taken not to destroy the microstructure but still obtain a leak-tight seal. Thus, the bonding techniques have to be adapted to meet the special demands of microstructure technology.

Microstructure devices are often made from metal foils. These foils are stacked, thereby obtaining passages. These passages have to be sealed against each other and against the environment. If there is one foil only, a straightforward way to seal the structure against the environment is to include the foil into an adapter using sealing such as O-rings or gaskets. This has been a standard method since the beginning of microstructured flow devices [51, 52].

However, the use of gaskets or O-rings between foils is at least tedious, if not impossible. The alternative method is compression of foils by the external casing as demonstrated by IMM Mainz using graphite foils for sealing [52] or compressing metal on metal [53]. Leak tightness is difficult to achieve, however, due to roughness of the microstructured platelets.

Most other methods of joining are nonremovable, since they include alteration of the materials connected. These can again be differentiated into those needing additional material to form the connecting layer, such as brazing, soldering and a technique called intermetallic bonding, and those techniques where no other material is introduced, like electron-beam welding, laser welding and diffusion bonding. Here, conventional welding and soldering or all the other quite well-known techniques will not be discussed, but two processes will be described more detailed: diffusion bonding of metals and laser microwelding. Other techniques are described briefly in the following.

For the use of brazing to join aluminum there is a patent application by Atotech [54]. In this application, soldering of foils by stacking aluminum foils that have been layered with nickel, silver and tin is described. The metallic phase that is responsible for the joint is an alloy of silver and tin and is said to be more temperature resistant than the starting material. During the "Micromotive" project a soldering technique using foils from solder material was developed [55, 56].

Using different materials for joining metals, however, introduces corrosion problems due to the formation of electrochemical local elements. Furthermore, in most cases the newly formed phase will not have the same corrosion behavior as the construction material. Thus, the joining areas will probably be those most attacked by corrosion.

The formation of strong intermetallic phases from foils made from precursors, here aluminum and nickel has been described as 'intermetallic bonding' by Paul et al. [57].

Electron-beam welding has been used to join microstructured, diffusion-bonded metal stacks into macroscopic housings [58]. Experiments have been performed to join foils by laser welding [59], but it proved to be difficult to create a joint stable to a second laser welding operation in a distance of only one foil thickness. The use of this method is currently restricted to the gasket-sealing method or in joining other devices. More details can be found in the following.

In general, it is concluded that for foil joining techniques that affect the whole joining area will be better suited than those techniques that join only the outer circumference. This is due to the fact that the foils are comparatively thin and lack the stability needed to transfer the force from the outer perimeter to the central area. However, by using diffusion bonding and brazing or soldering methods devices from microstructured foils can be built that withstand high external forces.

10.4.2.1 Diffusion Bonding

Diffusion bonding has been introduced as the joining technique for microstructured devices by Schubert et al. [58] in the beginning of the technology. It is a common process to generate microstructure devices to withstand high temperatures and high pressures and is therefore often used.

Diffusion bonding is achieved by pressing a stack of metal foils in a controlled atmosphere at temperatures in excess of 2/3 of the melting temperature of the material for some hours. Other groups use diffusion bonded also, sometimes under the label '(multi-) lamination' [59].

In principle, the cleaned, assembled parts are fixed in the desired arrangement. This complete arrangement is placed into a furnace, mostly including the fixation device. The furnace itself is either evacuated or set under pressure with an inert gas. A strictly defined temperature and mechanic force program is applied, which highly depends on the microstructure design, the material, the surface quality and the designated application. This leads to a diffusion process at the connecting surfaces of the single parts, following the known laws of diffusion. If the diffusion-bonding process is done correctly, a very stable and insolvable interconnection between the single parts of the microstructure device will be reached. The microstructures will obtain a shrinkage of some per cent, depending on the material and the bonding parameters. In Fig. 10.28, the principle of the diffusion-bonding process is shown on the example of a crossflow microstructure heat exchanger made of stainless steel. The grain structure of an alloy grown across the boundary layer of microstructured foils is also shown.

Some restrictions have to be taken into account by using diffusion bonding. It is, e.g., extremely difficult to bond materials that generate protective layers on top of the surface (like aluminum alloys or alloys with a high chromium percentage). Some metals simply do not bond by diffusion bonding because of the high temperature needed (like tantalum or tungsten). Bonding alloys has to be done very carefully, because the process may lead to a local change of the alloy composition and therefore to an undesired material change.

Beside this, precise parameters for correct diffusion-bonding processes are not easy to obtain. Depending on the material, a set of parameters may work or may fail, because the alloy composition was changed in one point by a few tenths of a per cent. It is definitely necessary to have certified material qualities

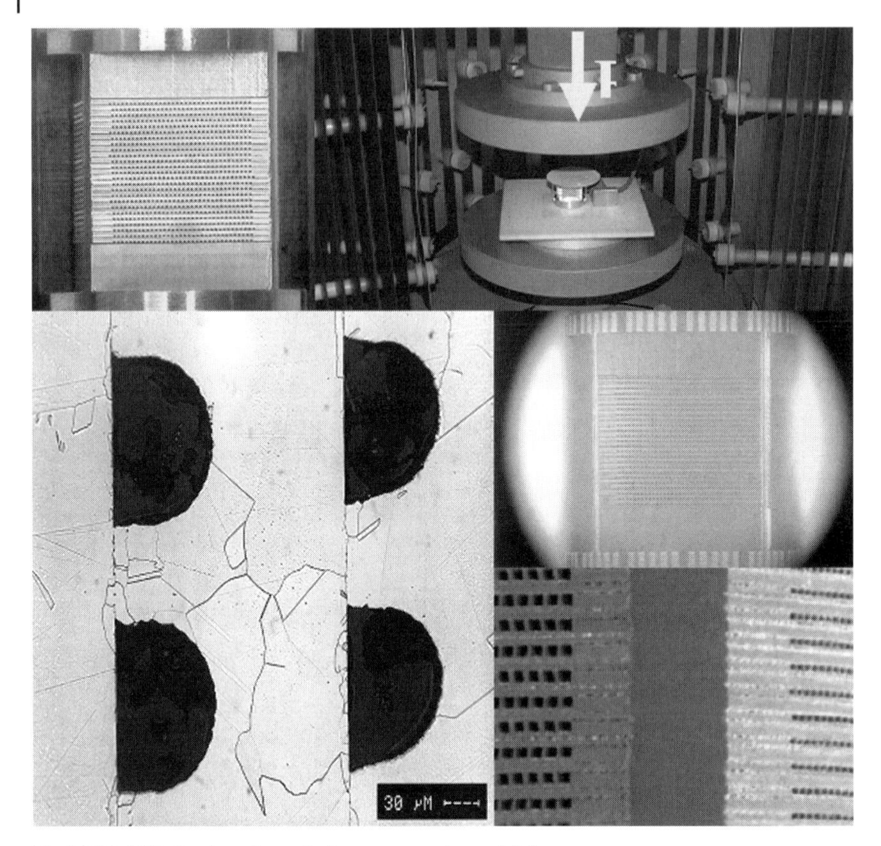

Fig. 10.28 Diffusion bonding of microstructured metal foils. Clockwise: Stacked foils, diffusion bonding furnace, glowing device while bonding, generated channel system, boundary-layer grain structure of diffusion-bonded microstructure foils.

and use only these for each set of bonding parameters. Moreover, depending on the microstructuring technique, the generated microstructure parts may be useless for direct diffusion bonding. If large burrs form on top of microstructure designs (e.g. generated by mechanical micromachining not done properly), diffusion bonding will not work, because the interconnection surface between the single microstructures is not large enough.

10.4.2.2 **Laser Microwelding**

Although laser welding may not be applicable for the joining of single foils (see above), the technology itself is very interesting and gives some opportunities that makes a more precise description suitable.

Two process strategies in laser welding are well established: laser deep penetration welding and laser conduction welding.

For high laser power densities (> 10^6 W cm^{-2}), a plasma and a so-called keyhole is formed that enables a large aspect ratio of the weld seam (→ deep penetration welding). But in general the volume of the weld seam is too large for applications in microsystems. Furthermore, the dynamic of the laser-generated melt can lead to a significant contamination of the surrounding material by melt ejection.

Conduction welding is performed at lower laser power intensities (< 10^6 W cm^{-2}). The width and depth of the weld seams are in the range of 100 to 300 μm. Because of the highly localized thermal impact, the conduction mode is more appropriate for micropackaging applications. Figure 10.29 gives some examples.

In chemical-reaction technologies, the demand of improved microchannel increases. While not working properly for high pressures with stainless steel yet, laser microwelding is of interest if conventional welding technologies such as diffusion bonding or electron-beam welding can no longer be applied. This is the case for special materials such as aluminum alloys, tantalum and high alloyed steel materials, as is shown in Fig. 10.30.

Laser welding of materials with high affinity concerning oxidation, pore and crack formation is still a great challenge in microsystem technology. Another challenge is to combine the main advantage of heat-conduction welding (small heat-affected zone) with the advantage of deep penetration welding (aspect ratio of weld seam >1). In order to overcome this limitation, a new technical approach uses temporal laser pulse shaping (see Fig. 10.31). The first pulse is used in order to overcome the high reflectivity and high heat conduction of the material. When a melt phase is formed, the reflectivity is reduced significantly and a second laser pulse can be absorbed with high efficiency. Figure 10.31 reveals that the weld penetration depth can be increased significantly up to aspect ratios of about 5, while the lateral dimension of the weld seam is kept unchanged.

With this process strategy, complete laser-based manufacturing of microheat exchangers made of high alloyed steels were performed. These microappara-

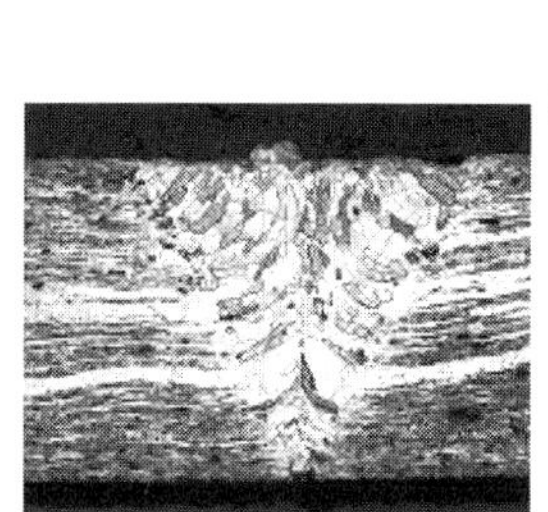

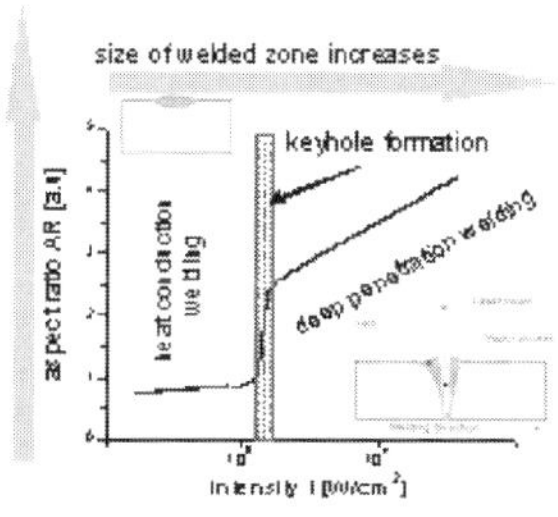

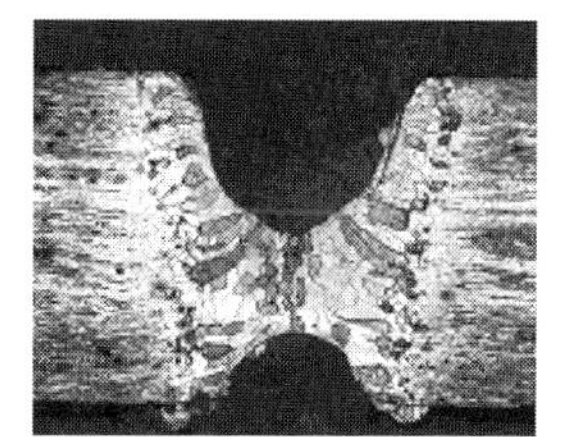

Fig. 10.29 Cross sections of butt-welding geometries with low (left) and high (right) laser intensity. Middle: General dependence between aspect ratio of weld seam as a function of laser intensity.

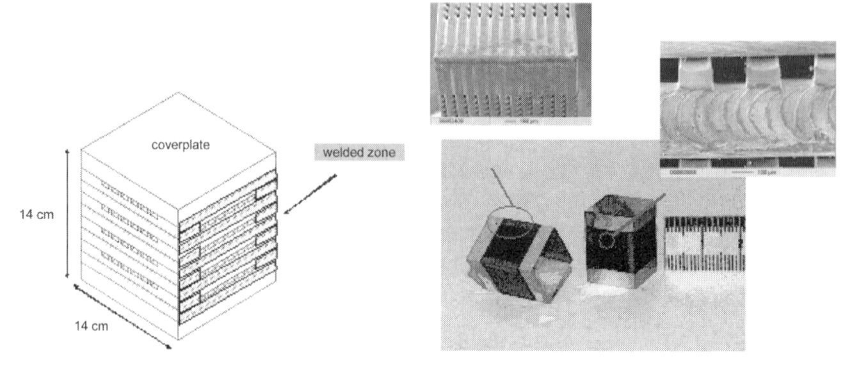

Fig. 10.30 Left: Schematic view of a microheat exchanger; right: photo and SEM of laser welded microheat exchanger made of high alloyed steel.

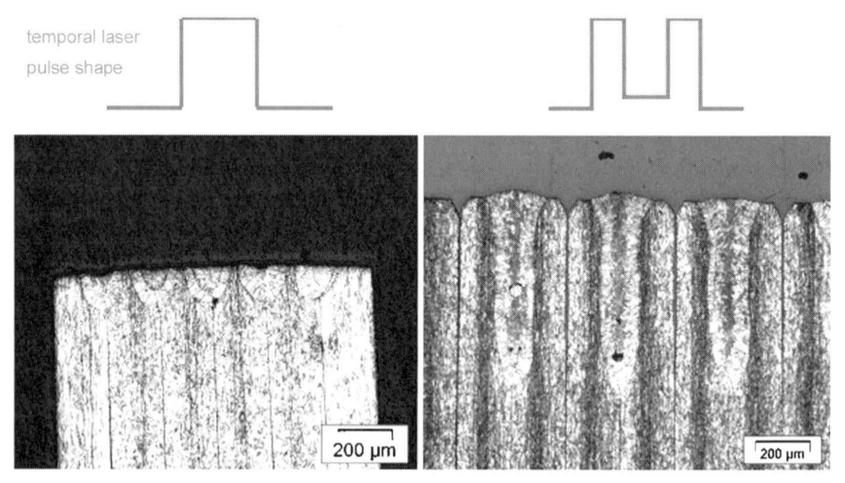

Fig. 10.31 Cross sections of laser-welding zones in steel without (left, rectangular pulse shape) and with pulse shaping (right, double-pulse technique).

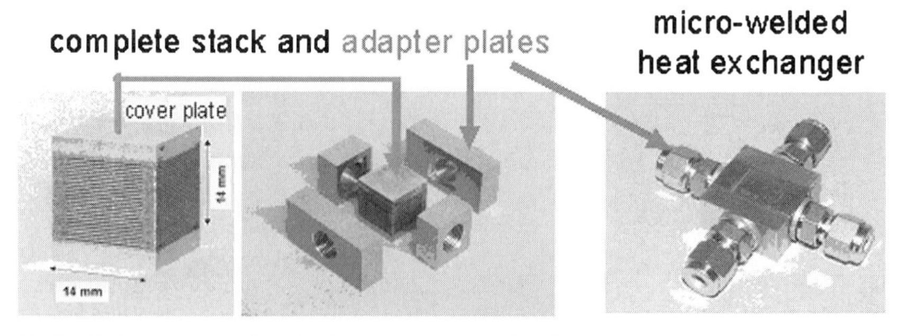

Fig. 10.32 Components of a microheat exchanger made of steel.

tures are tested successfully. The quality of the weld seams enables a high vacuum density of the components (He leakage $<10^{-9}$ mbar l s^{-1}). Furthermore, the laser welded stack and adapter plates are mechanically stable against internal fluid pressures of about 30 bar. Examples are shown in Fig. 10.32.

10.4.3 Bonding of Polymers

The methods of bonding microstructured polymers can be described with a standard issue in the mounting for microfluidic devices. Most manufacturing processes do not allow production of tubes – the first step is the production of a groove, that has to be covered with a top cover to get a pipeline for liquid handling. For all the principles described in the following, schematic sketches are given in Fig. 10.33.

10.4.3.1 Gluing

One of the easiest methods is the use of glue (Fig. 10.33 A). There may be two serious problems:

Some polymers have no adhesive surface. PTFE, POM and the polyolefines are known for not being suitable for gluing. There are etching agents for each material to modify the surface and make it possible to glue, but then an extra process step with contamination problems is being added. For polyolefines, there is an alternative in oxidizing the surface, either by plasma oxidation or by an electrostatic corona.

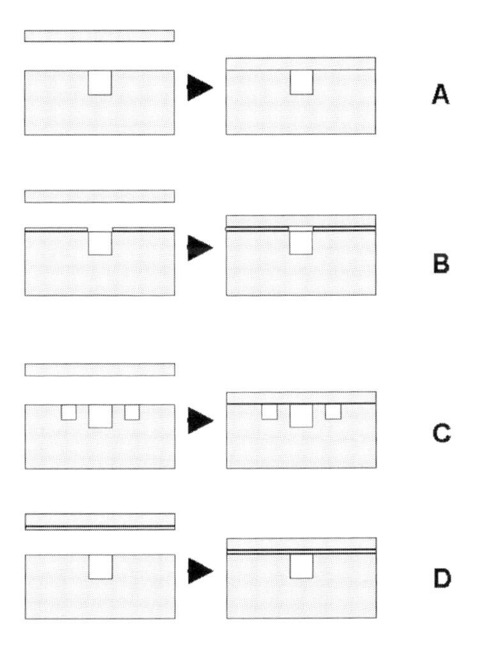

Fig. 10.33 A: Seal a molded groove to form a pipeline, B: structured gluing by screen printing, C: gluing with extra cavities for the adhesive, D: sealing with a sealing foil.

Capillary force is the second issue. The glue tends to flow into the microstructures and clogs the drain. One proven solution is to dose the adhesive only on top of the substrate by screen or tampon printing (Fig. 10.33 B). Another solution is to provide extra cavities with higher capillarity, which will be filled by the glue (Fig. 10.33 C) [60].

The last gluing option is to prepare the adhesive on top of the nonstructured parts. Foils with such an adhesive (tapping or heat sealing) are available commercially. The process is simple and easy to be automated. Two major drawbacks have to be handled: The semiliquid adhesive tends to fill very small cavities as well as a real liquid. The smaller the cavities the more controlled should be the sealing process. Moreover, the sealing material, mostly unknown and often chemically reactive, gets into intimate contact with the medium (see Fig. 10.33 D).

10.4.3.2 **Welding**

Since any additional material in contact to the media can cause dysfunction, welding processes for polymers have found great interest. Without any adhesive, the polymer material itself is to be activated in order to diffuse into the counter part. Due to the molecular chains more difficulties appear than for other materials. The polymer chains must have sufficient time and mobility to build entanglements by reptation, a snake-like moving process, whereas one chain molecule moves forward while moving to both sides. There are some well-known methods in the technique, but most of them practically do not work properly with microstructured parts. Since a general molding would destroy the microstructures, the issue is to enhance the polymer chain mobility only on the bonding sites.

Only three welding methods are proven to work as yet: Solution welding (often named solution gluing), ultrasonic welding and laser welding.

Solution Welding A solvent for the polymer is dropped onto the structured part and a cover is placed over this. The whole structure is set into vacuum to evaporate the solvent. The mobility of the chains is enhanced by the fast diffusion of the solvent into the surface, the fast evacuation saves the bulk of the parts from deformation. Mixtures of solvents can be used to adjust the swelling velocity [61].

Ultrasonic welding Energy feeding by friction is known as an initiator of the polymer-welding process. In rotation and vibration welding, the movement of both parts against each other is too large to get a well-positioned welding. Only a vibration with a very small amplitude is suitable, e.g. the heating of the joining lines with ultrasonic power. Some modifications have to be made, since the

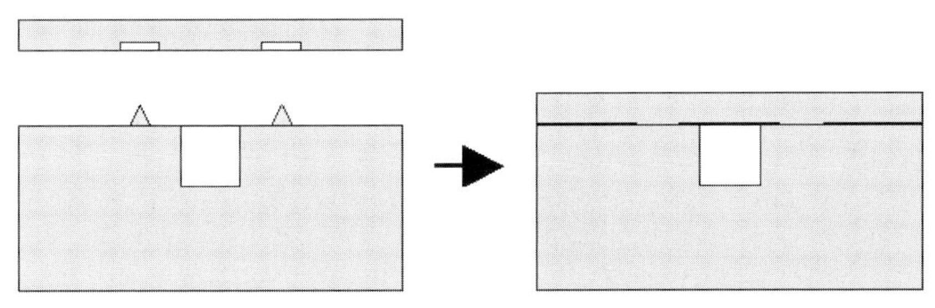

Fig. 10.34 Ultrasonic process with energy directors and melt traps on top of the substrate.

power has to be directed to the joining lines providing a reservoir for excess melt. Another drawback (see gluing with extra cavities) is the space that is needed for the bonding process, inhibiting very tightly packed channels as used in microreaction systems. Nevertheless, ultrasonic welding is a very fast process with a good prospect for automation [61]. The principle is schematically shown in Fig. 10.34.

Laser Welding The laser-transmission welding of transparent and opaque polymers is a new technology but still well investigated [62, 63]. Several applications in the "macroworld" such as filter housings for the automotive industry are well established. The state-of-the-art concerning laser-transmission welding of transparent-transparent polymers is represented by the so-called Clearweld^{TM} [64] technology. Another similar process is called Microclear [65]. In both cases, resins filled with dye or pigments absorbing near-infrared radiation are deposited in general via spin-coating on polymer surfaces. Two polymer substrates, one of which is transparent and another that is coated with the absorbing layer, are overlapped, and the laser beam is radiated from the transparent one. The layer thickness may be in the range of several micrometers up to several 100 µm. Nevertheless, the welding of completely transparent components is still a challenge in microsystem technology, especially if microstructures are included. In order to meet these requirements a new laser-based process technology was developed: "Laser-transmission welding with nanolayers". For this objective absorbing layers with a thickness of about 5 nm to 15 nm are used in order to establish a welding process between transparent and micropatterned polymers [66]. An example of a welded microchannel structure is shown in Fig. 10.35.

Laser-transmission welding with nanolayers is a powerful tool for welding applications in microsystem technology with structural details down to 10 µm. If the absorbing layer is to be patterned, e.g., by contact mask deposition or via UV-laser ablation, then it is expected that microstructures with smaller features can be bonded without any distortions.

Microchannel devices such as capillary electrophoresis chips, heat exchangers or static mixers for liquids and gases consist of different micropatterned sheets with structural details in the range of a few µm. The laser-assisted welding of

Fig. 10.35 Left: SEM of a channel structure in polyvinylidene fluoride (PVDF, sheet thickness 250 µm; channel width 100 µm); right: cross section of a covered PVDF channel structure after laser-transmission welding.

transparent and micropatterned polymers such as polyvinylidene fluoride (PVDF), polymethylmethacrylate (PMMA), polycarbonate (PC), Topas® COC, polyamide (PA) and polypropylene (PP) was successfully developed. Devices are built up by stacks of micropatterned sheets that have to be joined to each other. These sheets are patterned by micromilling, CO_2-laser cutting or hot embossing. Figure 10.36 shows an example of the process sequence used.

In comparison to mechanical micromilling or excimer laser ablation the direct patterning of the polymer sheets by CO_2-laser cutting seems to be a more powerful tool for rapid manufacturing in combination with multilayer welding. Therefore, the CO_2-laser writing of channels in polymer sheets was optimized with respect to the roughness and material damage. For a three-layer stack one welding cycle only was necessary. The surface quality at the cross section of the

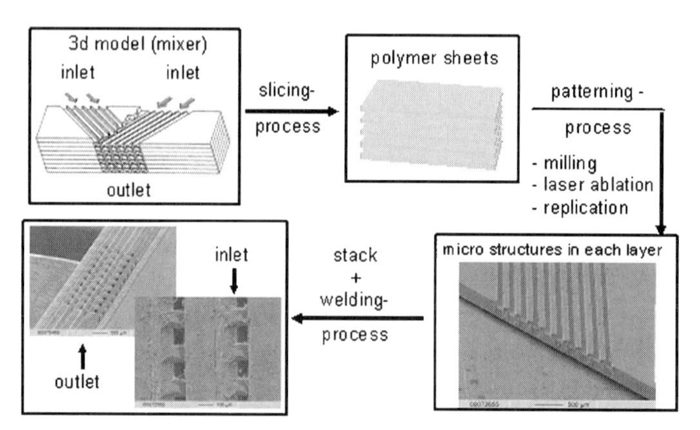

Fig. 10.36 Process sequence for a rapid manufacturing process using multilayer welding: 3D model design, slicing process, patterning process and laser-transmission welding with nanolayers.

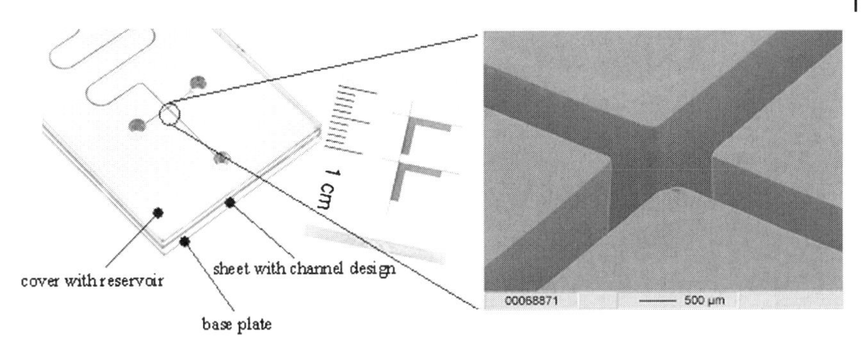

Fig. 10.37 Left: Chip structure in PMMA; right: detail view of the crossing channels (sheet thickness 1 mm, laser power 4 W, velocity 500 mm min^{-1}, argon gas flow 100 l h^{-1}).

polymer chip is shown in Fig. 10.37 (right). At the sidewall a surface roughness of $R_a = 20$ nm was obtained. Other demonstrators made of PMMA were fabricated with more than 6 micropatterned sheets. During laser patterning aspect ratios of about 30 were possible.

10.4.4
Packaging and Sealing

After manufacturing and bonding, packaging and sealing has to be performed to make devices connectable to conventional process-engineering equipment. It depends on several parameters, how easy or complicated those steps may be, and whether conventional sealing techniques and off-the-shelf housings can be used or not. One is the manufacturing process itself, another is the material, others are the demands of the application the device is used in. With a few exemplary points the complexity of this stage of the manufacturing process will be explained.

10.4.4.1 **Packaging**

Packaging means to insert a microstructure arrangement (mostly completely bonded and ready-to-use) into a housing and therefore make a interconnection to either other microscale or macroscale devices possible. Materials used have to be chosen carefully (thermal behavior, generation of local electrochemical elements, change of alloy composition and more), for temperature and pressure stability, for resistance against corrosion and for the possibility to connect on the one hand the microstructure device parts to the housing, on the other hand to connect the housing to the rest of the process equipment, like gas or liquid handling systems, pumps, etc. Eventually, sensors or actuators needed for the process control (thermocouples, optical equipment, pressure transducers, heating or cooling elements, stirrers, etc.) may have to be integrated into the housing, and interconnections to electrical systems like data acquisition, power sources or analytical systems (GC, MS, spectroscopy systems, etc.) may be demanded.

Therefore, generation of modular, combinational systems right from the first design step of the microstructure device as well as correctly defined interfaces is of great importance. Modular systems make it much easier to build multipurpose devices without major redesign work and reduce development time as well as costs.

Packaging for metallic devices normally is done with metallic housings also – a housing made of the same material as the microstructure device can be taken. Depending on the desired pressure resistance, clamping, thermal shrinking, hard

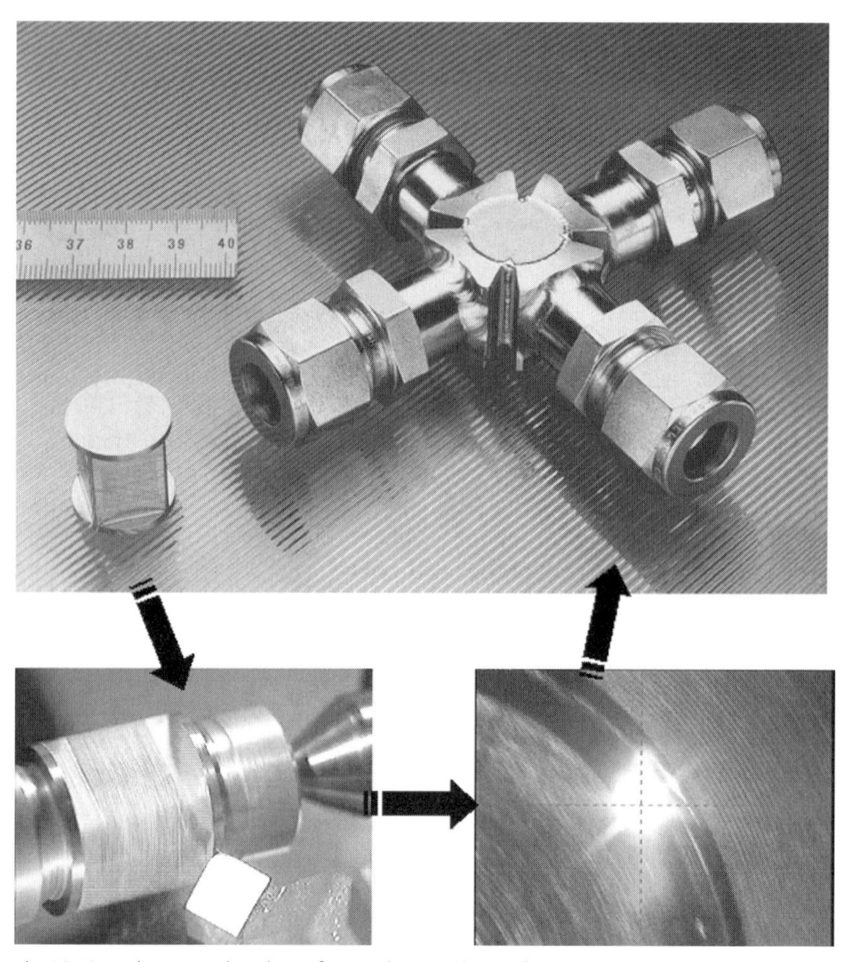

Fig. 10.38 Packaging and sealing of a stainless steel crossflow microstructure heat exchanger. From top-left to top-right (counterclockwise): Central core body, mechanical machining to smooth the edges, electron-beam welding, ready-to-use device welded into a housing.

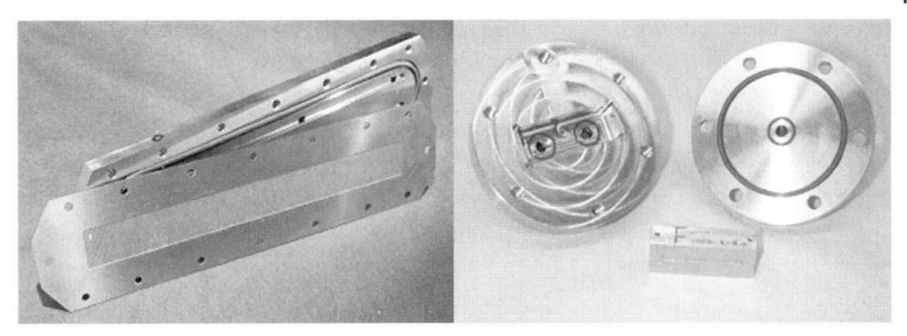

Fig. 10.39 Microstructure devices connected by clamping. Left: Lab-module with exchangeable microstructure foil. Right: Flange with exchangeable microstructure mixer inlay.

soldering or conventional welding techniques like WIG (wolfram inert gas welding), laser or electron-beam welding may be used (see above) [1, 30]. In Fig. 10.38, preparing and packaging of a crossflow microstructure heat exchanger made of stainless steel is shown. Starting with the central core of such a device, the edges are machined to produce smooth surfaces for, in this particular case, electron-beam welding into the housing equipped with standard fittings.

When using welding methods, normally leak-tight devices are achieved, which – depending on the thickness, the shape and the chosen material – will withstand very high pressures of up to 100 MPa [27]. Considerations for the material strength as well as for the wall thickness of the housing can be taken from commonly used design rules.

Mostly, it is easier to package polymer devices. Here, we have two opportunities: Either to manufacture a housing and to mount the device parts in the housing or to generate one (or more) combined housing-assembled-microstructure-device unit(s) that might then be connected to each other. Anyway, polymer devices often can be packaged with gluing methods with or without additives as well as with isotropic and nonisotropic gluing. Details of these processes can be found in [1, 5, 30]. Beside these processes, different welding techniques like diffusion bonding, laser welding and ultrasonic welding are possible (see Section 10.4.3).

Clamping of microstructures is useful if the microstructure parts of a device are to be exchangeable (see Fig. 10.39).

10.4.4.2 **Sealing**

Some aspects of sealing of microstructures have already been described briefly (see Section 10.4.2). No more details will be given here, they may be found in [1, 30], but a few more points will be noted.

As in conventional technology, sealing in microstructure devices can be done by using gaskets or O-ring sealing. These methods may be applied to single-foil devices or to the adapters to conventional process equipment (and with the drawback that the process fluid might come into contact with the sealing), but to generate internal sealing between, e.g., single microstructure foils to generate leak-tight passages, the use of gaskets in any design is not feasible in all cases. Here, bonding methods like the ones described previously are the methods of choice.

Another point is the connection of microstructure devices to other devices, either microstructure or conventional devices. Here, sealing with metal, carbon or polymer gaskets, O-rings or any other known technique to generate pressure- and leak-tight connections like standard HPLC fittings or other process equipment fittings may be used.

A point to think about is the ratio between the passive and active volumes of such combinations. In Fig. 10.40 (top), the size of a crossflow microchannel heat exchanger is shown in comparison to the size of the four standard fittings used to connect it to the test rig. In this case, the ratio between passive (or dead) volume and active volume is in the order of 12:1. This relatively large ratio may lead to residence-time effects or even destruction of the device, if a highly exothermic chemical reaction is starting inside the noncooled adapter fitting.

Fig. 10.40 Top: Size comparison between the central core of a crossflow microchannel heat exchanger (1 cm^3 active volume) to three standard fittings used for adaptation (about 2.9 cm^3 each). Bottom left: micromixer and microchannel reactor/heat exchanger with flange system to avoid large dead volumes. Bottom right: combined micromixer/reactor/heat exchanger with flange interconnection at the outlet side and polymer O-ring sealing.

To avoid this high value for the dead volume, at least in those parts where a large dead volume might be a disadvantage, flange systems can be used for interconnections between single microstructure devices. In Fig. 10.40 (bottom left), a microchannel mixer and a microchannel reactor/heat exchanger connectable with flanges is shown. Figure 10.40 (bottom right) shows a combined mixer/reactor with integrated heat transfer passage and flange system to avoid large dead volumes. The polymer O-ring sealing is clearly seen. With these devices, the dead volume of sealed systems can easily be reduced.

10.5 Characterization and Quality Assurance

Any fabrication process, whether in a research and development or in a commercial production environment, has to be complemented by a set of suitable characterization methods. In research and development, their objective is primarily to identify the parameters that influence the desirable as well as the undesirable properties of the devices being manufactured, thus aiding in the empirical optimization of the devices and in the development of simulation tools for the fabrication process. In a commercially driven fabrication environment, characterization methods are indispensable as part of a quality assurance (QA) system.

At different stages of the microstructure device-fabrication process, different parameters are of primary interest, and hence different tools are used at these times. The most important stages at which characterization methods come into play are

1. after the microstructuring, before the joining,
2. after the joining, before the packaging, and
3. after the packaging, at the end of the process chain.

10.5.1 Characterization of Microstructured, Nonjoined Devices

The main properties of interest at this stage are

1. geometrical dimensions and their deviation from the design,
2. surface quality, and
3. the quality of edges, e.g. the presence of burrs.

Geometrical dimensions may deviate from design due to inhomogeneities and drift introduced in wet chemical etching processes, or, in the case of mechanical micromachining, due to drift in the control of the machinery and the wear and tear of tools. The most important of the surface properties is the surface roughness, which is influenced by parameters like the material grain structure and by the interaction between the material and the tool's edge, or, in the case of microstructuring by wet chemical etching, by different etching selectivities of grains and intergrain boundaries. Finally, the quality of edges is often degraded by the presence of burrs that are to a certain degree unavoidable in any cutting process. In microstructuring (as opposed to conventional machining), burrs are especially detrimental since any efficient deburring process (like electropolishing) will significantly affect the entire microstructure.

While the measurement of geometrical dimensions in the two-dimensional (x–y) plane is relatively simple, the third dimension (z coordinate) poses more problems. Inspection tools for 2D usually consist of a motorized stage, a camera mounted on a microscope, and software combining image analysis with control of the motor stage. Such systems have been in use in the semiconductor industry for decades and allow highly automated inspection of microstructures. Unlike in the semiconductor industry, however, there is no narrow set of standard sample sizes (like wafer diameters in planar technology) in mechanical and wet chemical etching microstructuring, so that the loading and unloading of such inspection tools has to be done manually.

Along with the development of high aspect ratio microstructures and bulk micromachining since the late 1980s, special tools have been created that add varying degrees of third-dimensional functionality to the 2D-measurement process. White light interferometry can provide z coordinate measurements over an entire field of vision in a short time and with an accuracy of below one micrometer. The angle of almost vertical sidewalls, which is inaccessible to this method, can be measured with an accuracy of about 10° by specialized machines that scan such walls along the z direction with a very small ball mounted on the tip of a glass fiber. The resilience of the fiber keeps the ball in contact with the wall, and variations of the ball's location in the x and y directions during the z scan are then read out by the usual 2D-image analysis procedure.

Surface roughness on areas that are not too much inclined can be measured either by the analysis of data generated by white-light interferometry, or by stylus methods in which a very sharp needle is dragged over the surface in a line scan, and the variation of its z coordinate is read out electromechanically.

All properties of interest mentioned above can be measured to varying degrees or at least assessed by scanning electron microscopy (SEM). In the semiconductor industry, fully automated inspection systems based on SEM have been standard for years, while in the fabrication of metallic, ceramic or plastics microstructures, SEM is still a work-intensive and slow process, practically restricted to research and development activities and possibly the fabrication of individual high-end demonstrator devices. SEM offers a resolution over many orders of magnitude and can deliver both overviews of structures several square centimeters in area as well as images of surface structures measuring only a few nanometers. With additional modules like energy dispersive X-ray analysis (EDX), SEM can also provide some information about the chemical composition of microstructure surfaces.

10.5.2
Characterization of Joined, Nonpackaged Devices

Characterization at this stage is essentially limited to measuring several geometrical parameters, like the compression a stack suffered during diffusion bonding, or the potential skew of such a stack. The resulting deformation of the microchannels can be determined by optical or scanning electron microscopy of the joined microstructure block. Relatively large faults in the joining process can also be detected at this point.

10.5.3
Characterization of Finished, Packaged Devices

At the end of the fabrication process, after the packaging of the microstructure device (e.g. by welding into a housing with suitable fittings), the most interesting properties of the device include

1. the leak tightness and
2. the resistance to pressure, both of which are strongly dependent on the quality of the joining,
3. the homogeneity of the flow distribution among the individual channels (especially when the packaging step was followed by a coating of the microstructures), and
4. any application specific properties, such as
   - heat-transfer properties,
   - the quality of mixing (for micromixers),
   - surface enhancement by and catalytic effect of a coating.

Leak tightness is usually checked with a helium leak detector, consisting of a vacuum pump and a mass spectrometer. The microstructure device is closed using caps and blank flanges on all ports but one, which is connected to a high vacuum pump. After the device has been evacuated, helium gas is applied to the outside of the device, by spraying or by wrapping the device in a helium-filled bag. Even through relatively small leaks, helium will enter easily because of its small molecules, and then be detected by the mass spectrometer. When a microstructure comprises polymeric components like gaskets, a certain diffusivity of the helium through these materials will determine a lower limit of the size of leaks that can be detected. Leak rates are usually given as the quotient of the driving pressure difference divided by the volume of substance of test gas that leaked through and the time it took.

The resistance to pressure can be tested by pressurizing the device with an incompressible fluid (to minimize the energy released in case of a failure), and by either recording whether this pressure is maintained over a certain period of time, or by depressurizing the device and conducting another leak-tightness test as described above.

The homogeneity of the flow distribution can in principle be measured by microanemometry. Air or another gas is forced through a passage of the device by a certain pressure difference, and the resulting flow speed at the microchannel exits is then measured with the help of miniaturized hot-wire resistor-bridge sensors. The faster the gas flows past such a hot wire, the more electrical power is needed to maintain a certain temperature of this wire. There are, however, two important limitations to this technique:

- even a miniaturized hot-wire sensor is relatively large compared to a microchannel cross section, so that backaction of the sensor on the gas flow to be measured can occur;
- microchannel exits can be difficult to access for the hot-wire sensor via adapter fittings and through the outlet plenum.

If microanemometry cannot be applied, at least a rough estimate of the channel diameters resulting from the fabrication process can be reached by recording the overall fluid flow rate as a function of the driving pressure difference and comparing this dependence to theoretical calculations. This is especially important if relatively thick coatings are applied on the inside of the finished devices.

The variety of application specific characterization methods is beyond the scope of this book. Mostly, these methods will have to rely on the measurement of integral parameters like temperatures, pressures at inlets and outlets, and mass flow rates. While temperature and mass-flow sensors have to a certain degree successfully been integrated into microstructure devices, the integration of pressure sensors and chemical sensors is still the subject of active research and development activities.

10.6 From Prototype to Series Production

Series production of microstructure devices for applications in reaction engineering or process engineering is, for certain applications like microelectronics processing, a running task in industry. Other applications like microreaction engineering are not covered with series production yet. Here, most companies are in a research and development state, only a small number of facilities exists worldwide to produce a chemical or pharmaceutical product within microstructure devices.

The requested number of microstructure devices for reaction or process engineering per year is not very high so far (in the range of several thousand, taking all device types together, status: May 2005). At this point, microstructure devices for reaction engineering are mainly used as tools in research and development, and therefore the throughput range is not suitable for industrial production in most cases. Thus, before talking about series production, the question has to be asked.

10.6.1 Scaling, Numbering-up or Equaling?

Currently, the mass-flow range for microstructure devices suitable for reaction engineering reaches from some mg h^{-1} up to about some kg h^{-1}. Rare examples exist to handle mass-flow ranges up to some thousand kg h^{-1}.

To make microstructure devices feasible for industrial production, several questions have to be asked. The first and most important (which will not be discussed here) is whether it makes sense to run the process in microstructure devices. In many cases, conventional devices will do.

When the decision is made for microstructure devices, another important question is how to reach the desired mass-flow range. Here, three different opportunities have been discussed so far.

10.6.1.1 Scaling

Scaling means to increase the size of a device (and therefore also the devices characteristic lengths) to a range to accomplish the demands of the mass flow, which means to obtain the proof of principle with a very tiny microstructure device, and then enlarge it to the desired size.

Unfortunately, most of the microstructure device advantages in reaction engineering are based on the very small characteristic lengths of the devices. By increasing these lengths, the effects used mostly disappear. When the microstructure characteristic lengths are kept but the outer dimensions of the microstructure device (and therefore the noncharacteristic dimensions of the microstructures, too) are increased, other effects will dominate the behavior of the devices and therefore in most cases lead to major disadvantages of microstructure technology. As an example, microheat exchangers can be taken. They are extremely efficient tools in process engineering providing high transferable thermal power combined with a reasonably small pressure drop [24, 27, 67]. The characteristic length is the small hydraulic diameter of the channels, and an important dimension for the functionality is the overall channel length. By increasing the hydraulic diameter (scaling of characteristic length), the efficiency of the devices is decreased tremendously. By changing the overall length of the microchannels to reach larger heat-transfer surface areas, the pressure drop is increased. Thus, the advantages of the microstructures are lost by scaling.

10.6.1.2 Numbering-up

Another possibility to increase the mass flow of a process is to increase the number of processing units, e.g. microstructure reactors. Several limitations have to be taken into account.

First, the flow distributions to the single units have to be as equal as possible to provide the same pressure drop, the same reaction behavior and the same residence-time behavior for all devices. This is not easy to achieve but is solvable.

A more serious problem will arise from measurement and control of the devices. Assuming a single microstructure device to be monitored and controlled will need four temperature sensors, four pressure sensors and two mass-flow controllers, this is a reasonable number of 10 sensors and controllers that can easily be handled. Numbering-up by a factor of 100, the number of sensors and controllers will also increase by this factor: 1000 sensors and controllers have to be monitored at the same time. This number might still be possible, but the tendency is clear: The larger the number of devices, the larger is the number of independent sensors and controllers. Sooner or later this will lead not only to a problem in measurement and control but also to a cost factor in building the process application.

10.6.1.3 Equaling

A possible solution for the problems described briefly above is equaling (sometimes also called internal numbering-up). Here, the number of the single devices is kept as small as possible, but the number of the single microstructures, having their characteristic lengths and reasonable outer dimensions to fit the process demands, is increased. Thus, the properties (and advantages) of the microstructure systems are kept, the measurement and control efforts are limited, and the mass-flow range is efficiently increased. The problem of equal flow distribution is serious but solvable by design changes in the device adaptation system [68, 69].

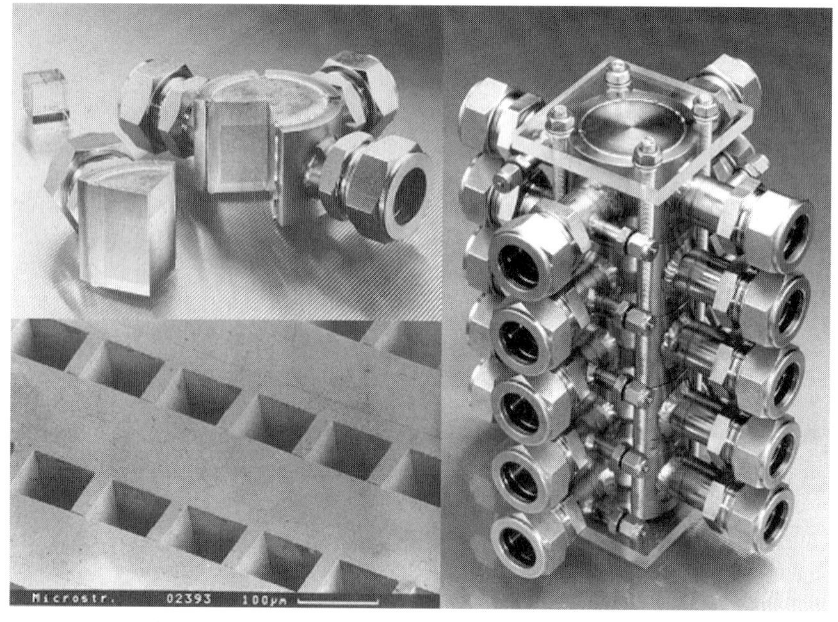

Fig. 10.41 Equaling of microstructure devices. In this example, five crossflow microchannel heat exchangers have been parallelized.

An example is shown in Fig. 10.41. The photo top-left shows a microstructure crossflow heat exchanger suitable for a mass-flow range of up to 7000 kg h^{-1}, measured with water. The SEM bottom left shows a detail of the integrated microchannel system: Several thousands of microchannels have been parallelized.

By taking five of those devices in parallel (Fig. 10.41, right), the mass-flow range is increased to 35000 kg h^{-1}, which means about 300000 tons per year mass flow.

10.6.2
Metal-component Series Production

Having in mind the considerations mentioned in Section 10.6.1, series production of metallic microstructure components becomes another quality. With regard to the equaling of devices, series production of thousands of metallic *components* bonded to a considerably smaller number of microstructure devices may be a possible solution. In Fig. 10.42, an example of large-number production of stainless steel microchannel foils by wet chemical etching is shown.

Therefore, only a limited number of manufacturing methods are feasible. While all the methods described in Section 10.2 are useful to obtain prototype devices, only the low-cost manufacturing methods like wet chemical etching may be used for series production of large numbers of metallic components.

Fig. 10.42 Series production of numerous microstructure stainless steel foils for counterflow microstructure heat exchangers by wet chemical etching.

Methods that are costly in working time or with high tooling costs will not be considered suitable for mass production [1, 30, 56].

It is the same for the assembling, bonding and packaging methods. The techniques have to be adequate not only to the technical demands but also to the pricing of the device. Techniques like flow-through diffusion bonding or laser welding for the packaging and standardized fitting and sealing of components should be used as often as possible.

10.6.3
Polymer-component Series Production

While prototyping is not that easy and mostly rather expensive in metal technology, prototyping in polymers is, within certain limitations, easy to obtain due to the rapid prototyping technologies like microstereolithography (see Section 10.3.5). Here, in most cases it is only a small step from prototype to series product.

Due to the main application fields of polymer microstructure devices (healthcare, analytic systems and others, see examples below), the mass-flow range is considerably smaller than for metallic devices. Hence, equaling is not of the same importance as for metallic devices. Moreover, in microstructure reaction technology and microfluidics with polymer devices, the use of *disposable* devices is feasible due to the very cheap production technology, and often necessary due to the main industrial applications.

Polymer-component series production is mainly done using injection molding or hot embossing techniques. Both have been described in detail before (see Sections 10.3.2 and 10.3.3). The main task is to generate one or more mold inserts to transfer the microstructure generated in prototype numbers into series production. Here, the point is not only to reproduce the microstructure with sufficient precision but also to demold them and to generate polymer parts that are easy to assemble and package.

Nevertheless, series production of polymer components is well established in the industry, not only in the macroscale but also in the microscale. As an example, watch industry all over the world can be taken, where billions of toothed wheels and other components are produced (see, e.g., [71, 72]). Another example is medical technology industries, where pipettes and many single-use components with microstructure dimensions are produced from polymers. Last, but not least, semiconductor and electronics are dealing with polymer microstructure components in huge numbers – mostly used as packages or sealing against the environment.

Microfluidic devices are not that common so far, some can be found in the healthcare sector or in analytical systems for online process monitoring or similar processes (see Bio Disk, Fig. 10.22). Here, industry is about to start with series production. For chemical-reaction engineering, polymer devices will probably be a niche market only. The reason is the limited temperature stability and corrosion resistance of most polymers combined with the difficulties of integrating catalytically active materials.

10.6.4
Long-term Stability and Quality Control

Users of process-engineering devices have a well-founded interest in the longevity of those. At least they will want to know the time during which the appliances can safely be used. Several factors contribute to this stability: Foremost, of course, is the resistance of the construction and joining materials against corrosive attacks by the fluids processed. Then the alteration of the materials under outer influences such as ultraviolet radiation or corrosion by the laboratory atmosphere have to be taken into account. Sadly, the literature does not provide much information on these topics, other than stating that a given device has been used for more or less prolonged periods of time.

As with the more general topic of corrosion, estimations from experience gained with standard devices will not carry too far, except for upper limitations in longevity. There are well-known standard tests for estimating long-term stability of materials in current technology, which have to be applied to microstructured components as well.

Quality control is also an issue with microstructured devices. The specific problems are the number of microchannels for devices bigger than lab-on-a-chip ones and the relative thinness of the material used. While there are methods to test chips automatically, there is no analogous method for process devices.

Testing of single devices, made from a multitude of metal foils, is no problem as long as one is talking about prototypes. However, if series production is performed, criteria by which the quality can be assured are needed.

One straightforward criterion is leak tightness of the whole device against the outer atmosphere. A second is the leak tightness between passages (if there are different fluids to be processed into the device). A third criterion for process devices is the integral pressure drop, which can be measured quite easily. The pressure drop can be calculated within a reasonable accuracy for most devices using the formula found in standard textbooks, such as the VDI-Wärmeatlas [73]. The pressure drop gives a rough estimate of the usefulness of the channels within the device. A more detailed description about the measurement principles is given in Section 10.5.

After those more general criteria, the actual properties of the function of the device can and should be checked by simply using the device for the intended objective.

10.7
Conclusion

Many requirements of processes in microreaction engineering may be fulfilled with microcomponents or devices made out of metals or polymers. Although there is a lot of experience from silicon and polymer processing as well as from mechanical machining, it is a very complex task to manufacture those devices.

As was described briefly in this chapter, a large variety of different limitations have to be considered. Thus, the efficient manufacturing of metallic or polymer microstructure devices, either for R&D, pilot-plant scale or in series production, is a key for success in microreaction engineering.

References

1 Madou, M., Fundamentals of Microfabrication, CRC Press, London, **1997**

2 ABI. EG Nr. L 181 from 9th of July **1997**

3 AD 2000-Regelwerk (Ed.: Verband der Technischen Überwachungs-Vereine e.V., Essen), Carl Heymanns Verlag KG, Köln, Germany, **2003**

4 Dechema Corrosion Handbook (Ed.: D. Behrens), VCH, Weinheim, Germany, **1987**

5 Menz, W., Mohr, J., Mikrosystemtechnik für Ingenieure, VCH, Weinheim, Germany, **1997**

6 Slocum, A. H., Precision Machine Design: Macromachine Design Philosophy and its Applicability to the Design of Micromachines, Proc. IEEE MEMS 1992, Travemünde, Germany, **1992**

7 Boothroyd, G., Knight, W. A., Fundamentals of Machining and Machine Tools, Marcel Dekker, Inc., New York, **1989**

8 Evans, C., Precision Engineering: An evolutionary View, Cranfield Press, Cranfield, Bedford, England, **1989**

9 Snoeys, R., Non-Conventional Machining Techniques, The State of the Art, Advances in Non-Traditional Machining, Anaheim California, USA, **1986**

10 Shaw, M. C., Metal Cutting Principles, Clarendon Press, Oxford, **1984**

11 DeVries, W. R., Analysis of Material Removal Processes, Springer, New York, **1992**

12 Schaller, T., Bohn, L., Mayer, J., Schubert, K., Microstructure Grooves with a width of less than 50 μm cut with ground hard metal micro end mills, Prec. Eng. 23 (**1999**), 229–235

13 Chryssolouris, G., Laser Machining, Springer, New York, **1991**

14 Kohl, M., Just, E., Pfleging, W., Miyazaki, S., SMA microgripper with integrated antagonism, Sens. Actuators 83 (**2000**), 208–213

15 Fischer, H., Vogel, B., Pfleging, W., Besser, H., Flexible distal Tip of nitinol (NiTi) for a steerable endoscopic camera system, Mater. Sci. Eng. A273–275 (**1999**), 780–789

16 Gillner, A., Laser Micro Machining, Proc. of the International Seminar on Precision Engineering and Micro Technology EUSPEN **2000**, Aachen, Germany, 105–112

17 Pfleging, W., Hanemann, T., Meier, A., Surface Modifications of Metallic Mold Inserts during Laser Treatment, Surface Engineering, VCH, Weinheim, EUROMAT, **1999**, 11, 455–460

18 Pfleging, W., Meier, A., Hanemann, T., Gruhn, H., Zum Gahr, K.-H., Laser micromachining of metallic mold inserts for replication techniques, Mater. Res. Soc. Symp., **2000**, 617, J5.5.1–J5.5.6

19 Pfleging, W., Hanemann, T., Torge, M., Bernauer, W., Rapid Fabrication and Replication of Metal, Ceramic and Plastic Mold Inserts for Application in Microsystem Technologies, J. Mech. Eng. Sci., 217 (**2003**), 53–63

20 Pfleging, W., Bernauer, W., Hanemann, T., Torge, M., Rapid fabrication of microcomponents – UV-laser assisted prototyping, laser micro-machining of mold inserts and replication via photomolding, Microsyst. Technol. 9 (**2002**), 67–74

21 Pfleging, W., Hanemann, T., Bernauer, T., Torge, M., Laser micromachining of mold inserts for replication techniques – State of the art and application, Laser Applications in Microelectronic and Optoelectronic Manufacturing VI, Jan 22–24, 2001, San Jose, USA, Proc. SPIE 4274, **2001**, 331–345

22 Petzow, G., Metallographisches, Keramographisches und Plastographisches Ätzen, Gebrüder Bornträger Berlin, **1994**

23 Harris, T.W., Chemical Milling, Clarendon Press, Oxford, **1976**

24 Brandner, J.J., Bohn, L., Schygulla, U., Wenka, A., Schubert, K., Microstructure devices for thermal and chemical process engineering. Yoshida, J.I. [Ed.], Microreactors: Epoch-Making Technology for Synthesis; Proc. Int. Workshop on Micro Chemical Plant Techn. / JCII, May 24, (MCPT) **2001**, Tokyo, J, CMC Publ. Co., **2003**, 75–87, 213–223

25 Brandner, J.J., Entwicklung von Mikrostrukturreaktoren zum thermisch instationären Betrieb chemischer Reaktionen. Forschungszentrum Karlsruhe, Wissenschaftliche Berichte FZKA 6891, Diss. Universität Karlsruhe (TH), **2003**

26 Henning, T., Brandner, J.J., Schubert, K., Characterisation of electrically powered microheat exchangers. Chem. Eng. J. Vol. 101/1–3, **2004**, 339–345

27 Schubert, K., Brandner, J.J., Fichtner, M., Linder, G., Schygulla, U., Wenka, A., Microstructure devices for applications in thermal and chemical process engineering, Microscale Thermophys. Eng., 5, **2001**, 17–39

28 Brandner, J.J., Emig, G., Liauw, M.A., Schubert, K., Fast temperature cycling in microstructure devices, Chem. Eng. J. Vol. 101/1–3, **2004**, 217–224

29 Henning, T., Brandner, J.J., Schubert, K., High speed imaging of flow in microchannel array water evaporators, Microfluidics Nanofluidics No. 1, **2005**, 128–136

30 Eigler, H., Beyer, W., Moderne Produktionsprozesse der Elektrotechnik, Elektronik und Mikrosystemtechnik, expert-Verlag Renningen, **1995**

31 Anurjew, E., Brandner, J.J., Hansjosten, E., Schygulla, U., Schubert, K., Mikroverfahrenstechnik und Rapid Prototyping im Forschungszentrum Karlsruhe, GEO-Siberia, Intern. Exhibition of Geodesy, Geoinformation Systems, Environment Analysis and Instrument Engineering, Novosibirsk, Russia, April 27–29, **2005**, to be published

32 Vansteenkiste, G., Boudeau, N., Leclerc, H., Barriere, T., Celin, J.C., Carmes, C. R., Millot, C., Benoit, C., Boilat, C., Investigations in direct tooling for microtechnology with SLS, Proc. LANE2004, Erlangen, Germany, **2004**, 425–434

33 Fischer, P., Blatter, A., Romano, V., Weber, H.P., Highly precise pulsed selective laser sintering of metal powders, Laser Phys. Lett., **2004**, 1–8

34 Giselbrecht, S., Gottwald, E., Schlingloff, G., Schober, A., Truckenmüller, R., Weibezahn, K.F., Welle, A., Highly adaptable microstructured 3D cell culture platform in the 96 well format for stem cell differentiation and characterization, 9th Int. Conf. on Miniaturized Systems for Chemistry and Life Sciences μTAS 2005, October 9–13, 2005, Boston, MA, USA, to be published

35 Ehrenstein, G.W., Erhard, G., Konstruieren mit Polymerwerkstoffen ein Bericht zum Stand der Technik, Hanser, München, **1983**

36 Mohr, J.A., Last, A., Hollenbach, U., Oka, T., Wallrabe, U., A modular fabrication concept for microoptical systems, J. Lightwave Technol., 21 (**2003**), 643–647

37 Ruprecht, R., Benzler, T., Holzer, P., Müller, K., Norajitra, P., Piotter, V., Ulrich, H., Spritzgießen von Mikroteilen aus Kunststoff, Metall und Keramik, Galvanotechnik, 90 (**1999**) 2260–2267

38 Heckele, M., Schomburg, W.K., Review on micro moulding of thermoplastic polymers, J. Mikromech. Mikroeng. 14 (**2004**) R1–R14

39 Pfleging, W., Finke, S., Gaganidze, E., Litfin, K., Steinbock, L., Heidinger, R., Laser-assisted fabrication of monomode polymer waveguides and their optical characterization, Mat.-wiss. u. Werkstofftech. 34 No. 10/11 (**2003**), 904–911

40 Pfleging, W., Böhm, J., Finke, S., Gaganidze, E., Hanemann, Th., Heidinger, R., Litfin, K., Fabrication of functional polymeric prototypes for microfluidical and micro-optical applications, CLEO/Pacific Rim 2003: Proc. 5th Pacific Rim Conf. on Lasers and Electro-Optics, Taipei, Taiwan, December 15–19, 2003, IEEE CD-ROM Vol. II, 739, **2003**

41 Pfleging, W., Böhm, J., Finke, S., Gaganidze, E., Hanemann, Th., Heidinger, R., Litfin, K., Direct laser-assisted processing of polymers for microfluidic and micro-optical applications, Photon

Processing in Microelectronics and Photonics II, 27–30 January 2003, San Jose, USA, Proc. SPIE 4977, **2003**, 346–356

42 Pfleging, W., Hanemann, Th., Bernauer, W., Torge, M., Laser Micromachining of Polymeric Mold Inserts for Rapid Prototyping of PMMA-Devices via Photomolding, Laser Applications in Microelectronic and Optoelectronic Manufacturing VII, 21–23 January 2002, San Jose, USA, Proc. SPIE 4637, **2002**, 318–329

43 Gaganidze, E., Litfin, K., Böhm, J., Finke, S., Henzi, P., Heidinger, R., Pfleging, W., Steinbock, L., Fabrication and characterization of single-mode integrated polymer waveguide components, Proc. Int. Conf. Integrated Optics and Photonic Integrated Circuits, Strasbourg, F, April 27–29, 2004 Bellingham, Wash., Proc. SPIE 5451, **2004**, 32–39

44 Pettit, G. H., Sauerbrey, R., Pulsed Ultraviolet Laser Ablation, Appl. Phys. A 56, **1993**, 51–63

45 Pfleging, W., Hanemann, Th., Bernauer, W., Torge, M., Laser micromachining of mold inserts for replication techniques – State of the art and application, Proc. SPIE 4274, **2001**, 331–345

46 Cheng, J.-Y., Wie, C.-W., Hsu, K.-H., Young, T.-H., Direct-write laser micromachining and universal surface modification of PMMA for device development, Sens. Actuators B 99 (**2004**), 186–196

47 Gebharth, A., Rapid Prototyping, Hansa, München, **1996**

48 Ikuta, K., Hirowatari, K., Ogata, T., Three Dimensional Integrated Fluid Systems (MIFS) Fabricated by Stereo Lithography, Proc. IEEE Int. Workshop on Micro Electro Mechanical Systems, MEMS '94, Osio, Japan, **1994**, 1–6

49 Ikuta, K., Hasegawa, T., Adachi, T., Maruo, S., Proc. IEEE Int. Workshop on Micro Electro Mechanical Systems (MEMS '2000), **2000**, 739 ff.

50 Ikuta, K., Module Micro chemical Device based on Biochemical IC Chips – From 3D micro/nano fabrication toward biomedical Applications, Proc. 1st Int. Workshop on Micro Chemical Plant Techn., Feb 3–4, Kyoto, Japan (MCPT 1), **2003**, 54–65

51 Ehrfeld, W., Gärtner, C., Golbig, K., Hessel, V., Konrad, R., Löwe, H., Richter, T. and Schulz, C., IMRET 1, **1997**, 72–90

52 Kolb, G., Cominos, V., Drese, K., Hessel, V., Hofmann, C., Löwe, H., Wörz, O. and Zapf, R., IMRET 6, **2002**, 61–69

53 Ziogas, A., Löwe, H., Küpper, M. and Ehrfeld, W., IMRET 3, **2000**, 136–150

54 Meyer, H., Crämer, K., Kurtz, O., Herber, R., Friz, W., Schwiekendick, C., Ringtunatus, O. and Madry, C., Patent Application DE 10251658 A1

55 Pfeifer, P., et al., "Micromotive", **2004**, unpublished results

56 Pfeifer, P., Görke, O., Schubert, K., Martin, D., Herz, S., Horn, U., Gräbener, Th., Micromotive – Development and Fabrication of Miniaturised Components for Gas Generation in Fuel Cell Systems, IMRET 8, **2005**

57 Paul, B. K., Hasan, H., Dewey, T., Alman, D., Wilson, R. D., IMRET 6, **2002**, 202–211

58 Bier, W., Keller, W., Linder, G., Seidel, D. and Schubert, K., in: Symposium Volume, DSC-Vol. 19, ASME, New York, **1990**, 189–197

59 Pfleging, W., Lambach, H. et al., unpublished results

60 Bacher, W., Saile, V., LIGA and AMANDA technologies for the fabrication of advanced micro-devices, Proc. 2003 JSME-IPP/ASME-ISPS Joint Conf. on Micromechatronics for Information and Precision Equipment, Yokohama, Japan, June 16–18, **2003**, 133–137

61 Truckenmüller, R., Ahrens, R., Bahrs, H., Cheng, Y., Fischer, G., Lehmann, J., Micro Ultrasonic Welding – Joining of Chemically Inert, High Temperature Polymer Microparts for Single Material Fluidic Components and Systems, Proc. of DTIP, June 1–3, **2005**, Montreux, CH, to be published

62 Bader, R., Jacob, P., Volk, P., Moritz, H., Process for joining microstructured plastic parts and component produced by this process, European Patent No. WO 99/25783

63 Bachmann, F., Russek, U., Laser welding of polymers using high power diode lasers, SPIE 4637, **2002**, 505–518

64 Sato, K., Kurosaki, Y., Saito, T., Satoh, I., Laser welding of plastics transparent to near-infrared radiation, SPIE 4637, **2002**, 528–536

65 http://www.clearweld.com

66 Teubner, U., Klotzbuecher, T., MICRO-CLEAR – a novel method for diode laser welding of transparent microstructured polymer chips, Proc. of ICALEO 2004, Laser Microfabrication Conference, San Francisco, CA, USA, October 4–7, **2004**, in press

67 Pfleging, W., Baldus, O., Bruns, M., Baldini, A., Bemporad, E., Laser assisted welding of transparent Polymers for Micro Chemical Engineering and Life Science, Proc. Int. Conf. "Laser and Applications in Science and Technology (LASE)", Photonic West 2005, San Jose, CA, USA, Jan 22–27, **2005**, to be published in Proc. SPIE Vol. 5713

68 Hessel, V., Löwe, H., Mikroverfahrenstechnik: Komponenten – Anlagenkonzeptionen – Anwenderakzeptanz, Chemie Ingenieur Technik (74) **2002**, Vol. 1/2, 17–30, Vol. 3, 185–207, Vol. 4, 381–400

69 Commenge, J. M., Falk, L., Corriou, P., Matlosz, M., Optimal Design for Flow Uniformity in Microchannel Reactors, IMRET 4, **2000**, 23–30

70 Walter, S., Frischmann, G., Broucek, R., Bergfeld, M., Liauw, M. A., Fluiddynamische Aspekte in Mikrostrukturreaktoren, Chemie Ingenieur Technik (71) Vol. 5, **1999**, 447–455

71 http://www.swatchgroup.com/home.php

72 http://www.eta.ch/

73 VDI Wärmeatlas, Berechnungsblätter für den Wärmeübergang, VDI-Verlag, **2000**

11
Silicon Microfabrication for Microfluidics

Frank Goldschmidtböing, Michael Engler, und Alexander Doll,
Laboratory for Design of Microsystems, Department of Microsystems Engineering (IMTEK), University of Freiburg, Germany

Abstract
This chapter gives a brief introduction into the silicon fabrication processes used in microreaction technology. Special emphasis is given to the overall fabrication process and not to specific details of certain process steps. It is demonstrated that the silicon material itself has advantageous thermal and mechanical properties for many microreaction devices and that the possibility of fabricating highly complicated structures with high accuracy in a batch process makes silicon technology the technology of choice for some classes of devices. In particular, those devices with moving parts (e.g. micropumps, valves or dispensers) or devices with multiple complex structures like certain micromixers are preferably fabricated by silicon technology. Some typical cases are discussed in the applications section.

Keywords
Silicon microfabrication, lithography, etching, deposition, micropumps, microdispensers, micromixers

Advanced Micro and Nanosystems Vol. 5. Micro Process Engineering. Edited by N. Kockmann

ISBN: 3-527-31246-3

11.1 Silicon as a Design Material

Historically, silicon was used as a bulk material for most electronic devices and microchips. In the beginning of the 1980s, not only the electrical, but also other physical properties of silicon bit by bit came into focus and were widely studied, see especially the benchmark paper by Petersen in 1982 [1], which was the first to point out the mechanical rather than the electrical properties of silicon and that is regarded by many as the birthplace of microsystems technology or MEMS.

In particular, the outstanding mechanical properties of silicon, like high durability and low mechanical abrasion, have made it suitable for applications other than electronic. The mature knowledge of the fabrication and process technologies of silicon wafers in the dimensions of micrometers has since been the most important driving force of the emerging new discipline of microsystems technology. It can even be said that microsystems technology is mainly derived from the results of microelectronics.

There are several advantages, but also some disadvantages of silicon in comparison with other materials, which have to be balanced when using silicon for the fabrication of micro process engineering devices.

Advantages include:
- fabricated structures are very fine and differ only little from device to device,
- the parallelization of the fabrication of silicon structures allows a cheap and fast mass production of highly complex devices,
- complex structures are easily possible,
- channel walls are usually very smooth,
- silicon, silicon oxide and silicone nitrid is chemically inert to many substances,
- a variety of properties can be given to the surface of the silicon structures, e.g. chemical activation for surface reactions,
- silicon has a high thermal conductivity, which is especially important when dealing with highly exothermic reactions or reactions that have to take place at defined temperatures.

Disadvantages include:
- For the fabrication of silicon structures, a cleanroom and special micromachinery is necessary, which raises investment costs,
- rapid prototyping of silicon structures is – up to now – usually difficult to achieve which raises design costs,

- due to the peculiarities of the fabrication technologies, the desired silicon structures are not arbitrary, but are limited depending on the chosen technology,
- the brittle silicon chips are easily broken which can complicate the handling and packaging of the devices, especially concerning fluidic interfaces and sealing problems,
- up to now, there is a lack of standardized peripheral equipment to connect the silicon chips with the outer world.

Basically, silicon can be divided into poly- and monocrystalline silicon. Polycrystalline silicon has isotropic properties, which means that its properties are independent of the direction they act inside the silicon. In contrast, the properties of monocrystalline silicon change depending on the direction they are acting inside the silicon. Important for micro process engineering is monocrystalline silicon, as many of the processes for the structuring of silicon are not usable on polycrystalline silicon.

11.1.1
The Fabrication Process

Silicon is always provided in the form of flat and circular "wafers" with diameters ranging from 3 up to 8 inch and thicknesses of 300, 525, 700 or 1000 µm. With different techniques, explained later in this chapter, the surface of the wafers is lithographically structured and etched using various chemical and physical methods. Further processes concerning surface modification or surface activation can be applied. With these methods, a multitude of nearly identical complex chips can be structured on one wafer. Bonding techniques for the connection of two or more wafers are further used to improve the functionality. The chips are then separated by fine saws and can be used (Fig. 11.1).

11.1.2
Housing and Connections

A difficult problem is the packaging of silicon chips and the connection with fluidic conductors. It is always necessary to provide a housing processed by standard machining so that common ports can be connected (e.g. standard fluidic connectors). Adequate materials are metals and polymers. For the connection of the silicon chip with the housing, several possibilities are imaginable, the most important of which are listed here:
- splicing of the silicon chip onto the housing with two-component glue,
- pressing the silicon chip onto a rubber sealing – for this, two-component silicones are suitable,
- casting of polymers to serve for a tight connection between silicon chip and housing (e.g. with epoxy resin).

Fig. 11.1 A fully processed silicon chip with a diameter of 4 inch consisting of 12 T-shaped micromixers.

Since no standardized solutions for housing and connecting silicon exist, there is still major design work needed when designing components for micro process engineering out of silicon. The situation is likely to change when the use of silicon is being more and more intensified in future devices.

11.1.3
Important Material Properties

When using silicon as a design material for micro process engineering, knowledge of certain material properties is necessary. These include the thermal properties for the calculation of heat conduction, mechanical properties for proper housing and other properties that might be useful for special applications. They can be found in Table 11.1 for monocrystalline pure and undoped silicon. Electrical properties are not given, since they are firstly not very important in micro process engineering, and secondly they can be adjusted over a wide range by using doped silicon. The reader is referred to a number of publications that deal with these terms.

11.2
Processes and Structures

Silicon microtechnology is often referred to as a planar technology. While microelectronic devices only use the top few micrometers of a silicon substrate as the active material, this is definitely true for microelectronic technologies. Microfluidic devices, however, have structure depths of a few tens of micrometers up to the whole wafer thickness, to avoid the enormous viscous pressure losses that go along with micrometer-sized cavities. These technologies that are used to structure the silicon wafer in depth are called "bulk technologies". Though there are a few microfluidic devices fabricated by so-called "surface technologies", that is by structuring thin surface layers, this chapter is restricted to bulk technologies.

In the following, the terms *process* and *process step* will be distinguished. The term process refers to the whole fabrication from substrate to device, for example "Fabricating a channel with rectangular cross section and transparent cover", whereas process steps are distinct fabrication steps, i.e. "depositing a metal layer by sputtering".

Most books on silicon technologies [8, 9] treat all the different methods of depositing layers, lithography, bonding and packaging to enable the reader to find the right process steps for his individual process. Due to the character of this book as an overview we take a different approach. We first introduce the differ-

Table 11.1 Important material properties of monocrystalline silicon at room temperature (except value with an asterisk).

Properties	Numerical values
General	
Density ρ [kg/m^3]	2328 [2]
Melting point* [°C]	1412 [2]
Thermal	
Heat conductivity λ [W/m · K]	148 [3]; $100/(0.1598 + 1.532\times10^{-3}\,T + 1.583\times10^{-6}\,T^2)$* ($T$ in K) [4]
Specific heat capacity c_p [J/kg · K]	702 [3]
Thermal expansion coefficient* α_{th} [1/K]	$2.362\times10^{-6} + 1.026\times10^{-8}\,\vartheta - 2.988\times10^{-11}\vartheta^2 + 3.948\times10^{-14}\,\vartheta^3$ (ϑ in °C) [5]
Mechanical	
Young's modulus E_{mech} [GPa]	122–168 (dep. on crystal orientation) [6, 8]
Poisson number ν_{mech} [–]	0.06–0.28 (dep. on crystal orientation) [6, 8]
Fracture toughness K_{Ic} [MPa · m$^{1/2}$]	0.83–0.95 (dep. on crystal orientation) [7]
Other	
Electric permittivity ε_r [–]	11.7 [2]
Magnetic susceptibility μ_r [–]	-4.9×10^{-5} (diamagnetic) [3]

ent possible structures and the general processes involved, before we go into some details of the process steps. Once the reader has identified silicon technology as an appropriate technology for his means, we recommend the above books for more detailed information on the process steps.

Devices for microreaction applications generally contain closed cavities and channels. Thus the two main fabrication steps are structuring of one or more wafers (see Section "Structuring") and bonding of two or more wafers to enclose the cavities and channels (see Section "Bonding Techniques"). Furthermore, some deposition techniques to fabricate hard masks for structuring, bonding layers or to functionalize the surface for certain reactions are performed (see Section "Surface Coating and Modification").

A comparably complex dummy structure is depicted in Fig. 11.2. It is made of two silicon wafers. The graph shows only one chip. The cambered edges indicate that several instances of the same chip adjoin on both sides. The wafers are denoted as top wafer and bottom wafer. Both wafers have a front and a back side. The front sides are the bonding surfaces. This nomenclature seems confusing, because the front side of the top wafer is the bottom one in the picture. This definition for the front side is nevertheless useful because the bonding surface has to be treated with special care to enable proper bonding. Thus the term front side always reminds the process designer to take care in the surface quality.

Processes are typically documented in process slips and visualized by process charts. As an example we briefly describe the subprocess for fabricating the top wafer (Fig. 11.3).

This process slip is a very brief description of the process with the purpose to demonstrate the general process outline. A real process slip contains much more details on the process parameters and many cleaning steps that are omitted here for clarity. The detailed process slip typically contains much intellectual property of the fabrication facility and is often confidential.

Process charts are used during the design of a process. The possible interactions between different process steps can be judged from these graphs. The graphs are not to scale. While the dimensions of silicon chips for fluid handling are typically in the order of centimeters, the thickness of a typical 6-inch wafer is about 700 μm. Typical hard masks are about 100–300 nm thick, while photoresists are spun on in a thickness of the order of one micrometer.

The outline of the following three sections is as follows. First, the different methods of depositing surface layers on silicon or similar substrates will be briefly summarized. After that, some structuring methods are discussed. The last technology section deals with bonding methods. These three sections are intended to impart a general idea of the prospects and limits of these process steps and do not claim completeness.

In the application chapter some silicon devices are presented in detail to demonstrate the advantage of silicon technology especially for devices with moving parts, for example pumps, valves and droplet dispensers and those devices that need a high geometrical accuracy or contain complex structures, for example certain passive micromixers.

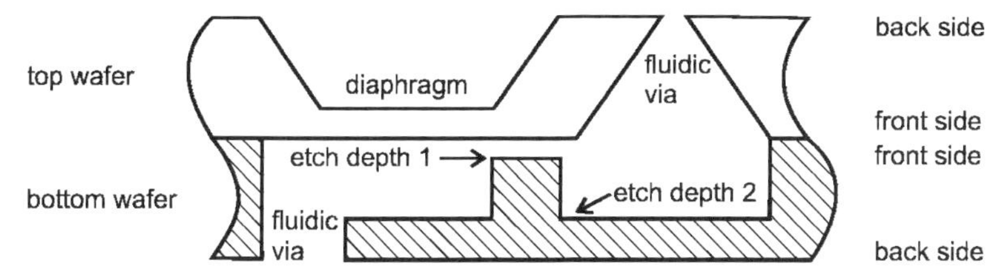

Fig. 11.2 Cross section of a "dummy" device.

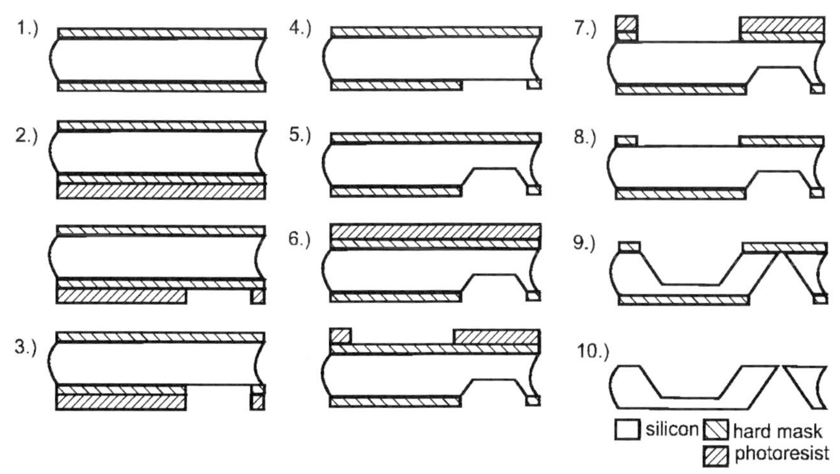

Fig. 11.3 Process chart of the "dummy" device. 1) Deposition of the hard mask (both sides). 2) Lithography (front side). 3) Etching of the hard mask (front side). 4) Removal of the photoresist (front side). 5) Anisotropic wet etching (1st step). 6) Lithography (back side). 7) Etching of the hard mask (back side). 8) Removal of the photoresist (back side). 9) Anisotropic wet etching (2nd step). 10) Removal of the hard mask (both sides).

11.2.1 Surface Coating and Modification

One of the advantages of silicon micromachining is the possibility to modify the surface and therefore enhance the design variability. When using lithography, the geometry on which the surface is modified can even be limited. With this, several design variations are possible, for example:

- Coating of channels with passivating or insulating layers like SiO_2 and Si_3N_4,
- deposition of chemically active layers for the implementation of surface reactions,
- deposition of thermally active layers for heating purposes,
- variation of the porosity of the silicon surface.

In particular, the ability to coat the silicon surface with different materials gives a variety of additional possibilities. One may think of coating materials with poor adhesion, coating of special components reacting with the species inside the channels or simply increasing or decreasing the roughness of the channel walls. Many possibilities are imaginable and there will surely be much research work to be done.

Several methods for the modification and coating of the silicon surface are possible. They are mainly taken from [9], [10] and [11]. Depending on what properties the surface should obtain, the appropriate process method must be chosen.

Evaporation

Evaporation belongs to the physical vapour deposition (PVD) processes. It can only be applied to materials that can be thermally evaporated and is therefore mainly used with pure metals. The material that is to be deposited is located in a special pot and is heated by different devices (electron beam, electric current or radiation heating). The material evaporates and, since the process is accomplished inside vacuum, deposits onto the silicon wafer that is positioned above the heating device. Evaporation gives very homogeneous layers and has a high coating rate, but, since the kinetic energy of the evaporated atoms is not very high (approx. 0.1 eV), the adhesion might become problematic.

Sputtering

Like evaporation, sputtering is a PVD process that is mainly applied to metals. The advantages of sputtering against evaporation are first that the adhesion is much stronger, and secondly that the materials that are to be deposited do not have to pass the melting phase, which means that also materials with high melting points like tantalum or tungsten can be attached to the silicon.

The process is also accomplished inside vacuum. In contrary to evaporation, the material is not heated but is bombarded by argon ions, which are created inside a plasma discharge and accelerated towards the material by an electric field. The atoms sputtered off therefore have a 10 to 100 times higher kinetic energy than when using evaporation, but the coating rates are lower. Reasonable coating thicknesses are in the range of several nanometers to some micrometers.

Chemical Vapor Deposition (CVD)

In this process, a chemical reaction takes place at the heated surface of the silicon substrate (or on the layer material of the silicon). During this reaction, which is accomplished by gaseous reactants, the resulting solid component is deposited on the substrate, while the resulting gaseous component is removed from the process.

An important example is silicon nitride, which is very often used as protection and insulation layer. A substrate consisting of a silicon bulk covered with silicon oxide is heated to a temperature of about 1000 °C and flooded with a mixture of silane SiH_4 and ammonia gas NH_3. The ongoing chemical reaction is of the form

$$3SiH_4(g) + 4NH_3(g) \rightarrow Si_3N_4(s) + 12H_2(g)$$

Silicon nitride is a solid that is deposited on the surface of SiO_2, while hydrogen is removed from the process.

The process can be improved for certain applications when it is either carried out inside a low vacuum (low-pressure CVD or LPCVD) or by using a plasma to carry out the reaction at lower temperatures (plasma-enhanced CVD or PECVD). The first one has advantages mainly in faster deposition rates, while the second one can be used for surfaces that are not capable of holding the high temperatures of over 1000°C because of the additional energy provided from the plasma discharge.

Oxidization

Pure silicon reacts with oxygen coming from the ambient air and forms a very thin film of silicon oxide of several nanometers. However, it is often useful to have a much thicker oxide layer on the silicon that is used as protection, insulation or adhesion layer. For example, silicon nitride needs a layer of silicon oxide, with thicknesses of 100–300 nm, to safely adhere on silicon. Therefore, another very important process is the oxidization process, which is performed on the silicon surface. Basically, it is a special CVD process accomplished at temperatures of 900–1200°C with oxygen or steam. The first one is called dry and the second one wet oxidization. The reaction schemes are as follows:

$$Si(s) + O_2(g) \rightarrow SiO_2(s) \qquad \text{dry oxidization}$$

$$Si(s) + 2H_2O(g) \rightarrow SiO_2(s) + 2H_2(g) \qquad \text{wet oxidization}$$

Dry oxidization gives an oxide with better quality than using wet oxidization, but the deposition rate is much slower. The deposition rate is also strongly temperature dependent. To give an example, to build an oxide thickness of 100 nm at 900 °C, about 30 min of wet, but 60 min of dry oxidization is needed. Exact tables can be found in [10].

11.2.2 Structuring

The whole process step from bare substrate to the final structure is denoted as structuring. It generally consists of defining the structure by lithography and etching the substrate.

The lithography step (Fig. 11.4) begins with depositing photoresists by spinning or spraying. A mask (a glass substrate with a thin structured metal layer) is placed onto the resist. Then the resist is exposed to light. Those parts of the resist that are covered by the metal layer of the mask are not exposed. The light induces a chain-breaking or, depending on the resist, a chain-linking reaction in the polymeric resist. This reaction alters the solubility of the resist in the developer, a solvent for the resist. In positive resists, the exposed resist undergoes a

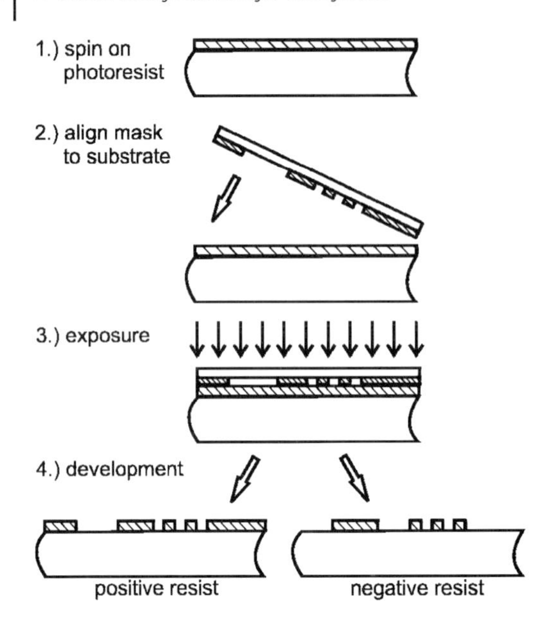

Fig. 11.4 Process chart of lithography process.

chain-breaking reaction and the exposed areas are more soluble in the developer. Thus the exposed areas are dissolved in the developer while the resist remains in the protected areas. For negative resists, the development works vice versa, the exposed (linked) areas remain, while the protected areas are dissolved away.

Resolutions of 1 μm can be achieved by standard equipment. This resolution is sufficient for microreaction devices, higher resolutions up to about 100 nm are obtained in the microelectronic industry. For more information on lithography methods and equipment, see [9].

Several etching methods are available in bulk micromachining. These methods differ in terms of etch rate, etch-rate uniformity, selectivity and etch profile. The etch rate is defined as etch depth per unit time. The uniformity is the relative variation of the etch depth over the whole wafer. The selectivity is characteristic of the etch method. It is the ratio of the etch rate of the substrate and of the masking layer. The etch profile is determined by different aspects and has to be treated separately for different technologies. Generally there are three different etch-profile types that are generated by different technologies, these are anisotropic dry etching (Fig. 11.5 a), anisotropic wet etching (Fig. 11.5 b) and isotropic dry or wet etching (Fig. 11.5 c).

Anisotropic Dry Etching

The simplest method to achieve vertical sidewalls is physical etching. Inert ions are accelerated towards the substrate and impact it with a high momentum, resulting in the removal of material. This is the basic principle of sputter etching and ion milling. Purely physical etching suffers from some limitations that prevent

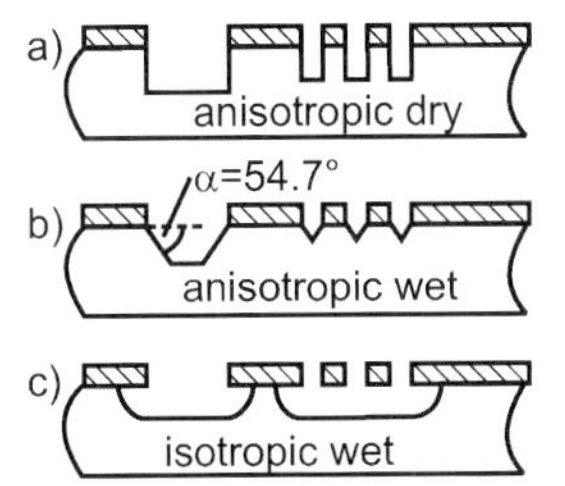

Fig. 11.5 Structures defined by different etch technologies.

Fig. 11.6 Structure fabricated by the anisotropic ICP process.

the usage of these technologies for bulk micromechanics. The selectivity is poor and the etch rate is very low (several tens of nanometers per minute [8]). Thus, a chemical mechanism has to be added to increase selectivity and etch rate.

The most common physiochemical etching method is deep reactive ion etching (DRIE). It is a combined process of etching the bottom of the cavity and passivating its sidewalls in a high-density inductively coupled plasma (ICP) reactor. The Teflon-like passivating layer is deposited from fluorocarbon plasma (CHF_3). Then the process is switched to SF_6 plasma with a negatively biased substrate. SF_x^+ ions are accelerated towards the substrate and remove the passivation layer by ion bombardment allowing the etching reaction to take place at the bottom of the etch cavity but not on the still passivated sidewalls. These steps are alternately repeated, to provide nearly perfect anisotropy. The alternating process steps of etching and passivating can be seen as small corrugations in the sidewalls (see Figs. 11.6 and 11.7).

As the ICP reactor yields a very dense plasma the etch rate is very high (up to 6 µm/min). Selectivity to masking material (more than 100 : 1 for photoresist or silicon oxide) is also very high because the etching mechanism is mainly chemical, as the ion bombardment is of much lower energy than in purely physical etching. The etching process is diffusion limited, which leads to problems regarding etch-rate uniformity. Areas of the wafer with much exposed area consume more reactants than areas with less exposed area and the diffusion of

Fig. 11.7 Magnified image of corner of the structure from Fig. 11.6.

reactants in narrow cavities is slower than in broad ones. Both effects alter the local etch rate. If high uniformity is needed, these effects have to be taken into account in the process design.

Though the DRIE technology has very high technological potential, its use is limited by economic factors. The drawbacks of the DRIE technology are its high investment cost and the long processing times. Etching times may exceed two hours for etching through a 6-inch wafer and only one wafer can be etched in the reactor at a time.

Structures with intentionally different etch depths are fabricated by two process steps. Two alternative processes can be used: (A) sequential lithography and etching and (B) generation of steps in the hard mask. These processes are sketched in Fig. 11.8.

Process A is the straightforward method for producing structures with multiple etch depths. Both etch depths can be adjusted individually in the etch steps A3 and A6, offering good etch-depth control. The critical process step in this scheme is the second lithography (A5). The resist has to be spun on a prestructured substrate, which is only possible for relatively shallow structures. The alternative process B circumvents this problem by patterning the hard mask with a step pattern (B1–B3) prior to the first DRIE step (B4). The thinned areas in the hard mask are then removed by a maskless etching step. A thin hard mask layer remains in those areas that have not been structured before (B5). In the second DRIE step (B6), both structures are etched, thus both etch depth are interrelated. A disadvantage of this method is the relatively thick hard mask that is needed to produce the step pattern.

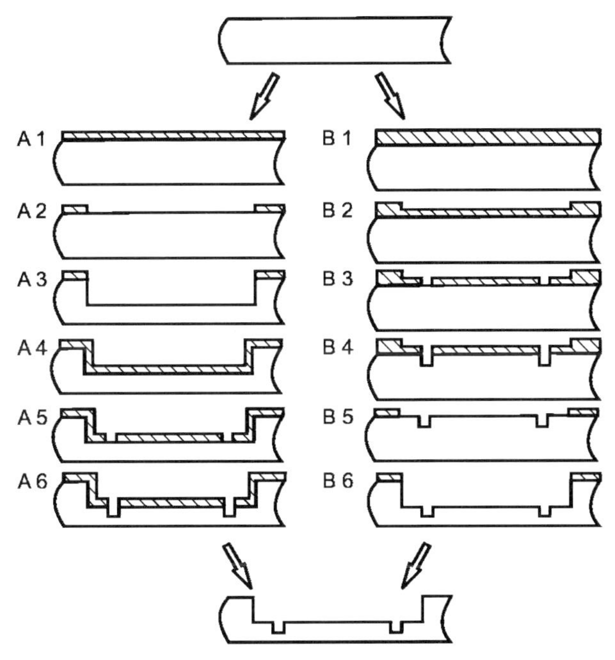

Fig. 11.8 Alternative processes to fabricate two-level structures.

Anisotropic Wet Etching

Anisotropic wet etching utilizes the single-crystal properties of silicon wafers. Some liquid etchings have different etch rates for different crystallographic directions. Thus the sidewalls of an etch cavity align to those planes with the lowest etch rate. Common anisotropic etchings are potassium hydroxide in water (KOH), ethylene diamine pyrocatechol (EDP) and tetramethyl ammonium hydroxide (TMAH).

Crystallographic directions are typically denoted by their Miller indices [*hjk*]. For the definition of the Miller indices see for example [12]. Those directions relevant for anisotropic wet etching of silicon are defined in Fig. 11.9. The crystallographic planes are denoted by the Miller indices of their normal direction. Due to the cubic symmetry of the silicon crystal, the three Cartesian directions ([100], [010] and [001]) are all referred to as [100]. In anisotropic wet etching those directions with the lowest etch rates have to be considered. These are in ascending order the [111], [100] and [110] directions. For KOH etching, the ratio of these etch rates can be as high as 1/200/400 [8]. For determining the cavity shape of anisotropic etching the orientation of the silicon surface has to be considered. This chapter is restricted to the most common [100]-oriented silicon wafers. Their surface is in the [100] plane. The wafer flat is aligned to the [110] direction. Figure 11.10 shows the cavities that evolve from square windows and channel-like openings in the hard mask. The etch front propagates in the [100] direction and

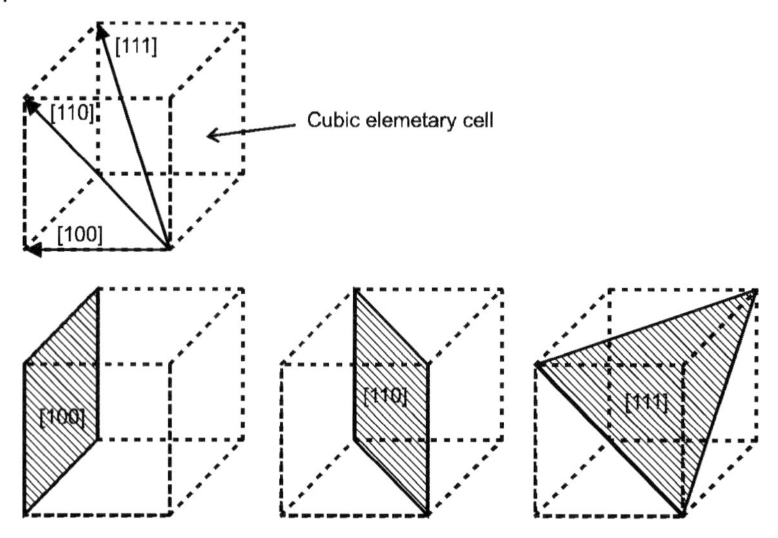

Fig. 11.9 Important crystallographic planes of silicon.

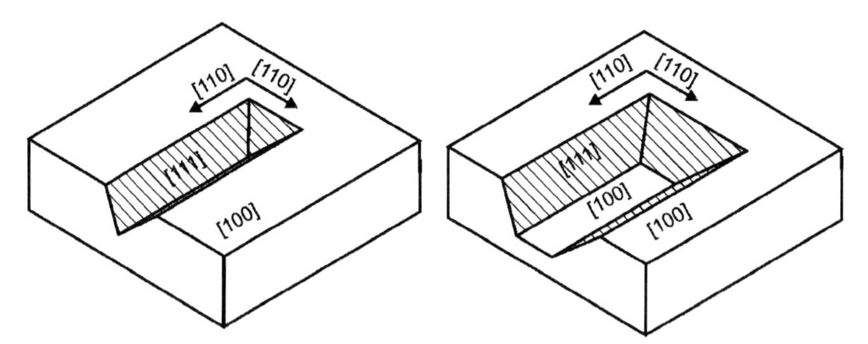

Fig. 11.10 KOH-etch cavities a) V-groove, b) trapezoidal cavity.

stops as soon as it hits an etch resistant [111] plane. If the structure is narrow enough a so-called V-groove (Fig. 11.10a) forms. A V-groove is bordered solely by [111] planes and is thus virtually etch resistant in all the following anisotropic etch steps. If the etch time does not suffice to build a V-groove, the bottom of the etch cavity remains in the [100] plane (Fig. 10b). Now the etch depth is not solely determined by the mask pattern but also by the etch time. This method is used to fabricate channels with trapezoidal cross sections or thin silicon diaphragms (in the [100] plane). Anisotropic wet etching is a reaction-rate-limited process, thus the diffusion path of reactants to the substrate does not influence the etch rate. Therefore a very high etch-rate uniformity over the wafer can be obtained. The etch rate has an Arrhenius-type temperature dependence, so a high temperature stability is necessary to obtain a reproducible etch rate. Very thin diaphragms are often fabricated by an etch-stop technique [8]. The etch front proceeds until it reaches an etch stop layer (for example a highly boron doped silicon layer).

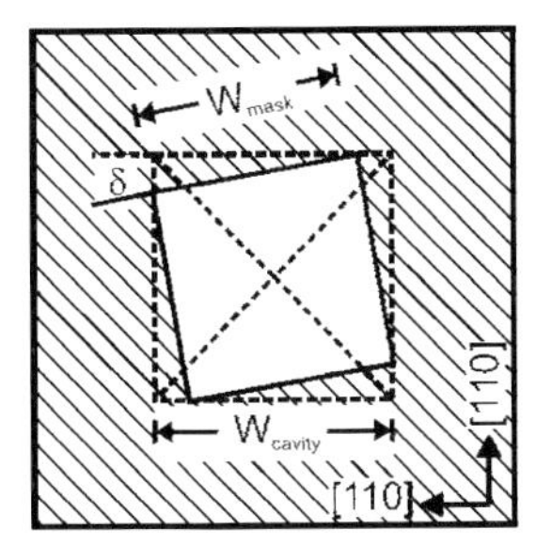

Fig. 11.11 Structure widening by misalignment.

The self-alignment of the sidewalls to the [111] plane restricts the possible structures to 90° angles, which is a disadvantage compared to anisotropic dry etching that enables free-forms in the lateral plane.

The accuracy of the etched structures is mainly limited by the alignment of the mask pattern to the crystallographic [110] direction. The effect of misalignment is demonstrated in Fig. 11.11. The mask opening is misaligned to the crystallographic [110] direction by an angle of δ (measured in radians). The etch cavity aligns perfectly to the [110] direction and thus undercuts the hard mask (shaded area). The etch cavity is therefore enlarged by factor of $1+\delta$. Typical silicon wafers have an alignment tolerance of 0.5° of the flat to the [110] direction. By introducing special alignment structures the misalignment can be reduced to about 0.1° [8].

Another important issue in anisotropic wet etching is the underetching of convex corners. Convex corners (for example the corners of a mesa in an etched cavity) may in principle be built by two [111] planes. The corners are nevertheless rapidly underetched (Fig. 11.12). Convex corners have to be protected from underetching by compensation structures. A compensation structure itself is underetched and just disappears and exposes the corner to the etch, when the desired etch depth is reached. A simple compensation structure is illustrated in Fig. 11.12. More sophisticated structures are explained in [13]. Compensation structures generally consume some chip area, and in some case leave a certain degree of rugged non-[111] surface.

Isotropic Etching

Isotropic etching may be performed either by dry plasma etching (provided that ion bombardment is negligible), gas-phase etching (XeF_2) or wet etching (for example HNA, a dilute mixture of HNO_3, HF and CH_3COOH).

The gas-phase etching with XeF_2 does not need any plasma source, the silicon reacts with the gaseous XeF_2 that sublimes from its solid-state form in a low-pressure chamber (at 1–5 Pa). The etch rate is quite high (typically 1–3 μm/min) [8] and may reach values of up to 40 μm/min. The selectivity to the masking materials silicon dioxide, silicon nitride, aluminium and photoresist is very high. Pure XeF_2 etching produces a very rough surface with granular features of about 10 μm size.

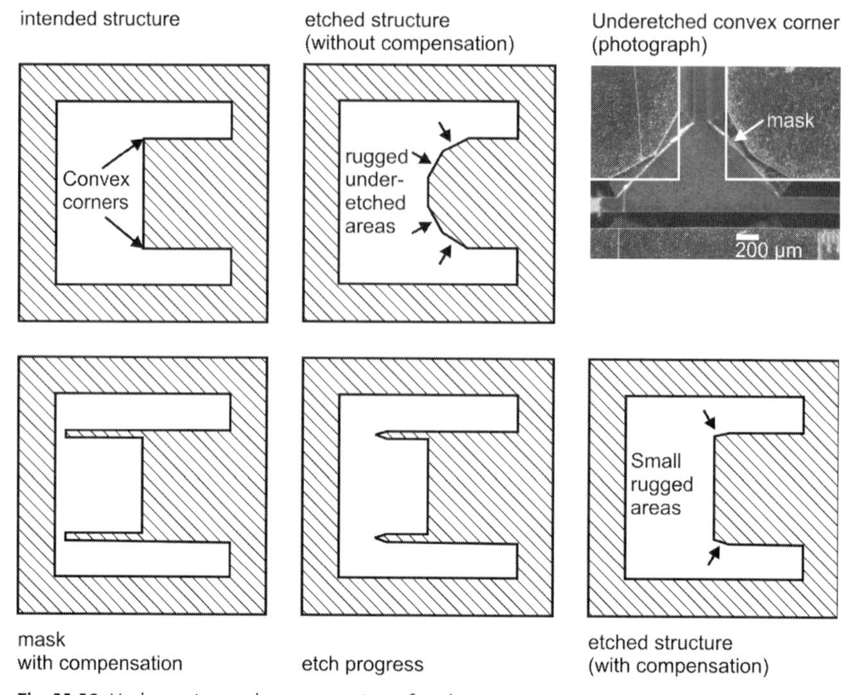

Fig. 11.12 Undercutting and compensation of undercutting.

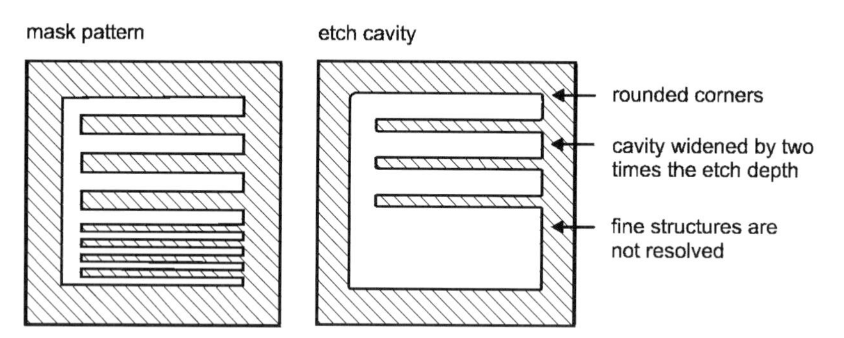

Fig. 11.13 Isotropic underetching.

HNA etching does not need any special equipment, a reaction vessel and an extractor hood is sufficient. The etch rate of about 4 to 7 µm/min [8] is very high, but masking with oxide is a severe problem for deep structures (oxide etch rate up to 70 nm/min [8]). The silicon etch rate is very sensitive to temperature, silicon doping level and even illumination, thus the repeatability of this process step is a problem.

The isotropic character reduces the minimum feature size of the etched structures to values much above the etch depth (Fig. 11.13). Sophisticated structures with well-controlled vertical dimensions cannot be fabricated. Therefore isotropic etching is not used very often for microfluidic devices.

11.2.3 Bonding Techniques

In the previous section, typical micromachining techniques for the structuring of blank silicon wafers were introduced. The structured wafers can be assembled to form multilayer superstructures. Depending on the complexity of the device, several combinations of bond partners and materials are possible. Using the whole range of available bonding techniques, a vast variety of sophisticated devices can be realized. In this section, a short overview of available bonding methods will be given and briefly discussed for their abilities for micro process engineering.

Adhesive Bonding

Adhesive bonding enables a low-budget bonding possibility for almost all substrate materials. The equipment costs are low as well as the demands on the wafer surface condition. Usually, a thin polymer layer, e.g. polymethylmethacrylate (PMMA) [14] or a glass layer is spun or stencilled onto the wafer surface. The bond partners are brought into contact and heated. An advantage of glass intermediate-layer bonding (also called "glass frit" bonding) compared to polymer-layer bonding is the higher physical stability of the intermediate layer against chemicals in devices used for, e.g., liquid handling. However, the homogeneous deposition of the intermediate layer onto a structured substrate seems to be a technological challenge. Moreover, channels, trenches and cavities are easily filled by the adhesive that results in malfunction of the device.

Eutectic Bonding

The gold–silicon phase diagram shows a eutectic point at 363 °C at 19 at.% silicon. This low eutectic point can be used to form bonds between silicon wafers when cooling down the molten gold–silicon interfacial composite. To prepare wafers for bonding, gold (e.g. 500 nm) is deposited onto the bonding site of the silicon wafer. The bond partners are aligned and a thermal treatment above 363 °C is performed while the wafers are compressed by a weak force. The drawback of eutectic bonding at low temperatures is the strong void formation when bonding larger areas. A very interesting approach is localized eutectic bonding [15]. In this technique, the subsequent bonding sites are patterned into the gold layer (e.g. bonding frames, meanders) and locally heated by current flowing through the resistance of the patterns. The heated structures will consequently form eutectic bonds after the current is turned off. Temperatures above 800 °C can thus be achieved locally at the bonding sites, while the substrate temperature remains low. The high process temperature results in strong bonds and low void density at short process times ($\sim$5 min).

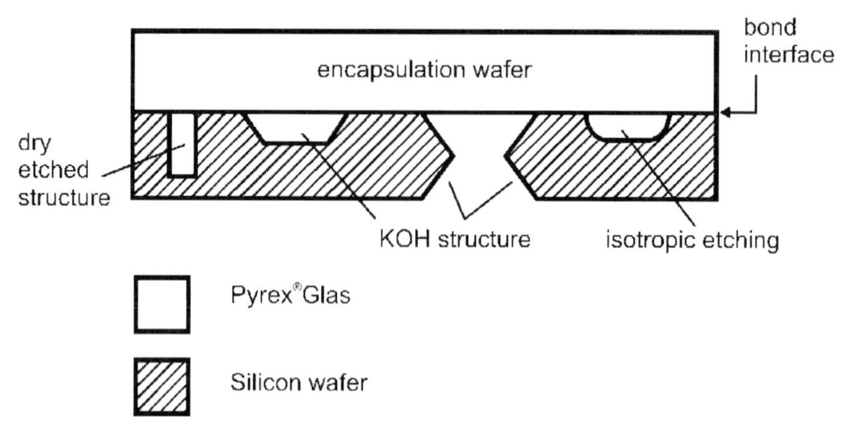

Fig. 11.14 A bonded Pyrex®-glass silicon stack.

Anodic Bonding

This method is one of the oldest and most established bonding techniques. Borosilicate glass wafers, e.g. Pyrex® glass, can be joined to silicon wafers by anodic bonding. The physical effect lies in a permanent ion exchange between the glass and the silicon wafer. Therefore, a strong electric field (1 kV) has to be applied to the wafer pair at elevated temperatures (450 °C). Sodium ions (+) are depleted at the bonding interface of the borosilicate glass, while oxygen ions (–) react with the silicon wafer to form silicon oxide. The static imbalance results in strong, permanent bonds. Since the structuring of the glass wafer can only be done isotropically (e.g. using hydrofluoric acid) and with less precision compared to silicon technologies, the glass wafer is commonly applied as an encapsulation of structures fabricated in the silicon wafer as shown in Fig. 11.14.

Fusion Bonding

Silicon fusion bonding can be subdivided into high-temperature and low-temperature bonding. Low-temperature bonding processes usually deploy temperatures below 400 °C. The mechanism of bonding is a cross-linking of silanol groups (Si–OH) over the bond interface by hydrogen bridges. At elevated temperatures, the bridged silanol groups are converted into water and silicon oxide (Si–OH + HO–Si → Si–O–Si + HOH). This results in a cross-oxidation of the silicon over the bonding interface. Consequently, strong permanent bonds are achieved. To enhance the initial number of hydrogen bonding sites the native oxide or thermal oxide on the wafer surface has to be activated, i.e. the closed Si–O–Si chains on the surface have to be broken up by special activation techniques. Chemical cleaning and/or plasma treatment followed by a deionized water dip are typical surface activation process steps. More details on the mechanism of silicon wafer bonding can be found in [16].

To provide bondable wafer surfaces for fusion bonding, a high-grade cleanroom with a strict particle control is necessary. The achievable bond strength can be as high as the fracture toughness of silicon and the bond area yield is very high and reproducible. The silicon and oxidized silicon wafers can be bonded in all combinations and therefore enables choices between insulation or electrically conductive stacking of the bond partners during the design process. Moreover, surface properties, e.g. the contact angle of a fluidic device can also be adjusted.

The bonding can be performed utilizing standard cleanroom equipment [17]. Recently, commercial low-temperature plasma bonding equipment has also become available.

Characterization of the Bond Quality

The strength of a bond, i.e. the degree of adhesion between the wafers is commonly quantified by the bond strength [Pa] or the surface energy [J/m^2] of a bond. For its determination only a few measurement methods exist. One of the most-cited measurement strategies is the crack-opening method by Maszara et al. [18]. Here, a blade is inserted between the bonded wafer pair and the length of the initiated crack is measured. From the crack length, the surface energy can be calculated. Other, also destructive tests, focus on the separation of the wafer pair by an applied outside force, e.g. diverse pull tests or pressurized cavities (blister test).

For the examination of bonded silicon wafers, infrared transmission imaging is typically used (Fig. 11.15). Figure 11.16 shows an infrared transmission image

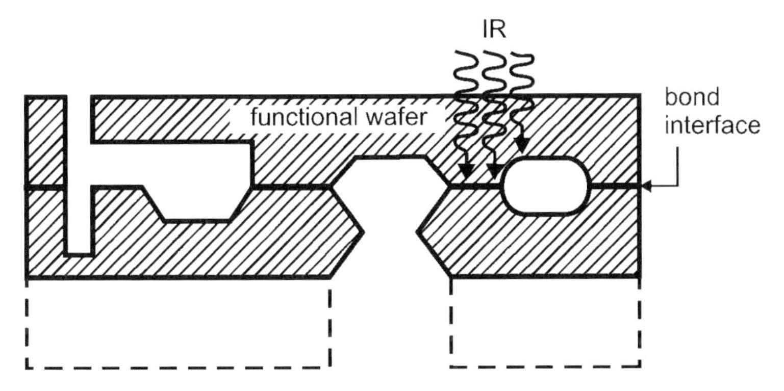

Fig. 11.15 Principle of IR inspection.

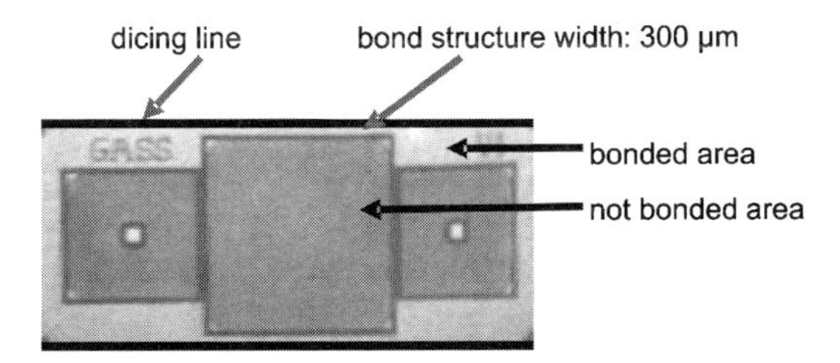

Fig. 11.16 IR photograph of a silicon micropump.

of a bonded silicon micropump [17]. Here, the bright regions are bonded. The darker regions are nonbonded, functional membrane areas. If a higher imaging resolution is necessary, surface acoustic microscopy (SAM) or X-ray imaging can be applied, accompanied with higher equipment cost.

11.3 Applications

Because of the relatively high equipment and processing costs of silicon microfabrication, applications are found in areas where the advantageous properties of silicon microfabrication are essential for the device performance or in those areas where very high lots of small chips are fabricated (e.g. pressure sensors). Here, the concentration is on microfluidic applications for which the advantageous properties of silicon microfabrication lead to this choice of technology.

The high fracture strength and the perfect elastic behavior of silicon as well as etch-rate uniformity of anisotropic wet etching enable the fabrication of thin flexible silicon diaphragms. These are often used in liquid- or gas-handling devices. They are deflected by different means of actuation, i.e. electrostatic, piezoelectric, pneumatic or thermopneumatic actuation. The advantage of silicon diaphragms as driving elements is their chemical inertness, their mechanical stability and the absence of fatigue. As an example we will discuss a silicon nanolitre disperser [19] and a silicon micropump [20–22] from our laboratory. Many groups have fabricated similar liquid- or gas-handling devices, e.g. micropumps [23], microvalves [24–27] and dispensers [28–30].

Another application field for silicon microstructures are micromixers. Mixing stages are cascaded to increase mixing quality and several mixing paths are parallelized to increase throughput. Therefore many structures occur repeatedly on a chip. These structures are preferably fabricated by lithographic steps and not by sequential mechanical fabrication. If mixing is accompanied by an exothermic chemical reaction the high thermal conductivity of silicon permits a good temperature control. The chemical inertness of silicon or its coatings allows for a variety of chemicals. We present a micromixer from our laboratory and give citations to many other micromixers.

11.3.1 The Nanoliter Dispenser

The nanolitre dispenser (Fig. 11.17) is a good example for the advantages of anisotropic wet etching of silicon for the fabrication of microfluidic devices. The working principle is demonstrated in Fig. 11.18. A silicon diaphragm is displaced by a pressure pulse and a liquid jet is ejected through the nozzle. The two critical elements for a proper function of the dispenser are the diaphragm and the nozzle. The diaphragm must have a precisely fabricated thickness, because the dosing volume depends to the third power on the diaphragm thick-

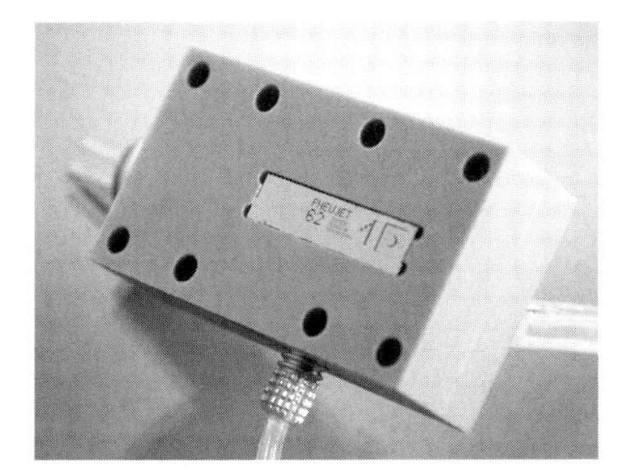

Fig. 11.17 Photograph of the nanoliter dispenser.

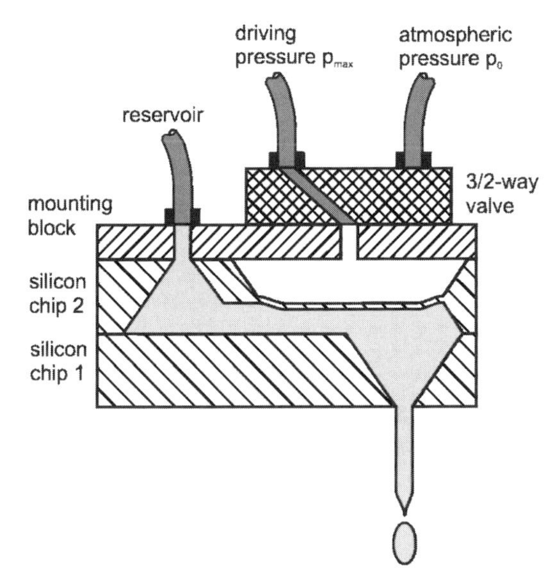

Fig. 11.18 Working principle of the nanoliter dispenser.

ness. Therefore wet anisotropic etching is the preferred technology because of its good etch uniformity. Silicon is well suited for this kind of actuation because it stands many load cycles without fatigue or plastic deformation. The nozzle should have a small orifice (50 μm width) but the viscous losses should be minimized. A tapered cross section is thus the optimal choice. This tapered shape is the natural shape of wet anisotropically etched [100] silicon. The high [100]/[111] etch-rate ratio allows for a good control of the orifice width, though the etch pit is opened on the opposite wafer face. The inlet restriction should be of the same shape as the nozzle and should therefore be fabricated with the same technology. All other elements do not significantly influence the device performance and may thus be fabricated without special care.

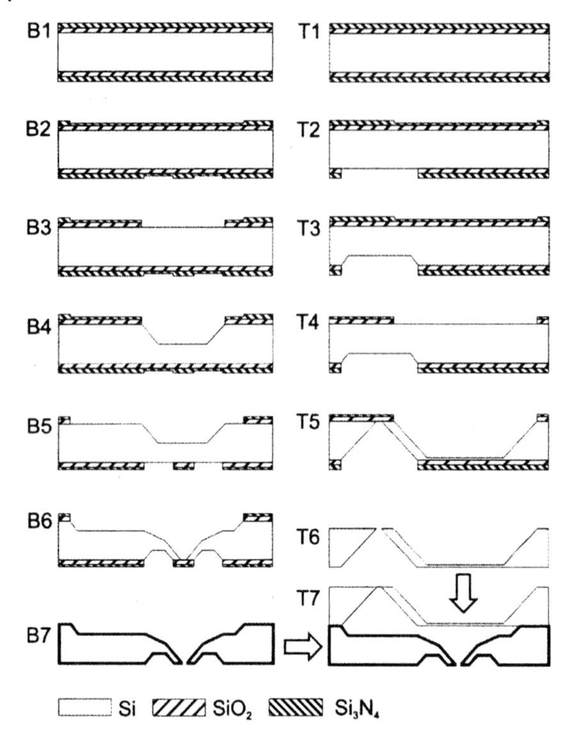

Fig. 11.19 Process chart of the nanoliter dispenser.

These considerations lead to the following process (Fig. 11.19, Table 11.2) using two wafers, "bottom" and "top", structured with KOH from both sides. The nozzle is fabricated in two etch stage processes. Here the nozzle inlet cannot be protected against the etching in the process step B6, therefore the nozzle inlet is attacked and shows a rugged surface, but this does not influence the overall performance.

11.3.2 Silicon Micropumps

Figure 11.20 shows the cross section of a peristaltic micropump fabricated in silicon and a photograph of the device embedded in a fluidic housing. The outer two diaphragms are used to seal a circular valve lip when actuated (see also Fig. 11.15). The pump diaphragm in the middle enables adjustment of a pressure difference between the pump chamber and its peripheral. The micropump is fabricated similar to the above discussed nanolitre dispenser. Here, different etching technologies are being used. The top wafer is structured on both sides by KOH etching. The inlet and outlet of the bottom wafer are deep etched by KOH etching. The valve seats – circular, very thin structures with high aspect

Table 11.2 Process flow of the nanoliter dispenser.

	"Bottom"		"Top"
B1	Material: double-sided polished [100] silicon 1. Coating: Oxidization 2. Coating: LPCVD-nitride	T1	Material: double-sided polished [100] silicon 1. Coating: Oxidization 2. Coating: LPCVD-nitride
B2	1. Litho: "Top" side 2. Dry etch: Step on "top" side 3. Litho: "Bottom" side 4. Dry etch: Step on "bottom" side	T2	1. Litho: "top" side 2. Dry etch: Step on "top" side 3. Litho: "bottom" side 4. Dry etch: Open hard mask on "bottom" side
B3	1. Litho: "top" side 2. Dry etch: Open hard mask on "top" side	T3	1. KOH etch: Cavity in "bottom" side
B4	1. KOH etch: Cavity in "bottom" side	T4	1. Dry etch: Open hard mask on "top" side
B5	1. Dry etch: Open hard mask on "Top" side	T5	1. KOH etch: Up to hard mask from "bottom" side, to desired diaphragm thickness from "top" side
B6	1. KOH etch: Up to hard mask from "top" side, to desired cavity depth from "bottom" side	T6	1. Strip hard mask layers
B7	1. Strip hard mask layers 2. Oxidization for hydrophilic bond process	T7	1. Hydrophilic bonding of top onto bottom wafer

ratios – are etched by deep reactive ion etching (DRIE). Because of its flow-limiting properties, the quality of the valve lips mainly determines the pump performance. Using an optimized DRIE process, valve lips with a thickness of 3 μm and a height of 40 μm can be realized. Figure 11.21 (left) shows a valve seat that was fabricated as described above. The through-etched KOH square and the circular valve lip can be clearly seen. The structures inside the ring are employed to absorb the impacting valve membrane. In Fig. 11.21 (right), a magnified segment of the valve lip is shown. A human hair is also shown as a structure size reference. After structuring, the top and the bottom wafer are bonded by a low-temperature plasma-assisted fusion-bonding process.

For the nanolitre dispenser pneumatic actuation was used. This micropump employs piezoelectric actuators. The actuator plates (mostly made of lead zirconate titanate PZT) are adhered onto the silicon diaphragms. A cross section of the piezo-adhesive-silicon compound is shown in Fig. 11.22. Applying an electric field in the polarization direction of the piezoelectric material results in a contraction of the piezoelectric plate and consequently in a downward movement of the piezo-silicon bending actuator. Respectively, a negative electric field re-

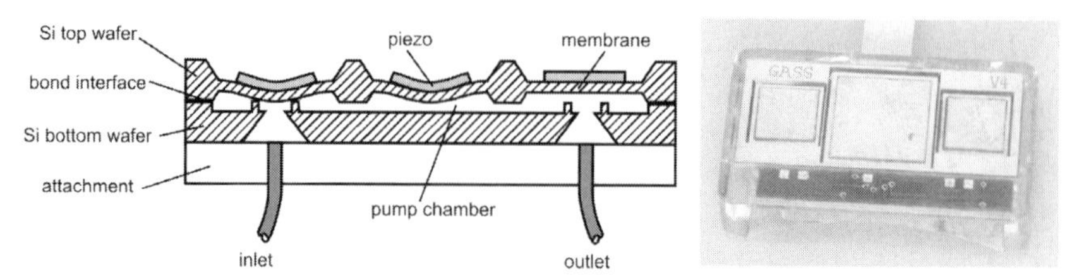

Fig. 11.20 Cross section and photograph of a peristaltic silicon micropump.

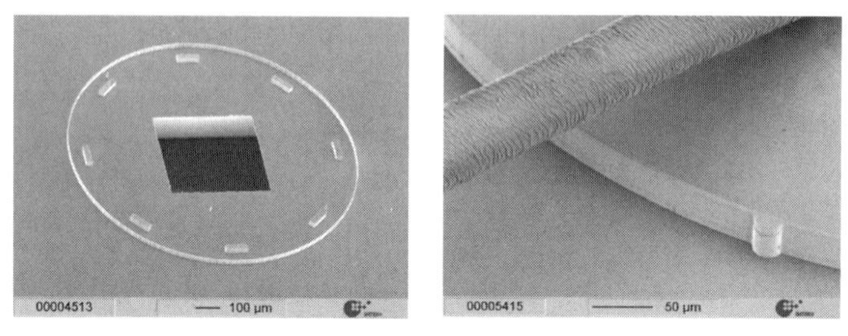

Fig. 11.21 Scanning Electron Microscopy (SEM) image of the valve lip of the peristaltic silicon micropump.

Fig. 11.22 Cross section through the piezoactuator of the micropump.

sults in an upward bending of the diaphragm. This effect is used to seal the valves and to generate a pressure gradient inside the pump chamber. The actuation scheme of the different actuators determines the net flow direction of the micropump [20]. In Fig. 11.23, a typical four-phase driving scheme is shown. Here, a portion of gas/fluid is transported from left to right during one pump cycle. Switching the driving algorithm of the valves changes the flow direction. Thus, this pump can be operated bidirectionally.

A typical micropump using this working principle and having an overall size of about $30\times11\times1$ mm^3 can pump water at a flow rate of 1.8 ml/min. The flow rate linearly decreases with the backpressure exerted at the outlet of the pump. Our micropump was specially designed for a medical application [21] and can build up a backpressure of up to 60 kPa [22]. A very comprehensive review of micropumps can be found in [23].

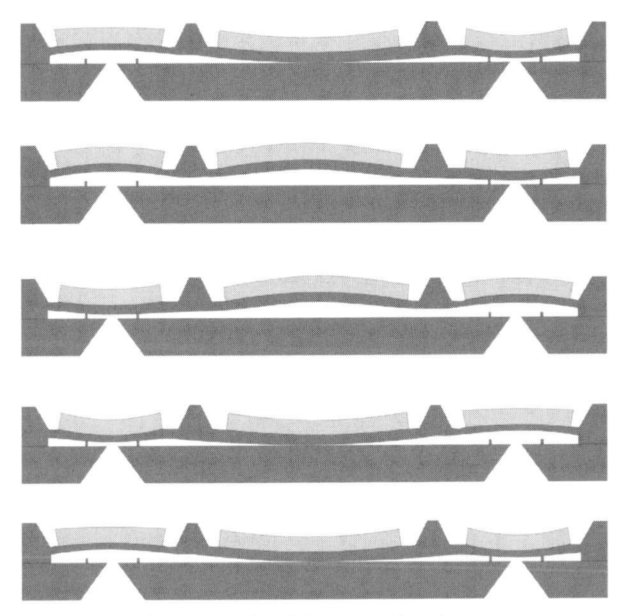

Fig. 11.23 Working principle of the peristaltic micropump.

11.3.3 Silicon Micromixers

Since mixers are among the most important devices for process engineering, a vast variety of micromixers have been reported up to now. Among these are also many micromixers made of silicon or based on silicon processes.

There exist more passive than active mixers. This is understandable, since it is far more difficult to design proper active micromixers than passive ones, whereas the advantages of active mixers prevail about the disadvantages only in special cases. Implemented are mixing devices using electrokinetic dielectrophoresis in combination with time-pulsed in-flow [31], ultrasound [32], or flow provided by micropumps [33, 34].

Passive micromixers play a more important role in micromixing engineering. The reasons for this are mainly the easier and reliable design and fabrication on the one hand, and on the other hand the small difference between the performance of active and passive mixing. There is often no advantage in using active mixers.

Many passive silicon micromixers make use of the fact that silicon channels can be made very thin and can consist of very complicated structures. This is, e.g., implemented in a capillary-force, self-filling micromixer [35], in micromixers using very narrow fluidic networks [36, 37], or in different mixers using complex channel structures [38–40]. These mixers are mainly provided for μTAS-applications. Furthermore, some multilamination mixers have also been presented [41–43], the latter of which being especially interesting, since the channels have additionally been fabricated not in silicon, but in SU8 resist within common silicon fabrication techniques.

In recent years, more focus has been placed on the research of the underlying processes inside micromixers. For this, silicon structures are suitable since therewith a vast variety of different designs can easily be fabricated. This has mainly been accomplished using T-shaped micromixers [44–46], where vortex effects arising at high Re numbers lead to a very good mixing performance. The very fast mixing performance of T-shaped micromixers at high Re numbers has also been shown in an older contribution, without detailing the underlying processes [47]. The creation of vortices in microchannels and the associated enhancement of mixing due to convective effects has been examined in Y-mixers with bent mixing channels [48] or in S-shaped mixing channels [49].

The advantages of silicon technology are now presented on a special silicon micromixer. These mixers consist of 16 mixing elements arranged in parallel. The mixing elements are arranged on two silicon wafers, with the first wafer building up the inlet and combining layer where the two fluids are conducted into and then combined, while the second layer builds up the mixing and outlet layer, see Fig. 11.24 for details.

For the single mixing elements, various designs have been implemented (asymmetric 3D T mixers, tangential mixers, double T mixers and cascading T mixers). To ensure that the pressure losses and therefore the flow conditions are equal at each mixing element, the inlet and mixing channels have been broadened. This has been analytically calculated. The transition from the lower to the upper layer is achieved by transition holes being positioned directly after the fluids are combined, see Fig. 11.25. Therewith, the flow is three-dimension-

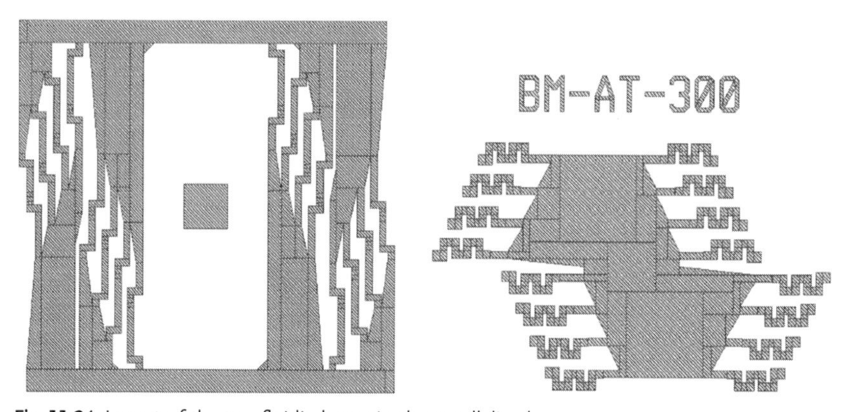

Fig. 11.24 Layout of the two fluidic layers in the parallelized convective micromixers. The outlet is situated in the middle, the inlets are on the top and bottom edge of the lower layer (left side). The mixing elements consist of asymmetric T-shaped mixing elements.

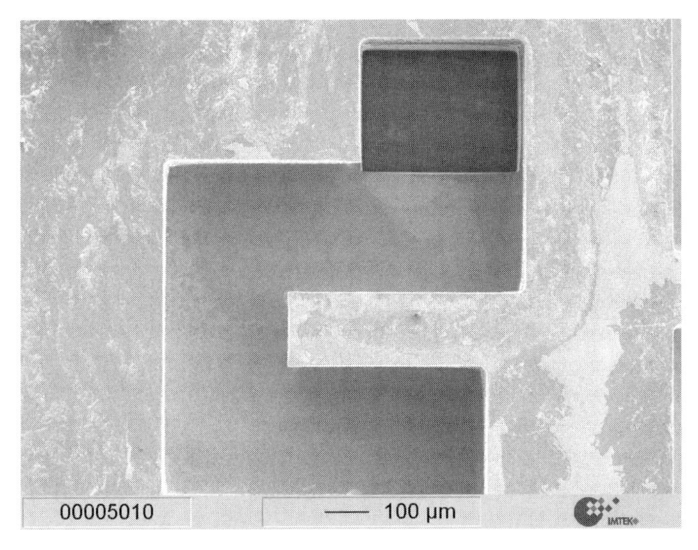

Fig. 11.25 SEM image of the mixing channel (upper layer) with the transition hole to the lower layer.

ally bent, leading to chaotic mixing by the resulting convective effects. This is further enhanced by a multiple bending of the mixing channel.

The channels are fabricated by DRIE, leading to channels with smooth walls, see also Fig. 11.25. By using reactive ion etching, the channels may have almost any shape on the surface of the wafer, only the third dimension is fixed. The two silicon chips are bonded together by silicon fusion bonding explained earlier in this chapter, and sealed by a lid made from pyrex glass that is anodically bonded. The inlets and the outlet are positioned at the downside of the chip. The fully processed chip has dimensions of only $20\times20\times1.5$ mm^3 and is shown in Fig. 11.26. The mixing devices are described in more detail in [50].

Compared to these dimensions, the performance data are surprising. For aqueous solutions, a maximum throughput of 22 kg/h has been measured, at a full mixing time of about 0.5 ms and a pressure loss of about 1 bar. This is achieved by a high porosity of the device through the parallelization of the single mixing elements and the broadening of the channels. Such properties are only possible by using lithography, etching processes and the bonding processes mentioned earlier in this chapter.

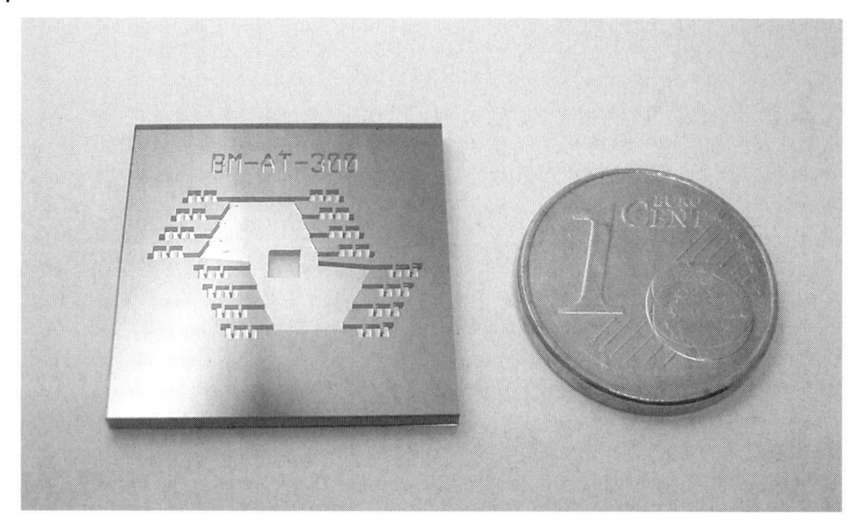

Fig. 11.26 One of the fabricated micromixers compared to the size of 1 Euro cent.

11.4 Conclusion

Silicon technology offers a variety of opportunities to microreaction technology. Its mechanical and thermal properties denote silicon as a good design material for moderately high temperatures ($<1000\,^{\circ}C$) especially when moving parts are needed. The technically mature process technology enables the fabrication of highly accurate structures with a high degree of complexity.

Two main strategies for a commercial implementation of silicon devices in micro process engineering have to be pursued. The first strategy rests on the huge complexity of silicon devices, combining complex channels with a high functionality of the surfaces or bulk structures. This can lead to new applications that are economically not reasonable or that cannot be fabricated with other technologies. For example, chemical reactors including two or more consecutive reactions on one single chip are thinkable, leading to a vast variety of possible new reaction schemes and chemical products.

The second strategy uses the high degree of parallelization of silicon devices. The mass fabrication of silicon devices is relatively cheap, leading to low unit costs for the single product. This is already true for microelectronics and many applications of micromechanics (computer chips, airbag sensors, pressure sensors, etc.). Similar developments in the mass production of microfluidic components made of silicon and used in process engineering are thus expected in the near future.

Finally, the designed silicon structures can also be used as masters for any casting techniques. With this, a certain number of structures made from any cheap material, e.g. plastics, can be casted from only a small amount of silicon structures. This leads to very cheap devices that are designated for single-use only.

References

1 K. E. Petersen, Silicon as a Mechanical Material, *Proc. IEEE* **1982**, *70(5)*, 420–457.

2 D. K. Schröder, *Semiconductor Material and Device Characterization*, John Wiley & Sons, Inc., New York, **1990**.

3 D. R. Lide (ed.), *Handbook of Chemistry and Physics, 76th Edition*, CRC-Press, New York, **1995**.

4 W. Fulkerson, J. P. Moore, R. K. Williams, R. S. Graves, D. L. McElroy, *Phys. Rev.* **1968**, *167*, 765.

5 V. Ziebart, O. Paul, H. Baltes, *Mater. Res. Soc. Symp. Proc.* **1999**, *546*, 103–108.

6 V. L. Spiering, S. Bouwstra, R. Spiering, On chip decoupling zone for package-stress reduction, *Sensors and Actuators A* **1993**, *39*, 149–156.

7 F. Ericson et al., Hardness and Fracture Toughness of Semiconducting Materials Studied by Indentation and Erosion Techniques, *Mater. Sci. Eng. A* **1988**, *105/106*, 131–141.

8 M. Madou, *Fundamentals of Microfabrication*, CRC Press, Boca Raton, **1997**.

9 W. Menz, J. Mohr, O. Paul, *Microsystems Technology*, VCH, Weinheim, **2001**.

10 S. Sze, *VLSI Technology*, McGraw-Hill, 1988.

11 C. Y. Chang, S. Sze, *ULSI Technology*, McGraw Hill, **1996**.

12 C. Kittel, *Introduction to Solid State Physics*, John Wiley & Sons, New York, **1996**.

13 H. L. Offreins, Compensating Corner Undercutting of (100) silicon in KOH, *Sens. Mater. 3, 127–144.*

14 W. P. Eaton, S. H. Risbud, R. L. Smith, Silicon Wafer-to-Wafer Bonding at T < 200 °C with PMMA, *Appl. Phys. Lett.* **1994**, *65(4)*, 439–441.

15 Y. T. Cheng, Liwei Lin, K. Najafi, Localized Silicon Fusion and Eutectic Bonding for MEMS Fabrication and Packaging, *J. Microelectromech. Sys.* **2000**, *9 (1)*.

16 Q.-Y. Tong, U. Gösele, Semiconductor Wafer Bonding, Science and Technology, *The Electrochemical Society Series* **1999**.

17 A. Doll et al., Versatile Low Temperature wafer Bonding and Bond Strength Measurement by a Blister Test Method, *J. Microsystem Technologies* (in press)

18 W. P. Maszara, G. Goetz, A. Caviglia, J. McKitterick, Bonding of Silicon Wafers for Silicon-on-Insulator, *J. Appl. Phys.* **1988**, *64*, 4943.

19 F. Goldschmidtböing and P. Woias, A Reliable Nanoliter Dispenser Based on Pneumatic Actuation, *Proceedings of Actuator 2004*, June 14–16, Bremen Germany, 569–572.

20 F. Goldschmidtböing, A. Doll, M. Heinrich, P. Woias, H. J. Schrag, U. T. Hopt, A Generic Analytical Model for Micro-Diaphragm Pumps with Active Valves, *Micromech. Microeng.* **2005**, *15(4)*, 673–683.

21 A. Doll, F. Goldschmidtböing, M. Heinrichs, P. Woias, H. J. Schrag, U. T. Hopt, A Piezoelectric Bidirectional Micropump for a Medical Application, *IMECE04*, Anaheim, **2004**, 61083.

22 A. Doll, F. Goldschmidtböing, S. Reimers, H.-J. Schrag, U. T. Hopt, P. Woias, A High Performance Bidirectional Micropump For A Novel Artificial Sphincter System, *Transducers*, Seoul, **2005**, 188–191.

23 D. J. Laser, J. G. Santiago, A Review of Micropumps, *J. Micromech. Microeng.* **2004**, *14*, R35–R64.

24 A. K. Henning, J. Fitch, E. Falsken, D. Hopkins, L. Lilly, R. Faeth, M. Zdeblick, A Thermopneumatically Actuated Microvalve for Liquid Expansion and Proportional Control, *Transducers*, Piscataway, **1997**, 825–828.

25 S. Kluge, G. Klink, P. Woias, A fast switching, low power pneumatic microvalve with electrostatic actuation made by silicon micromachining, *American Laboratory* **1998**, *30(6)*, 17.

26 J. Schaible, J. Vollmer, R. Zengerle, H. Sandmaier, T. Strobelt, Electrostatic Microvalves in Silicon with 2-way-Function for Industrial Applications, *Proc. Transducers*, Munich, **2001**, 928–931.

27 G. Hahm et al., Microfabricated, Silicon Spring Biased, Shape Memory Actuated Microvalve, *Proceedings of the IEEE Solid-State Sensor and Actuator Workshop*, **2000**, 230–233.

28 B. de Heij, C. Steinert, H. Sandmaier, R. Zengerle, A Tuneable and Highly Parallel Picolitre-Dispenser Based on Direct Liquid Displacement, *Sens. Actuators A* **2003**, *103*, 88–92.

29 P. Koltay, G. Birkle, R. Steger, H. Kuhn, M. Mayer, H. Sandmaier, R. Zengerle, Highly parallel and Accurate Nanoliter Dispenser for High-Throughput-Synthesis of Chemical Compounds, *Proc. International MEMS Workshop (I-MEMS)*, Singapore, **2001**, 115–124.

30 T. Laurell et al., Design and Development of a Silicon Microfabricated Flow-through Dispenser for On-line Picolitre Sample Handling, *J. Micromech. Microeng.* **1999**, *9*, 369–376.

31 Y.-K. Lee, J. Deval, C.M. Ho, P. Tabeling, Chaotic Mixing in Electrokinetically and Pressure Driven Micro flows, *IMRET5*, Springer, Berlin, **2001**, 185–191.

32 Z. Yang, S. Matsumoto, H. Goto, M. Matsumoto, R. Maeda, Ultrasonic Micromixer for Microfluidic Systems, *Sens. Actuators A* **2001**, *93*, 266–272.

33 Z. Yang, H. Goto, M. Matsumoto, T. Yada, Micromixer Incorporated with Piezoelectrically Driven Valveless Micropump, in *Micro Total Analysis Systems*. Kluwer, Dordrecht, **1998**, 177–180.

34 P. Woias, K. Hauser, E. Yacoub-George, An Active Silicon Micromixer for μTAS Applications, in *Micro Total Analysis Systems*. Kluwer, Dordrecht, **2000**, 277–282.

35 R.U. Seidel, D.Y. Sim, W. Menz, M. Esashi, A Capillary Force Filled Automixing Device, *IMRET3*. Springer, Berlin, **2000**, 506–513.

36 F.G. Besoth, A.J. DeMellow, A. Manz, Microstructure for Efficient Continuous Flow Mixing, *Anal. Commun.* **1999**, *36*, 213–215.

37 D. Lim, Y. Kamotani, B. Cho, J. Mazumder, S. Takayama, Fabrication of Microfulidic Mixers and Artificial Vasculaires Using a High-Brightness Diode-Pumped Nd:YAG Laser Direct Write Method, *Lab on a Chip* **2003**, *3*, 318–323.

38 M.K. Jeon, J.-H. Kim, H.J. Yoon, J. Noh, S.H. Kim, E. Yoon, H.G. Park, S.I. Woo, Design and Characterization of a Passive Recycle-Micromixer, in *Proc. 7th Int. Conf. on μTAS*, Squaw Valley, **2003**, 109–112.

39 J. Voldman, M.L. Gray, M.A. Schmidt, Liquid Mixing Studies with an Integrated Mixer/Valve, in *μTAS* (Eds.: J. Harrison, A. van den Berg), Kluwer, Dordrecht, **1998**, 181–184.

40 S.-J. Park, J.K. Kim, J. Park, S. Chung, C. Chung, J.K. Chang, Rapid Threedimensional Passive Rotation Micromixer Using the Breakup Process, *J. Micromechm. Microeng.* **2004**, *14*, 6–14.

41 S. Böhm, K. Greiner, S. Schlautmann, S. de Vries, A. van den Berg, A Rapid Vortex Micromixer for Studying High-Speed Chemical Reactions, in *μTAS* (Eds.: J.M. Ramsey, A. van den Berg), Kluwer, Dordrecht, **2001**, 25–27.

42 M. Koch, D. Chatelain, A.G.R. Evans, A. Brunnschweiler, Two Simple Micromixers Based on Silicon, *J. Micromech. Microeng.* **1998**, *8*, 123–126.

43 R.J. Jackman, T.M. Floyd, R. Ghodssi, M.A. Schmidt, K.F. Jensen, Integrated Microchemical Reactors Fabricated by Both Conventional and Unconventional Techniques, *IMRET4*, Atlanta, **2000**, 62–70.

44 M. Engler, N. Kockmann, T. Kiefer, P. Woias, Numerical and Experimental Investigations on Liquid Mixing in Static Micromixers, *Chem. Eng. J.* **2004**, *101*, 315–322.

45 S.H. Wong, M.C.L. Ward, C.W. Wharton, Micro T-Mixer as a Rapid Mixing Micromixer, *Sens. Actuators B* **2004**, *100*, 359–379.

46 M. Engler, N. Kockmann, T. Kiefer, P. Woias, Convective Mixing and its Application to Micro Reactors, *ICMM2004*, ASME, Rochester, **2004**, 781–788.

47 D. Bökenkamp, A. Desai, X. Yang, Y.-C. Tai, E.M. Marzluff, S.L. Mayo, Microfabricated Silicon Mixers for Submillisecond Quench Flow Analysis, *Anal. Chem.* **1998**, *70*, 232–236.

48 V. Mengeaud, J. Josserand, H. Girault, Mixing Processes in a Zigzag Microchannel: Finite Element Simulations and Optical Study, *Anal. Chem.* **2002**, *74*, 4279–4286.

49 F. Schönfeld, S. Hardt, Simulation of Helical Flows in Microchannels, *AIChE J.* **2004**, *50*, 771–778.

50 M. Engler, T. Kiefer, N. Kockmann, P. Woias, Effective Mixing by the Use of Convective Micromixers, *IMRET8*. AIChE, Atlanta, **2005**, TK 128d.

12
Microfabrication in Ceramics and Glass

Regina Knitter, Institute for Materials Research III, Forschungszentrum Karlsruhe, Germany
Thomas. R. Dietrich, mikroglas chemtech GmbH, Mainz, Germany

Abstract
The use of ceramics and glass as reactor materials is very promising for high-temperature and corrosive applications in microprocess engineering. Ceramic materials in particular demand more effort in the fabrication processes and joining techniques than metals or silicon. Notwithstanding these obstacles, several components have already been developed and successfully applied for chemical or thermal applications. Glass microreactors have been developed and manufactured over recent years by several companies. Several techniques, such as sand blasting, standard photolithography with wet etching have been used, as well as mechanical methods. A very interesting material is a photostructurable glass like FOTURAN, which allows the very accurate manufacture of fine channels and chambers with high aspect ratio. The manufactured structures are connected by diffusion bonding to form three-dimensional channel structures.

Keywords
Ceramic injection molding, tape casting, stereolithography, coating, joining, photoetchable glass, FOTURAN

Advanced Micro and Nanosystems Vol. 5. Micro Process Engineering. Edited by N. Kockmann

ISBN: 3-527-31246-3

12.1 Introduction

Microcomponents made of ceramics and glass are of particular interest in microtechnology applications when their properties qualify them for uses that cannot be covered by metals or polymers. While ceramics are valued in microtechnology in general because of their good mechanical and tribological properties or because of special physical characteristics, like dielectricity or piezoelectricity, their high thermal and chemical resistance is the most interesting feature for applications in microprocess engineering. Glass components also offer the advantage of a high chemical stability and a good resistivity at elevated temperatures. In addition, glass features optical transparency, which is of particular interest for analytical or photochemical applications and an asset in mixing processes. Tables 12.1 and 12.2 give an overview of the most relevant physical properties of selected ceramic and glass materials.

Besides the thermal and chemical resistivity, ceramic and glass materials have some characteristics that distinguish them from metals. Ceramics excel metals in their compression strength, but are inferior in tensile strength and in elasticity, that is, they are brittle and do not exhibit ductility. Furthermore, it should be emphasized that the properties of ceramics do not only depend on the chemical composition, but are greatly influenced by the fabrication process. Large grain sizes and defects like pores or agglomerates may considerably decrease their properties.

Glass is more stable than one would expect. Glass tends to break starting from microscratches, which are induced during the manufacturing process such as the grinding and polishing processes. Therefore, the last production step is always an HF etch, which heals the scratches and makes the remaining glass structure mechanically very stable.

Table 12.1 Physical properties of selected ceramic materials.

Property	Al_2O_3 [1]	SiC [2]	LTCC 951 [a] [3]
Density (g/cm^3)	3.98	>3.10	>3.10
Porosity (%)	<2	<2	n/a
Vickers hardness (GPa)	15 ± 2	25.5	n/a
Compressive strength (GPa)	3.0 ± 0.5	2.5	n/a
Bending strength (MPa)	380 ± 50	400	320
Young's modulus (GPa)	416 ± 30	n/a	152
Thermal expansion (10^{-6}/K)	7.1 (0–500 °C) 8.1 (0–1000 °C)	4.1 (20–500 °C) 5.2 (500–1000 °C)	5.8 (20–400 °C)
Heat conductivity at 20 °C (W/m K)	33 ± 2	125	3.0
Specific heat capacity at 20 °C (J/kg K)	755 ± 15	600	989

a) DuPont 951 Green Tape™, low-temperature cofiring ceramics in the sintered state.

Table 12.2 Physical properties of FOTURAN® glass and glass ceramic [4].

Property	FOTURAN® glass	FOTURAN® glass–ceramic
Density (g/cm^3)	2.37	2.41
Porosity (%)	0	n/a
Knoop hardness (GPa)	4.6	5.2
Bending strength (MPa)	60	150
Young's modulus (GPa)	78	88
Thermal expansion at 20–300 °C (10^{-6}/K)	8.6	10.5
Heat conductivity at 20 °C (W/m K)	1.35	2.73
Specific heat capacity (J/kg K)	880	920
Transformation temperature (°C)	465	–
Electrical resistivity at 25 °C (Ω cm)	8.1×10^{12}	5.6×10^{16}
Electrical resistivity at 200 °C (Ω cm)	1.3×10^{7}	4.3×10^{7}
Dielectric constant at 20 °C, 1 MHz	6.5	5.7

The application of a material, however, is not only determined by its properties but also by the costs involved in development and production. While microshaping techniques of metals, silicon, polymers and glass have benefited from micropatterning methods developed for semiconductor technology, micromolding of ceramics evolved only in the 1990s. Only for the first mentioned materials a comprehensive know-how on micropatterning is available and their assembly and joining techniques have already been largely established. This is why ceramic components emerged later in applications of microprocess engineering.

This chapter will focus on the most important techniques for the microfabrication of ceramic and glass microcomponents and will give a survey on developments and applications of such components in microprocess engineering.

12.2 Microfabrication of Ceramics

Most structural or so-called engineering ceramics are characterized by a high thermal and chemical stability so that they are predestined as reactor materials for chemical reactions at high temperatures and/or with corrosive reactants involved. Yet they have a significant drawback: Due to their mechanical properties and their high melting point, molding processes and joining techniques known from other materials cannot be transferred to ceramics. Furthermore, conventional shaping techniques used for the mass production of ceramic parts, like dry pressing or slip casting, are hardly suited for microshaping or micropatterning. This is why many recent R&D activities have been dedicated to the development of molding processes down to the micrometer or even nanometer range [5].

While some shaping processes, such as laser machining, electrical discharge machining or special micromechanical methods, are applied to micropattern ceramics in the sintered state [6–8], most processes developed or adapted for the microshaping and micropatterning of ceramic parts are based on a powder-technological molding process in the green, unfired state. Processes like slip pressing [9, 10], centrifugal casting [11], gel and sol-gel casting [12, 13], direct coagulation casting [14], tape casting and embossing [15, 16], injection molding [17, 18], electrophoretic deposition [19, 20] and soft molding [21, 22] differ in terms of design restrictions, achievable part and structure size, aspect ratio (ratio of structure height to structure width), accuracy, manufacturing expenditure, and the economic lot size, but have in common that a negative, micropatterned mold is required. The ceramic powder is dispersed with the help of various amounts of mostly organic additives and consolidated into the mold to obtain the desired shape. After demolding from the mold, the green body has to be heat treated to remove the additives during debinding and annealed for densification in a sintering step (Fig. 12.1 a). For all these molding processes a mold is necessary that exhibits the desired micropatterning to be replicated into the ceramic component. Depending on the demands of the component to be produced and the requirements of the shaping technique to the mold material and its properties, the mold may be directly fabricated by milling or micromachining, indirectly via one or more replication steps from a master model or even by the LIGA technique (German acronym for X-ray deep lithography, electrodeposition and polymer molding). Thus, the mold fabrication can be a costly and time-consuming step, especially during product development where several redesigns may be necessary. In addition, the dimensions of the mold have to take into account the shrinkage of the ceramic part during sintering, and the molding technique has to be suited for near net-shape fabrication as remachining or finishing is either impossible or at least uneconomic for micropatterned parts.

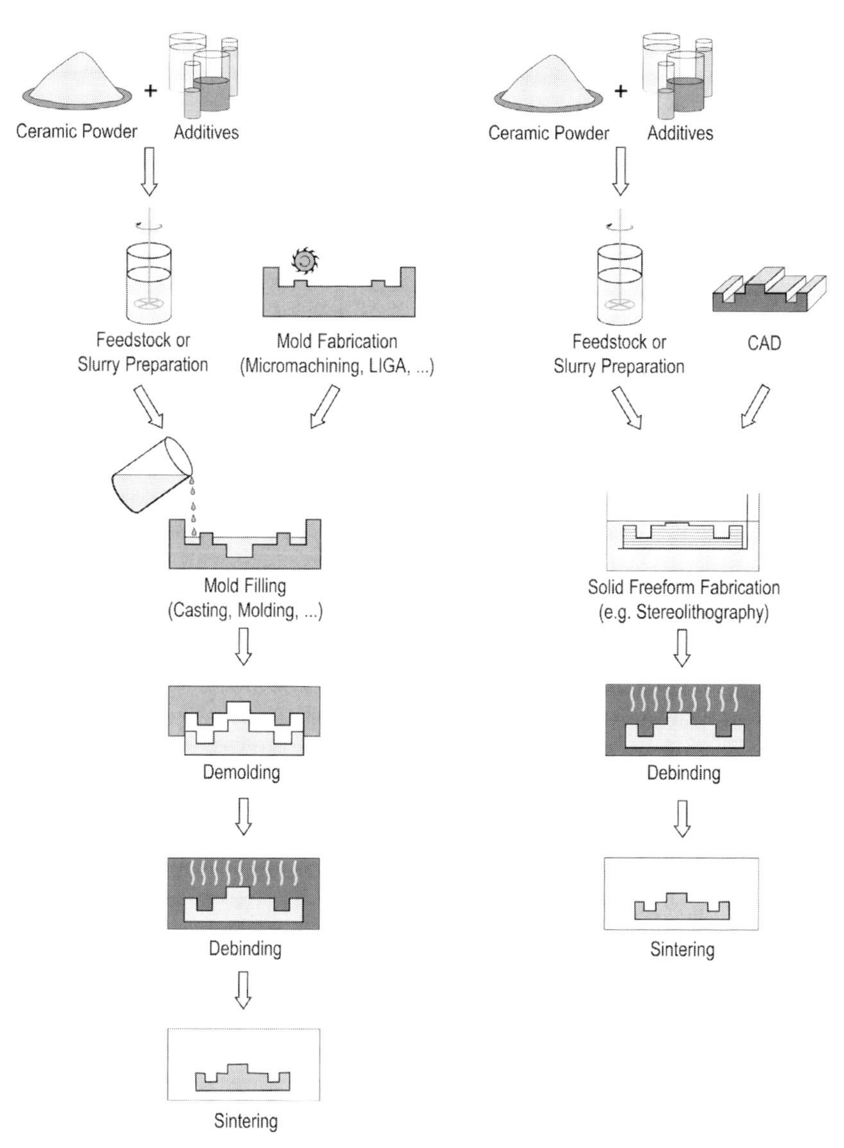

Fig. 12.1 Schematic representation of the process steps of (a) fluidic or plastic molding of ceramics and (b) solid freeform fabrication of ceramics.

A different approach is made by rapid prototyping or so-called solid freeform fabrication (SFF) techniques of ceramics, which enable the realization of a solid, three-dimensional model starting from computer-aided design (CAD) data and using moldless, generative fabrication processes [23, 24]. In these techniques, for example stereolithography, printing techniques or fused deposition of ceramics [25–27], the costly mold fabrication is avoided, the process steps of slurry preparation and thermal treatment, however, correspond to the above-mentioned molding techniques (Fig. 12.1 b). This also means that the considerations and requirements that are specific to ceramic technology are similar for all powder-based processes.

The ceramic powder used for microfabrication must have a grain size small enough to reproduce the details of the micropatterning and to allow for relatively smooth surfaces. Even after sintering, which is usually accompanied by grain coarsening, the grain size should be at least one order of magnitude smaller than the smallest structural detail to be replicated. Dealing with micropatterns in the range of 50–100 μm, the ceramic grain size should be only in the range of 1–2 μm, and for details in the micrometer range, nanosized powders have to be used. The powder is usually added to a solvent or binder and dispersed with the help of surfactants (dispersants) or other process-specific additives to form a stable suspension without agglomerates. The amount of necessary additives to get a homogeneous, low-viscous suspension may be material specific but is dominated by the specific surface area of the powder, so that small-grained powders usually need more additives. In most casting or (injection) molding processes as well as in SFF techniques, like stereolithography or printing techniques, the viscosity of the suspension is a very crucial point, and is limited by the requirements of the particular fabrication process.

Unlike, for example, in slip pressing, printing techniques or electrophoretic deposition, in those fluidic or plastic ceramic-shaping processes, where no solvent is evaporated and no densification is obtained during or directly after shaping, however, the solid volume fraction of the suspension equals the green density of the ceramic part. A certain amount of the density in the green compact is yet needed to prevent warping or cracking during debinding and sintering and to achieve a sufficient densification of the part. Inhomogeneities in the green part will lead to anisotropic shrinkage, and the dimensional accuracy will be improved by a low shrinkage, that is a high green density. For typical green densities in the range of 50–65% of the theoretical density (TD), the linear shrinkage will result in 21–13%, assuming a homogeneous and complete densification during sintering.

In fluidic and plastic forming processes, the additives also play an important role during mold filling as well as during demolding. The viscosity has to be sufficiently low to enable degassing and removal of bubbles from the fluid and to ensure a proper mold filling even of the tiniest details of the mold. The binders have to provide for a sufficient strength in the green body to allow for a defect-free demolding.

The removal of the additives usually takes place during debinding at temperatures of up to 600 °C. In the case of high amounts of additives, this is the most crucial and time-consuming step of thermal treatment, as it has to be carried out very carefully and slowly, possibly with one or more dwell times, to avoid distortion of the green body and the formation of cracks. Yet, the small dimensions facilitate the debinding of microparts. The densification is achieved during a sintering step at higher temperatures and typical dwell times of about 30–90 min, for example at 1600–1700 °C for alumina, or at approx. 1500 °C for zirconia.

Besides the processing issues, it is important to consider the design during the outline of a new component. The design has to be adequate to the material and its properties and has to consider the shaping and joining possibilities and their specific requirements. This means that the design of metallic microcomponents cannot simply be transferred but has to be adapted to ceramic components, and a more challenging assembly and joining technique has to be faced. Furthermore, design guidelines for the micrometer range are still missing and the properties in microdimensions cannot always be inferred from experiences gained in macrodimensions.

12.2.1 Fabrication Processes

As mentioned above, various shaping and molding techniques have been developed or adapted to fabricate ceramic microcomponents. It is beyond the scope of this chapter to give a complete overview on all these processes and their variants. This chapter will rather concentrate on those shaping techniques that have actually been used to fabricate ceramic components for applications in microprocess engineering. In addition, some considerations on ceramic coatings and joining will be given.

12.2.1.1 Ceramic Injeciton Molding

Ceramic injection molding (CIM) is a well-established technique for the mass production of parts with complex shapes [17, 28]. In the last fifteen years it has been adapted to the high-precision molding of parts with outer diameters in the millimeter range or of structural details of several micrometers [18]. In contrast to many other microfabrication techniques of ceramics, an advantage of CIM is the possibility to mold relatively large parts of several centimeters but with micropatterns in the micrometer range. For high-pressure injection molding (HPIM) the ceramic powder is dispersed in high-melting, thermoplastic polymers like polyoxymethylene, polyethene, polypropylene or polyamide, which may be combined with waxes and other additives. At pressures of >50 MPa and molding temperatures of 120–200 °C the ceramic feedstock is injected into the mold. After cooling down and solidification, the green body is demolded from the mold for subsequent debinding and sintering. Owing to the high pressures and the relatively high temperatures involved, metallic molds are needed that have to be fabricated by expensive micromachining or even LIGA techniques. The high cost for mold fabrication will amortize in mass production with large series, but is uneconomic during product development or for small series.

A process variant in which less-expensive molds, even soft molds made of silicone rubber, can be used is the low-pressure injection molding (LPIM) that is

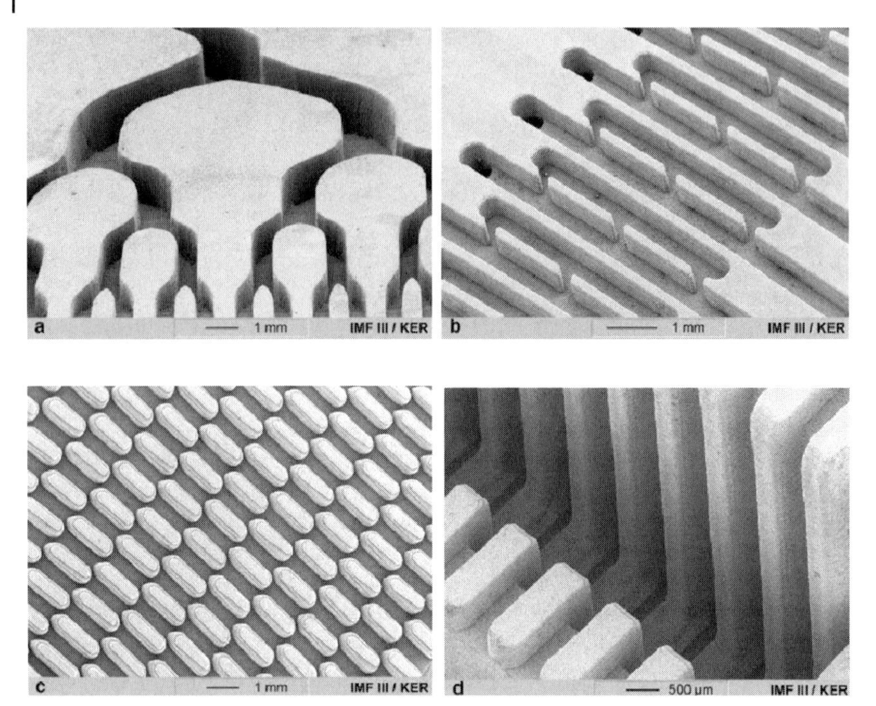

Fig. 12.2 Alumina inserts for a modular microreactor system fabricated by LPIM. Detailed views of (a) a gas distributor, (b) a part of a static mixer, and (c, d) catalyst carrier plates ([33] – reproduced by permission of The Royal Society of Chemistry).

also called hot molding. Here the ceramic powder is dispersed in low-melting paraffins and waxes, which allow for molding at temperatures of 70–90 °C and pressures of <1 MPa [18, 29, 30]. Especially for the fabrication of prototypes, functional models and small series, a favorable approach is a rapid prototyping process chain (RPPC), where the silicone rubber mold is obtained by replication of a polymer master model made by rapid prototyping [31, 32]. Examples of microreactor inserts fabricated via the RPPC are shown in Fig. 12.2 (see also Section 12.2.4) [33]. Ceramic injection molding of microparts and the process details of different variants have comprehensively been described recently and readers are referred to [18] for further details.

12.2.1.2 **Tape Casting**

The casting of ceramic tapes by the doctor-blade process was developed to satisfy the demands of microelectronics for thin substrates [34, 35]. It is widely used for the fabrication of multilayered ceramic packages and multilayer capacitors. In this process the ceramic powder is dispersed in a solvent, mixed with various additives and then the slurry is cast through a gating device ("doctor blade") onto a carrier film to thin flexible tapes. The selection of additives plays a decisive role for the properties of the tape. While dispersants are used to stabilize the slurry and to control its viscosity, binders must provide for a sufficient strength in the green tape after evaporation of the solvent and plasticizers impart the necessary flexibility. After drying, the tape can be machined and cut into parts by punching or laser machining. For multilayer applications the tapes may be equipped with metallic pattern by screen printing and are laminated to a stack and subsequently cofired to form a monolithic structure. To enable the cofiring with the metallic pattern, low-temperature cofiring ceramic (LTCC) tapes are used, where the sintering temperature is reduced by an addition of a glass phase.

For the micropatterning of ceramics a combination of tape casting and embossing or stamping was implemented [15, 16]. For this purpose, the plasticity and flow behavior of the green tapes had to be enhanced and the shear strength reduced compared to commercially available tapes. A relatively thick tape or a stack of a few laminated tapes are embossed with a micropatterned mold to obtain micropatterned parts with aspect ratios of up to ~ 5 [36]. To achieve a complete filling of the molds, the ceramic loading of the tapes has to be adjusted as well as the amount of suitable binders and plasticizers. Because of the thermoplastic behavior of most binders, an increase of the embossing temperature can also be used to facilitate the mold filling [37, 38]. A related approach is the embossing of viscous-polymer-processed ceramic tapes [39].

12.2.1.3 **Stereolithography**

Stereolithography for the generative fabrication of polymer parts was the first rapid prototyping technique to be invented and launched [40]. In spite of a large variety of other rapid prototyping techniques subsequently developed, stereolithography is still the most precise fabrication process with the highest resolution and the best-suited technique for microfabrication [41]. The principle of stereolithography consists in the in-layer-solidification of low-viscosity polymer reactive resins by a laser beam (see also Chapter 10).

Like most of the other rapid prototyping techniques, stereolithography has also been applied to the rapid fabrication of ceramics [25, 42]. For this purpose the resin is replaced by a suspension of ceramic powder dispersed in a UV-curable resin. The adjustment of the ceramic loading poses a challenge also in this process, as a low viscosity is needed for a proper recoating of the cured layer when lowering the platform to fabricate the next layer, while a ceramic loading of about 50 vol.% is needed to prevent distortion and cracking and to gain sufficient densification during thermal treatment. At the same time the cured layer depth is inversely proportional to the volume fraction of solids. Thus, an increase of ceramic loading is favorable to increase the resolution but a sufficient curing depth is necessary to ensure a good cohesion between the layers. Examples of fluidic microdevices fabricated by stereolithography of alumina suspen-

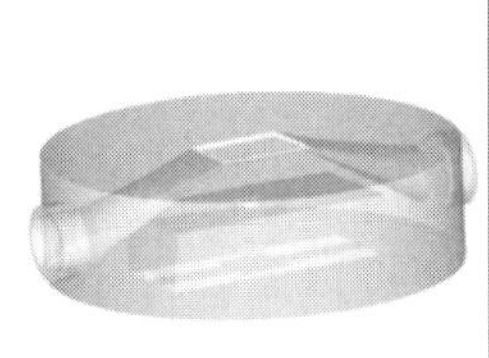

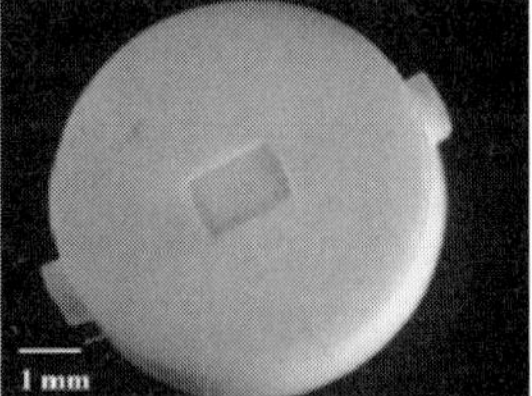

Fig. 12.3 Microfluidic devices made by stereolithography of alumina suspensions: (a) CAD view and (b) sintered component, (c) central microstructure of a micro-exchanger ((a) by courtesy of CNRS, Nancy, France, (b) reprinted from [45], Copyright 2003, Wiley-VCH, (c) reprinted from [43], Copyright 2001, with permission of Springer).

sions are shown in Fig. 12.3 (see also Section 12.2.4) [43–45]. The advantage of stereolithography is a nearly complete design freedom to build three-dimensional structures with undercuts or cavities. Various rapid prototyping techniques for the microfabrication of ceramics as well as of other materials have recently been reviewed in detail [24].

12.2.2
Coatings and Foams

Ceramic coatings of components in microprocess engineering are either applied as catalyst carrier or catalytic layers or as a protective layer to improve the corrosion resistance of metals [46]. The techniques for the coating of microparts do not differ in principle from those employed for macrodimensions. Methods applied for the coating of microsystems are CVD processes [47], sputtering [48], electrophoretic deposition [49, 50], sol-gel methods in combination with dip or spin coating [51, 52], washcoating of support layers to enhance the surface area or of catalytically active layers [49, 53, 54]. While coating techniques on ceramic and metal substrates may be similar, adhesion of ceramic films on ceramic substrates is easier to achieve and the thermoshock resistance is higher due to similar thermal expansion coefficients. In addition, a thermal treatment of the deposited layer can usually be performed at higher temperatures on ceramics than on metal substrates because of the higher thermal resistance. This is an asset especially for protection coatings, where fully dense layers have to be achieved. The problem may be simplified by special coating compositions, laser treatment or by using nanosized powders [55–58].

For the sake of completeness it should be mentioned that the use of ceramic foams as catalysts or catalyst supports is a well-established technique in conventional reactors [59, 60], and they have also been used as inserts of microreactors for catalytic [61] as well as for heating purposes (Fig. 12.4, see also Section 12.2.4) [62]. These foams are usually fabricated by infiltration and subsequent pyrolysis of polyurethane preforms but may also be prepared by self-foaming processes [63].

Fig. 12.4 Alumina microreaction module fabricated by LPIM with a silicon carbide foam insert for heating purposes (by courtesy of FhG IKTS, Dresden, Germany).

12.2.3
Joining and Sealing

First of all, there is no all-purpose solution for joining or sealing of ceramic components or of ceramic to metal components. The intended application and the materials involved have to be carefully taken into account and the joint has to meet their demands. As emphasized before, the joining should already be considered during design of the component and avoided wherever possible. In general there are three different techniques, the mechanical joining or fastening, the direct joining by diffusion bonding or fusion welding, and the indirect joining by intermediate joining materials [64, 65]. The most crucial point in joining is the mismatch of thermal expansion coefficient of the materials involved, leading to thermomechanical stresses, either during later use or already during the joining process.

When ceramic components are used in microprocess engineering, they are mostly chosen because of their high thermal and chemical resistance. This means that a possible joint should also be able to endure the same conditions, so that ceramic–metal joints made by brazing or gluing of parts with organic-based adhesives are excluded or at least limited to those areas of the component that are exposed to less-demanding conditions. Connections of ceramic parts to metals, for example to media supplies, are preferably moved to areas far from the heating zone, where they may be achieved by conventional fittings or clamping.

An ideal joining of ceramics should only involve materials with similar properties. For parts of the same material or materials with at least matching sintering and thermal expansion characteristics this can be achieved by joining in the green state and subsequent cofiring [66, 67]. While diffusion bonding of sintered parts is possible for nonoxide ceramics, this process has to be facilitated for oxide ceramics by an intermediate layer (powder, slurry or green tape) preferably of the same material to avoid high bearing pressures. The successful joining of sintered alumina parts was also achieved by two-laser-beam welding in the liquid phase [68].

A different approach is the soldering of ceramic parts by vitreous or glass–ceramic sealants, which have been successfully applied as the thermal expansion coefficient of glass or glass–ceramic composites can quite easily be tailored to match those of the ceramic components. Starting from a mixture of oxides, a stable glass or glass–ceramic is formed during heat treatment [38, 69]. Especially with glass–ceramic sealants high-temperature resistant joints may be gained, but the working temperature of the joint will, nevertheless, be limited to temperatures below the softening or melting point of the glass phase.

While joining leads to irreversible bonding of parts, there may also be the need for reversible assembly by sealing. As in the case of joints, the use of materials with inferior properties will lead to critical weak points in the component or to a decrease of the working temperature. In principle, metal, polymer or ceramic gaskets or seal rings can be used as in metallic components, but the sealing should already be considered during design to minimize any tensile stresses.

12.2.4 Applications

Ceramic components were the last to be used in microprocess engineering. For a long time ceramics were only employed as single parts of microreactors such as heater supports [70], catalytic inserts or supports [71–73], or reactor housings [74, 75]. The first developments of ceramic microcomponents intended to be used as heat exchangers, reactors or as microfluidic devices were based on the lamination of green tapes [76–79].

Using LTCC tapes with reduced sintering temperatures as in multilayer circuits, screen-printed metal patterns can be added before firing that may act as heaters or electrodes [80, 81]. An assembly of four layers, each formed by the lamination of a few tapes, was used for a microreactor with a single reaction channel and heated by a resistor. It was successfully tested for the oxidation of n-butane at temperatures up to 430 °C [80]. A similar approach was used to assemble a ceramic electrochemical microreactor made of five layers with screen-printed platinum electrodes beneath the reaction channels (Fig. 12.5) [81]. The reactor was directly connected to a mass spectrometer to investigate the effect of residence time on the methoxylation of methyl-2-furoate.

The development of a plate heat exchanger for heterogeneous gas-phase reactions made by the lamination of coarse-grained alumina tapes was reported recently [82]. The device was built up from seven layers, resulting in two cooling passages and one reaction passage. The remaining porosity in the sintered assembly

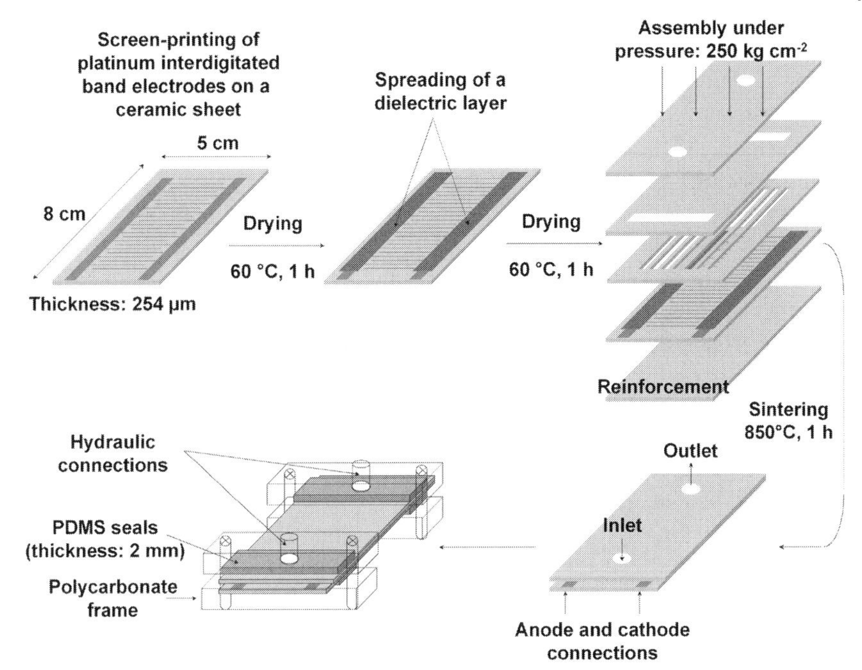

Fig. 12.5 Schematic representation of the ceramic electrochemical microreactor assembly ([81] – reproduced by permission of The Royal Society of Chemistry).

was favored for the catalytic coating, however, a glass paste had to be used to seal the outside of the component. As preliminary experiments showed leakage between the cooling and reaction channels, in addition, CVD with TEOS and ozone was investigated to achieve an internal sealing of the component. After washcoating with a noble metal catalyst, the device was used for the selective methanization of carbon monoxide between 180–280 °C. As the connection to the external equipment was accomplished by O-ring seals in a metallic clamping device, the maximum operating temperature was limited to approx. 320 °C.

Small microfluidic devices with complex internal structures for microreactor or bioanalytical applications were realized by stereolithography of alumina suspensions with ceramic loadings of 24–50 vol.% [43–45]. Using a liquid crystal display for the simultaneous curing of one layer, components were built up from 100–200 layers with a thickness of 25–35 µm each. While problems occurred during sintering of components made from resins with low ceramic loadings, components with a green density of 50% TD could be successfully sintered with only slightly inhomogeneous shrinkage [45]. Examples presented so far are a crossed microexchanger with a wall thickness of 75 µm and fluidic devices with dimensions of only a few millimeters (see Fig. 12.3 in Section 12.2.1). The latter were used for the detection of single molecules of cyanine 5 by fluorescent correlation spectroscopy [43]. Stereolithography has also been used for the development of alumina microreactors for power generation [83].

For a microheat exchanger, silicon carbide (SiC) plates micropatterned by milling or laser machining were stacked to obtain a cross-flow heat exchanger that may also be used as a microreactor [2]. Hermetic sealing of the 3.5-mm thick plates was achieved by diffusion bonding at temperatures above 1700 °C. Each plate contained 13 channels having a length of 50 mm and a cross section of 1.5×1.5 mm^2 (Fig. 12.6). The ceramic body was pressed against a steel housing and stainless steel tubes were used for the external connections. The development aims at applications well above 700 °C, due to restrictions caused by the experimental setup, however, the first results of heat-exchange experiments with nitrogen have only been reported up to 340 °C.

The first ceramic microdevice actually used for gas-phase reactions at high temperatures was a modular microreactor system fabricated by LPIM of alumina [33, 61, 84]. By using stereolithography for a fast and inexpensive supply of master models for the replication of silicone molds, different designs and various inserts could be fabricated that allow the system to be rapidly adapted to special require-

Fig. 12.6 Si microheat exchanger: (a) heat-exchanger plates, (b) cross section of diffusion-bonded monolith, (c) SiC monolith inside steel housing (reprinted from [2], Copyright 2005, Wiley-VCH).

ments (Fig. 12.7) [33, 85]. Differently shaped, multichannel catalyst carrier plates may be used for a rapid screening of catalysts, and flow distributors or different static mixers can be exchanged in the reactor housing (see also Fig. 12.2 in Section 12.2.1). Sealing of the reactor housing was accomplished for normal pressure by pressing on a polished lid with a ceramic clamping device. The cover plate may contain a sapphire window or an inlet for a fiber optic device for visual or spectroscopic control. For media supply and discharge, ceramic tubes were sealed to the housing by a glass–ceramic solder well suited for gas-tight applications at 1000 °C [69]. The connections to the peripheral equipment were positioned far outside the reaction zone, so that conventional fittings could be used. In addition, the reactors can be equipped with internal or external heating elements made either by screen-printing or LPIM and that allow the reactor to be heated to 1000 °C by inductive coupling or from direct current flow [66, 86]. For this purpose, electroconductive elements were made from alumina/titanium nitride compositions, which are particularly suitable as both phases have very similar thermal expansion coefficients and the electrical resistivity can be adjusted within several orders of magnitude by the volume fraction of TiN. Furthermore, TiN exhibits a metallic electroconductivity that is easily controlled due to a linear increase of the conductivity with increasing temperature and may even act as a temperature sensor. The microreactors were successfully used for the oxidative coupling of methane at temperatures of 850–

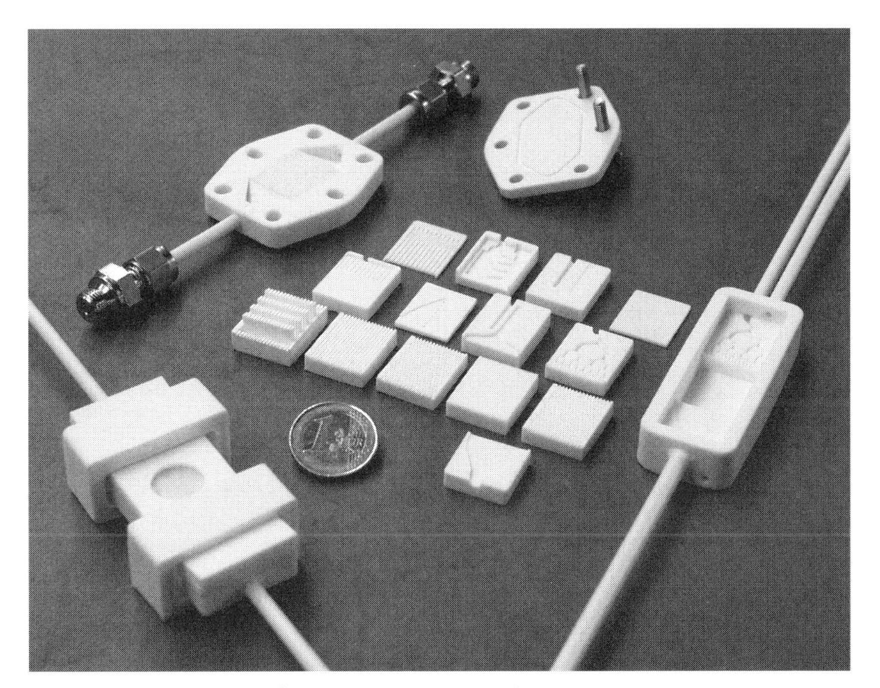

Fig. 12.7 Different designs of ceramic microreactors with a variety of exchangeable inserts ([33] – reproduced by permission of The Royal Society of Chemistry).

1000 °C, as well as for the selective oxidation of isoprene in the range of 300–500 °C [33, 61]. As a further advantage of these alumina microreactors, it has been proven that they are catalytically inactive, so that the yields of reactions carried out in these microreactors are not affected by the housing or catalyst support. Aiming at reactions at higher pressures, the microreactors can either be sealed with the glass–ceramic solder or used with seal rings or gaskets. Depending on the materials used as seals, however, the operating temperature may be reduced.

Based on the same fabrication technique, an alumina microheat exchanger was developed aiming at high-temperature gas-phase applications [87]. Sintered, micropatterned plates were assembled by a glass solder and the monolith could subsequently be applied at pressures of 8 bar. In preliminary heat-exchange experiments with water, heat-transfer coefficients of up to 15 kW/m^2K were reached. Simulation of high-temperature gas-phase applications evidenced high heat-transfer efficiencies according to [88].

A modular microreaction system with modules made from different materials like metals, polymers, silicon, and ceramics was particularly designed for liquid applications in the temperature range from –20 °C to 150 °C [89] (see also Chapter 9). Various, standardized, hexagonal microfluidic modules (with a distance of 40 mm between opposite sides of the hexagon) can be mounted on a baseplate and are interconnected for example via PTFE tubings at the underside, which may be tempered with a fluid. For this toolkit several ceramic microcomponents such as reaction or mixing modules were made from alumina by LPIM with optionally further patterning by laser machining (Fig. 12.8) [62]. For the ceramic reaction modules different multichannel or ceramic foam inlays are available that may be additionally washcoated to enhance the surface area for catalytic applications. Special, adjustable mixing units allow for precision dosing of two liquids. Heating of the modules can be accomplished by screen-printed silver-palladium pattern or by SiC foams, that is by surface or volume heating, respectively (see also Fig. 12.4 in Section 12.2.2). In principle, fluid temperatures of 800–1000 °C can be achieved by volume heating of semiconducting SiC foams [90]. The modules are

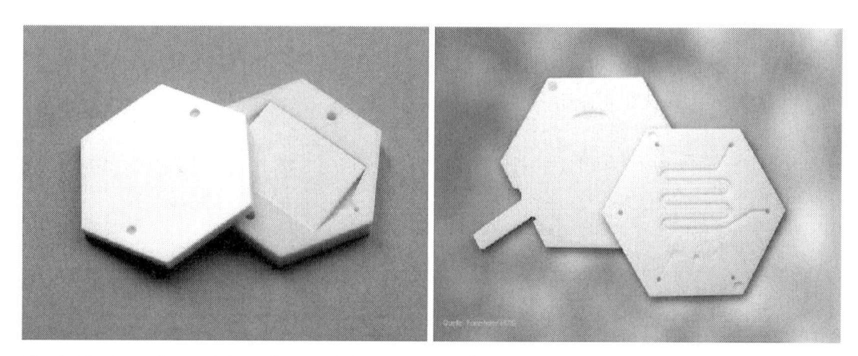

Fig. 12.8 Ceramic modules of a modular system with (a) a multichannel inlay, and (b) an adjustable mixing unit (by courtesy of FhG IKTS, Dresden, Germany).

closed by a lid, which may be equipped with a sapphire window to allow visual or spectroscopic monitoring. Depending on the application, they can be sealed by adhesive bonding, metal brazing or glass or glass–ceramic solders.

12.3 Microfabrication of Glass

Various glass types have been used for different applications in microtechnology. Among these are glass bondable with silicon (Borofloat (Schott) or Pyrex (Corning)) or quartz glass, which is used if a high optical (UV) transparency is necessary.

In microsystems technology, glass and glass–ceramic components are required with well-defined shapes and strict tolerances. Often, conventional molding methods used in the manufacturing of glass cannot fulfill these requirements. Mechanical operations such as drilling or milling are very expensive and limited in their possibilities.

Therefore photolithography, as known from silicon micromachining, is mainly used to manufacture microstructures in glass. A special technology is offered by glass from the basic Li_2O/Si_2O family containing traces of noble metals. After exposure to UV light and subsequent heat treatment, this glass will partially crystallize. The crystalline phase is lithium metasilicate, which is much more soluble in hydrofluoric acid than the surrounding unexposed amorphous glass. This makes the production of complicated and high-precision components possible via an etching process.

12.3.1 Isotropic Etching of Glass

Usually, glass is microstructured with methods coming from semiconductor technology, as shown in Fig. 12.9. A well-polished glass surface is coated with an adhesion layer (approx. 100 nm Ti or Cr) and then a photoresist is spin coated. This resist is exposed through a mask, which contains the required structures. The protection resist is then removed from the exposed areas to open the glass surface for attack of the etchant.

In most cases, hydrofluoric acid (HF) is used as an etchant. With a 10% hydrofluoric acid concentration, etching rates of 0.5–3.0 μm/min can be achieved with possible etching depths of 10–300 μm. This limit of a maximum etching depth is due to the fact that the hydrofluoric acid also attacks the adhesion layers. After a while the photoresist is underetched and removed by the HF. If very flat channels of a few micrometers are required (for example for biochips), a very slow, but very precise etching rate is necessary. This can be achieved by a buffered hydrofluoric acid (HF/NH_4F). Under these conditions, etching rates are within the range of 100–1000 nm/min with possible etching depths of 1–20 μm.

Also plasma etching processes are suitable for the production of structures in glass. Here, a photoresist protection layer is used to protect the areas that

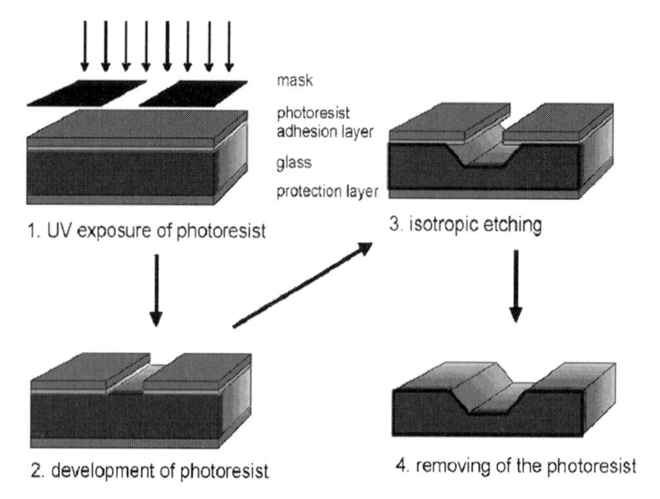

Fig. 12.9 Microstructuring of glass using photoresists (reprinted from [91], Copyright 2005, Wiley-VCH).

should not be etched by a SF_6/O_2 plasma. The etching rates are up to 1 μm/min with possible etching depths on a 1–20 μm scale.

Glass is an amorphous material. This means that it is etched in all directions with the same speed. Therefore, the etchant does not only etch into the depth, but also attacks the side walls. This leads to an underetching of the protection layer. The resulting structure is therefore at least twice as wide as deep and the aspect ratio is always small.

12.3.2
Microstructuring of Photoetchable Glass

With FOTURAN (Schott) it is possible to induce an anisotropic behavior by first exposing the glass to UV light through a chromium mask with the required structures. The exposed parts of the glass then crystallize during a heat-development step, while the nonexposed parts remain in their vitreous form. In a solution of hydrofluoric acid the crystallized parts are etched faster than the glass parts. By etching a substrate from both sides it is possible to get holes with an aspect ratio of 30. The size of the developed crystals during the heat treatment (3–5 μm) defines both the roughness of the etched walls ($r_a \approx 1$ μm) and the smallest possible hole diameters of approx. 25 μm. After finishing the etching process the substrates can be bonded together to build closed channels or can be combined with other materials (Fig. 12.10). For details see [4, 92].

Photoetchable glass is based on lithium aluminum silicate with some additives that define the specific properties of this material. The exact composition varies from glass type to glass type but is in the range given in Table 12.3.

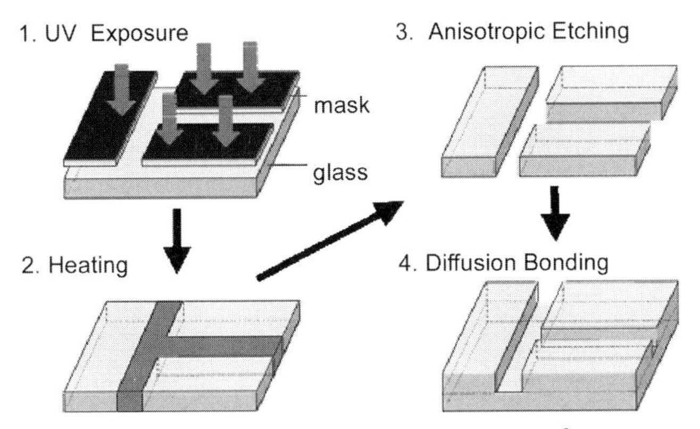

Fig. 12.10 Principal processes during the structurization of photoetchable glass using photoresists (reprinted from [91], Copyright 2005, Wiley-VCH).

Table 12.3 Composition of photoetchable glass.

Compound	wt.%
SiO_2	75–85
Li_2O	7–11
K_2O	3–6
Al_2O_3	3–6
Na_2O	1–2
ZnO	0–2
Sb_2O_3	0.2–0.4
Ag_2O	0.05–0.15
CeO_2	0.01–0.04

When high operating temperature (max. 750 °C) is required, microstructured FOTURAN components can be modified to the ceramic state by a subsequent exposure and heat treatment without losing the shape of the microstructures. The physical properties of the glass as well as the glass–ceramic structures are summarized in Table 12.2 (see Section 12.1).

With this technology, any two-dimensional structure can be etched into the glass, because there is no preferred etching direction like in silicon due to the crystal structure.

There are also different ways to obtain a required 3D channel system:

- several subsequent structurization steps in one glass plate
- combining the FOTURAN process with an isotropic etching process
- laser processing of glass
- bonding of different structured layers.

12.3.3 Laser-patterning Process

The laser-patterning process leverages the characteristics of photostructurable glass–ceramics [93]. When such glasses are exposed to modest levels of pulsed UV laser energy, in excess of the critical fluence F_c, a photochemical reaction occurs within the glass that creates a density of nanocrystals within the critically exposed volume as shown in Fig. 12.11 [94].

The density of nanocrystals is directly proportional to the fluence within the critically exposed volume. By controlling the depth of focus and numerical aperture of the UV laser optics, submicrometer features can easily be created. Or, when larger critically-exposed volumes are required, for example for large-diameter holes, optics with a larger depth of focus can be employed [95, 96]. The critically exposed volume can be extended by employing successive pulses of laser energy along a specified path, defined by a computer-controlled three-axis nanopositioner.

The heat treatment and subsequent etching process is similar to the 2D structurization described above. With the laser process it is possible to reduce the size of the formed crystals, to obtain structure sizes and resolutions below 5 μm [95]. With this technology real 3D structures [97], even embedded in the material [98] can be manufactured. Because no mask is needed, the laser allows a rapid prototyping [99].

Lasers can also be used to structure other glasses that are not photosensitive. Due to the fact that no mask is needed, the laser structurization especially allows a fast redesign in the development phase. Laser structurization is often used for the structurization of polymers. Unfortunately, laser structurization of glass requires high laser power, which is mainly transformed into heat leading to a local ablation of the material. A new crack-free direct-writing technology has been developed by a Taiwanese group using a 266-nm laser with low power (20 mW) but high repetition rate (6 kHz). Complex microfluidic structures leading to an integrated microreactor/concentrator have been demonstrated [100].

To increase the absorption of the laser light special glasses have been developed for the direct laser ablation. In these glasses different dopants are added to absorb the laser energy leading to an effective ablation process [101].

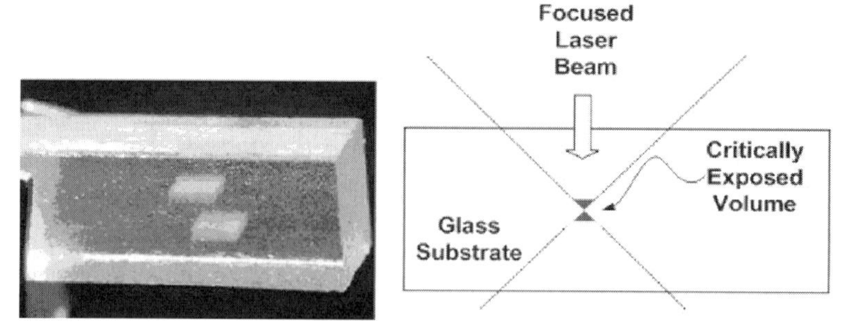

Fig. 12.11 Laser exposure of photostructurable glass (by courtesy of Invenios Inc., Santa Barbara, USA).

12.3.4
Sandblasting

Already for decades, sandblasting is known as an easy and cheap process for cleaning and modification of surfaces. It has recently also been used for the manufacturing of microreactors [102]. The new microsandblasting process provides the simple opportunity to produce various microstructures out of materials like glass or silicon. A mask containing holes for the required structures is positioned above the glass. A beam of compressed air, saturated with fine grains of SiC, is directed through this mask. This leads to ablation of the surface.

With this technology it is possible to structure glass plates up to a thickness of 6 mm, with aspect ratios up to 3. The smallest achievable structures have a size of 300 μm, with spacing in the same range and tolerances of ± 30 μm. The walls have an angle of about 20° due to reflection of the SiC particles; for through-holes this can be reduced to 5°. A surface roughness of 2 μm can be achieved.

The microsandblasting process can be used for all glass types, like special alkaline-free glass, quartz glass, or cheap window glass. Even ceramic and semiconductor materials can be structured. By processing at room temperature the finished part is free of distortion. This is the major advantage over processes, where the material goes through structural changes or high temperatures. With this technology different microreaction modules have been manufactured and used (Fig. 12.12) [102].

Fig. 12.12 Micromixer with reaction channel (by courtesy of Little Things Factory, Ilmenau, Germany).

12.3.5 Bonding Technologies

Microfluidic components are usually built out of several layers, each of them containing different functions. These layers are connected by bonding processes to three-dimensional channel structures. The following processes are mostly used to bond glass layers:

- *Diffusion bonding*: Two or more glass layers with well-polished surfaces are adjusted on top of each other and bonded under pressure and at temperatures around the glass transition temperature. For this process it is necessary that the different layers have the same coefficient of thermal expansion (CTE). The main advantage of such a connection is that it works without intermediate layers (e.g. glue) and therefore leads to a chemically very stable structure.
- *Glass soldering*: In some cases a direct bonding without intermediate layers cannot be used. For example, it can be necessary to manufacture a window from a different material than the microstructured channel plate. Different materials with different CTEs cannot be diffusion bonded directly. Glass solders as an intermediate layer help to overcome the mismatch in CTEs, by having a CTE in between and by using lower processing temperatures. Glass solder is a mixture of a low melting glass powder and an organic solvent. It is printed as a paste onto the surface of one of the plates.
- *Gluing*: If chemical needs are not too critical, standard glues can be a less-expensive alternative to connect different structured layers.
- *Other processes*: To connect silicon or metals other connecting techniques like anodic bonding, sputtering, evaporation, or electroforming can be used.

12.3.6 Applications

Applications for photoetchable glass are products where the special glass properties of this material are necessary. Like other glass, FOTURAN is very resistant against most chemicals. It shows good heat resistance up to 750 °C in the ceramic form. Therefore, it is ideal as a component or a substrate material in different applications, like environmental engineering, medicine, biotechnology, chemical engineering, or analytics [91].

12.3.6.1 Microreactors

The above-described technologies were used for manufacturing different microfluidic components, like micromixers, microheat exchangers, and so-called microreactors, a combination of both [103, 104].

The first FOTURAN microreactor was realized for a nitration reaction [105, 106] (Fig. 12.13). Conventional nitration of aromatic compounds exhibits some principle problems: The reactions are strongly exothermic and can even lead to explosive mixtures. In order to control the reactions, very expensive cooling of

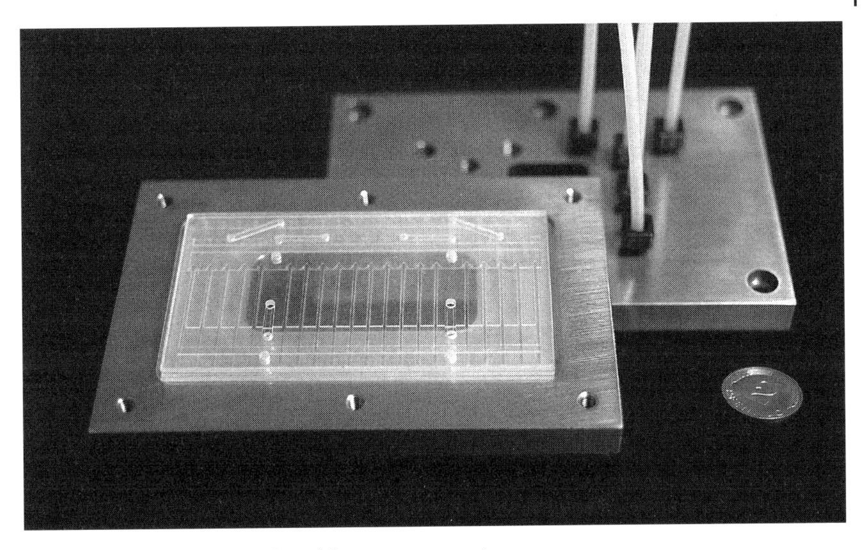

Fig. 12.13 Microreactor developed for nitration reactions.

the reaction mixture is necessary. Different secondary, side, and competitive reactions occur.

By carrying out the nitration of naphthalin with nitrogen pentoxide in glass microreactors it could be shown that all these issues can be controlled. Compared to conventional technology, in microreactors the reaction temperature can be increased from –50 °C to +30 °C, which leads to substantial savings of energy. The reaction parameters can be controlled to suppress side reactions.

Since the nitrating reagents are chemically very aggressive, glass was the material of choice. This fluidic system consists of 7 structured layers with thicknesses between 200 and 700 μm, which contain the different functions (e.g. mixing, heat transfer) of the reactor. The surface of the wall for the heat exchange is 12.6 mm^2 for one channel. In this system 16 channels run parallel and the total area is 201 mm^2. A flow rate for the total system of 36 ml/min has been used.

The mixing performance of the reactor was characterized and the experiments clearly showed a laminar flow behavior of the educt streams and thus a diffusion-controlled mixing of the reactants. Depending on the flow rates complete mixing was achieved at the end of the reaction channel or much earlier. These experimental data could be confirmed by CFD simulations (Fig. 12.14) [107, 108].

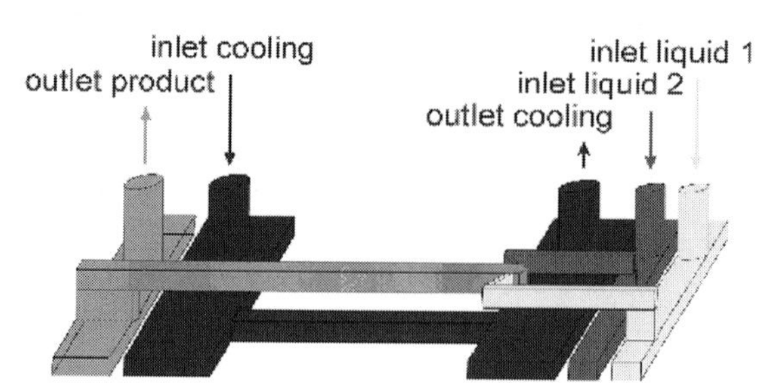

Fig. 12.14 Sketch of one reaction chamber of the microreactor (reprinted from [91], Copyright 2005, Wiley-VCH).

12.3.6.2 **Micromixers**

To optimize the mixing properties of micromixers, it is necessary to increase the contact area between two liquids. This can be done by splitting up the two liquids into a number of smaller streams and combine them again interdigitally. Even though a microfluidic channel system shows laminar flow, the mixing only by diffusion leads to very short mixing times. Due to the interdigital design, the diffusion length is only several micrometers, which leads to total mixing within milliseconds [107].

Using the synthesis of benzaldehyde from benzal chloride and sulfuric acid as an example, it could be shown that even aqueous and organic phases can be mixed easily in a suitable glass micromixer for an effective reaction. Standard reaction times of several hours can be reduced to minutes by this approach.

For this kind of reaction a multilamination mixer was developed and built (Fig. 12.15) [109, 110]. Within this reactor, two liquid streams are split up into a row of smaller streams, which are brought together again "interdigitally" (=A B A B ...).

Glass was used as the material for the microfluidic device, because it permits the observation of the mixing behavior directly in a microscope. Different studies could be performed to improve the design of such a micromixer. With the support of these investigations the "Superfocus Mixer" has been developed: 124 channels (2×62) lead into a mixing chamber, which is focused into the 500-μm wide reaction channel. The resulting lamellae have a width of only 4 μm. Using typical flow rates of 8 l/h of aqueous solutions a complete mixing takes place within the channel length of 50 mm.

12.3.6.3 **Gas–Liquid Mixing**

For gas–liquid mixtures a glass component called the cyclone mixer has been developed (Fig. 12.16). Within this component, the two reactants are directed by tangential aligned nozzles (50 μm×150 μm) into a central mixing chamber (diameter 10 mm, height of 2.15 mm). This leads to a "cyclone" formation with a very homogeneous gas bubble size (~150 μm) in the liquid phase. Thus the

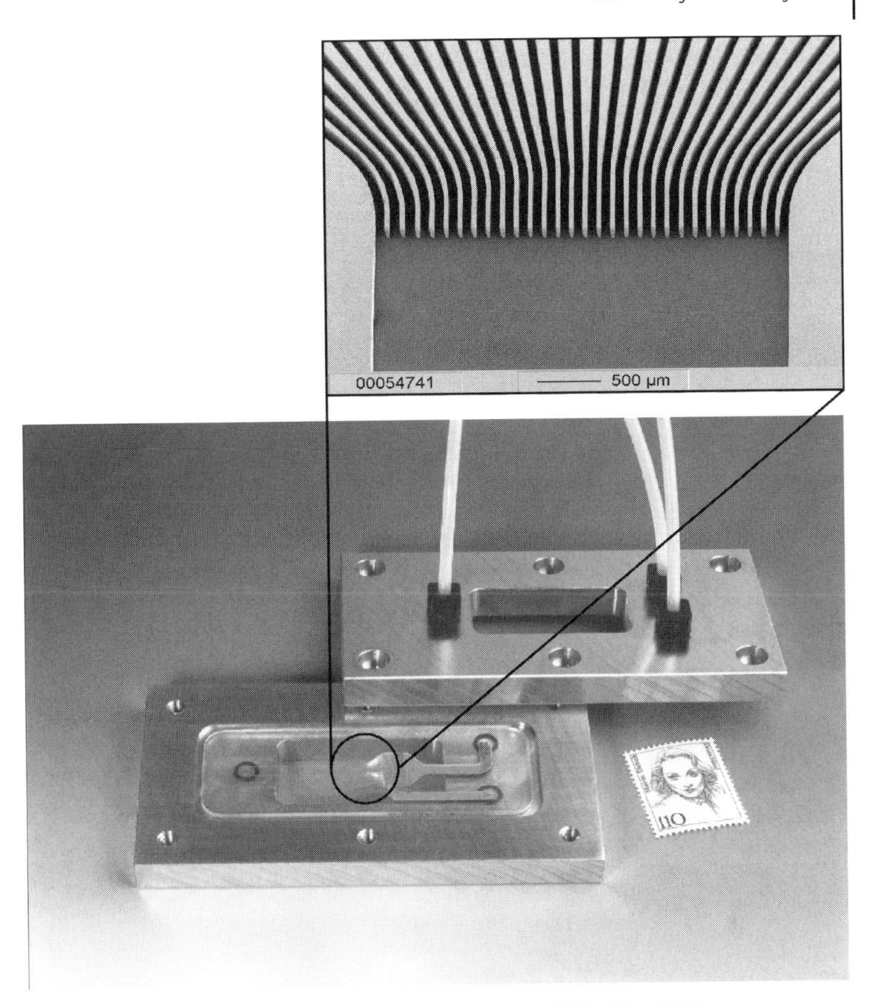

Fig. 12.15 Interdigital mixer (reprinted from [91], Copyright 2005, Wiley-VCH).

solubility of oxygen in water/isopropanol, for example could be increased to 35 g/l [107, 111].

12.3.6.4 **Photoreactors**

Glass is a perfect material for photoreactions. Photons have only a very short depth of penetration (several 10 µm), before they are absorbed. Therefore, microreactors offer a perfect opportunity to lead the reactants through the volume that is exposed by the light source and increase the photon yield. This has been demonstrated for 2+2 cycloadditions in glass microreactors [112]. Reaction time could be decreased, while the yield and selectivity has been increased.

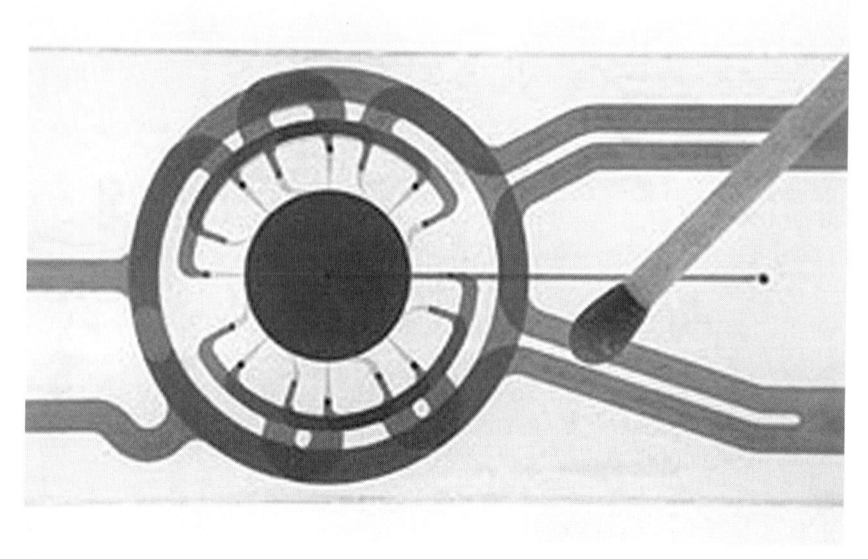

Fig. 12.16 Cyclone mixer for gas–liquid mixing (reprinted from [91], Copyright 2005, Wiley-VCH).

Also, reactors are available having only a glass window for transmission of light. The chlorination of alkylaromatics in a gas–liquid reaction has been demonstrated using a falling film reactor with a glass window, again showing a reduced reaction time and an increased yield and selectivity comparing to the conventional batch reactor [113].

12.3.6.5 **Heat Exchanger Made from Glass**

Heat exchangers made of glass can be important for special applications. While for conventional heat exchangers materials with high heat conductivity are favored, it was confirmed by numerical simulations that in the case of countercurrent micro heat exchangers the heat-transfer efficiency can be optimized using materials with low heat conductivities like glass or some ceramics [88]. In metal reactors the heat is conducted mainly within the walls of the channel system instead of getting through the walls. In the case of microdimensions the axial heat conduction in the channel wall will influence the heat-transfer efficiency. While for very low heat conductivities the efficiency tends to zero, the efficiency for very high heat conductivities approaches 50%, corresponding to the maximum value of parallel-flow heat exchangers. In between there is a maximum in efficiency for optimum heat conductivity. Glass with a heat conductivity λ of 0.67–1.38 W/m K shows a substantially more effective heat transfer [88] (Fig. 12.17). The material glass appears in the maximum of this efficiency curve. Yet this optimum depends on geometry and flow rates.

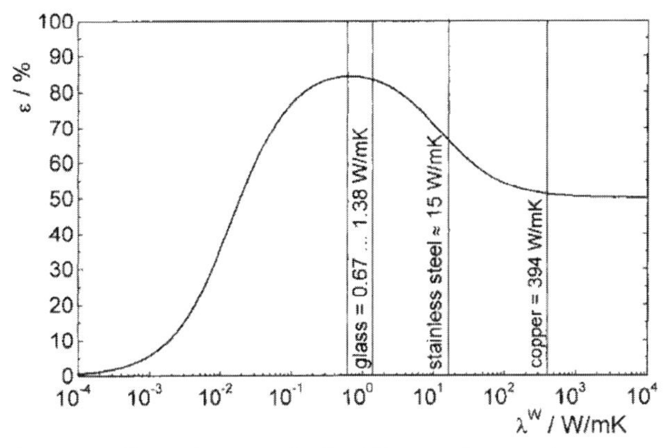

Fig. 12.17 Effective heat transfer for different wall materials with different heat conductivities (reprinted from [88], Copyright 1999, Wiley-VCH).

12.3.6.6 **Microfluidic Interfaces**

Usually, the glass reactors are connected with Teflon tubes. Figure 12.18 shows a metal frame that mechanically protects the glass reactors and that, on the other hand, presses the Teflon tubes with standard HPLC fittings on the glass surface. This connection has the main advantage that only chemically stable materials (glass and Teflon) get in contact with the media. In addition, the tubing can be disconnected at any time to change or to clean the microfluidic devices. These connectors can be used up to a pressure of 20 bar.

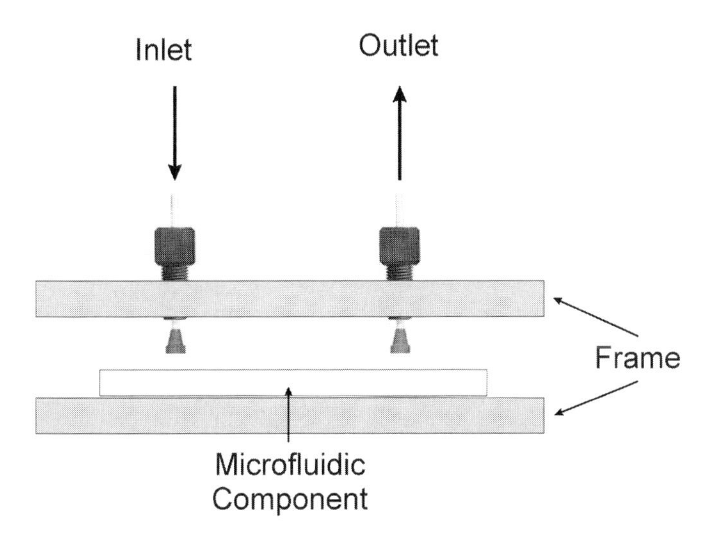

Fig. 12.18 Connection technology for microreactors made of glass (reprinted from [91], Copyright 2005, Wiley-VCH).

12.4 Conclusions

Ceramic and glass materials offer a high thermal and chemical resistivity so that they outmatch polymers, metals or silicon in many high-temperature or corrosive applications in microprocess engineering. While in the last ten years considerable advances have been achieved in microfabrication and many processes now meet high standards, the joining of components still poses a challenge in high-temperature or high-pressure applications. Due to the mismatch of mechanical and thermal properties and due to the absence of any standardized fittings, particularly the joining to metal parts and the connection to the peripheral equipment often limits the range of possible applications. Recently, a consortium of mainly German companies defined standardized interfaces for microreaction modules [114]. Several companies already produce respective modules. Glass components are now commercially available with connections to Teflon tubings as well as metal tubings (e.g. by Swagelok fittings) [115]. Pressures up to 15 bar and temperatures up to 200 °C can be applied. Together with the good chemical, optical, and electrical properties, glass and ceramic microreactors are suitable for a large variety of chemical applications.

References

1 R.G. Munro, Evaluated material properties for a sintered α-alumina, *J. Am. Ceram. Soc.* **1997**, *80*, 1919–1928.

2 F. Meschke, G. Riebler, V. Hessel, J. Schürer, T. Baier, Hermetic gas-tight ceramic microreactor, *Chem. Eng. Technol.* **2005**, *28*, 465–473.

3 http://www.dupont.com/mcm/gtapesys/part4.html, DuPont, June 2005.

4 T.R. Dietrich, Photostrukturierung von Glas, in *Handbuch Mikrotechnik* (Ed.: W. Ehrfeld), Hanser, Wien, **2002**, 407–429.

5 M. Heule, S. Vuillemin, L.J. Gauckler, Powder-based ceramic meso- and microscale fabrication processes, *Adv. Eng. Mater.* **2003**, *15*, 1237–1245.

6 T.B. Thoe, D.K. Aspinwall, M.L.H. Wise, Review on ultrasonic machining, *Int. J. Machine Tools Manuf.* **1998**, *38*, 239–255.

7 D.T. Pham, S.S. Dimov, C. Ji, J.V. Petkov, T. Dobrev, Laser milling as a rapid micromanufacturing process, *Proc. Instn. Mech. Engrs. Part B, J. Eng. Manuf.* **2004**, *218*, 1–7.

8 Z.Y. Yu, K.P. Rajurkar, A. Tandon, Study of 3D micro-ultrasonic machining, *J. Manuf. Sci. Eng., Trans. ASME* **2004**, *126*, 727–732.

9 F. Nöker, E. Beyer, Herstellung von Mikrostrukturkörpern aus Keramik, *Keram. Z.* **1992**, *44*, 677–681.

10 W. Bauer, H.-J. Ritzhaupt-Kleissl, J. Haußelt, Micropatterning of ceramics by slip pressing, *Ceram. Int.* **1999**, *25*, 201–205.

11 W. Bauer, H.-J. Ritzhaupt-Kleissl, J. Haußelt, Mikrostrukturierung von keramischen Bauteilen im Wachszentrifugalgußverfahren, *Keram. Z.* **1998**, *50*, 411–415.

12 O.O. Omatete, M.A. Janney, R.A. Strehlow, Gelcasting – a new ceramic forming process, *Am. Ceram. Soc. Bull.* **1991**, *70*, 1641–1649.

13 C.M. Chan, G.Z. Cao, T.G. Stoebe, Net shape ceramic microcomponents by modified sol-gel casting, *Microsyst. Technol.* **2000**, *6*, 200–204.

14 L.J. Gauckler, T. Graule, F. Baader, Ceramic forming using enzyme cata-

lyzed reactions, *Mater. Chem. Phys.* **1999**, *61*, 78–102.
15 R. Knitter, E. Günther, U. Maciejewski, C. Odemer, Preparation of ceramic microstructures, *cfi/Ber. DKG* **1994**, *71*, 549–556.
16 M. L. Griffith, A. R. Barda, N. Taylor, J. W. Holloran, Micromolding of ceramics using photolithographic polyimide patterns, in *Ceramic Transactions Vol. 51, Ceramic Processing and Technology* (Eds.: H. Hausner, G. L. Messing, S. Hirano), The American Ceramic Society, Westerville, Ohio, USA, **1995**, 321–325.
17 B. C. Mutsuddy, R. G. Ford, *Ceramic Injection Molding*, Chapman & Hall, London, GB, **1995**, p. 368.
18 W. Bauer, J. Haußelt, L. Merz, M. Müller, G. Örlygsson, S. Rath, Microceramic injection molding, in *Adv. Micro and Nanosystems Vol. 3, Microengineering of Metals and Ceramics* (Eds.: H. Baltes, O. Brand, G. K. Fedder, C. Hierold, J. Korving, O. Tabata), Wiley-VCH, Weinheim, Germany, **2005**, Chapter 12.
19 P. Sarkar, P. S. Nicholson, Electrophoretic deposition EPD: mechanisms, kinetics and application to ceramics, *J. Am. Ceram. Soc.* **1996**, *79*, 1987–2002.
20 J. Tabellion, R. Clasen, Electrophoretic deposition from aqueous suspensions for near-shape manufacturing of advanced ceramics and glasses – applications, *J. Mater. Sci.* **2004**, *39*, 803–811.
21 E. Kim, Y. Xia, G. M. Whitesides, Micromolding in capillaries: Applications in materials science, *J. Am. Chem. Soc.* **1996**, *118*, 5722–5731.
22 D. Zhang, B. Su, T. W. Button, Microfabrication of three-dimensional, free-standing ceramic MEMS components by soft moulding, *Adv. Eng. Mater.* **2003**, *5*, 924–927.
23 B. Y. Tay, J. R. G. Evans, M. J. Edirisinghe, Solid freeform fabrication of ceramics, *Int. Mater. Rev.* **2003**, *48*, 341–370.
24 T. Hanemann, W. Bauer, R. Knitter, P. Woias, Rapid prototyping and rapid tooling techniques for the manufacturing of silicon, polymer, metal and ceramic microdevices, in *MEMS/NEMS Handbook: Techniques and Applications Vol. 3, Manufacturing Methods* (Ed.: C. T. Leondes), Springer, Heidelberg, Germany, **2005**, Chapter 4.
25 M. L. Griffith, J. W. Halloran, Freeform fabrication of ceramics via stereolithography, *J. Am. Ceram. Soc.* **1996**, *79*, 2601–2608.
26 P. F. Blazdell, J. R. G. Evans, M. J. Edirisinghe, P. Shaw, M. J. Binstead, The computer aided manufacture of ceramics using multilayer jet printing, *J. Mater. Sci. Lett.* **1995**, *14*, 1562–1565.
27 M. K. Agrarwala, A. Bandyopadhyay, R. van Weeren, A. Safari, S. C. Danforth, N. Langrana, V. R. Jamalabad, P. J. Whalen, FDC, rapid fabrication of structural components, *Am. Ceram. Soc. Bull.* **1996**, *75*, 60–65.
28 J. R. G. Evans, Injection moulding, in *Materials Science and Technology Vol. 17a, Processing of Ceramics* Part I (Ed.: R. J. Brook), VCH, Weinheim, Germany, **1996**, Chapter 8.
29 J. A. Mangels, Low-pressure injection molding, *Am. Ceram. Soc. Bull.* **1994**, *73*, 37–41.
30 R. Lenk, Hot moulding – an interesting forming process, *cfi/Ber. DKG* **1995**, *72*, 636–642.
31 R. Knitter, W. Bauer, D. Göhring, J. Haußelt, Manufacturing of ceramic microcomponents by a rapid prototyping process chain, *Adv. Eng. Mater.* **2001**, *3*, 49–54.
32 W. Bauer, R. Knitter, Development of a rapid prototyping process chain for the production of ceramic microcomponents, *J. Mater. Sci.* **2002**, *37*, 3127–3140.
33 R. Knitter, M. A. Liauw, Ceramic microreactors for heterogeneously catalysed gas-phase reactions, *Lab Chip* **2004**, *4*, 378–383.
34 J. C. Williams, Doctor-blade process, in *Treatise on Materials Science and Technology Vol. 9, Ceramic Fabrication Processes* (Ed.: F. F. Y. Wang), Academic Press, New York, USA, **1976**, 173–198.
35 R. E. Mistler, The principles of tape casting and tape casting applications, in *Ceramic Processing* (Eds.: R. A. Terpstra, P. P. A. C. Pex, A. H. de Vries), Chapman & Hall, London, GB, **1995**, Chapter 5.
36 H.-J. Ritzhaupt-Kleissl, H. von Both, M. Dauscher, R. Knitter, Further ceramic

replication techniques, in *Adv. Micro and Nanosystems Vol. 4, Microengineering of Metals and Ceramics* (Eds.: H. Baltes, O. Brand, G.K. Fedder, C. Hierold, J. Korving, O. Tabata), Wiley-VCH, Weinheim, Germany, **2005**, Chapter 15.

37 M. Stadel, H. Freimuth, V. Hessel, M. Lacher, Abformung keramischer Mikrostrukturen durch die LIGA-Technik, *Keram. Z.* **1996**, *48*, 1112–1117.

38 V. Hessel, W. Ehrfeld, H. Freimuth, V. Haverkamp, H. Löwe, T. Richter, M. Stadel, A. Wolf, Fabrication and interconnection of ceramic microreaction systems for high-temperature applications, in *IMRET1*, **1998**, 146–157.

39 B. Su, T.W. Button, A. Schneider, L. Singleton, P. Prewett, Embossing of 3D ceramic microstructures, *Microsyst. Technol.* **2002**, *8*, 359–362.

40 C.H. Hull, Apparatus for production of 3D objects by stereolithography, 3D Systems, *US Patent 4.575.330*, **1984**.

41 V.K. Varadan, X. Jiang, V.V. Varadan, *Microstereolithography and other Fabrication Techniques for 3D MEMS*, Wiley & Sons, Chichester, UK, **2001**, p. 260.

42 C. Hinczewski, S. Corbel, T. Chartier, Ceramic suspensions suitable for stereolithography, *J. Eur. Ceram. Soc.* **1998**, *18*, 583–590.

43 C. Provin, S. Monneret, H. Le Gall, H. Rigneault, P.-F. Lenne, H. Giovanni, New process for manufacturing ceramic microfluidic devices for microreactor and bioanalytical applications, in *IMRET5*, **2001**, 103–112.

44 C. Provin, S. Monneret, Complex ceramic-polymer composite microparts made by microstereolithography, *IEEE Trans. Electron. Packag. Manuf.* **2002**, *25*, 59–63.

45 C. Provin, S. Monneret, H. Le Gall, S. Corbell, Three-dimensional ceramic microcomponents made by stereolithography, *Adv. Mater.* **2003**, *15*, 994–997.

46 M. Fichtner, W. Benzinger, K. Haas-Sato, R. Wunsch, K. Schubert, Functional coatings for microstructure reactors and heat exchangers, in *IMRET3*, **2000**, 90–101.

47 M. Janicke, H. Kestenbaum, U. Hagendorf, F. Schüth, M. Fichtner, K. Schubert, The controlled oxidation of hydrogen from an explosive mixture of gases using a microstructured reactor/heat exchanger and Pt/Al_2O_3 catalyst, *J. Catal.* **2000**, *191*, 283–293.

48 A. Kursawe, R. Pilz, H. Dürr, D. Hönicke, Development and design of a modular microchannel reactor for laboratory use, in *IMRET4*, **2000**, 227–235.

49 P. Pfeifer, O. Görke, K. Schubert, Washcoats and electrophoresis with coated and uncoated nanoparticles on microstructured metal foils and microstructured reactors, in *IMRET6*, **2002**, 281–287.

50 K. Haas-Sato, O. Görke, P. Pfeifer, K. Schubert, Catalyst coatings for microstructure reactors, *Chimia* **2002**, *56*, 605–610.

51 S. Zhao, R.S. Besser, Selective deposition of supported platinum catalyst hydrogenation in a microchannel reactor, in *IMRET6*, **2002**, 289–296.

52 K. Haas-Sato, M. Fichtner, K. Schubert, Preparation of microstructure compatible porous supports by sol-gel synthesis for catalyst coatings, *Appl. Catal. A: Gen.* **2001**, *220*, 79–92.

53 M. Valentini, G. Groppi, C. Cristiani, M. Levi, E. Tronconi, P. Forzatti, The deposition of γ-Al_2O_3 layers on ceramic and metallic supports for the preparation of structured catalysts, *Catal. Today* **2001**, *69*, 307–314.

54 P. Pfeifer, K. Schubert, M.A. Liauw, G. Emig, PdZn catalysts prepared by washcoating microstructured reactors, *Appl. Catal. A: Gen.* **2004**, *270*, 165–175.

55 Z. Wang, J. Shemilt, P. Xiao, Fabrication of ceramic composite coatings using electrophoretic deposition, reaction bonding and low temperature sintering, *J. Eur. Ceram. Soc.* **2002**, *22*, 183–189.

56 H. Exner, A.M. Reinecke, N. Nieher, Laser beam sintering of thin alumina coatings on metals, *J. Ceram. Proc. Res.* **2002**, *3*, 66–69.

57 X. Wang, P. Xiao, M. Schmidt, L. Li, Laser processing of yttria stabilised zirconia/alumina coatings on Fecralloy substrates, *Surface & Coatings Technol.* **2004**, *187*, 370–376.

58 A. Pfrengle, H. von Both, R. Knitter, J. Haußelt, Electrophoretic deposition and sintering of zirconia layers on microstructured steel substrates, *J. Eur. Ceram. Soc.*, in press.

59 W. M. Carty, P. W. Lednor, Monolithic ceramics and heterogeneous catalysts: Honeycombs and foams, *Curr. Opin. Solid State Mater. Sci.* **1996**, *1*, 88–95.
60 J. T. Richardson, Y. Peng, D. Remue, Properties of ceramic foam catalyst supports: pressure drop, *Appl. Catal. A: Gen.* **2000**, *204*, 19–32.
61 D. Göhring, R. Knitter, P. Risthaus, St. Walter, M. A. Liauw, P. Lebens, Gas-phase reactions in ceramic microreactors, in *IMRET6*, **2002**, 55–60.
62 T. Moritz, R. Lenk, J. Adler, Ceramic components in a modular microreaction system, *cfi/Ber. DKG* **2003**, *80*, E47–E48.
63 F. Scheffler, A. Zampieri, W. Schweiger, J. Zeschky, M. Scheffler, P. Greil, Zeolite covered polymer derived ceramic foams: novel hierarchical pore system for sorption and catalysis, *Adv. Appl. Ceram.* **2005**, *104*, 43–48.
64 M. M. Schwartz, *Ceramic Joining*, ASM Int., Materials Park, Ohio, USA, **1990**, p. 196.
65 M. G. Nicholas, Joining of ceramics, in *Materials Science and Technology Vol. 17b, Processing of Ceramics* Part II (Ed.: R. J. Brook), VCH, Weinheim, Germany, **1996**, Chapter 19.
66 R. Knitter, R. Lurk, M. Rohde, S. Stolz, V. Winter, Heating concepts for ceramic microreactors, in *IMRET5*, **2001**, 86–93.
67 L. Wang, F. Aldinger, Joining of advanced ceramics in green state, *Mater. Lett.* **2002**, *54*, 93–97.
68 A. M. Reinecke, H. Exner, A new promising joining technology, *J. Ceram. Proc. Res.* **2001**, *2*, 45–50.
69 N. H. Menzler, M. Bram, H. P. Buchkremer, D. Stöver, Development of a gastight sealing material for ceramic components, *J. Eur. Ceram. Soc.* **2003**, *23*, 445–454.
70 H. Löwe, W. Ehrfeld, K. Gebauer, K. Golbig, O. Hausner, V. Haverkamp, V. Hessel, T. Richter, Microreactor concepts for heterogeneous gas phase reactions, in *IMRET2*, **1998**, 63–74.
71 W. L. Allen, P. M. Irving, W. J. Thompson, Microreactor systems for hydrogen generation and oxidative coupling of methane, in *IMRET4*, **2000**, 351–357.
72 U. Rodemerck, P. Ignaszewski, M. Lucas, P. Claus, M. Baerns, Parallel synthesis and testing of heterogeneous catalysts, in *IMRET3*, **2000**, 287–293.
73 V. Hessel, W. Ehrfeld, K. Golbig, C. Hofmann, St. Jungwirth, H. Löwe, Th. Richter, M. Storz, A. Wolf, O. Wörz, J. Breysse, High temperature HCN generation in an integrated microreaction system, in *IMRET3*, **2000**, 151–164.
74 G. Veser, G. Friedrich, M. Freygang, R. Zengerle, A modular microreactor design for high-temperature catalytic oxidation reactions, in *IMRET3*, **2000**, 674–686.
75 J. Mayer, M. Fichtner, D. Wolf, K. Schubert, A microstructured reactor for the catalytic partial oxidation of methane to syngas, in *IMRET3*, **2000**, 187–196.
76 R. Knitter, E. Günther, C. Odemer, U. Maciejewski, Ceramic microstructures and potential applications, *Microsyst. Technol.* **1996**, *2*, 135–138.
77 R. Knitter, W. Bauer, C. Fechler, A. Winter, H.-J. Ritzhaupt-Kleissl, J. Haußelt, Ceramics in microreaction technology: materials and processing, in *IMRET2*, **1998**, 164–168.
78 P. M. Martin, D. W. Matson, W. D. Bennett, D. C. Stewart, C. C. Bonham, Laminated ceramic microfluidic components for microreactor applications, in *IMRET4*, **2000**, 410–415.
79 M. R. Gongora-Rubio, P. Espinoza-Vallejos, L. Sola-Laguna, J. J. Santiago-Avilés, Overview of low temperature co-fired ceramics tape technology for meso-system technology (MsST), *Sens. Actuators* **2001**, *A 89*, 222–241.
80 X. Wang, J. Zhu, H. Bau, R. J. Gorte, Fabrication of micro-reactors using tape casting, *Catal. Lett.* **2001**, *77*, 173–177.
81 V. Mengeaud, O. Bagel, R. Ferrigno, H. H. Girault, A. Haider, A ceramic electrochemical microreactor for the methoxylation of methyl-2-furoate with direct mass spectroscopy coupling, *Lab Chip* **2002**, *2*, 39–44.
82 C. Schmitt, D. W. Agar, F. Platte, S. Buijssen, B. Pawlowski, M. Duisberg, Ceramic plate heat exchanger for heterogeneous gas-phase reactions, *Chem. Eng. Technol.* **2005**, *28*, 337–343.

83 J. Vican, B.F. Gajdeczko, F.L. Dryer, D.L. Milius, I.A. Aksay, R.A. Yetter, Development of a microreactor as a thermal source for micromechanical systems power generation, *Proc. Combus. Inst.* **2002**, *29*, 909–916.

84 R. Knitter, D. Göhring, M. Bram, P. Mechnich, R. Broucek, Ceramic microreactor for high-temperature reactions, in *IMRET4*, **2000**, 455–460.

85 R. Knitter, D. Göhring, P. Risthaus, J. Haußelt, Microfabrication of ceramic microreactors, *Microsyst. Technol.* **2001**, *7*, 85–90.

86 S. Stolz, W. Bauer, H.-J. Ritzhaupt-Kleissl, J. Haußelt, Screen printed electro-conductive ceramics, *J. Eur. Ceram. Soc.* **2004**, *24*, 1087–1090.

87 B. Alm, R. Knitter, J. Haußelt, Development of a ceramic micro heat exchanger – design, construction, and testing, *Chem. Eng. Technol.*, in press.

88 T. Stief, O.-U. Langer, K. Schubert, Numerical investigations of optimal heat conductivity in microheat exchangers, *Chem. Eng. Technol.* **1999**, *21*, 297–303.

89 S. Löbbecke, W. Ferstl, S. Panić, T. Türke, Concepts for modularization and automation of microreaction technology, *Chem. Eng. Technol.* **2005**, *28*, 484–493.

90 T. Moritz, R. Lenk, J. Adler, M. Zins, Modular micro reaction system including ceramic components, in *Proc. of the 1st Int. Conf. and Exhibition on Ceramic Interconnect and Ceramic Microsystems Technologies (CICMT)*, April 10–13, Baltimore, Maryland, USA, IMAPS/ACerS, USA, **2005**, on CD-ROM.

91 T.R. Dietrich, A. Freitag, R. Scholz, Production and characteristics of microreactors made from glass, *Chem. Eng. Technol.* **2005**, *28*, 477–483.

92 T.R. Dietrich, W. Ehrfeld, M. Lacher, M. Krämer, B. Speit, Fabrication technologies for microsystems utilizing photoetchable glass, *Microelectronic Eng.* **1996**, *30*, 497–504.

93 M. Masuda, K. Sugioka, Y. Cheng, N. Aoki, M. Kawachi, K. Shihoyama, K. Toyoda, H. Helvajian, K. Midorikawa, 3-D microstructuring inside photosensitive glass by femtosecond laser excitation, *Appl. Phys. A* **2003**, *76*, 857–860.

94 R.M. Karam, R.J. Cassler, A New 3D, Direct-Write, Sub-Micrometer Microfabrication Process that Achieves True Optical, Mechatronic and Packaging Integration on Glass–ceramic Substrates, presented at *Sensors 2003 Conference*, Rosemont, IL, USA, June 2–5, **2003**, and R.M. Karam, Invenios Inc., private communication.

95 F.E. Livingston, W.W. Hansen, A. Huang, H. Helvajian, Effect of laser parameters on the exposure and selective etch rate in photostructurable glass, *Proc. SPIE* **2002**, *4637*, 404–412.

96 F.E. Livingston, H. Helvajian, True 3D volumetric patterning of photostructurable glass using UV laser irradiation and variable exposure processing: Fabrication of meso-scale devices, *Proc. SPIE* **2003**, *4830*, 189–195.

97 W.W. Hansen, S.W. Jansen, H. Helvajian, Direct-write UV laser microfabrication of 3D microstructures in lithium alumosilicate glass, *Proc. SPIE* **1997**, *2991*, 104–112.

98 P.D. Fuqua, D.P. Taylor, H. Helvajian, W.W. Hansen, M.H. Abraham, UV direct-write approach for formation of embedded structures in photostructurable glass–ceramics, in *Materials Development for Direct-Write Technologies* (Eds.: D.B. Chrisey, D.R. Gamota, H. Helvajian, D.P. Taylor), *Mater. Res. Soc. Symp. Proc.* **2000**, *624*, 79–86.

99 H. Helvajian, 3D microengineering via laser direct-write processing approaches, in *Direct-Write Technologies for Rapid Prototyping Applications* (Eds.: A. Piqué, D. B. Chrisey), Academic Press, New York, **2002**, 415–474.

100 J. Cheng, M. Yen, C. Wei, T. Young, Crack-free direct-writing in glass using a low-power UV laser in the manufacturing of a microfluidic chip, *J. Micromech. Microeng.* **2005**, *15*, 1147–1156.

101 M. May, B. Straube, D. Hülsenberg, M. Eckstein, Laserstrukturiertes Glas für die Mikrosystemtechnik, *F&M* **1995**, *103*, 309–311.

102 www.lft-gmbh.de, Little Things Factory GmbH, June **2005**.

103 A. Freitag, T.R. Dietrich, R. Scholz, Glass as a material for microreaction technology, in *IMRET4*, **2000**, 48–54.

104 T.R. Dietrich, A. Freitag, R. Scholz, Herstellung und Eigenschaften von Mikroreaktoren aus Glas, *Chemie Ingenieur Technik* **2004**, *76*, 575–580.

105 J. Antes, T. Türcke, E. Marioth, K. Schmid, H. Krause, S. Löbbecke, Use of microreactors for nitration processes, in *IMRET4*, **2000**, 194–200.

106 J. Antes, T. Türcke, J. Kerth, E. Marioth, F. Schnürer, H.H. Krause, S. Löbbecke, Investigation, analysis and optimization of exothermic nitrations in microreactor processes, in *IMRET5*, **2001**, 446–454.

107 K. Jähnisch, V. Hessel, H. Löwe, M. Baerns, Chemistry in microstructured reactors, *Angew. Chem. Int. Ed.* **2004**, *43*, 406–446.

108 S. Löbbecke, J. Antes, T. Türcke, E. Marioth, K. Schmid, H. Krause, The potential of microreactors for the synthesis of energetic materials, in *Proceedings of 31 Int. Annu. Conf. ICT Energetic Materials – Analysis, Diagnostics and Testing*, 33, June 27–30, **2000**, FhG-ITC, Karlsruhe, 33-01–33-16.

109 T. Herweck, S. Hardt, V. Hessel, H. Löwe, C. Hofmann, F. Weise, T. Dietrich, A. Freitag, Visualization of flow patterns and chemical synthesis in transparent micromixers, in *IMRET5*, **2001**, 215–229.

110 V. Hessel, H. Löwe, F. Schönfeld, Micromixers – a review on passive and active mixing principles, *Chem. Eng. Sci.* **2005**, *60*, 2479–2501.

111 S. Hardt, T. Dietrich, A. Freitag, V. Hessel, H. Löwe, C. Hofmann, A. Oroskar, K. vanden Bussche, Radial and tangential injection of liquid/liquid and gas/liquid streams and focusing thereof in a special cyclone mixer, in *IMRET6*, **2002**, 329–342.

112 T. Fukuyama, Y. Hino, N. Kamata, I. Ryu, Quick execution of [2+2] type photochemical cycloaddition reaction by continuous flow system using a glass-made microreactor, *Chem. Lett.* **2004**, *33*, 1430–1431.

113 H. Ehrlich, D. Linke, K. Morgenschweis, M. Baerns, K. Jähnisch, Application of microstructured reactor technology for the photochemical chlorination of alkylaromatics, *Chimia* **2002**, *56*, 647–653.

114 www.MicroChemTec.de, DECHEMA, June **2005**.

115 mikroglas chemtech GmbH, *Product brochures: microreactors*, **2005**.

13
Industrial Applications of Microchannel Process Technology in the United States

Daniel R. Palo [1,2], *Victoria S. Stenkamp* [1], *Robert A. Dagle* [1], *and Goran N. Jovanovic* [2,3]
1) Pacific Northwest National Laboratory, Richland, Washington, USA,
2) Microproducts Breakthrough Institute, Corvallis, Oregon, USA
3) Oregon State University, Corvallis, Oregon, USA

Abstract
An overview is provided of the industrial applications of microchannel process technology in the United States. Major application areas include thermal processing, separations, mixing and emulsification, chemicals production and the supporting catalyst development, fuel processing, and other integrated systems that require the coordination of multiple unit operations. Significant industrial and preindustrial activities are detailed in each application area, and future directions are discussed.

Keywords
Heat exchange, mixing, separations, chemicals production, two-phase systems, reforming, microtechnology

Advanced Micro and Nanosystems Vol. 5. Micro Process Engineering. Edited by N. Kockmann

ISBN: 3-527-31246-3

13.1
Introduction

Research in microchannel process technology (microtechnology) for industrial applications has been ongoing in the United States since the 1990s [1–8]. Since that time, the number of investigators and the number of applications has grown significantly, to the point where the United States has many university research departments, national laboratories, and small and large companies investigating the merits of and developing microtechnology for various applications that extend beyond development of the devices for analytical, laboratory, or screening purposes. Although much has been published on the use of microfluidics for discovery, sensing, and analysis, the subject will not be taken up here [9–11]. Current industrially relevant topics of microtechnology range from hydrogen production for fuel cells, to the processing of explosive gas mixtures in the production of hydrogen peroxide, to medical applications that promise to drastically improve the performance of portable and stationary dialysis units.

While microtechnology has not been extensively deployed in North American industry, progress is moving in that direction, with the first adoption of the technology likely to be in military applications. The military has seen the promise of using microreactors for generating hydrogen for portable fuel cells, for destroying chemical and biological agents in field operations, for cooling soldiers and shelters, and for producing synthetic fuels in-theatre. In each of these cases, the technology has strict performance demands, must be small and lightweight, and is not as price sensitive as many commercial applications.

The increased industrial interest in microtechnology in recent years is evidenced by the number of small companies formed that have microtechnology as a significant part of their portfolio (Table 13.1). This suite of small companies and start-ups is joined by several larger companies that have identified microtechnology for potential process or product enhancements. Larger companies in the United States that are investigating microtechnology include Dow, Motorola, Modine, Bristol-Myers Squibb (BMS), Hydro Aluminum, FMC Corporation (FMC), and UOP, LLC (UOP). Other companies likely have ongoing microchannel component development activities, but have not reported them publicly.

Several universities have formed centers or departments focused exclusively on microtechnology, such as the Institute for Micromanufacturing at Louisiana Tech University (LTU), Microsystems Technology Laboratories at the Massachusetts Institute of Technology (MIT), the Microtechnology-Based Energy, Chemical and Biological Systems Program at Oregon State University (OSU), and the New Jersey Center for MicroChemical Systems at Stevens Institute of Technology (Stevens). Also evident is the extensive teaming taking place around microreactor research and development. Stevens has been enlisted by both FMC and BMS to investigate promising applications of microtechnology for chemical processing. UOP has enlisted the help of the German Institut für Mikrotechnik Mainz, GmbH (IMM) in its pursuit of microchannel reactor technology, and Pacific Northwest National Laboratory (PNNL) and OSU have teamed to form the

Table 13.1 Small companies conducting microtechnology development in the United States.

Company	Microtechnology focus areas
Ceramatec	Heat exchangers
Exergy, LLC	Heat exchangers
Innovatek Inc.	Fuel processing
International Mezzo Technologies	Manufacturing
Makel Engineering Inc.	Device materials
MER Corporation	Porous materials for microchannel reactors
Mesoscopic Devices, LLC	Fuel processing and heat exchange
MesoSystems	Fuel processing and air purification
UltraCell	Fuel processing
Vellocys Inc.	Chemicals, energy, heat exchange, catalysis

Microproducts Breakthrough Institute, with significant assistance from HP Corporation. Additionally, the Center for Process Analytical Chemistry at the University of Washington, which represents a consortium of industry, academia, and national laboratories, is investigating the usefulness of microreactors in process monitoring, analysis, and high-throughput experimentation [12, 13].

In each application detailed in this chapter, the end user – whether military or industry – has identified a significant role that microchannel reactors, heat exchangers, and contactors can play in intensifying processes, increasing efficiency, enabling new applications, or providing a platform for novel chemical-synthesis routes. The industrial applications of microtechnology in North America are detailed in this chapter according to application, in order of increasing complexity, and starting with single-unit operations. The advantages, disadvantages, and appropriate use of microtechnology have been identified, and some other less-developed areas of microchannel research are summarized with an eye toward future applications.

13.2 Thermal Processing

Thermal processing represents the first area of development that was targeted for microtechnology [2–4]. The very small characteristic transport distances of microchannels were clearly identified as advantageous for highly effective heat exchangers, vaporizers, and other thermal processors. Flow-channel dimensions ranging from micrometers to millimeters significantly reduce the distance for heat and mass transfer, leading to very high transfer coefficients and resulting in compact thermal processing devices. These devices can be quite robust, capable of operating under extreme pressures and temperatures, given proper materials selection.

A number of companies, such as Modine, Hydro Aluminum, Exergy, and Ceramatec have developed microchannel heat exchangers, some of which are for commercial sale, with custom-fabrication options available. The material of fabrication depends on the expertise of the company as well as the targeted application (Table 13.2), but each offers high heat transfer in compact units through the use of microtechnology. As examples, Exergy offers the capability of transferring 88 kW of heat in a 1.3-L stainless steel heat exchanger, while Ceramatec offers the capability of transferring 107 kW in a 14-L ceramic heat exchanger.

Development of microchannel heat exchangers is a logical outgrowth of the trends seen in American industry. Automobile radiators with channels on the size of micrometers to millimeters enable the use of less efficient refrigerants without adding size or weight to the refrigerant system [14]. The movement towards environmentally friendly systems has also resulted in other developments such as heat exchangers for fuel-cell systems and for carbon-dioxide residential heat pumps [15]. Other drivers for the development of microchannel heat exchangers include the desire for cheaper systems [16] or the targeting of specific processing conditions such as Exergy's developments in low flow rate (less than 115 L/min) and high-pressure (34.5 to 51.7 bar) applications.

Increasing performance requirements and the impetus to offer advantages over existing technology have resulted in numerous innovations specific to the microtechnology industry. Such innovations include Ceramatec's recuperator that could be used to increase system efficiencies of microturbines from the 30% of current systems to the United States Department of Energy (USDOE) goal of 40% using microchannel ceramic heat exchangers [17]. Ceramatec created a laminated object manufacturing method that permits fabrication of these heat exchangers at low cost and in mass production by using inexpensive ceramic materials laminated in the green state.

Groups such as PNNL and Velocys, LLC (Velocys) have demonstrated methods of achieving the high-efficiency heat exchange of microchannels, while simultaneously achieving uncharacteristically low pressure drops. This combined

Table 13.2 Examples of companies and their commercial microchannel heat-exchanger applications.

Company	Materials	Application areas
Modine	Metals	Vehicles, HVAC equipment, industrial equipment, refrigeration systems, fuel cells and electronics
Hydro Aluminum	Aluminum	HVAC equipment, refrigeration systems
Exergy	Stainless steel	Aerospace, pharmaceutical, semiconductor, petrochemcial, and fuel cell
Ceramatec	Ceramics	Recuperators for high-temperature applications

set of attributes is attained through the use of massively parallel channels with short path lengths. Velocys has targeted liquefied natural gas (LNG) liquefaction cycles since the capacity is generally limited by the capacity of compression equipment, which in turn is constrained by pressure losses in the traditional heat-exchanger network. It is anticipated that increasing capacity by 10% or more with the equivalent capital expenditures positively alters the economics allowing stranded gas to be monetized via liquefaction [18].

More recently, PNNL has delivered an economizer to the Gas Technology Institute/Cleaver-Brooks, designed to support a 880-kW boiler. This highly effective, thin panel heat exchanger built for recuperating heat from a hot gas stream into a liquid stream has only inches of water pressure drop on the gas side. The completed economizer assembly (Fig. 13.1) contains 16 individual heat exchanger panels, with a combined heat duty of over 60 kW, and an expected effectiveness of more than 94% [19].

Instead of short path lengths, OSU has used fractal devices (Fig. 13.2) in desorbers to decrease the pressure drop and flow surges normally associated with

Fig. 13.1 Microchannel economizer based on PNNL panel design with low air-side pressure drop (courtesy of G. Whyatt, PNNL).

Fig. 13.2 Example of a fractal-based desorber unit designed by OSU for heat pump applications (courtesy of D. Pence, OSU) (from ref. [20]).

two-phase flow in heat exchangers [20, 21]. In this design, fluid is fed to the center of the device and is heated as it flows through the increasing number of bifurcations.

The use of microchannels can alter the physics of phenomena such as homogenous (flame) combustion, normally used to enhance heat transfer, as demonstrated by researchers at the University of Pittsburgh [22]. In related work, PNNL has developed compact devices heated by catalytic combustion, such as the 350-mL gasoline vaporizer they developed to support a 50-kW fuel-cell system [23].

13.3 Separations

Separations encompass a variety of processes ranging from adsorption to physical phase separation to multiphase processing involving mass transfer. Separations are enhanced in microchannel architecture through heat- and mass-transfer advantages, and the increased importance of interfacial phenomena. Microchannel separators are already seeing application in military, space, and portable systems, where compact processing in an orientation-independent architecture is required, or in biomedical or waste-recovery applications where rapid heat and mass transfer enable purification to unprecedented levels or at unprecedented rates.

In the case of adsorption, thermal-swing processing has become more attractive in a microchannel architecture since the high heat-transfer rate allows cycle times of seconds or minutes, rather than hours. When multiple beds are linked, this allows for semicontinuous processing in a compact architecture [24]. Current efforts at Velocys include recovery of methane from coal mines and landfills to produce pipeline-quality gas from a low-quality stream that would otherwise be flared or emitted. Commercial deployment of this technology has the potential to reduce annual US greenhouse gas emissions by 21.4 million metric tons of carbon dioxide equivalent, while simultaneously recovering 100 billion standard cubic meters of natural gas. At PNNL, microchannel TSA is being used as a vapor chemical compressor as part of an integrated propellant production system for a space application as discussed later in this chapter.

The small channel sizes used in microtechnology enhance capillary, surface, and hydrodynamic forces, which in turn enhance a variety of multiphase processes. For instance, this results in phase separators that are compact, passive, and gravity independent, unlike many of their conventional counterparts. In one case, a multichannel separator (Fig. 13.3) was designed, built and tested for recovery of water from the cathode side of a 5-kW PEM fuel cell being developed for the National Aeronautics and Space Administration (NASA) [25]. The most compact embodiment of this technology showed a size decrease factor of 3 to 4 times and a weight decrease factor of 6 to 7 times over a conventional centrifugal separator.

Incorporation of heat exchange into a phase separator can result in a unique single device that has the capacity for partial condensation with simultaneous phase separation in a process that is orientation independent. Results from tests performed on a device flown in microgravity show complete separation, with heat fluxes ranging from 1 to 7 W/cm^2, mean overall heat transfer coefficients of 500 to 2000 W/m^2 K, and power densities ranging from 2 to 12 W/cm^3 [26]. Comparisons with conventional condensers coupled to cyclone separators suggest a 20-fold reduction in weight. This reduction in weight is critically important to applications such as water management in environmental life-support systems and space suits, and portable fuel-cell systems.

In multiphase separation processes such as pervaporation, reverse osmosis, absorption, desorption, and distillation, the rate of separation is typically controlled by the diffusion time of the component of interest in the liquid phase, since liquid diffusivities are lower than gas diffusivities by orders of magnitude. Typically the liquid-phase thickness is controlled by wicks, membranes, or surface forces to enhance process intensification. The time for diffusion is characterized by the square of the thickness of the fluid film divided by the diffusion coefficient of the component in that phase. A study from LTU [27] has shown that the diffusion time for ethanol in water varies by orders of magnitude as

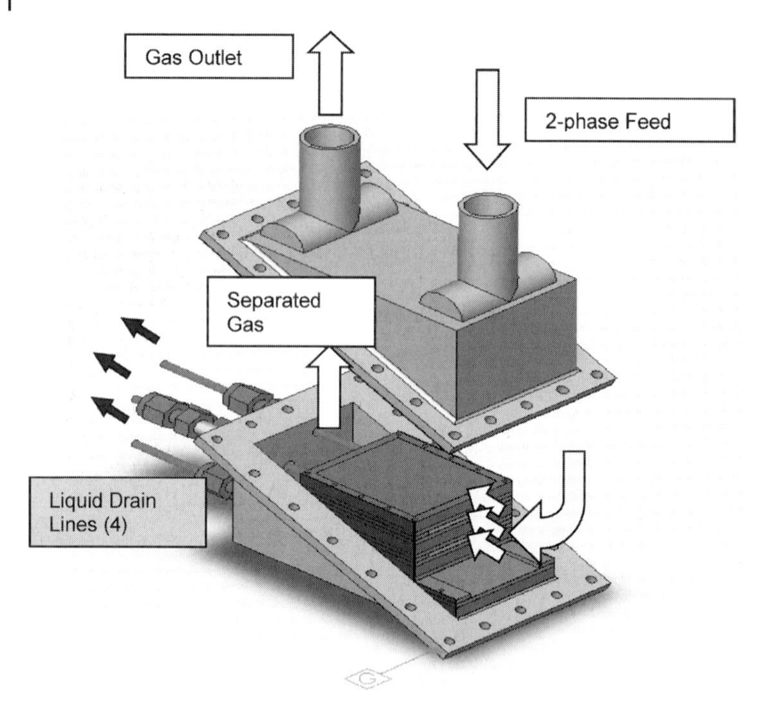

Fig. 13.3 Schematic showing operation of the multichannel phase separator (used with permission from ref. [25], copyright 2005, American Chemical Society).

the channel dimensions vary from 50 to 400 μm. Using this advantage, LTU was able to show between 55% and 97% recovery of ethanol from an ethanol/water mixture that was processed in a pervaporation device using a polydimethylsiloxane membrane with 98 channels 50 μm wide and 2.7 cm long (Fig. 13.4). It is anticipated that applications for microchannel pervaporation include dehydration of alcohols and solvents, removal of trace organics from water, and separation of isomers.

Conventional distillation remains a highly energy-intensive process, representing consumption of 5 quadrillion kJ annually, and is thus a target of great interest for programs such as the USDOE Industrial Technologies Program. Conventional distillation systems could potentially increase efficiency by incorporating reboilers at multiple stages, but the resultant capital costs are prohibitive. In contrast, microtechnology can dramatically decrease equipment size and capital intensity, while integrating heat transfer for better energy efficiency. Early studies show that the height equivalent of a theoretical plate (HETP) is decreased to about 2.5 cm compared to the 30 to 45 cm achieved in commercial packings (Fig. 13.5). Although conventional laboratory-scale devices show HETP in the range of several centimeters, the scaling processes typically do not result in similar HETP in commercial systems. In contrast, the scaling processes of microtechnology automati-

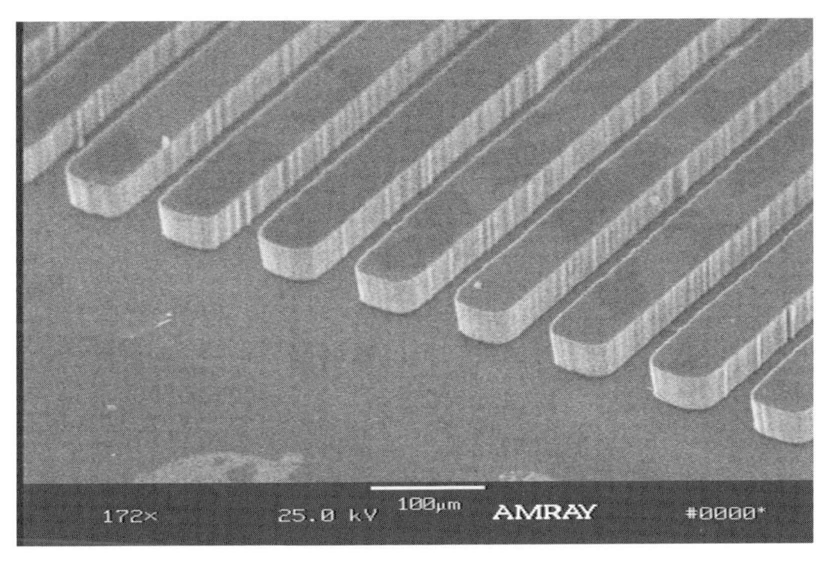

Fig. 13.4 Detail of a LTU pervaporation device for the separation of ethanol/water mixtures (used with permission from ref. [27]).

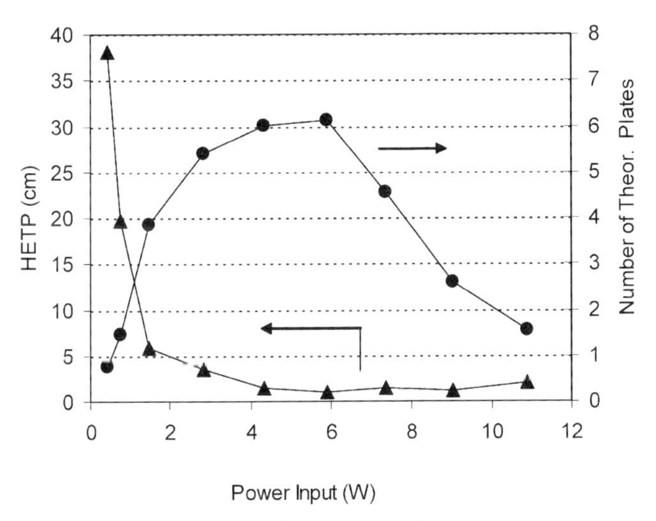

Fig. 13.5 Height equivalence of a theoretical plate and number of theoretical plates as functions of heat input for the separation of acetone and methanol (used with permission from ref. [29]).

cally result in the same HETP. Commercial applications currently targeted for microchannel distillation include process components with low relative volatility, requiring a large number of stages. Velocys has targeted the separation of ethane from ethylene, where microtechnology has the potential to reduce energy consumption by up to 20% for the entire process [28]. PNNL has targeted the separation of C_4 hydrocarbons for similar reasons [29].

In the area of biological separations, the development of the hemodialysis cartridge is a success story in the implementation of microscale features. Modern hollow-fiber technology, as deployed in hemodialyzers has made a huge difference in the treatment of patients with renal failure. The fact that a single hollow fiber has an internal diameter of approximately 200 μm and that the wall of the fiber is an order of magnitude smaller shows how microscale features improve operation in devices where mass transfer is the prevailing phenomenon. Improvement over the current devices has been afforded through formal microtechnology research, which indicates that improvements to the uniformity of the microscale features can result in an increase in the overall efficiency of mass transfer in the dialysate side of the hollow-fiber bundle. OSU, along with HD+ of Portland, OR, has demonstrated a novel multichannel unit that shows substantial improvement in urea removal from a simulated blood stream containing serum albumin and 0.7 g/L of urea. The multilayer architecture provides a platform for future developments that, according to HD+, will change the current paradigm in hemodialysis from clinical to home hemodialysis and allow for the creation of a wearable hemodialyzer in the future [30].

13.4 Mixing and Emulsification

Mixing creates new challenges in microchannels, especially for liquids, since the flow is laminar and, if left unperturbed, occurs by diffusion alone. Several methods to address this issue have been developed within the United States. In some cases, mixing is performed by creating disturbances to the flow through patterned surface charges or through features in the microchannel such as grooves or posts [31, 32]. Other approaches rely on decreasing the diffusion distance as in hydrodynamic focusing or the use of side flows to force the inlet flow into a thin stream [33, 34]. Most of these approaches originally focused on extremely low flows for applications in analysis or extremely small biological applications, but some have moved into the realm of larger-scale mixing such as a PNNL-developed gas–gas mixer created for a 10-kW steam reformer [35]. In this work, the start-up time of the steam reformer from room temperature to 650 °C was reduced to 20–30 s from the previous time of 15–20 min. The mixer rapidly and uniformly mixes hydrogen and air as part of a spark ignition system that replaces a slower catalytic combustion system.

Emulsification is the mixing of two fluid phases to obtain a dispersed phase of small droplets within a continuous phase. Emulsions are ubiquitous in both industrial and consumer products, as are the expensive surfactants and other additives that are often needed to form and stabilize emulsions. Agglomeration of droplets is the most common way for an emulsion to fail, and this instability is primarily caused by nonuniform droplet size. Typically, emulsions are prepared using highly energetic systems such as high-pressure or ultrasound homogenizers. In contrast, emulsions can be prepared in microchannels by either pushing the discontinuous phase through a membrane or other porous material [36, 37], or by controlling shear at the junction of two phases [38].

Forming emulsions in microchannels offers many advantages, such as low bulk shear to protect fragile components, narrow droplet size distribution that enhances mixture stability, and in-line heating and cooling that permits gentle reagent heating, rapid product quenching, and hence very stable emulsions. Studies in the United States have demonstrated that the use of junctions with controlled shear or channel geometry allows the production of nonspherical droplets such as slugs or disks [38]. In a Velocys study for the production of hand lotion using a microchannel membrane, the resulting droplet sizes were an order of magnitude smaller than in the emulsion prepared by conventional hardware, and the size distribution was much narrower, although substantially lower levels of surfactants were employed [39, 40]. However, in order for these processes to become commercially viable on an industrial scale, the performance issues associated with traditional membrane processes must be overcome or the unique advantages offered by the microchannel processes must give the product an economic advantage.

13.5 Chemical Reactions

Chemical processing represents another large area of potential application for microtechnology. Any process that would benefit from enhanced heat transfer or mass transfer, better thermal control of unit operations, or the ability to directly scale down, scale up, or modularize is a candidate for microreactor application. In the case of exothermic reactions, efficient heat removal leads to prolonged catalyst life and lower operating costs. Interleaved microchannel structures allow coupling of endothermic and exothermic reactions, or the removal of products such as hydrogen through an adjacent membrane [41]. Additionally, because microreactors have at least one short characteristic dimension, radical-based reactions are quickly quenched before runaway or explosion can occur. This enables operation, for instance, with explosive mixtures that would not be possible in conventional systems.

Production of hydrogen peroxide by direct combination of hydrogen and oxygen is one reaction of great interest in light of the above benefits of microtechnology, and is being studied by UOP [42] and Stevens [43, 44]. UOP indicates that they expect a simple, direct production process to reduce peroxide production costs to as low as 0.22–0.33 US dollars per kg. This represents a significant cost saving compared to the current, very complicated process utilizing anthraquinone. The potentially lower cost of peroxide promises to enable new processes, such as the production of propylene oxide, phenol, and epichlorohydrin [42]. Stevens anticipates very large energy savings with reductions to 2300 kJ/kg from the 13 800 kJ/kg of conventional processes. As part of their goal to accelerate technology development, UOP is striving for direct scale up from laboratory to production, a magnitude of 10^5 for process hardware, and 10^6 for catalysts. This strategy requires precise quantification of the fundamental resistances and identification of the intrinsic kinetics.

The previously unattainable goal of direct production of hydrogen peroxide has been made possible with microtechnology since the enhanced heat transfer enables rapid wall quenching of free radicals, which in conventional-size reactors lead to thermal runaway conditions and concomitant explosion. This finding was confirmed by UOP through joint investigation with IMM, and by Stevens who is partnered with FMC, the world's largest producer of hydrogen peroxide. Single-channel studies by UOP and IMM operating at 300–600 psig, achieved up to 100% conversion, and up to 95% selectivity to product, with yields up to 0.80. Using explosive O_2/H_2 ratios ranging from 1.0 to 6.8 in channels 100 to 500 μm high resulted in a 5% H_2O_2 product [42]. Studies by Stevens employed a residence time of 2 to 3 s compared to 0.5 to 1.0 h in conventional reactors and resulted in peroxide product concentration of 1.5 wt% [43].

Numerous other examples exist where microchannel reactors are demonstrating advantages in production of commodity chemicals. For example, Velocys reports the use of microreactors for gas-to-liquids (GTL) processing. This reaction is of interest for the monetizing of remote natural-gas supplies into synthetic fuel through the Fischer–Tropsch (FT) process. Taking advantage of rapid heat and mass transfer allows preservation of catalyst integrity under aggressive conditions, resulting in higher yield than can be achieved with conventional reactors. Higher yield reduces the amount of methane formed as a side product and minimizes the amount of recycle required in the system, which in turn allows for compression equipment to be smaller, reducing plant-wide capital and operating costs [45]. Working with PNNL, Velocys reports 12 to 36-fold increases in reactor throughput, and a seven-fold increase in efficiency compared to conventional FT reactors [46, 47].

Velocys (along with partners Dow and PNNL) is investigating the use of microchannel reactors for the high intensity oxidation of ethane to ethylene [48]. Utilizing intensive heat integration and staged oxygen addition, the team hopes to increase ethylene yield in a process that typically suffers from several side reactions and requires intensive heat removal. Initial results of operation at 70 to 90% conversion have shown improved selectivity in the microchannel reactor (82 to 88% selectivity to ethylene) versus conventional technology (69 to 79% selectivity to ethylene).

Process intensification with microtechnology has shown promise to also expedite pharmaceutical process development, as shown by BMS. Microreactors are used to conduct feasibility studies with limited-availability materials to generate intense processes for chemistries that are highly hazardous or require the use of extreme operating conditions. Table 13.3 is a summary of chemistries that have been successfully tackled and subsequently scaled up by BMS in microreactors or minireactors [49]. In the case of catalytic hydrogenation, BMS partner Stevens has reported the reduction of residence times to 2 to 3 s compared to the conventional 4 h, which in turn results in fewer side reactions and fewer purification steps [50]. The greater heat-transfer control also removes safety concerns with this highly exothermic reaction involving hydrogen. In 2004 alone, processes initially demonstrated by microreactors have been scaled up to generate 0.75 metric tons of pharmaceutical intermediates or active pharmaceutical ingredients (API) funding clinical and toxicological studies.

The achievement of the ultimate cost and quality gains in API manufacturing efficiency necessitates integration of microreactors with process-analytical technology (PAT). PAT provides a fundamental understanding of process knowledge during development through implementation of various at-line or on-line sensors to monitor and control process or product attributes. Such integration represents a paradigm shift governing pharmaceutical regulation and constitutes the foundations for the new 21st Century Risk-based cGMP initiative of the United States Food and Drug Administration. BMS has constructed a miniplant that exemplifies 21st century pharmaceutical manufacturing and has a production capacity of 5 kg/day for a reaction with thermal runaway potential. The unit includes a minireactor, PAT, and downstream purification, yet takes up the space of less than half a laboratory hood.

Experience at BMS has shown that even with painstaking development efforts, there is no guarantee that microtechnology solutions circumvent all road-

Table 13.3 List of pharmaceutical reactions that have been successfully scaled up in microreactors and minireactors [49].

Chemistry	Issues	Results
Lithiation and coupling	Two cryogenic steps Capital investment	Eliminate both cryogenic conditions Capacity: 5 kg/h with pilot microreactor unit
Reduction	Process cost advantage	Process developed with kg-quantity product generated
Acidolysis	Fast reaction Product degradation	Process with consistent quality developed Scale up and tech transfer completed
Deprotonation oxidation	Explosive hazard	Inherently safer process developed Scale-up demonstrated
Cyclopropanation	Low yield on scale Hazardous	Yield improvement from $<30\%$ to 70% Scale-up and tech transfer completed
Oxidation	Thermal runaway hazard	Inherently safer process developed Scale-up demonstrated Tech transfer ongoing

blocks. Some of the crucial issues include the lack of compatibility with other process streams, the inability to satisfy cGMP regulations, lack of time to implement the technology, and unclear microtechnology advantages.

Bio-based fuel production is another application area where microtechnology can provide significant advantage, speeding up a typically slow chemical process. OSU recently finished a feasibility study of biodiesel synthesis in a microscale reactor system. Figure 13.6 illustrates typical experimental results obtained in a high aspect ratio (1:2000) thin-film reactor [51]. It is important to mention that the OSU reactor system did not use a micromixer unit, which one would normally anticipate to be useful in this two-phase process. In fact, premixing of reactants (soybean oil and methyl alcohol) in a micromixer proved to be counterproductive for overall process performance, which hinges on second-order, multistep equilibrium reactions and separation operations. The thin-film chemical reactor validated expectations obtained through CFD simulations that the overall process rate of biodiesel production in a thin-film microreactor with a high-aspect ratio architecture is substantially better than an ideal mixed flow microreactor. These results confirm that diffusion-limited reaction processes greatly benefit from the reduction of the diffusion time ($\tau_D = l^2/D$). Furthermore, the absence of a micromixer unit allows for additional in-situ reaction processing, which is not usually associated with commercially known biodiesel processes [51]. OSU and its partners, working through the Oregon Nanoscience and Microtechnologies Institute (ONAMI), are continuing the development of this technology.

As microtechnology continues to revolutionize the way unit operations are performed, a new area of catalyst development has also emerged. Due to decreased heat- and mass-transfer resistances offered by microtechnology, catalysts employed in microchemical reactors can be tailored to take full advantage of

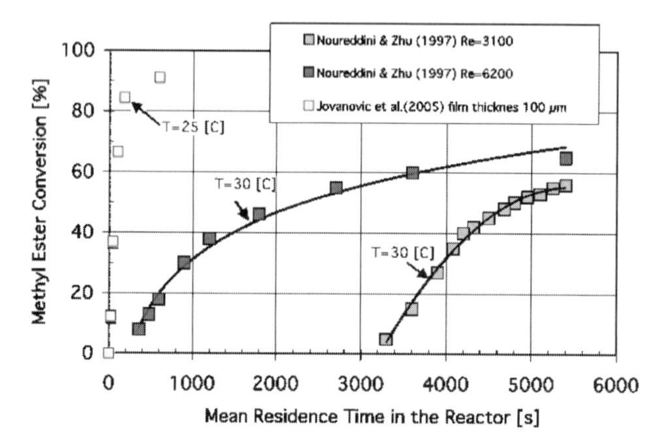

Fig. 13.6 Comparison of OSU microchannel biodiesel production results to those obtained using conventional means (used with permission from ref. [51]).

their fast intrinsic kinetics. For example, unlike conventional fixed-bed reactors systems, the use of more expensive precious-metal catalyst compositions can be made economical. Monolith-type substrates utilizing highly active catalyst species are integrated in microreactors to minimize heat- and mass-transfer resistances and maximize catalyst efficiency. The use of smaller, more efficient systems can compensate for the higher cost of the catalytic material. It is the development of highly active, selective, and stable catalysts in combination with microchannel reactor design that determines the efficiency of many chemical processing systems.

Researchers at University of Minnesota have pioneered the development of catalysts specifically for use in millisecond reactors. High-temperature catalytic oxidation reactions have been developed for the partial oxidation of hydrocarbons to more valuable products, such as syngas, olefins, and oxygenates [52, 53]. More recently, the development of water-gas-shift catalysts utilizing noble metals for use in fuel-reformation systems at fast contact times has also been achieved [52]. Researchers at the University of Pittsburgh have investigated the integration of millisecond contact time reactors and high-temperature catalysis for partial oxidation of methane to synthesis gas [54]. PNNL has developed catalysts for use in microreactor fuel-processing applications [55, 56] and for methane steam reforming microreactor chemical-production applications [57]. Catalyst coatings in microchannels (10 to 100 μm thickness) have been investigated by researchers at the University of New Mexico, yielding improved ceramic microreactors for methanol reforming to hydrogen [58].

13.6 Fuel Processing for Hydrogen Production

Beyond thermal processing and chemicals manufacture, much of the sustained interest in microchannel reactors has been in the area of fuel processing for hydrogen production for fuel cells [59, 60]. Hydrogen fuel cells hold great promise as clean, silent, efficient energy sources for various applications, including portable, mobile, and residential power production. However, since hydrogen is not easily stored or transported, much focus has been placed on the production of hydrogen "onboard", or in conjunction with the fuel cell. Since microchannel architecture provides an opportunity to produce highly efficient and compact devices, use of such reactors to produce hydrogen for portable, mobile, and even stationary applications is a natural fit. Great interest has been shown in the United States in this area by the USDOE, the Department of Defense, and various companies [59].

The fuel of choice depends on the application. For small power, simpler fuels have been of most interest, including methanol, ethanol, formic acid, and ammonia. Larger applications, such as mobile or stationary power, tend to be focused on fuels that are readily available and have extensive distribution networks, such as methane, propane, gasoline, and diesel.

Table 13.4 lists the major developers in microchannel fuel processing for hydrogen production for fuel cells, also indicating the fuels used, the reactor materials, the reactions employed, and the applications sought. Since there are many aspects to any fuel-processing system, such as preprocessing (e.g., desulfurization, cracking), primary conversion (e.g., steam reforming, partial oxidation),

Table 13.4 Major developers and activities in microchannel fuel processing for fuel-cell applications.

Developer	Fuels employed	Reactor materials	Reactions employed [a)]	Applications
Air Force Research Laboratory	Diesel	Metal	Thermal cracking, desulfurization, steam reforming	Mobile
Battelle	Synthetic diesel	Metal	Steam reforming	Mobile, stationary
Innovatek	Methane, diesel, gasoline	Metal	Steam reforming	Mobile
Makel Engineering	Methane	Metal, ceramic	Methane POx, methanation, WGS, PrOx	Portable, mobile, space
Massachusetts Institute of Technology	Ammonia	Silicon nitride	Catalytic cracking	Portable
Mesoscopic Devices	Various hydrocarbons	Metal	Steam reforming, partial oxidation	Portable, mobile
MesoSystems	Ammonia	Metal	Catalytic cracking	Portable
Motorola	Methanol	Ceramic	Steam reforming	Portable
Oregon State University	Diesel	Metal	Desulfurization	Mobile, modular
PNNL	Methanol, various hydrocarbons, alcohols	Metal	Steam reforming, WGS, PrOx, methanation	Portable, mobile, stationary
Ultracell	Methanol	Silicon, polymers	Steam reforming	Portable
University of Delaware	Ammonia	Metal	Catalytic cracking	
University of Illinois	Ammonia	Ceramic	Catalytic cracking	Portable
University of Michigan	Methanol	Ceramic	Steam reforming	Portable
Velocys	Methane	Metal	Steam reforming	Merchant

a) POx = partial oxidation, WGS = water-gas shift, PrOx = preferential oxidation.

secondary conversion (e.g., water-gas shift), and CO mitigation (e.g., preferential oxidation, methanation), various investigators are involved in some or all aspects of the process.

Work in fuel processing for fuel-cell applications has taken several different paths; methanol reforming at moderate temperatures, hydrocarbon reforming at high temperatures, ammonia cracking at high temperatures, and the challenging area of hydrocarbon desulfurization. Within these main areas, different approaches have been taken, including metal-, ceramic-, and silicon-based reactors. Most of the fuel-processing work has been conducted in metal reactors (generally stainless steel or nickel-based alloys like Inconel®, but advantages have also been realized in using ceramics or silicon.

13.6.1 Methanol Systems

Methanol reforming is generally accomplished at temperatures below 350 °C using a suitable catalyst – usually copper-zinc oxide. Many investigators are active in this area, but not all of them are employing microchannel architecture. The ceramic-reactor approach to methanol conversion has been employed by both Motorola and the University of Michigan. Motorola has taken the approach of directly coupling their methanol steam reformer to a high-temperature PEM fuel cell [61, 62]. The reactor includes a vaporizer, steam reformer, and either a catalytic combustor or thick-film heaters in a device that is 35 mm by 15 mm by 5 mm. Motorola reports high methanol conversion at 200 to 230 °C in this device. Researchers at the University of Michigan [63] are using ceramic tape-casting procedures to fabricate a microchannel methanol steam reformer heated by hydrogen combustion. Reported reactor characteristics include 5 to 10 channels, with aspect ratios of 0.7 to 2.0, and wall thicknesses of 500 to 800 μm.

Using silicon and polymer-based technology, UltraCell Corporation, of Livermore CA is currently under contract to the US Army to develop a soldier-portable fuel-cell power system based on methanol steam reforming, and delivering roughly 20 W of power. The base technology was originally developed at Lawrence Livermore National Laboratory [64, 65].

PNNL have demonstrated methanol reformer systems at both the subwatt and multiwatt power ranges using stainless steel reactors. In the former case, PNNL teamed with Case Western Reserve University (CWRU) to develop an integrated methanol steam reformer and fuel-cell system [59, 66]. The PNNL device, which contained several unit operations in a transistor-sized package, was operated at thermal efficiencies of up to 33%, and yielded a hydrogen-rich reformate stream. The reformate was fed to the CWRU high-temperature PEM fuel cell to produce power in the mW range [67].

Other work at PNNL yielded microchannel-based methanol fuel processors with design powers of 20, 50, 100, and 150 W. At each power level, the device contained catalytic combustion, catalytic steam reforming, fuel vaporization, and heat recuperation in one unit [59, 68]. The devices were demonstrated for methanol steam reforming at thermal efficiencies up to 85%. The team also incorporated a selective methanation reactor to reduce the CO content of the reformate below 100 ppm and demonstrated the processor with several PEM fuel cells [69]. The PNNL team assembled a demonstration unit based on the 150-W reformer, and although the reformer was not optimized, its size and weight represented a mere 4% of the volume and 7% of the mass of the demonstration unit [69].

13.6.2 Hydrocarbon Systems

The area of hydrocarbon reforming has seen much more microchannel reactor activity, and promises to have widespread application, ranging from portable and mobile military power to commercial automotive power, to merchant hydrogen production. As with methanol reforming, various reactor materials have been investigated. Makel Engineering, Inc., for instance, uses a parallel-plate microchannel design for fuel processing, employing both ceramic and metal reactors. Specific reactions investigated include methane partial oxidation, methanation of CO and CO_2, water-gas shift, and preferential oxidation of CO [70, 71].

Researchers at the United States Air Force Research Laboratory (USAFL) have developed a highly integrated microchannel-based fuel-processing device for hydrogen production from jet fuel. The group has developed a prototype system that performs upstream fuel desulfurization (99.98% removal) before feeding the preprocessed fuel to a microchannel-based steam reformer. The design uses channels of less than 500 μm, which are heated by a radiant burner. Estimates on the ultimate size of the device predict a 100-kW processor with a volume of about 28 L [72].

Mesoscopic Devices, LLC reports the development of steam reformers and partial oxidation reactors for hydrocarbon fuels. One application utilizes catalytic partial oxidation to produce a reformate stream for a 75-W solid oxide fuel cell (SOFC) [73, 74]. This processor has been demonstrated with multiple fuels at powers from 50 to 500 W, and with a pressure drop of about 0.07 bar. The company has demonstrated cold startup in less than four minutes, using about 3600 J of energy and enabled by the small size of the device (13 mm diameter by 18 mm long). In a second application, Mesoscopic Devices is utilizing steam reforming coupled with combustion in a normal flow reactor, which promises to have faster response time than traditional or parallel plate devices [74]. The company also reports greater than 90% thermal efficiency in their reformer device.

InnovaTek Inc. has reported the use of microchannel architecture for fuel reforming of complex hydrocarbons [75, 76]. The company employs a cross-flow heat-exchange system where hot combustion gas provides heat to the endothermic reforming reaction, which operates at high space velocities (100 000 h^{-1}) and roughly 850 °C. The device includes a burner, steam reformer, heat exchangers, and water vaporizer, and has a pressure drop of about 0.2 bar, processing either methane or diesel fuel.

In work funded by the USDOE, PNNL has developed hydrocarbon steam-reformer systems operating in the 200-W to 22-kW power range, and utilizing fuels such as methane, propane, ethanol, iso-octane, and synthetic diesel. The main focus of the work to date has been on-board automotive steam reforming, an application that requires extremely rapid reactor startup. Recently, the team at PNNL demonstrated a kilowatt-scale system with a cold-start time of 12 s, illustrating the advantage to be gained by utilizing microchannel architecture to reduce the thermal mass of a processing system [77].

Additional PNNL work in support of automotive fuel cells includes water-gas shift (WGS) and preferential oxidation (PrOx) reactor development. PNNL researchers used the precise temperature-control capabilities of microchannel devices to develop a differential temperature WGS reactor. By operating in a differential temperature mode, the PNNL team demonstrated that WGS could be performed in a single-stage reactor, replacing the traditional three-stage approach, which includes high-temperature WGS, cooling, and low-temperature WGS [78]. Development of a PrOx reactor in support of the PNNL system yielded a 2-kW unit that reduced CO content by 97 to 98%. The approach employed a four-stage reactor in which oxygen was injected in each of the four stages, leading to better temperature control and higher selectivity using both precious and non-precious metal commercial catalysts [79].

In related work, researchers at Stevens depicted the benefits of using a microchannel versus a fixed-bed approach for the PrOx reaction. Using model simula-

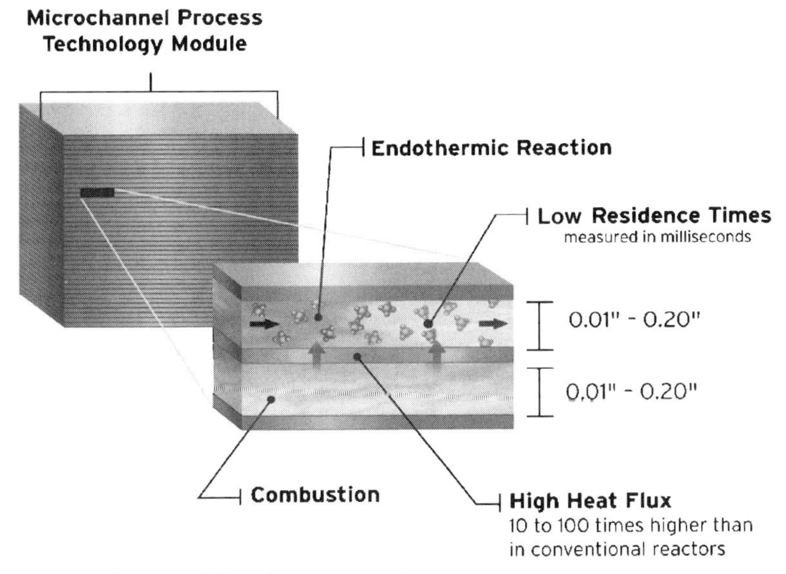

Fig. 13.7 Schematic of the Velocys integrated reactor system concept, coupling exothermic combustion with endothermic reforming (used with permission from ref. [83]).

tions, as well as empirical arguments, the efficient heat removal of the microchannel system was shown to eliminate temperature gradients and effectively prevent the onset of the undesired reverse WGS reaction, thus creating a much larger window of useful operation [80].

In conjunction with the United Defense Limited Partnership and the US Army National Automotive Center, Battelle has demonstrated a fully integrated steam-reformer-based polymer electrolyte membrane (PEM) fuel-cell system for use on the Bradley Fighting Vehicle for silent-watch applications [81]. The fuel-processing technology (developed by Battelle researchers at PNNL), along with the fuel cell, balance of plant, and controls were integrated into a package that fit in an existing utility box on the back of the vehicle (0.52 m by 0.44 m by 0.91 m), and produced 2.2 kW of net power in demonstrations conducted for the US Army.

On an even larger scale, Velocys is employing microchannel steam methane reforming (SMR) devices to produce merchant hydrogen. The device contains three key unit operations, an SMR reactor, a combustor, and a recuperating heat exchanger (Fig. 13.7), allowing for a significantly smaller process footprint, capital cost savings of up to 20%, and efficiency improvement of 7% over traditional fired tube reactors [82, 83].

13.6.3
Ammonia Systems

The third group of fuel-processing systems is based on ammonia cracking. Ammonia cracking is desirable because it produces no CO, avoiding the need for downstream CO mitigation before feeding the fuel cell. Several teams have employed microchannel reactors in the ammonia-cracking reaction, all of them at less than 100 W, but using differing approaches, as detailed below.

A group at MIT has been investigating silicon-based microchannel device fabrication and operation, including ammonia cracking [84]. Most notably, they have demonstrated a silicon-nitride suspended-tube reactor and heat-exchanger system that operates at up to 900 °C and is heated by an embedded platinum resistance heater or by combustion. Using special geometries, thermal losses were minimized, despite the high operating temperature.

MesoSystems has reported the development of a 50-W stainless steel ammonia cracker that operates at 575 °C to 625 °C and 1 to 4 bar. Microchannel architecture was employed for both reactor and heat exchanger, which were integrated with the necessary additional operations to produce a fuel-cell system demonstrator. Ammonia conversion is reported at 99.5 to 99.8% [85, 86].

Researchers at the University of Illinois, Urbana-Champaign (UIUC) have demonstrated ammonia cracking for hydrogen production at the 60 W scale [87]. The group at UIUC utilizes alumina-based microreactors to decompose ammonia at 600 °C with 99% conversion [88, 89]. Similarly, at the University of Delaware research has been conducted on integrated microdevices for the production of hydrogen from ammonia decomposition over Ru catalyst, heated by homogeneous propane combustion [90–92].

13.6.4 Desulfurization

Desulfurization is a key element to using fuels like gasoline, diesel, and natural gas, since many parts of the fuel-cell system cannot tolerate sulfur compounds. Sulfur readily poisons fuel processing and fuel cell catalysts, rapidly decreasing performance. Two approaches can be taken to the sulfur problem. One approach is to make the fuel-processing system tolerant to sulfur. The other approach is to desulfurize the fuel, either as part of the fuel-processing system or somewhere along the supply chain. Logistically, the latter approach is the most likely to succeed.

The most challenging types of sulfur compounds found in diesel, gasoline, and jet fuel are thiophenes. These compounds readily survive the hydrodesulfurization process and are the most recalcitrant compounds for removal. Several processes have been suggested for the removal of thiophenes, including hydrodesulfurization with a novel catalyst that only mildly reduces octane number of fuels, oxidative desulfurization, membrane separation with and without desulfurization, adsorption, and liquid–liquid extraction with and without oxidative desulfurization. In all of these processes microscale reactors and microseparation unit operations may have a potential role to play. As mentioned above, work has been conducted at USAFL on desulfurization of diesel fuel as part of a microchannel fuel-processing system [72]. The USAFL approach includes a cracking and distillation process.

Using an oxidative desulfurization method, researchers at OSU have demonstrated a continuous, cocurrent, thin-film, photochemical microreactor. In this device, an aqueous stream containing hydrogen peroxide (3 to 30%) was contacted with a hydrocarbon stream containing thiophenes and exposed to UV radiation at $\lambda = 254$ nm. Thiophene levels were reduced much more rapidly than in conventional batch reactors, reaching 60% removal in seconds rather than hours [93], illustrating the enhanced performance of microreactors relative to conventional laboratory batch reactors (0.5 to 1.0 L).

13.7 Integrated Systems

Integration of microchannel systems is a logical outgrowth of the requirement to meet particular performance criteria with numerous unit operations. Integration requires weighing the gains in performance of one device versus the loss in performance of another, in addition to considering heat integration of the various devices to allow increases in overall energy efficiency. The area where integration is most developed is in fuel processing, as discussed above, but integration has started to occur in other applications as discussed in the examples below.

Recent research by MesoSystems for the US military incorporated microchannel heat exchange with a microchannel catalytic reactor for air purification in field operations where chemical or biological agents may be present [94]. The resultant recirculating thermocatalytic air purifier (TCAP) avoids the use of short-lived filters and absorbers by destroying the offending agents over an oxidation catalyst developed by Honeywell. An internal combustor provides the high temperature environment (250 to 400 °C). Because the TCAP requires operation at high temperatures, but cannot dump high-temperature air into military shelters, highly effective heat exchangers are required. The system is further constrained by the size and weight requirements (< 110 L and < 30 kg), the processing rate (> 200 L/s), the required purification efficiency (> 90%), and the limited allowable power consumption (< 2 kW). These constraints lead to a heat-exchange effectiveness requirement of 95%, a value not attainable with conventional technology.

Working within the above constraints, MesoSystems' TCAP system achieved heat-exchange effectiveness up to 97% using a microchannel architecture with wall thicknesses on the order of 25 μm and channel thicknesses of 700 to 800 μm. The final device parameters were established using a rigorous design that investigated channel and wall dimensions, operating temperatures, flow rates, and materials of construction. One of the resultant MesoSystems TCAP prototypes is shown in Fig. 13.8. The device is less than 50 L in volume, less than 30 kg in mass, consumes about 1 kW of power, and has a heat-exchanger effectiveness of 95%.

One of the larger-scale systems incorporating mass transfer in multiphase microchannel processing is the absorption heat pump being developed for the US Army by OSU and PNNL [95]. The application is for portable cooling at 250 W

Fig. 13.8 A TCAP prototype developed by MesoSystems (used with permission from ref. [94]).

for man portable applications and 5000 W for cooling mobile command units and shelters. Target weights are 1.8 kg and 132 kg, respectively. A 250-W breadboard system has been successfully demonstrated that uses a single effect, absorption cycle heat pump with a heat-rejection temperature close to 50 °C. Multiphase processing is performed in microchannel wicking absorbers, microchannel wicking rectifiers, and fractal desorbers. The development of microchannel technology is expected to allow the realization of portable heat-actuated heat pumps, since conventional technologies are too large in size and are highly orientation dependent. The use of heat-actuated heat pumps significantly reduces the noise and reduces the electrical load by a factor of ten when compared to conventional compressor-driven heat pumps.

NASA has funded the development of a system that produces propellant from resources found on Mars, while simultaneously regenerating breathable air for life support. In this in situ propellant production system (ISPP), CO_2 and H_2O (from life support, the Martian atmosphere and landscape) will be converted to CH_4 and O_2 via microchannel Sabatier and reverse WGS reactors. The resultant CH_4 and O_2 can also be used to generate electrical power for vital life-support systems, making this capability central to a manned outer space infrastructure [96]. In order to achieve concentrations of CO_2 that can be reacted, a microchannel thermochemical compressor is being employed to increase the CO_2 partial pressure from $\sim$10 to $\sim$100 kPa, while microchannel separators are being employed for water recovery.

13.8 Future Directions

As can be seen from the above analysis, the application of microtechnology in North American industry has just begun. As research continues on many fronts, more opportunities will be identified where microtechnology can add value to industrial operations through increased efficiency, increased yields, or higher throughput, all methods of adding value in conventional systems. However, due to the orders of magnitude change in controlling mechanisms, microtechnology has shifted the paradigms of engineering and enables less conventional methods of achieving value. The unprecedented ability to control hazardous thermal runaway has made some reactions safer and allowed previously untenable reaction pathways to be pursued, such as in the production of hydrogen peroxide. The compact nature of microtechnology allows distributed production, lower capital costs, and when coupled with the low residence times, on site just-in-time manufacturing. For some applications, such as in the pharmaceutical industry, the compact nature allows easier production at extreme operating conditions. The method of scaling by "numbering up" the channels provides for increased safety due to the massive parallelism, while for some applications it provides a means of avoiding costly pilot plant and scale-up studies. In multiphase systems, the small channels allow interfacial forces to dominate, enabling processes such as phase separation, distillation, and pervaporation to become orientation independent. For military, space, and portable applications, the compact nature, increased safety, and orientation independence allow the deployment of unit operations and systems such as onboard fuel reformers. Ultimately, economics will determine the usefulness of microreactors in industrial applications, but as we have seen already, for certain applications, the payoff is obvious.

Just as in the microelectronics industry, it is anticipated that microtechnology will allow shifts in operations of large-scale systems. One such system is health care. Health-care delivery to large segments of the population in the United States and worldwide is under great financial strain due to an unprecedented rise in health-care costs in traditional health care institutions (i.e., hospitals and clinics). Current trends in healthcare delivery in Japan, Europe, and the United States point toward an ever-increasing fraction of the health care services delivered in the home environment to help potentially tame the overall cost of health care. Biomedical and pharmaceutical industries have noticed this trend and are developing business approaches that are now producing new products for testing, diagnostics, and simple health-care procedures that are specifically targeted for home use. Microscale devices can play an important role in this developing market, which is estimated at over $10 billion in the US alone. There are, however, uncompromising attributes that new products slated for home health-care delivery must have. These products have to be simple to use, inexpensive, readily available, self-calibrating, and equipped for network information sharing. Virtually none of these attributes apply for products that are used in traditional institutions of health care. This is an enormous challenge and opportunity for microscale devices since real needs and markets already exist, and these product attributes are not new for microscale devices.

At the same time, it must be noted that microtechnology is not a panacea for all industry. Care must be exercised when determining the applicability of microtechnology, because all systems do not benefit equally. For instance, mildly exothermic or endothermic reactions may be better pursued in conventional reactors. The entire costs and the entire scope of functionality must be considered. Functionality considerations include items such as the purity of feed components or potential for solids precipitation, which can in some instances result in channel plugging. Consideration of the entire costs include fabrication methods, which are currently expensive and can take a substantial amount of time to complete. As a result, in situations where size and weight are not factors, it may be more economical to use standard equipment for completing the same task at a fraction of the cost, even though it may be an order of magnitude larger. These cost considerations will become less of an issue as microtechnology is implemented and manufacturing costs decrease through improved fabrication techniques.

References

1 A. L. Y. Tonkovich, C. J. Call, D. M. Jimenez, R. S. Wegeng, M. K. Drost, *AIChE Symposium Series*, **1996**, *310*, 119–125.

2 T. S. Ravigururajan, J. Cuta, C. E. McDonald, M. K. Drost, *Proceedings of the 31st National Heat Transfer Conference*, **1996**, *7*, 157–166.

3 T. S. Ravigururajan, J. Cuta, C. E. McDonald, M. K. Drost, *Proceedings of the 31st National Heat Transfer Conference*, **1996**, *7*, 167–178.

4 A. L. Y. Tonkovich, C. J. Call, D. M. Jimenez, R. S. Wegeng, M. K. Drost, *Proceedings of the 31st National Heat Transfer Conference*, **1996**, *7*, 167–178.

5 C. J. Call, M. K. Drost, R. S. Wegeng, *AIChE Symposium Series*, **1996**, 310.

6 W. E. TeGrotenhuis, R. Cameron, M. G. Butcher, P. M. Martin, R. S. Wegeng, 10th *Sympos. Sep. Sci. Technol. Ener. Appl.*, Gatlinburg, TN, October, **1997**.

7 M. K. Drost, M. Friedrich, *Proceedings IECEC* (**1997**), 32nd, 1271–1274.

8 D. W. Matson, P. M. Martin, A. L. Y. Tonkovich, G. L. Roberts, SPIE *Proceedings – The International Society for Optical Engineering* (**1998**), 3514 (Micromachined Devices and Components IV), 386–392.

9 C. J. Cullen, R. C. R. Wootton, A. J. de Mello, *Curr. Opinion in Drug Discov. Dev.* **2004**, *7*(6), 798–806.

10 K. B. Mogensen, H. Klank, J. P. Kutter, *Electrophoresis* **2004**, *25*(21/22), 3498–3512.

11 C. K. Fredrickson, Z. H. Fan, *Lab Chip* **2004**, *4*(6), 526–533.

12 P. G. Vahey, S. A. Smith, C. D. Costin, Y. Xia, A. Brodsky, L. W. Burgess, R. E. Synovec, *Anal. Chem.*, **2002**, *74*(1), 177–184.

13 M. V. Koch, *Abstracts of Papers, 222nd ACS National Meeting*, Chicago, August 26–30, **2001**.

14 R. Babyak, in *Appliance Design*, June 6, **2000**.

15 Y. Zhao, M. M. Ohadi, R. Radermacher, *Microchannel Heat Exchangers with Carbon Dioxide*, final Report for the Air Conditioning and Refrigeration Technology Institute, ARTI-21CR/10020-01.

16 Modine Manufacturing Company, *ASHRAE Winter Meeting*, Feb. 5–9, **2005**.

17 M. A. Wilson, K. P. Recknagle, K. P. Brooks, *Proceedings of the AIChE Spring Meeting*, Atlanta, GA, April 10–14, **2005**.

18 L. Silva, R. Arora, A. Y. Tonkovich, *Proceedings of the AIChE Spring National Meeting*, New Orleans, LA, **2004**.

19 G. Whyatt, Pacific Northwest National Laboratory, *personal communication*.

20 A. Y. Alharbi, D. V. Pence, R. N. Cullion, *J. Fluids Eng. – ASME Trans.*, **2003**, *125* (6), 1051–1057.

21 A. Y. Alharbi, D. V. Pence, R. N. Cullion, *J. Heat Transfer – ASME Trans.*, **2004**, *126* (5), 744–752.

22 G. Veser, *Chem. Eng. Sci.*, **2001**, *56*, 1265–1273.

23 A. L. Y. Tonkovich, S. P. Fitzgerald, J. L. Zilka, M. J. LaMont, Y. Wang, D. P. VanderWiel, R. S. Wegeng, *Proceedings of the International Conference on Microreaction Technology, 3rd (IMRET 3)*, Frankfurt, Apr. 18–21, **1999**, 364–371.

24 R. S. Wegeng, S. D. Rassat, V. S. Stenkamp, W. E. TeGrotenhuis, D. W. Matson, M. K. Drost, V. V. Viswanathan, *US Patent*, 6,630,012.

25 W. Tegrotenhuis, S. Stenkamp, and A. Twitchell, in *Microreactor Technology and Process Intensification* (Eds.: Y. Wang and J. Holladay), American Chemical Society, Washington, DC, **2005**.

26 W. E. Tegrotenhuis, V. S. Stenkamp, *Proceedings of the First Internal Conference on Microchannels and Minichannels (ICMM)*, Rochester, NY, April 24–25, **2003**.

27 S. Ramprasad, S. Forrest, J. Palmer, *Proceedings of the AIChE National Meeting*, New Orleans, LA, April 1, **2003**.

28 A. L. Tonkovich, L. Silva, R. Arora, T. Hickey, *Proceedings of the Eighth International Conference on Microreaction Technology (IMRET 8)*, Atlanta, GA, **2005**.

29 W. E. TeGrotenhuis, R. A. Dagle, V. S. Stenkamp, *AIChE 2005 Spring National Meeting*, Atlanta, GA, April 10–14, **2005**.

30 G. Kleiner, *Medical News Today*, February 3, **2004**.

31 A. D. Stroock, G. M. Whitesides, *Acc. Chem. Res.*, **2003**, *36*, 597–604.
32 S. Deshmukh, D. Vlachos, *AIChE Spring National Meeting*, April 10–14, **2005**.
33 J. B. Knight, A. Vishwanath, J. P. Bordy, R. H. Austin, *Phys. Rev. Letters*, **1998**, *80*(17).
34 K. Jensen, *Nature* **1998**, *393*, 735–737.
35 G. A. Whyatt, C. M. Fischer, J. M. Davis, *AIChE 2004 Spring National Meeting*, April 25–29, **2004**.
36 T. Kawakatsu, Y. Kikuchi, M. Nakajima, *J. Am. Oil Chem. Soc.*, **1997**, *74*, 317–321.
37 C. Charcosset, H. Fessi, *Membrane Emulsification and Microchannel Emulsification Processes*, **2005**, *21*(1).
38 D. Dendukuri, K. Tsoi, T. A. Hatton, P. S. Doyle, *Langmuir*, **2005**, *21*(6), 2113–2116.
39 R. D'Aquino, *Chem. Eng. Progr.*, **2004**, *100*(8), 7–10.
40 J. McDaniel, Velocys, LLC, *personal communication.*
41 K. F. Jensen, *Chem. Eng. Sci.*, **2001**, *56*, 293–303.
42 K. M. VandenBussche, *Symposium on Modeling of Complex Processes*, College Station, TX, March 1, **2005**.
43 A. Lawal, E. Dada, R. Halder, S. Tadepalli, W. Lee, Y. Voloshin, *Proceedings of the AIChE Spring Meeting*, Atlanta, GA, April 10–14, **2005**.
44 M. Freemantle, *Chem. Eng. News*, **2004**, *82* (41), 39–43.
45 T. Mazanec, T. Dritz, F. Daly, *Proceedings of the AIChE Spring National Meeting*, Atlanta, GA, **2005**.
46 K. Jarosch, *Proceedings of the 226th ACS National Meeting*, September 9th, **2003**.
47 Y. Wang, J. Hu, J. Cao, D. Elliott, J. White, D. VanderWiel, T. Mazanec, *Catcon 2003*, Houston, May 5–6, **2003**.
48 T. Mazanec, *Proceedings of the AIChE Spring Meeting*, Atlanta, GA, April **2005**.
49 Y. Chan, Bristol-Myers Squibb, *personal communication.*
50 A. Lawal, D. Kientzler, L. Achenie, R. Halder, R. S. Besser, S. Tadepalli, W. Lee, *Proceedings of the AIChE Spring Meeting*, Atlanta, GA, April 10–14, **2005**.
51 G. N. Jovanovic, A. Al-Dhubabian, *Micro Nano Breakthrough Conference*, Portland, OR, July 25–28, **2005**.
52 K. Venkataraman, J. M. Redenius, L. D. Schmidt, *Chem. Eng. Sci.* **2002**, *57*, 2335–2343.
53 C. Weeler, A. Jhalani, E. Klein, S. Tummala, L. Schmidt, *J. Catal.*, **2004**, *223*, 199.
54 D. Neumann, M. Kirchhoff, G. Veser, *Catal. Today*, **2004**, *98*(4), 565–574.
55 J. Hu, Y. Wang, D. VanderWiel, C. Chin, D. Palo, R. Rozmiarek, R. Dagle, J. Cao, J. Holladay, E. Baker, *Chem. Eng. J.*, **2003**, *93*, 55.
56 G. Xia, J. Holladay, R. Dagle, E. Jones, Y. Wang, *Chem. Eng. Techn.* (in press).
57 Y. Wang, Y.-H. Chin, R. T. Rozmiarek, B. R. Johnson, Y. Gao, J. Watson, A. L. Y. Tonkovich, D. P. VanderWiel, *Catal. Today*, **2004**, *98*, 4.
58 J. Bravo, A. Karim, T. Conant, G. Lopez, A. Datye, *Proceedings of the 7th International Conference on Microreaction Technology (IMRET 7)*, Lausanne, Switzerland, September 7–10, **2004**.
59 J. D. Holladay, Y. Wang, E. O. Jones, *Chem. Rev.* **2004**, *104*(10), 4767–4789.
60 K. Shah, X. Ouyang, R. S. Besser, *Chem. Eng. Technol.* **2005**, *28*(3), 303–313.
61 R. Changrani, D. Gervasio, R. Koripella, S. P. Rogers, S. R. Samms, S. Tasic, *6th International Conference on Microreaction Technology (IMRET 6)*, New Orleans, LA, **2002**, p. 108.
62 D. Gervasio, S. Rogers, R. Koripella, S. Tasic, D. Zindel, R. Changrani, C. K. Dyer, J. Hallmark, D. Wilcox, *Ceramic Transactions*, **2002**, *127*, 157–166.
63 W. L. Johnson II, C. B. Phillips, Z. Chen, T. S. Ransom, L. T. Thompson, Jr., *Abstracts of Papers, 226th ACS National Meeting*, New York, NY, September 7–11, **2003**.
64 R. S. Upadhye, J. D. Morse, D. A. Sopchak, M. A. Havstad, R. T. Graff, *High aspect ratio chemical microreactor* PCT Int. Appl. **2005**.
65 H. G. Park, W. T. Piggott, J. Chung, J. D. Morse, M. Havstad, C. P. Grigoropoulos, R. Greif, W. Benett, D. Sopchak, R. Upadhye, *Conference Proceedings, Hydrogen and Fuel Cells Conference and Trade Show*, Vancouver, BC, Canada, June 8–11, **2003**, 269–277.

66 J. D. Holladay, E. O. Jones, R. A. Dagle, G. Xia, C. Cao, Y. Wang, In *Microreactor Technology and Process Intensification* (Eds.: Y. Wang and J. Holladay), American Chemical Society, Washington, DC, **2005**.
67 J. D. Holladay, J. Wainright, E. O. Jones, S. R. Gano, *J. Power Sources* **2004**, *130* (1/2), 111–118.
68 D. R. Palo, J. D. Holladay, R. A. Dagle, Y.-H. Chin, In *Microreactor Technology and Process Intensification* (Eds.: Y. Wang and J. Holladay), American Chemical Society, Washington, DC, **2005**.
69 D. R. Palo, J. D. Holladay, R. A. Dagle, *Fuel Cell Seminar,* San Antonio, TX, November 2–5, **2004**.
70 S. Carranza, S. D. Rosenberg, D. B. Makel, *Proceedings of the 36th Heat Transfer and Fluid Mechanics Institute* **1999**, 36th, 267–280.
71 S. Carranza, D. B. Makel, B. Blizman, B. J. Ward, *AIP Conference Proceedings* **2005**, *746*, 1229–1236.
72 T. J. Campbell, F. S. Thomas, M. L. Bostian, A. H. Shaaban, *5th Logistic Fuel Reforming Conference*, Panama City, FL, January 25–26, **2005**.
73 J. Martin, T. Armstrong, A. Virkar, *Fuel Cell Seminar,* San Antonio, TX, November 1–5, **2004**.
74 A. Kulprathipanja, J. C. Poshusta, E. N. Schneider, *Fuel Cell Seminar,* San Antonio, TX, November 1–5, **2004**.
75 Q. Ming, P. Irving, *Abstracts of Papers, 226th ACS National Meeting*, New York, NY, United States, September 7–11, **2003**.
76 Q. Ming, T. Healey, L. Allen, P. Irving, *Catal. Today* **2002**, *77*(1/2), 51–64.
77 G. A. Whyatt, C. M. Fischer, J. M. Davis, *Proceedings of the AIChE Spring National Meeting*, New Orleans, LA, April 25–29, **2004**, 932–939.
78 W. E. Tegrotenhuis, D. L. King, K. P. Brooks, B. J. Golladay, R. S. Wegeng, *6th International Conference on Microreaction Technology (IMRET 6)*, New Orleans, LA, March 10–14, **2002**.
79 D. L. King, K. P. Brooks, C. M. Fischer, L. R. Pederson, G. C. Rawlings, V. S. Stenkamp, W. E. Tegrotenhuis, R. S. Wegeng, G. A. Whyatt, In *Microreactor Technology and Process Intensification* (Eds.: Y. Wang and J. Holladay), American Chemical Society, Washington, DC, **2005**.
80 X. Ouyang, R. Besser, *J. Power Sources*, **2005**, *141*(1), 39–46.
81 http://www.battelle.org/news/05/02-16-05BradleyFuelCell.stm
82 A. Y. Tonkovich, S. Perry, Y. Wang, D. Qiu, T. LaPlante, W. A. Rogers, *Chem. Eng. Sci.*, **2004**, *59*(22/23), 4819–4824.
83 T. Dritz, J. McDaniel, E. Daymo, *Micro Nano Breakthrough Conference*, Portland, OR, July 28–29, **2004**.
84 L. R. Arana, S. B. Schaevitz, A. J. Franz, M. A. Schmidt, K. F. Jensen, *J. MEMS Systems*, **2003**, *12*(5), 600–612.
85 C. J. Call, M. R. Powell, M. Fountain, A. S. Chellappa, *3rd Annual International Symposium on Small Fuel Cells and Battery Technologies for Portable Power Applications*, Washington, DC, **2001**.
86 A. S. Chellappa, M. R. Powell, M. Fountain, C. J. Call, N. A. Godshall, *Abstr. Pap. Am. Chem. Soc.*, **2002**, *224*, 123-FUEL.
87 J. C. Ganley, E. G. Seebauer, R. I. Masel, *J. Power Sources* **2004**, *137*, 53–61.
88 E. Seebauer, J. Ganley, R. Masel, *Proceeding of the Eighth International Conference on Microreaction Technology (IMRET 8)*, Atlanta, GA, **2005**.
89 C. Miesse, R. Masel, S. Prakash, *Proceeding from the Eighth International Conference on Microreaction Technology (IMRET 8)*, Atlanta, GA, **2005**.
90 C. J. H. Jacobsen, S. Dahl, B. S. Clausen, S. Bahn, A. Logadottir, J. K. Norskov, *J. Am. Chem. Soc.*, **2001**, *123*, 8404.
91 P. Aghalayam, Y. K. Park, D. G. Vlachos, *AIChE J.*, **2000**, *46*, 2017.
92 A. Mhadeshwar, D. Viachos, S. Deshmukh, *Proceeding of the Eighth International Conference on Microreaction Technology (IMRET 8)*, Atlanta, GA, **2005**.
93 A. H. Al-Raie, *Desulfurization of Thiophene and Dibenzothiophene Peroxide in a Photochemical Microreactor,* Masters Thesis, Oregon State University, Corvallis, OR, March **2005**.
94 M. R. Powell, S. H. Hong, C. J. Call, *NBC Defense Collective Protection Conference* Oct. 29–31, **2002** Orlando, Florida.
95 V. S. Stenkamp, W. E. TeGrotenhuis, B. Q. Roberts, M. D. Flake, J. M. Davis, D. D.

Caldwell, *Proceedings of the AIChE Spring Meeting*, Atlanta, GA, April 10–14, **2005**.

96 K. P. Brooks, S. D. Rassat, R. S. Wegeng, V. S. Stenkamp, W. E. Tegrotenhuis, D. D. Caldwell, *Proceedings of the AIChE Spring National Meeting*, New Orleans, LA, March 10–14, **2002**.

14
Industrial Applications in Europe

Thomas Bayer and Markus Kinzl, Siemens AG, A&D SP Solutions Process Industries, Industriepark Höchst, Frankfurt am Main, Germany

Abstract

In recent years low-cost engineered components with structural features in the range of micrometers to millimeters ("microstructured objects") have become available in a wide variety of chemically resistant materials such as glass, stainless steel, ceramics, polymers, alloys and graphite (and no longer only silicon). These new objects offer stimulating perspectives for the development of a new generation of highly original process components and systems for chemical process development and production. When applied in laboratory plants, microstructured components allow the efficient gathering of experimental data for process development. They give access to new routes for process transfer into production scale. Finally, in many cases the application of these devices in production plants can improve quality, safety and effectiveness significantly. The traditional way of process development from lab to production (scale-up) is time-consuming and costly. Very often a lot of problems are linked to scale-up. This is due to a change in scale, e.g. from millimeters to centimeters resulting in dramatic different heat- and mass-transfer characteristics. Therefore in general the reaction design has to be adapted to the performance of the apparatus used, e.g. solutions are diluted or dosing is extended. Often reduced qualities or reduced selectivities cannot be avoided. Microstructured devices could overcome these drawbacks.

Keywords

Microreaction technology, industrial applications, microstructured devices, multiscale design

Advanced Micro and Nanosystems Vol. 5. Micro Process Engineering. Edited by N. Kockmann

ISBN: 3-527-31246-3

14.1
Overview of Current Activities in Europe

Over the last 15 years, substantial research on small-scale structured devices for chemical applications has been undertaken, and a host of academic studies, as well as the eight successful editions of the conference series IMRET (International Conference on Microreaction Technology), have established a solid scientific basis for the fabrication and analysis of individual (generally unconnected) units. A number of reference books are now available [1, 2] as a substantial contribution to the already well-established general area of process intensification [3]. Although much of the research effort in this area has been undertaken in Europe, significant contributions have also come from other regions, in particular the United States and Japan.

Talking about industrial applications of microreaction technology (MRT) the key question to be asked is "Why are there only so few large-scale industrial applications today?"

Generally, the development of new technologies before the occurrence of broad industrial applications takes about 20 to 25 years. Until now mainly physicists, microtechnicians, biologists, and chemists have contributed to the development of the technology and have worked with microstructured devices – but an adequate contribution of chemical engineers is only recently taking place.

The reason for the lack of comprehensive design rules and operational instructions is that the commercial availability of these devices started only recently. Industrial companies in Europe have just begun to use these devices for their daily work and not only for the evaluation of the new technology.

But there are more reasons retarding the breakthrough of microreaction technology. The continuous operation mode of microstructured devices is in contrast to the traditional batch- and semibatch-dominated experimental methods in chemical laboratories. Therefore, synthesis recipes have to be adapted. Furthermore, reactions with solids or solid formation are critical when microstructured devices are used (although they should not be excluded a priori). Also, the transport of highly viscous fluids through microchannels is limited due to the resulting pressure drops.

But, nevertheless, the advantages of microreaction technology are well known and emphasized by industrial users.

Components with structural features in the range of micrometers to millimeters ("microstructured objects") are now available in a wide variety of chemically resistant materials such as glass, stainless steel, ceramics, polymers, alloys and graphite. Therefore there is practically no restriction for chemical reactions with respect to material resistance in principle, but long-term experience still does not exist. A good overview of existing components and manufacturers, especially in Germany, can be found on the homepage of the Industrial Platform microchemtec: http://www.microchemtec.de.

The so far reported industrial applications of chemical microprocessing refer to scouting tests with bench-scale plants, extended test programs and screening studies, pilot studies with high-throughput plants, and a few reports on production plants. In addition, most of the large and medium chemical and pharmaceutical companies carry out tests with microstructured devices and systems at present.

The usage of microstructured devices offers a new possibility for the development of chemical processes without limitations in heat and mass transfer. Therefore, the scale-up risks that are well known in process development (see also Section 14.4) can be reduced. In many cases the transfer of MRT-based chemical processes from laboratory to technical scale can be realized without a numbering-up, since microstructured devices for high throughputs are commercially available today. For example, micromixers and heat exchangers designed for mass streams of several tons per hour can be purchased, e.g., from Heatric, Forschungszentrum Karlsruhe (FZK), Institut für Mikrotechnik Mainz (IMM) or Ehrfeld Mikrotechnik BTS.

14.2 Industrial Applications in Process Development

The motivation for microreaction technology activities in chemical laboratories is manifold. Predominantly, investigations aim at process intensification and the optimization of conventional batch processes; especially in terms of product selectivities and yields. In this chapter details on some selected laboratory projects are reported. The authors confine themselves to describe experimental studies executed at Siemens A&D SP, Frankfurt, Germany (formerly Axiva), since they were involved in these projects themselves. Of course, the selection is only a small extract of the many activities that have become known in recent years. A much more complete overview can be found in [2] with about 900 citations from the literature.

Microstructured devices have turned out to be applicable in a wide range of reaction conditions, as can be seen from Table 14.1. The table gives an overview of the characteristics of more than 20 microreaction technology laboratory projects that were executed at Siemens A&D SP in recent years.

A promising actual project at Siemens A&D SP deals with the production of a specialty chemical, which is synthesized by a very fast mixing-sensitive liquid-phase reaction. Characteristic reaction times are below 1 s. It is well known that the formation of undesired byproducts can be avoided only, if mixing is significantly faster than reaction, since reactants have to be contacted immediately in the optimum ratio. Thus, in the above-mentioned case mixing times of not more than some few 100 ms are required. In order to accelerate mixing, laboratory experiments were carried out with a micromixer instead of a conventional

Table 14.1 Overview on microreaction technology projects executed at Siemens A&D SP.

Nature of chemical reaction	Homogeneous liquid-phase reactions, gas–phase reactions, two-phase liquid–liquid reactions, gas–liquid reactions
Temperature	for reactions in liquid phase: –30 °C to +210 °C for reactions in gas phase: up to 600 °C
Pressure	up to 60 bars
Average residence times in microstructured devices	Between few milliseconds in micromixers and 20 minutes in micro heat exchangers

static mixer as used in the existing technical process. The experimental setup can be seen in Fig. 14.1. Since the first proof-of-principle experiments already showed a remarkable increase in product yield, a parameter study was started. By the variation of reaction temperature, reactant throughput and reactant concentrations, the optimum parameter set could be identified. Product yield could finally be increased by up to 15%. With the laboratory plant roughly 7 kg of the product solution can be produced per hour. The transfer from laboratory to technical scale requires an increase of the throughput by a factor of more than 1000. Micromixers for throughputs in the range of tons per hour are already available today and all experimental results indicate that a numbering-up of single mixers will not be necessary.

The importance of fast reactant mixing can also be demonstrated by means of the continuous radical polymerization of acrylates in solution [4]. At the beginning of process development at Axiva, monomers and initiator were mixed by a conventional static mixer. Unfortunately, the first experiments had to be stopped after few hours of polymerization, since insoluble deposits of high molecular weight polymers caused plugging of the reactor. In order to decrease fouling tendency, interdigital micromixers were used for reactant mixing. As a consequence, acrylate polymerization could be run for hours without plugging problems. The key was a change in mixing physics affected by multilamination. The improved mixing process influenced the whole polymerization, particularly the formation of the starting radicals. The pilot plant was operated with a throughput of about 8 kg/h, corresponding to a theoretical annual capacity of about 60 tons of polymer. This is a typical example for a multiscale plant, since only mixing is carried out in a micromixer, whereas polymerization takes place in a tube reactor with an inner diameter of some millimeters. With proceeding reaction and, therefore, reduced monomer concentration and reaction rate, less reaction heat is produced. Thus, heat removal becomes simplified and tube diameter can be increased.

A major part of laboratory projects dealing with microreaction technology aims at process intensification. Chemical processes can be intensified by raising space-time yields as a consequence of higher reaction rates, less reactant dilution or even the complete waiving of solvents.

Process intensification can be demonstrated by means of the production process of the industrial mass chemical monoethylene glycol (MEG) which is synthesized

Fig. 14.1 Laboratory plant for microreaction experiments at Siemens A&D SP, Frankfurt (Germany).

by the hydrolysis of ethylene oxide (EO), catalyzed by small amounts of sulfuric acid. In the technical scale the reaction is carried out at a temperature of 200 °C and pressures of more than 10 bars in continuously stirred vessels with a volume of several cubic meters. In order to prevent the formation of consecutive products, especially di-ethylene glycol (DEG) and tri-ethylene glycol (TEG), water and EO are fed in the molar ratio of 20 : 1 [5]. Since the separation of the excess water from the product is very costly, a reduction of the water excess would be beneficial but, unfortunately, it is known to be accompanied by a decrease in MEG selectivity. Siemens started an internal project in which the influences of reactant ratio and improved temperature control on MEG selectivity were investigated. Unfortunately, it was found that even the usage of microreactors did not allow a water-excess reduction without a loss in MEG selectivity and yield. But experiments demonstrated that space-time yield could be increased by a factor of 30. After all, this result is due to the outstanding heat-exchange performance of microstructured reactors. In the conventional process reaction rates must be limited in order to make the removal of reaction heat from the stirred reaction vessels possible. Reaction-rate limitation is realized by running the highly exothermic EO hydrolysis continuously at high conversion and, therefore, low reactant concentrations. A completely different situation is found when the process is carried out in a microreactor. Due to the effective heat removal from microstructured devices reaction rates can even be increased, e.g., by higher temperatures or sulfuric acid concentrations. As a consequence, characteristic reaction times can be reduced to several seconds and space-time yields increase significantly. Particularly when risky reactions such as the hydrolysis of EO are considered, the possibility of a reactor volume reduction is very important. As a consequence, fewer safety arrangements are necessary and simplified permit procedures could lead to a remarkable cost advantage.

Another field of micromixer application is emulsification. Various experimental studies deal with the influence of temperature, nature of surfactant, surfactant concentration and further parameters on emulsion stability and droplet-size distribution. Siemens A&D SP developed a laboratory-scale process for the continuous generation of a cosmetic emulsion that is traditionally produced in a stirred batch vessel. A micromixer was used for the emulsification of oil and water, before a thickener and perfume were added by means of conventional static mixers. Stability tests and the determination of droplet-size distribution showed that the emulsion quality was equal to that of the batch-manufactured product. Average droplet size (expressed in terms of the Sauter diameter) could be adjusted by a variation of throughput and mixing temperature. With a single micromixer (see Fig. 14.2) the production of up to 6 kg emulsion per hour was possible at a pressure drop of about 10 bar.

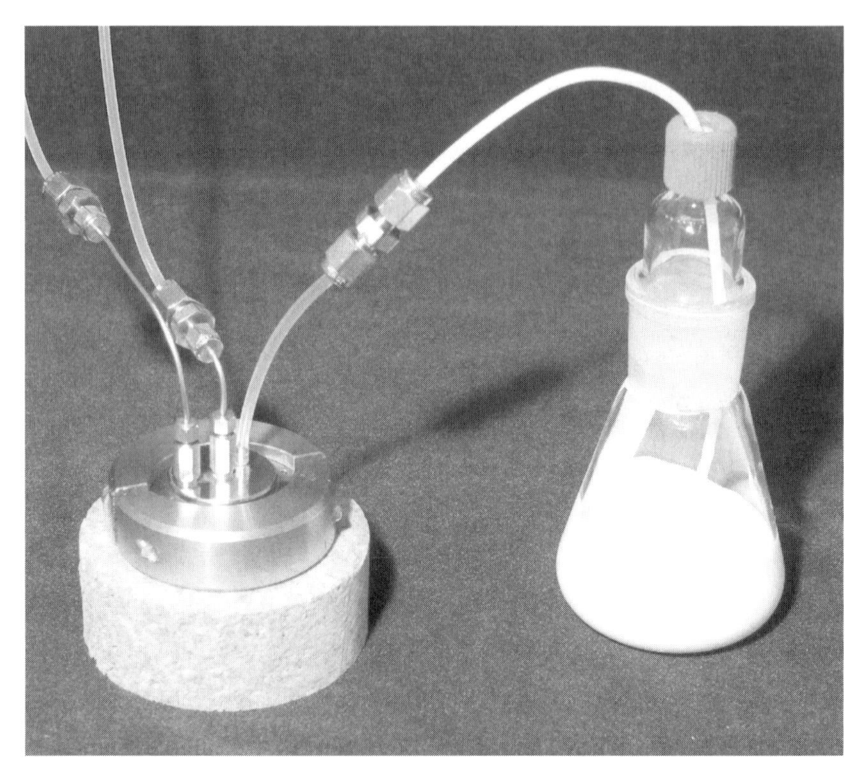

Fig. 14.2 Production of a cosmetic emulsion with a micromixer.

In another project, the necessary specific energy input for the generation of water–silicon oil emulsions was compared for the application of micromixing and conventional stirring. As can be seen from Fig. 14.3, micromixing requires a significantly lower specific energy input when emulsions with the same droplet Sauter diameter are produced [6].

The application of microstructured devices in gas-phase processes can be demonstrated by means of a radical gas-phase reaction that is run at temperatures of about 500 °C. Axiva compared the performance of a conventional preheater to that of a micro heat exchanger in a pilot plant unit for gas flow rates of about 1 m^3/h. The reaction mixture consisting of the reactants and a radical precursor is heated up to the desired reaction temperature. During heating, radicals are formed and undesired byproducts are generated. Therefore, fast heating is the key to low byproduct content. In the conventional process, preheating takes as long as the main reaction itself ($\approx$1 s). Furthermore, a temperature overshoot is observed. In the pilot plant trials at Axiva, two electrically heated microstructured heat exchangers (Fig. 14.4) were used in series for heating the gas stream.

The preheating took about 5 ms, which was 15 times faster than with the conventional equipment. Also, the desired temperature level was reached immediately without the occurrence of an overshoot (see Fig. 14.5). The microheat exchanger acts like a temperature switch – very fast and accurate. Conversion in the reaction zone could be enhanced by 25% for the investigated reaction and space-time yield was nearly doubled [7].

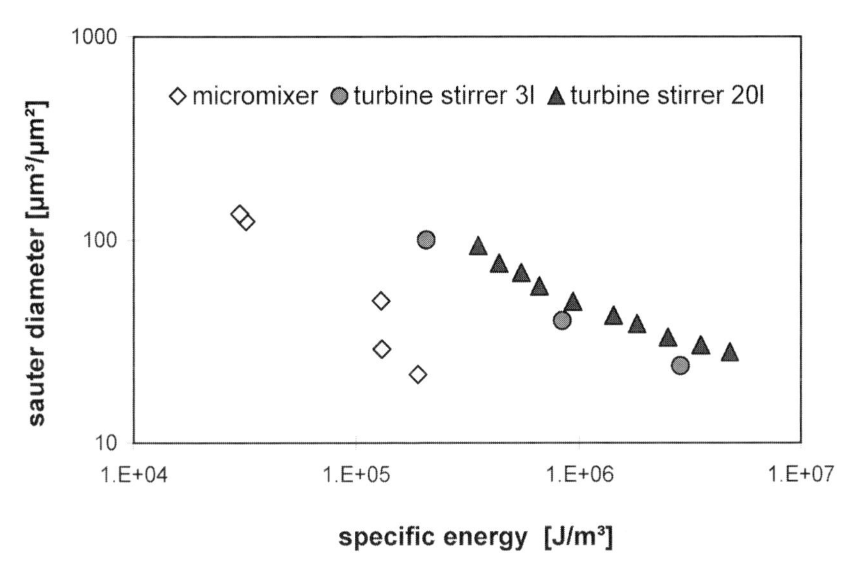

Fig. 14.3 Specific energy input for the generation of emulsions – comparison for micromixing and stirring.

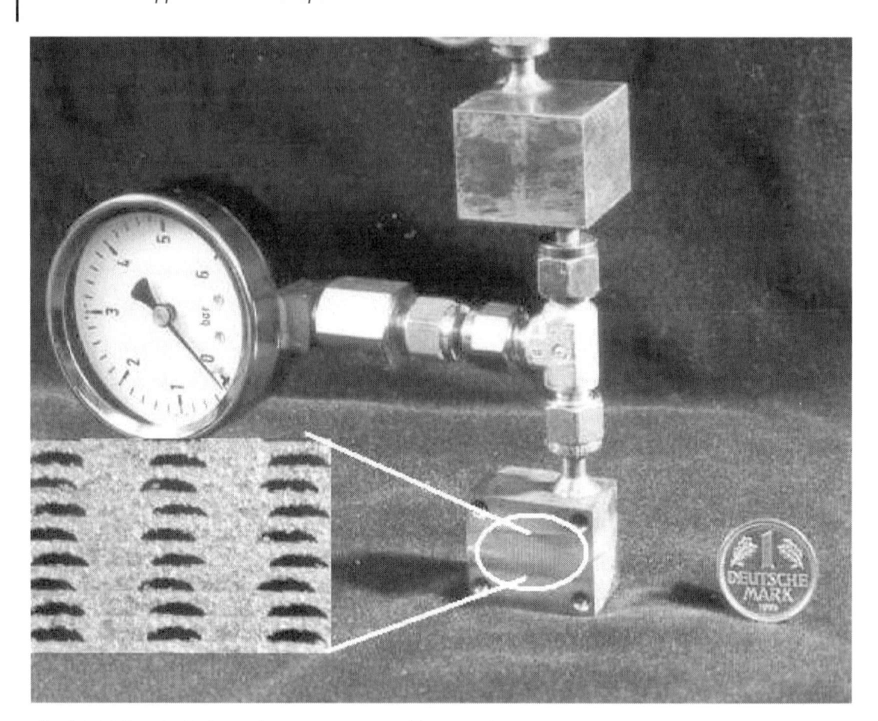

Fig. 14.4 Electrically heated microstructured heat exchangers.

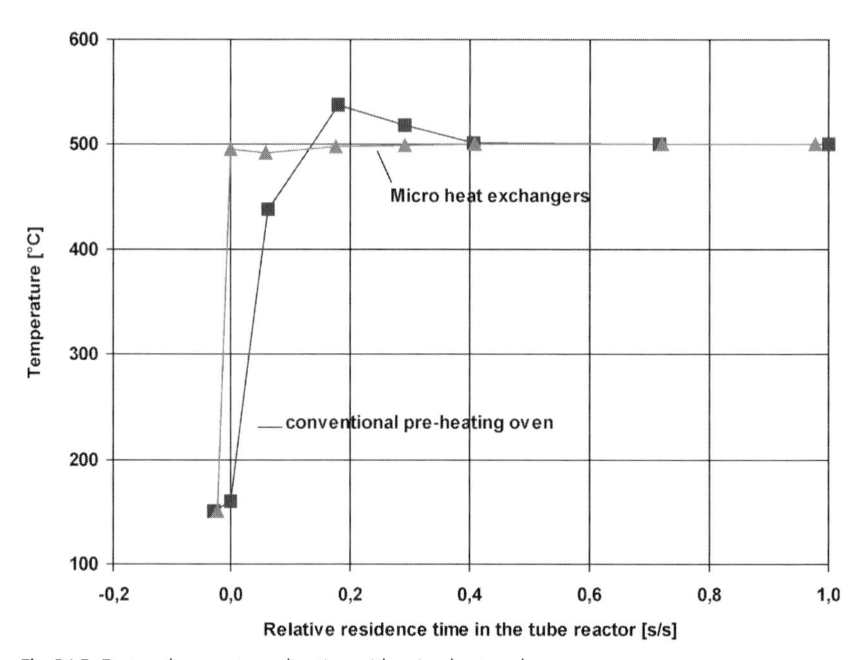

Fig. 14.5 Fast and accurate preheating with microheat exchangers.

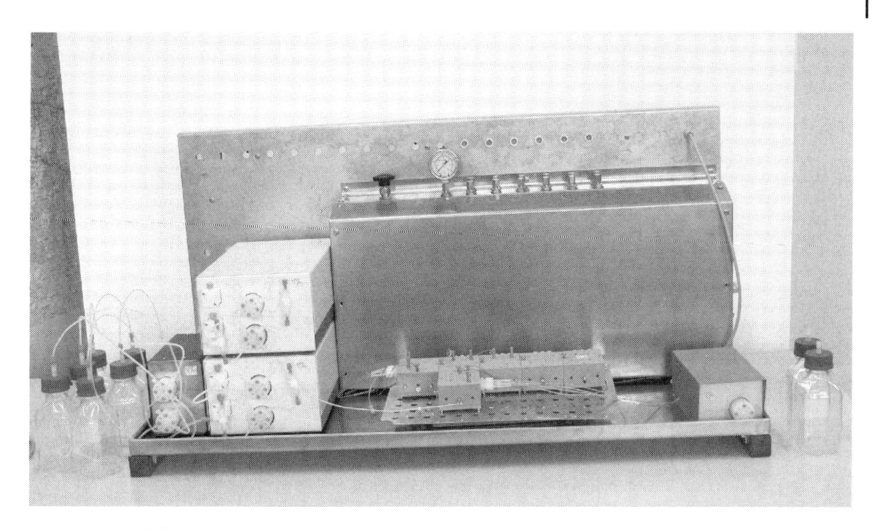

Fig. 14.6 Modular automated microreaction system.

When experiments are carried out in a continuously operated microtechnology laboratory setup, process development projects can be executed with very high efficiency. Extensive experimental studies can be completed within a short time, especially if laboratory equipment is automated. Currently, several companies and institutes are working on the development of these automated systems. Within a project funded by the German Government, Siemens, Merck (Darmstadt, Germany) and the Fraunhofer Institute for Chemical Technology (Pfinztal, Germany) developed a modular automated microreaction system for process development and the production of small product quantities in laboratory scale (several kg/h) [8]. The system, which is illustrated in Fig. 14.6, consists of modules for the dosage of liquid reactants, mixing, reaction, pressure control and online analytics. Due to its modular setup the system offers high flexibility. Data logging as well as the control of flow streams, temperature and pressure are managed by the Siemens process control system SIMATIC PCS7®.

Project results have shown the high potential of the modular system. In future, further effort will be necessary for the development of suitable microstructured sensors for the measurement of process data. They are necessary for the fully automated operation of the system that will allow the efficient gathering of experimental data. The first examples of sensors for the measurement of mass flows, temperatures and pressures have been developed within the project, but so far the integration into the microreaction system has not been completed.

14.3
Industrial Applications in Production Scale

Only a small number of examples on a production scale have so far been reported. Highlights from chemical applications and unit operations from other industrial applications are described in this chapter.

Merck, Darmstadt (Germany) was the first chemical company to use interdigital micromixers for bench-scale data gathering in order to build a production plant (see Fig. 14.7, [9]). Compared to the conventional batch process, the yield of a fast mixing-sensitive metal-organic reaction could be improved by 23% applying micromixing in laboratory-scale experiments. For production, home-made minimixers, matching rather classical static mixers than micromixers, were used due to the risk of fouling. The resulting fine-chemical production process ran successfully for five years. The yield was higher by 20% in comparison to the former batch production.

In 2001, Degussa AG, Düsseldorf and Uhde GmbH, Dortmund (both located in Germany) started the DEMiS project (Demonstration Project for the Evaluation of Microreaction Technology in industrial Systems) together with partners from academia (Max Planck Institute of Coal Research, Mülheim, Chemnitz University of Technology, Darmstadt University of Technology and University of Erlangen-Nuremberg). In this project, expertise in the areas of catalysis, process design, plant design and operation of chemical processes was brought together. With the support of the German Federal Ministry of Education and Research, DEMiS started up with a budget of about 4.5 million Euro. The main goal of the project is the investigation of the basic technical and operational feasibility of heterogeneously catalyzed reactions in microstructured reactors [10]. For scaling up to pilot scale, the gas-phase epoxidation of propylene to propylene oxide was selected. In this titanium silicate (TS-1) catalyzed reaction hydrogen peroxide acts as an oxidizing agent. Keeping in mind the difficult temperature control (reaction heat of –220 kJ/mol) and the possible decomposition of the labile oxidizing agent, the application of microreaction technology is undoubtedly advantageous for this challenging synthesis.

The reactor was realized in pilot-plant scale (see Fig. 14.8). The process concept provides a modular structure with regard to basic operations and capacity. It can also be used for other syntheses and allows the utilization of microreaction effects. Particularly (at least in principle), the safe operation in explosion regions is possible. The aim of further work is the operation of these microstructured reactors in such a way that for the development of other syntheses the pilot step becomes unnecessary and the direct transfer of the results from the laboratory scale into production is possible.

In the last three years, Clariant, Frankfurt (Germany) has established microreaction technology application know-how in an industrial environment. According to the concept of simultaneous engineering and processing, the development activities on this technology were rapidly expanded from laboratory to pilot-plant scale, and even a first concept for a production plant based on microreactors was planned at a very early stage.

Fig. 14.7 Production plant for the synthesis of a metal-organic chemical using minimixers (Merck, Darmstadt, Germany).

Clariant reported the use of an integrated chemical microprocessing system with a multilaminating mixer for azo pigment production in the pilot-scale range (see Fig. 14.9, [11]). The azo pigments produced in this plant displayed an improved color strength, while having the same transparency and brightness as the batch products. This result is due to the generation of much finer particles

Fig. 14.8 DEMiS reactor (Degussa AG, Düsseldorf, Germany).

Fig. 14.9 Pilot plant for azo pigment production with a multi-laminating mixer (Clariant, Frankfurt, Germany).

in the microreactor. In a second application, Clariant investigated a fast and mixing-sensitive reaction [12]. Bench-scale tests with an interdigital micromixer turned out to give yields that were higher by nearly 25% compared to the former batch-production process. Using the process parameters derived from these lab tests, a pilot-plant unit was built. The plant contains a caterpillar micromixer having larger internal dimensions than the interdigital device. As a result of the microdevice change the throughput could be increased up to 100 l/h at a pressure loss of 10 bar. The caterpillar micromixer provides high robustness, particularly regarding to low fouling sensitivity. Yields are still higher by 20% compared to the prior batch process.

For the above mentioned radical acrylate polymerization process, in the pilot-plant scale (see Section 14.2), Axiva prepared a concept study for a plant capacity of 2000 t/y on a prebasic engineering level using 28 parallel interdigital micromixers (Fig. 14.10). With the microstructured mixers available today only one mixer would be sufficient.

Much more is becoming visible today.

In March 2005, the Xi'an Huian Chemical Industrial Group China and the Institut für Mikrotechnik, Mainz (Germany) signed a R&D agreement for the investigation of the nitroglycerin production on a pilot-plant-scale level [13]. Since the manufactured nitroglycerin is expected to be used as a drug for acute cardiac infarction, product quality must be of the highest grade. The plant is ex-

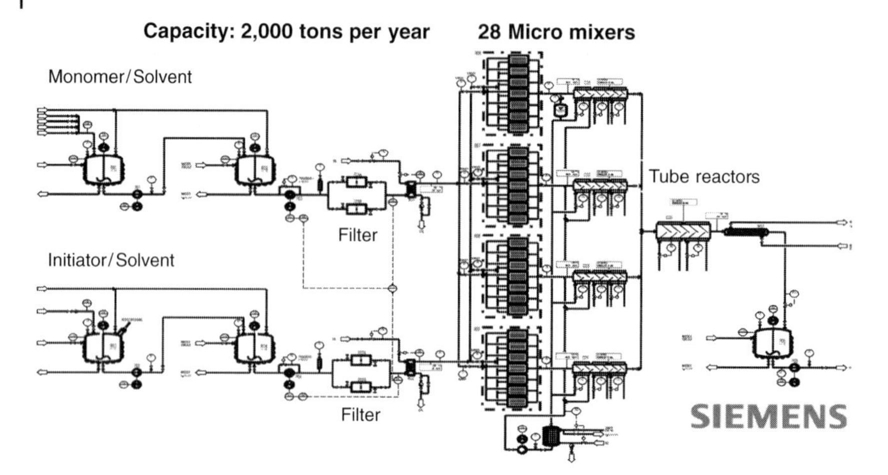

Fig. 14.10 Concept study for a polymerization plant with micromixers and a capacity of 2000 t/y (Siemens A&D SP, formerly Axiva).

pected to operate safely and fully automated. Environmental pollution is to be excluded by an advanced waste-water treatment and closed water cycle.

Synthacon GmbH, a sister company of CPC Systems GmbH, started a commercial multiproduct plant (capacity up to 20 tons/a) based on CPC's microreactor technology on the chemical industrial site of Leuna, Germany, in September 2005 [14].

Sigma-Aldrich, a major distributor and manufacturer of various chemical compounds for life-science research, uses a microreactor system for development and production of organic chemicals in its Switzerland facility Fluka [15].

There are also experiences with microstructured devices outside of chemical reactions.

The British company Heatric was established in 1985 to commercialize the design and manufacture of "micro/milli" scale heat-exchange core matrices called printed circuit heat exchangers (PCHEs), following several years of developmental work at the University of Sydney. PCHEs incorporate single-material (usually metal) matrices manufactured by diffusion bonding of plates. The inner diameters of the passages are typically in the range of some 100 micrometers to a few millimeters (see Fig. 14.11).

Compared to conventional heat exchangers PCHEs are compact, but they are not necessarily small, as can be seen from Fig. 14.12: single units of up to 100 tons have been manufactured, and clearly they are "compact" only in comparison to the 500-ton alternatives.

Since 1990, Heatric has supplied over 3000 tons of such matrix in hundreds of services. Many of them are used on offshore oil and gas platforms where the size and weight advantages of compact structures are of obvious major benefit. Other applications include liquefied natural gas (LNG), ethylene oxide, sulfuric

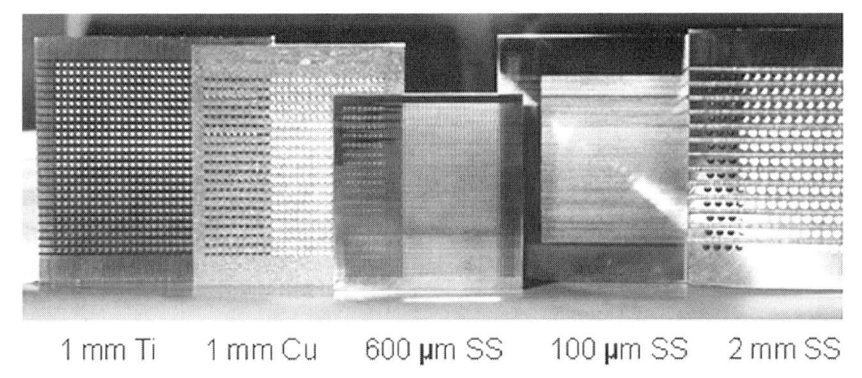

Fig. 14.11 Heat exchangers from Heatric (UK) with channel diameters in the range of micrometers to millimeters.

Fig. 14.12 Heat exchanger from Heatric (UK) with housing and connections.

acid, naphtha reforming, and caustic soda plants. The potential for application of PCHEs to temperature control in chemical reaction has also been recognized, with potential benefits anticipated in areas such as compactness, operability, efficiency, safety and cost.

In those duties where fouling does manifest itself, very successful cleaning procedures have been developed. These procedures include chemical cleaning, hydrokinetic cleaning, and so-called "gas puffing". In order to meet the needs of chemical industry, especially concerning chemical resistance, Heatric is developing heat exchangers made of stainless steel, titanium, Hastelloy as well as glass.

Another company working in the field of microreaction technology, especially the manufacturing of microstructured devices, is Atotech, Berlin (Germany). The company is a subsidiary of the French corporation Total and one of the world's largest suppliers of metallization production solutions for GMF (general metal finishing), PCB (printed circuit board) and semiconductor wafer applications.

Liquid-cooling units for microprocessors are an example for Atotech's prosperous products (see Figs. 14.13 and 14.14). Increasing performance of microprocessors (CPUs) goes along with a steep increase in thermal power evolution. Conventional air-cooling concepts for such high-performance devices have already reached their limits or will reach their limits in the near future. Liquid cooling is the most promising option, especially due to the high thermal capacity of, e.g., water.

The technological challenge was the combination of electroplating and PCB processing on the one hand with soldering as the assembly technology on the other hand. Combining these technologies a manufacturing process for microstructured components was developed. It is compatible for mass production needs and is based upon industry-proven elements. The process allows the production of high-tech products at competitive costs.

Due to the outstanding heat-removal performance of microstructured devices, Atotech expects microstructure technology to become a fundamental application also in chemical reaction and process engineering. Miniaturization is regarded as a strategy to success in these fields and requires small, high-performance, versatile, highly reliable and cost-efficient microstructured components.

14.4 Integration of Microdevices into Macroenvironment

Applications of microfabrication technology in microelectronics and micro mechanics have already led to a revolutionary development in numerous industries, and the structured multiscale chemical devices offer the perspective of significant innovation through the use of microfabrication in the chemical process industries as well. Miniaturized components made of robust materials can be exploited to create miniaturized sensors and actuators embedded in process systems and acting on the specific length scales (several tens to hundreds of micrometers) particularly relevant to chemical processing: boundary layers, transport processes, reaction and mixing zones.

To take full advantage of these advances, however, it is necessary to rethink the way in which chemical processes are designed, built and operated. The concept of structured multiscale process design proposed here is a direct result of this rethinking.

Fig. 14.13 Liquid-cooled server cluster (Atotech, Berlin, Germany).

Despite the well-known advantages of microstructured devices, development has been dominated by "technology-push", and industrial "market-pull" has not appeared to any extent to be comprehensive or systematic on the part of chemical producers. In a few limited cases, processes employing small-scale structured devices have been utilized for process intensification on the production scale. In most cases, the published examples are quite simple tasks: mixing of miscible liq-

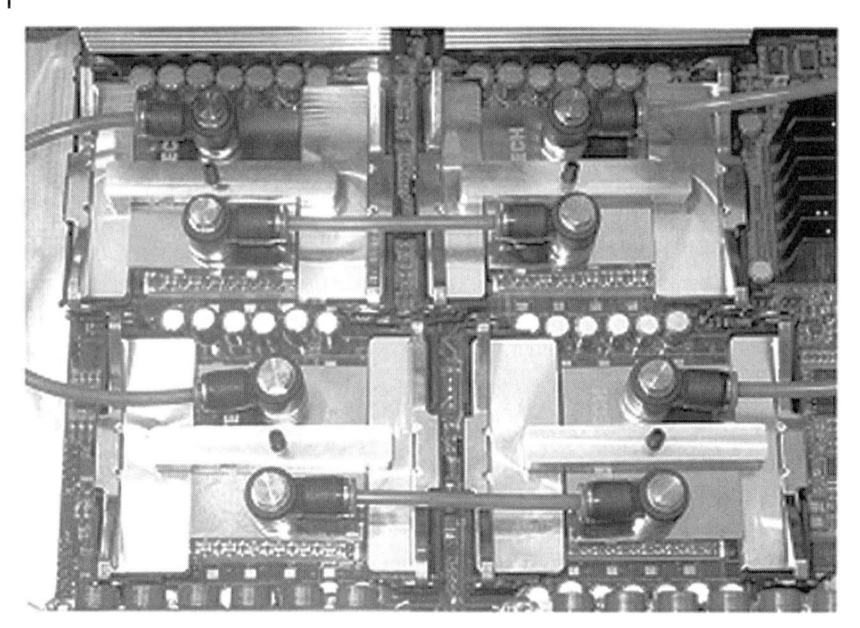

Fig. 14.14 Liquid-cooled Quad Opteron board (Atotech, Berlin, Germany).

uids, combined with effective heat transport and contact with immobile solids. There has also been some use of other mesostructured devices in production, including intensive rotating contactors [16] and oscillatory baffled reactors [17].

Given the very significant advances in the state-of-the-art in recent years on individual small-scale structured components and devices on the laboratory scale, the major challenge facing the chemical industry is not the further development of individual locally structured units but rather the effective integration of those units into complete production systems. In this connection, it should be noted that the seductively simple ideas of direct "numbering-up" or "scale-out" of microreactor systems initially envisioned in the 1990s are now being brought into question (see, for example, the critical remarks in this regard in [18] and [19]). It now appears clear that the principle of "numbering-up" through direct interconnection of individual small-scale units into large-scale production systems does not really "solve" the scale-up problem as initially intended. On the contrary, "numbering-up" displaces a (well-known) chemical engineering problem of process scale-up to a (for the moment essentially unsolved) problem of multiscale process interconnection. In a similar manner, the true impact of individual process-intensification units, such as spinning-disk reactors and similar devices, on whole-process performance (including reactant work-up and related process logistics) has not been completely explored. Basic design principles for process layout and process-performance evaluation are clearly needed and these issues will require substantial research efforts in view of the development of a truly generic multiscale design methodology.

As a complement to the detailed research and development efforts, it should be noted that sophisticated methodology for detailed equipment design, interconnection and layout is only of use for industrial application once an initial, preliminary decision has been made to explore new technological options. The availability of approximate, short-cut methods and principles, derived from the more complete, rigorous research results, is an additional challenge for emerging innovation that cannot be ignored.

Examination of the state-of-the-art reveals that for a thorough evaluation of true technological opportunities for the use of small-scale structured components in chemical production, a comprehensive and systematic protocol is required as an aid to decision making and for ultimate design and exploitation. Whether for retrofit of structured components into existing plant or for new design of future plant facilities, a new methodological approach is an urgent need and a clear contribution to future industrial competitiveness in chemical production technologies. Comparable to a pinch analysis for heat integration, or to HAZOP for safety issues, the structured multiscale design methodology should be developed in such a way as to permit reliable qualitative and quantitative technoeconomic evaluation of structured multiscale process systems for both existing and potential production processes.

Microstructured devices and process components set the stage for a true paradigm shift in the principles of chemical process engineering. Rather than adapting the operating conditions and chemistry to available equipment, the process structure, architecture and equipment can now be adapted to the physicochemical transformation. Production units can be created by integration and interconnection of diverse, small-scale structured units into large-scale macroproduction devices. A key feature of the resulting structured chemical devices is local process control (through integrated sensors and actuators), leading to enhanced global process performance.

In the last decade, a new alternative to traditional scale-up has been proposed in the context of "microreaction technology". The new approach has been coined "scale-out" or "numbering-up" and has attracted considerable academic interest.

With the "numbering-up" approach, the system of interest is studied only on a small scale in so-called "microreactors", and the final reactor design is simply a multiplication of (interconnected) small-scale devices acting independently. No attempt is made at large-scale optimization. Instead, the optimal functioning point is found for a small-scale device by empirical laboratory studies and then is simply reproduced by "replication" into a larger structure.

The obvious advantages of such an approach are that empirical measurements of "qualitative" performance can be used directly, without recourse to the precise mechanisms necessary for the detailed "modeling" of the traditional "scale-up" approach. The industrial user seeks optimal conditions for running the reactor or chemical device in microreactors, and that information is obtained quickly, with small quantities of reactants in a relatively short time.

It has even been claimed that scale-up "disappears" with the "numbering-up" approach, since the macrodevice can be obtained (in principle!) by simple multiplication (or replication) of the laboratory microreactors. By interconnecting large numbers of microreactors, macroproduction devices should be achievable for which each component microunit will run under the same optimum conditions as those found in the laboratory device.

In reality, although this approach may avoid "scale-up" of the reactor itself in the traditional sense, it does not "solve" the problem of scale-up but simply changes the nature of the problem. The true "scale-up" problem results in the numbering-up approach from the clear difficulty of optimal design for the numerous interconnections between reactors and for the essential connections of the "microreactor assembly" to the "macrostructure" necessary to feed the system.

Upon reflection, one concludes that it is rather unlikely that individual laboratory microreactors will be connected in this way in industrial designs. It is more likely that large-scale macrodevices will be created with internal microstructuring, and it is not evident that such devices (such as, for example, multiplate microchannel reactors) will truly operate under identical conditions at all points in the interconnected structure. Numbering-up is therefore not the complete answer to the scale-up problem, but it does provide a stimulating model for a totally new way to design and construct reactor devices.

Among the points that need to be addressed is the fact that even if direct numbering-up could be achieved, it is not clear that it would necessarily be desirable to have the same operating conditions at all points in a macroproduction device. In fact, when compared to a laboratory device, the "objective" for optimization of a production device is not the same: for a measurement device one is seeking "information", whereas for a production device, one is not seeking information but rather "performance" for a given throughput of matter and energy. The optimization criteria are generally different.

A laboratory reactor, for example, can frequently be operated at low conversion, whereas an industrial reactor is frequently operating at high conversion. It is therefore more likely that the optimal operating conditions for a true reactor will differ significantly from those found for the laboratory device. Here again, even if numbering-up were possible, it is not clear that one would want to have all points in a reactor operating at the same conditions. It might be preferable to modify operating conditions as a function of position in the macrostructure (see below: local process control).

With the multiscale assembly approach, diverse, interacting small-scale units are integrated into a large-scale device that is globally optimized with respect to standard industrial performance criteria. Local sensors and actuators permit local process control of operating conditions (that may vary with position and time) throughout the reactor unit. The extreme diversity of such devices promises particularly rich and innovative designs. The approach is not limited to any particular scale, but with the recent advances in microfabrication on the submillimeter scale and the fact that many important transport and transfer phenomena occur on length scales on the order of several tens to hundreds of micrometers, indicates that the time is ripe to look into new design methodologies that can benefit from the availability of low-cost microstructured components and devices in a variety of materials. These devices could be helpful to realize production on demand (avoid transportation of hazardous chemicals, elimination or reduction of ecological risks) and to speed-up time-to-market for new production processes.

In order to advance development of methodology on structured multiscale design, an industrial–academic consortium has been created with the support of CEFIC, the European Chemical Industry Council. The objective of the consortium research is enhanced performance through targeted, localized intensification.

IMPULSE (an acronym for "integrated multiscale process units with locally structured elements") aims at effective, targeted integration of innovative process equipment (such as microreactors, compact heat exchangers, thin-film devices and other micro and/or mesostructured components) to attain radical performance enhancement for whole processes (not simply individual components or unit operations), thereby contributing to significant improvement in supply-chain sustainability. The design concepts and process-development tools resulting from the IMPULSE research will be new and unfamiliar. To be accepted in industrial practice, the new approaches must be developed, analyzed, tested and proven on actual industrial processes. The collaborative effort of industrial and academic partners is a necessary element for success.

The strategy and rationale of the research is to move from optimization of individual equipment units to whole process design. As shown in the state-of-the-art described above, recent developments of intensified process equipment have mostly been limited to individual unit operations (such as mixers, reactors, separators or exchangers), but appropriate integration of these intensified devices into complete production units has not yet been fully addressed from an industrial perspective.

To respond to this challenge, the IMPULSE consortium proposes to orient research and development toward complete production systems. The design of such systems will involve integration and interconnection of diverse, small-scale structured components into large-scale macroproduction devices. The IMPULSE philosophy is to provide intensification locally only in those parts of a process where it is truly needed, and then to adapt interconnection of the locally intensified structures into a global macrodevice.

Furthermore, the IMPULSE approach is not only appropriate for new process designs but is also relevant for retrofit of innovative process equipment into existing plant structures and for modular plant designs.

To attain acceptance in industrial practice, it is necessary to provide a number of validated business cases for true commercial processes of industrial interest, in the framework of a multidisciplinary R & D program. Overall results should include:

- proof of principle in several major industrial supply-chain sectors
- validated business models, including technical economic analysis for each case
- generic design and optimization rules and software tools for their implementation
- decision criteria for appropriate choices of multiscale approaches in practice.

The project is structured around three broad subprojects covering products from three major supply chain sectors: pharmaceuticals, specialty chemicals and consumer goods. Each subproject sector includes one or more market segments. For each market segment, multidisciplinary research teams, led by an industrial producer, identify promising chemical process systems and undertake specific directed research, development and demonstration on those systems with respect to the IMPULSE rationale.

Parallel to this subproject structure, work packages on generic issues cover the common ground among the subprojects and provide cohesion and integration. Examples of generic issues include instrumentation (sensors and actuators), equipment and control devices, as well as tools, methodology, standardization, design rules and safety.

The following 20 companies and research institutes comprise the group of partners in the IMPULSE consortium: CNRS (F), Siemens (D), Degussa (D), Procter and Gamble (GB), GlaxoSmithKline (UK), Solvent Innovation (D), Britest Limited (CH), Dechema (D), INPL–Nancy (F), RWTH–Aachen (D), University of Manchester (UK), TNO (NL), IMM–Mainz (D), INERIS (F), FZ–Karlsruhe (D), ICPF–Prague (CZ), ICTP–Prague (CZ), Arttic (F), ETSEQ–Tarragona (ES) and WUT–Warsaw (PL). The IMPULSE Integrated R & D Project, programmed as a 4-year initiative, started officially in February 2005 as an Integrated Project (IP) within the European Commission's 6th Framework Program for Research and Technological Development (see: www.cordis.lu/fp6). The project has a budget of about 17 million Euros with a financial support of about 10.3 million Euros from the EC.

The methodology for structured multiscale chemical process design is an extremely promising area of scientific and technological investigation, rich in theoretical and experimental research and with significant potential for industrial competitiveness. In addition, the new approach should have a lasting impact on the teaching of chemical engineering and industrial chemistry, and should contribute to the attractiveness and acceptability of the chemical process industries as well.

Although relatively easy to imagine in principle, true industrial development and use of the concepts of structured multiscale devices requires further progress in research, combined with corresponding advances in the microfabrication methods required for the construction of the devices themselves. Targeted laboratory and pilot-scale demonstrator units are an urgent necessity in this regard, along with critical technical economic analysis of potential practical applications. Despite the difficulties, prospects and potential uses for these new chemical devices and production systems are considerable, and concerted research actions between industrial and academic institutions should lead to rapid advances and significant perspectives for their development in the near future. One such action, the IMPULSE consortium, has been founded on this basis in Europe and should encourage development of the necessary methodological tools required for future industrial implementation of the approach.

It has to be mentioned that it is not necessary to carry out all process steps in microstructured devices. Not "as small as possible", but "as small as necessary" should be the guideline for the application. And microprocess technology does not replace the classical way of chemical production, but it gives new possibilities for a lot of chemical processes, especially in the fine and specialty chemistry, but also in the pharmaceutical industry. Production-on-site and production-on-demand can become more important and therefore change the way of producing and distributing chemicals.

References

1 M. Matlosz, W. Ehrfeld, J.-P. Baselt, *Microreaction Technology: Proc. of the 5th Int. Conf. on Microreaction Technology, Strassbourg (IMRET 5)*, Springer, Berlin, **2002**.
2 V. Hessel, S. Hardt, H. Löwe, *Chemical Microprocess Engineering: Fundamentals, Modelling and Reactions*, Wiley-VCH, Weinheim, **2004**.
3 A. I. Stankiewicz, J. A. Moulijn, *Re-engineering the Chemical Processing Plant: Process Intensification*, Marcel Dekker, New York, **2004**.
4 Th. Bayer, D. Pysall, O. Wachsen in *Proc. of the 3rd Int. Conf. on Microreaction Technology, Frankfurt (IMRET 3)*, Springer, Berlin, **2000**, p. 165.
5 S. Rebsdat, D. Mayer, *Ullmann's Encyclopedia of Industrial Chemistry 6th edn.* **2002**, Electronic Release, Chapter "Ethylene Glycol".
6 Th. Bayer, H. Heinichen, T. Natelberg in *Proc. of the 4th Int. Conf. on Microreaction Technology, Atlanta (IMRET 4)*, AIChE, New York, **2000**, p. 167.
7 I. Leipprand, H. Heinichen, M. Kinzl, Th. Bayer, *mstnews* **2002**, *3*, 8.
8 W. Ferstl, S. Löbbecke, J. Antes, H. Krause, M. Grund, M. Häberl, H. Muntermann, D. Schmalz, J. Hassel, A. Lohf, A. Steckenborn, T. Bayer, M. Kinzl, I. Leipprand, *Chem. Eng. J.* **2004**, *101*, 431.
9 H. Krummradt, U. Kopp, J. Stoldt in *Proc. of the 3rd Int. Conf. on Microreaction Technology, Frankfurt (IMRET 3)*, Springer, Berlin, **2000**, p. 181.
10 G. Markowz, S. Schirrmeister, J. Albrecht, F. Becker, R. Schütte, K. J. Caspary, E. Klemm, *Chem. Eng. Technol.* **2005**, *28*.
11 U. Nickel et al., US Patent Appl. US2002/0014179, **2002**. C. Wille, V. Autze, H. Kim, U. Nickel, S. Overbeck, T. Schwalbe, L. Unverdorben in *Proc. of the 6th Int. Conf. on Microreaction Technology (IMRET 6)*, AIChE, New Orleans, **2002**, p. 7.
12 V. Hessel, H. Löwe, C. Hofmann, F. Schönfeld, D. Wehle, B. Werner, in *Proc. of the 6th Int. Conf. on Microreaction Technology (IMRET 6)*, AIChE, New Orleans, **2002**, p. 39.
13 http://www.imm-mainz.de (press release).
14 http://www.synthacon.biz.
15 C. Boswell, *Chemical Market Reporter* **2004**, 266.
16 D. H. Meikrantz, L. L. Macaluso, U.S. Patent 5,591,340.
17 X.-W. Ni, A. Fitch and Ph. Webster, Centre for Oscillatory Baffled Reactor Applications (COBRA), School of Engineering and Physical Sciences, Heriot-Watt University, Edinburgh – press release.
18 K. F. Jensen, *Chem. Eng. Sci.* **2001**, *56*, 293–303.
19 M. Matlosz and J.-M. Commenge, *Chimia* **2002**, *56*, 654–656.

15
Industrial Production Plants in Japan and Future Developments

Jun-ichi Yoshida and Hideho Okamoto, Department of Synthetic Chemistry and Biological Chemistry, Kyoto University, Kyoto, Japan

Abstract
The current activities directing toward microchemical plants in Japan are reviewed. Some examples of industrial applications are presented including the fields of particle synthesis, organometallic processes (Grignard exchange process, halogen–lithium exchange process), organic process (Swern oxidation, oxidation with H_2O_2, Friedel–Crafts alkylation reaction), and polymerization processes (radical polymerization, polycondensation). Significant progress in microprocess engineering has already been made to meet the future demands of the Japanese chemical industries.

Keywords
Gel particles, organic reactions, organometallic reactions, polymerization

Advanced Micro and Nanosystems Vol. 5. Micro Process Engineering. Edited by N. Kockmann

ISBN: 3-527-31246-3

15.1 Overview of Current Activities in Japan

Microchemical plants have emerged as an important new concept in the chemical industry, and are expected to make a revolutionary change to realize more compact and knowledge-intensive industries. The idea of chemical synthesis using microsystems is not very new. As a matter of fact, Bard had already proposed the concept of an integrated chemical synthesizer in his book entitled "Integrated Chemical Systems" [1]. The idea has been realized step-by-step for the past two decades. The research on design and fabrication of microstructured devices has witnessed steady progress of microtechnology and has opened up new opportunities of carrying out chemical reactions in such devices. This trend is accelerated by the increase in concern over the global environment that aims at the saving of natural resources and energy.

In Japan, the technical term of microchemistry was first used in the latter half of 1980s. The research in this field mainly focused on mechanistic studies of high-speed and dynamic chemical reactions in a micrometer-order reactor, using laser beams, which had been developed in the 1960s [2]. Afterwards, microfabrication processing in electronics, or the semiconductor industry, micromachining in mechanical engineering, MEMS (microelectromechanical systems), and microfluidics have developed independently and fused together in interdisciplinary approaches. With reactors having microstructures in hand, a new field of chemistry emerged in the latter half of the 1990s. This new chemistry was named microchemistry. The size of microstructure ranges from several μm to several hundred μm. Since then, the micro-total analysis system (μTAS) including lab-on-a-chip technology [3, 4] has received significant research interest in Japan and much effort has been devoted to realize "integrated chemistry" that focuses mainly on the integration of chemical systems for measurements and analysis on a microchip [5]. Inspired by the International Conference on Micro Total Analysis System (μTAS) (the 1st conference was held in Twente University, the Netherlands in 1994) [6], the corresponding domestic society, The Society of Chemistry and Micro-Nano Systems (CHEMINAS) was established in 2000, and has held a research meeting twice a year.

In the area of microchemistry, synthetic chemists and chemical engineers have realized the importance of microsystems for production, and since then enormous advances have been made in the study of microdevices, microunit operations, microproduction systems, and microchemical reactions [7]. In Germany, the 1st International Conference on Microreaction Technology (IMRET) was held in 1997 by the German Society for Chemical Engineering and Biotechnology, DECHEMA [8]. In Japan, the Group for Research on Automated and Microreactor Synthesis (GRAMS) started in 1997, in the Division of Synthetic Chemistry of the Kinki Chemical Society. The research group held its 6th Symposium (2nd Open Symposium) in August, 1998 in Osaka [9]. This was the first international workshop concerning microreaction technology in Japan. Other domestic societies in this field include the Research Group of Microchemical Process in The Society of Chemical Engineers, Japan SCEJ [10] (established in 2001), and the Research Group on Microreactors in the Catalysis Society of Japan (CatSJ) [11] (established in 2002).

In 2002, a national project, the Microchemical Technology for Productions, Analysis and Measurement System Project (Project leader: Professor H. Komiyama, University of Tokyo) started. This project was planned by the Ministry of Economy, Trade and Industry (METI) in Japan's Government and supported by NEDO (the New Energy and Industrial Technology Development Organization). The project is an academic–industrial–government complex, and is operationally managed by the Microchemical Process Technology Research Association (MCPT), which was organized by 30 chemical and instrumentation companies [12].

As shown in Fig. 15.1, the project consisted of three groups depending on the research field: the Microchemical Plant Technology Group (Group leader: Professor J. Yoshida, Kyoto University), the Microchip Technology Group (Group leader: Professor T. Kitamori, University of Tokyo), and the Systemization of Microchemical Process Technologies Group (Group leader: Professor C. Kuroda, Tokyo Institute of Technology). The project is to be terminated at the end of Fiscal 2005 (March, 2006). The R&D budget was 947, 1212, and 1150 Mio JPY for Fiscal 2002, 2003, and 2004, respectively.

The Microchemical Plant Technology Group is responsible for developing microchemical technology for industrial production and has an Intensive Research Center at Katsura Int'tech Center in Kyoto University. The group consists of 13 researchers from 13 chemical companies (2002) and more than 20 academic researchers in Kyoto University and Osaka Prefecture University. This Kyoto Intensive Research Center has collaborated with 10 different academic research groups in various districts in Japan, such as Tsukuba, Tokyo, Osaka, Kobe, Okayama, and Miyazaki (2002–2004). Major research areas of the plant group

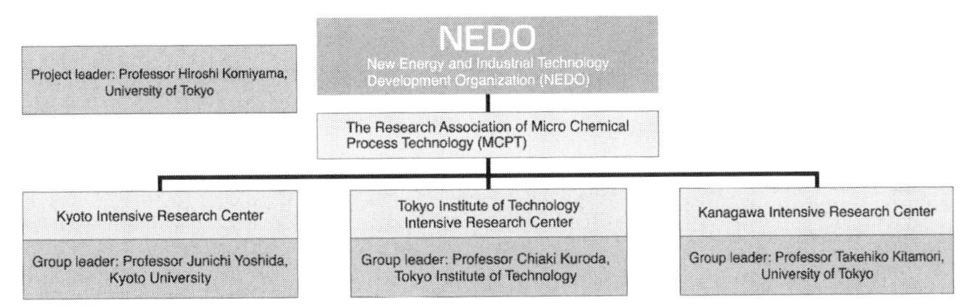

Fig. 15.1 Microchemical technology for productions, analysis and measurement system project managed by NEDO, Japan (2002).

include the study of microreactors, microunit operations, fundamental characteristics of microsystems, and synthetic reactions using microsystems, and the design and development of industrial microproduction systems. At present (April, 2005), three pilot plants (Grignard exchange process, halogen–lithium exchange process, and radical polymerization process) have already been developed, the details of which will be described in the next section.

In April, 2005, another project of an academic–industrial–government complex has just started in Okayama Prefecture in Japan as a part of the Cooperation of Innovative Technology and Advanced Research in Evolutional Area (CITY AREA), which is supported by the Ministry of Education, Culture, Sports, Science and Technology (MEXT) in Japan's Government. The project is under the initiative of the Okayama Prefecture Industrial Promotion Foundation. The association members include six universities, one public institute, and five companies. The project will be continued for three years from fiscal 2005. The R&D budget is 600 Mio JPY in total.

Furthermore, the Microchemical Initiative (MCI) started in May, 2004. This has been the consortium, consisted of six Japanese companies (eight companies in August, 2004), which have the technological base on micromachining and microfabrication. The activity of the development is expected to serve the manufacture of microreactor systems in the near future.

15.2 Industrial Production Plants in Japan

15.2.1 Particle Synthesis

15.2.1.1 Gel-particle Fabrication

Tosoh Corporation has developed a microreactor system based on their technological accumulation and the experience of R&D on magneto-optical disk. One of their targets is the fabrication of gel particles with uniform diameter, which do not need class selection [13a, b]. The fundamental structure is a Y-shaped mi-

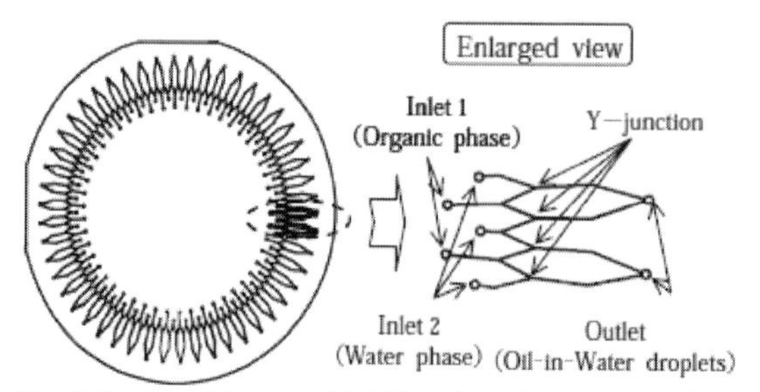

Fig. 15.2 Microchip disk with 100 Y-shaped microchannels (reproduced by permission of Tosoh Corporation).

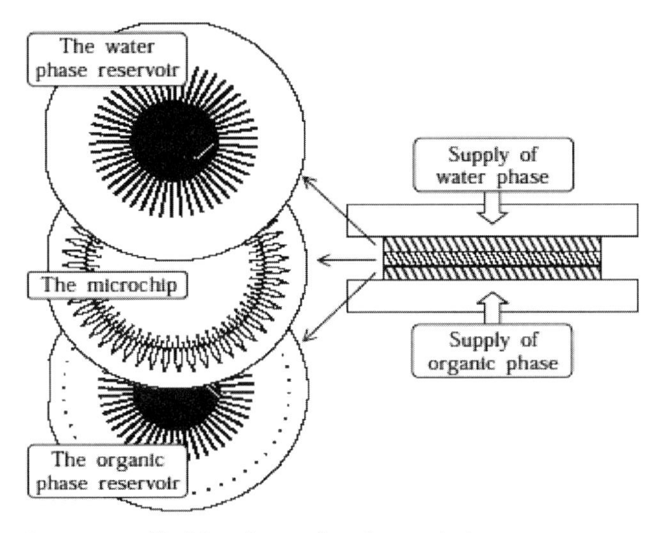

Fig. 15.3 Microblock based on uniform-flow method (reproduced by permission of Tosoh Corporation).

crochannel junction made of glass substrate by wet etching after optimizing its shape to increase the stability of droplets formed. Figure 15.2 shows the microchip disk with a 125-mm diameter having 100 microchannels, which are arranged in a radial pattern.

Aqueous phase of poly(vinyl alcohol) and organic phase of the mixture of styrene and divinylbenzene containing polymerization initiator are pumped into

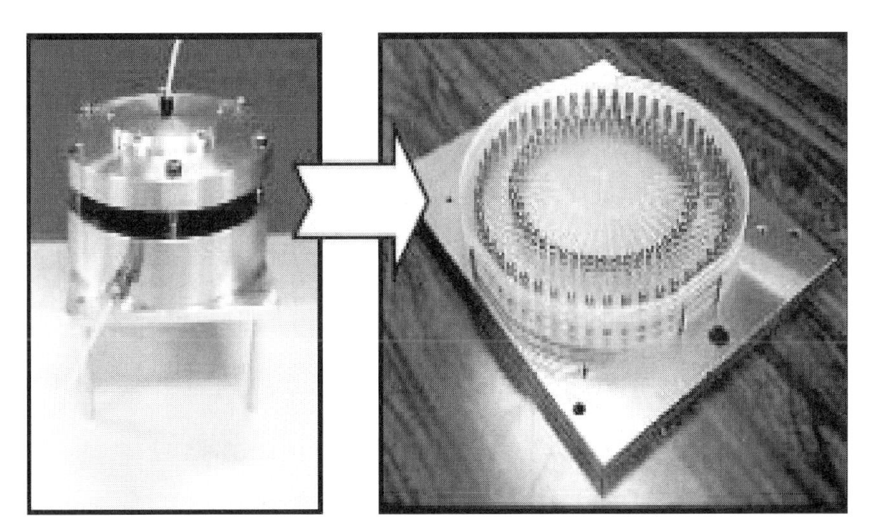

Fig. 15.4 Stacked blocks with three microchip disks (reproduced by permission of Tosoh Corporation).

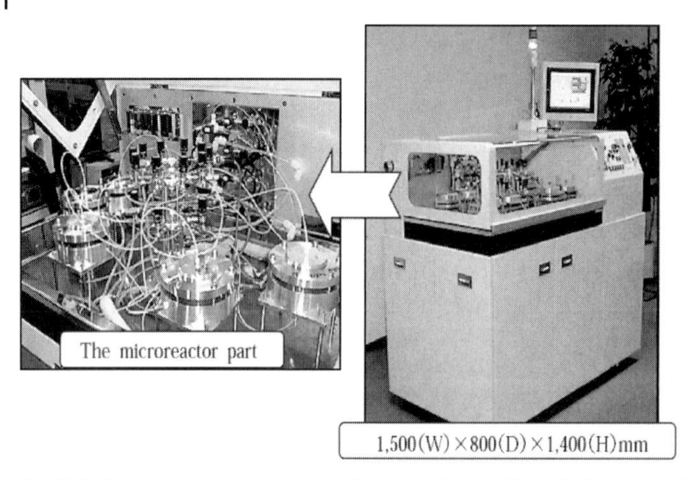

Fig. 15.5 Prototype system (reproduced by permission of Tosoh Corporation).

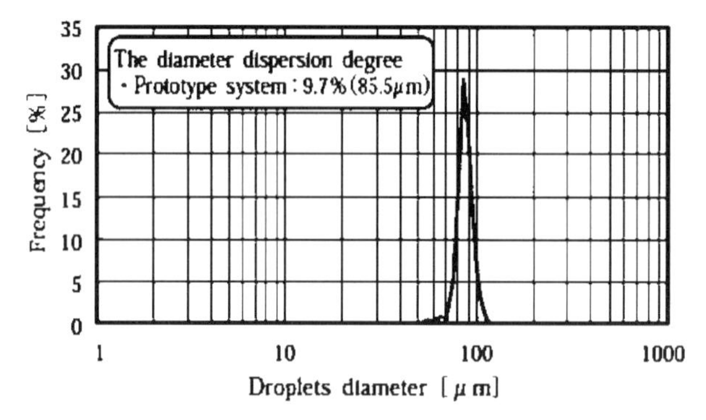

Fig. 15.6 Distribution of droplet diameter manufactured by prototype system (reproduced by permission of Tosoh Corporation).

Y-shaped microchannels to form organic droplets, nearly monodisperse, that are then polymerized in the succeeding batch reactor. The important point is to send solution uniformly into each microchannel. They developed two reservoir spaces for aqueous and organic phases, and then the microdisk was sandwiched between the two reservoir disks, as shown in Fig. 15.3. Three microchip disks are stacked into one block that contains 300 microchannels shown in Fig. 15.4. The prototype system is shown in Fig. 15.5, which can mount up to 10 blocks, namely to a maximum of 3000 microchannels. The maximum amount of production is 30 tons per year within design capacity. The distribution of the droplet diameter obtained was very sharp around 86 μm, which is illustrated in Fig. 15.6. The dispersion degree, CV value = (standard deviation)/(mean value), of the particle diameter was proved to be less than 10%.

15.2.1.2 Inorganic Fine Particle Crystallization

Photofunctional inorganic fine particles are generally used as materials for photographs or electronic displays. The performance of these inorganic fine particles is affected by the particle sizes, the distributions of the sizes and the crystal habits. Therefore, there is a strong need to control these features in industrial production of inorganic fine particles. On the other hand, the formation of nuclei of inorganic fine particles is an extremely fast reaction (completes within millisecond order). From this viewpoint, microreaction technology is very attractive to control the nucleus size, size distribution, and crystal habits, since microdevices have many inherent advantages such as fast mixing, rapid heat transfer and fast diffusion due to their small characteristic dimension. However, they also have disadvantages such as the clogging of microchannels by adhesion of precipitated nuclei on the wall of the channels [14].

A new microstructured device that prevents the clogging of the microchannel has been developed by MCPT (Fig. 15.7) [15]. This device enables crystallization of high-quality nuclei of inorganic fine particles.

The device consists of a cylindrical outer pipe with an inner diameter of 15.0 mm, three cylindrical partition pipes with different diameters for injecting four different fluids, and a central column with an outer diameter of 7.55 mm. These are arranged coaxially to create four layers of fluids. Two reactants and two inactive fluids are introduced from the inlets (2), (3) and (1), (4), respectively. The "position (A)" is the end of the three cylindrical partition pipes and the starting point to bring all four fluids into contact. From this point (A), the inner diameter of cylindrical outer pipe is reduced from 15.0 mm to 8.0 mm to decrease the thickness of the fluid layers, which causes the decrease of the diffusion distances of reactants to facilitate the reaction. The length of the mixing and reactant zone is 340 mm and the circumference of the surface contacting the reactants is 23.7 to 25.1 mm.

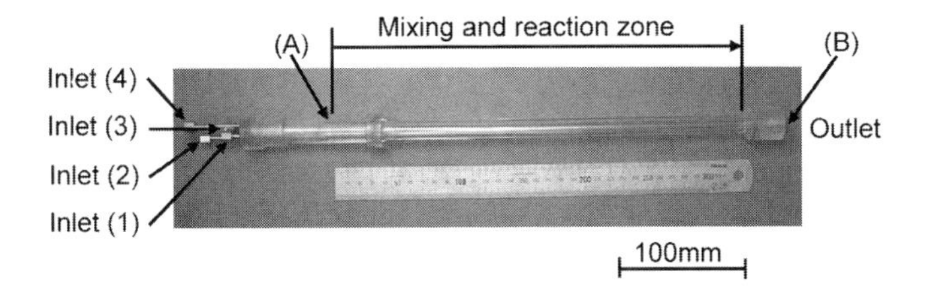

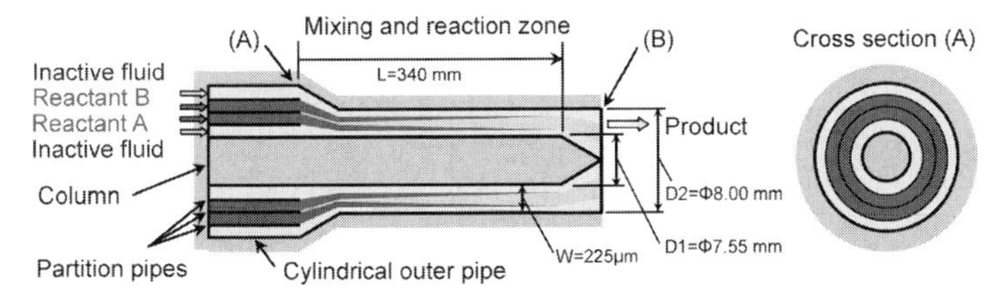

Fig. 15.7 Overview and schematic of flow in the microreactor.

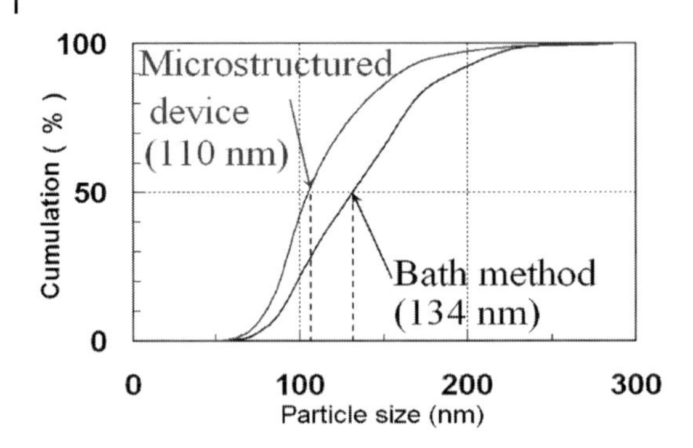

Fig. 15.8 Particle-size distributions.

The inactive fluid layers are arranged to cover the reactant layers to prevent the clogging of the microchannel by adhesion of precipitated nuclei on both inner and outer walls. By controlling the flow of the inactive fluid layers, the width of reactant layers can be varied.

Silver chloride (AgCl) nuclei crystallization, which is one of the most typical inorganic nuclei crystallizations, was carried out using this microdevice to examine the prevention of the clogging.

$$AgNO_3 + NaCl \rightarrow AgCl + Na^+ + NO_3^- \quad (1)$$

Silver nitrate solution (0.05 mol/l) and sodium chloride solution (0.05 mol/l) including low-molecular gelatin (about 20 000) as a protective colloid were used as reactants. Distilled water was used as an inactive fluid. The flow rate of each fluid was 3 ml/min. This means that the average velocity in the mixing and reaction zone was 36 mm/s and the residence time was ~9 s. The temperature of both reactants and inactive fluid was kept at 298 K. The average size of silver chloride nuclei thus obtained was smaller than that produced by the conventional batch method (134 nm) (see Fig. 15.8). In addition, the MCPT group could successfully run the continuous operation without clogging in the microchannel.

The present new microstructured device serves as one of the typical models for prevention of clogging in the microchannel by adhesion of precipitated nuclei on the wall together with crystallizing the high-quality nuclei of inorganic fine particles. It is hoped that the device will be used for crystallization in industrial production of various inorganic fine particles.

15.2.2 Organometallic Process

15.2.2.1 Grignard Exchange Process

Grignard reaction is one of the most fundamental reactions in organic syntheses and is widely utilized in laboratory syntheses and industrial production of fine chemicals. Although Grignard reagents are usually prepared by the reaction of organic halides with magnesium metal, Grignard exchange reactions (halo-

$$Et{-}MgBr + C_6F_5{-}Br \longrightarrow Et{-}Br + C_6F_5{-}MgBr \xrightarrow{BCl_3} (C_6F_5)_4B^- \; {}^+MgBr$$

Fig. 15.9 Grignard exchange reaction of ethylmagnesium bromide with BPFB.

gen–magnesium exchange reactions) are sometimes used for the preparation of Grignard reagents that are difficult to prepare by a conventional method.

Because Grignard exchange reactions are usually very fast and highly exothermic, they are difficult to control, especially in large-scale syntheses. However, a research group of MCPT found that Grignard exchange reactions can be conducted in a highly controlled manner by the use of a microsystem [16]. For example, the reaction of ethylmagnesium bromide with bromopentafluorobenzene (BPFB) proceeds smoothly in a microsystem to give pentafluorophenylmagnesium bromide; an intermediate for the production of tetrakis(pentafluorophenyl)borate, derivatives of which are useful compounds in metallocene-catalyzed polymerization and photopolymerization.

(a)

(b)

Fig. 15.10 Pilot plant of Grignard exchange process: (a): Full view, (b) enlarged view of reactor.

A microflow system (pilot plant) consisting of a micromixer connected to a shell and a tube microheat exchanger was constructed. The reaction temperature was automatically controlled at 20 °C by circulating water in the shell of the heat exchanger. Ethylmagnesium bromide solution in dibutyl ether (1.3 M) and BPFB (8.0 M, neat) were introduced into the micromixer at 20 °C. The residence time in the device was ~5 s. The product solution containing pentafluorophenylmagnesium bromide was immediately quenched by methanol. The commercial production, which is conventionally carried out using a batch reactor (volume size of 10 m^3), could be accomplished by numbering-up of only four microflow systems of the present scale, leading to a significant decrease in investment compared with current batch processes.

15.2.2.2 **Halogen–Lithium Exchange Process**

The halogen–lithium exchange has been used widely for making organolithium compounds in laboratory synthesis. This process should be, however, carried out at low temperatures such as –78 °C, because the reaction is usually highly exothermic and organolithium compounds thus produced are unstable and decompose at higher temperatures. Although laboratory-scale reactions can be easily carried out, industrial-scale batch reactions are problematic. The requirement of such low temperatures causes severe limitations in industrial use of this highly useful reaction.

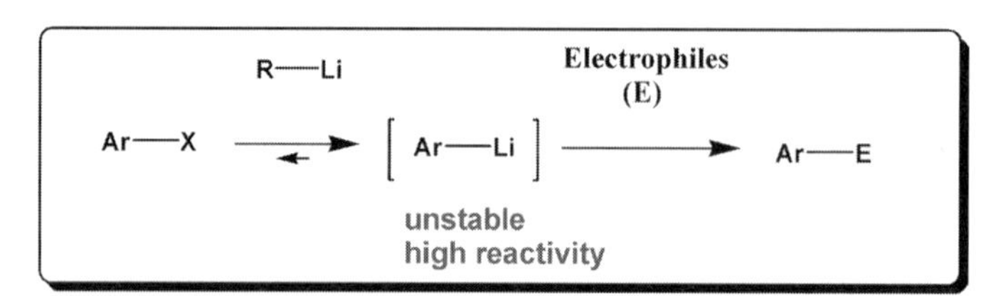

Fig. 15.11 Halogen–lithium exchange reactions.

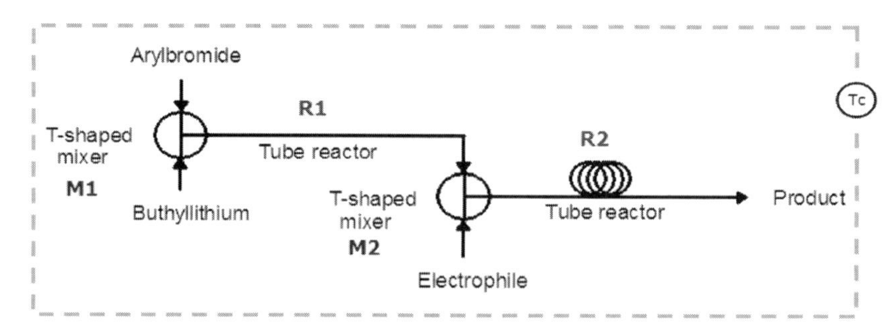

Fig. 15.12 Microflow system for halogen–lithium exchange reactions.

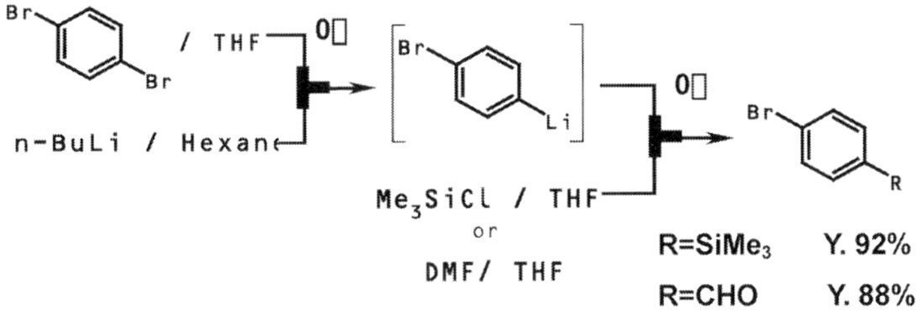

Br–C$_6$H$_4$–Br / THF (0□)
n-BuLi / Hexane
[Br–C$_6$H$_4$–Li]
Me$_3$SiCl / THF or DMF / THF (0□)
Br–C$_6$H$_4$–R

R=SiMe$_3$ Y. 92%
R=CHO Y. 88%

Fig. 15.13 Reaction scheme of halogen–lithium exchange.

Thus, the use of a microflow system was examined by MCPT to achieve halogen–lithium exchange process at higher temperatures. Efficient heat transfer was expected to prevent the decomposition of organolithium intermediates [17].

A microflow system consisting of micromixers and microtube reactors was developed. The halogen–lithium exchange reactions of aryl bromides and heteroaryl bromides with butyllithium followed by the trapping with an electrophile took place in a microsystem with high selectivity even at 0–20 °C to give the desired compounds in good yields. Efficient heat transfer of the microsystems seems to minimize the local deviation of temperature to avoid undesired side reactions. Precise control of the residence time may also be responsible for diminishing the decomposition of organolithium compounds.

Fig. 15.14 Pilot plant for halogen–lithium exchange process.

With the successful results in laboratory synthesis in hand, a pilot plant for the halogen–lithium exchange process has been built. Using this plant, the reaction can be conducted continuously without any problems, such as pressure drop.

15.2.3 Organic Process

15.2.3.1 Swern Oxidation

Moffatt–Swern-type oxidation, the oxidation using dimethyl sulfoxide (DMSO), is one of the most versatile and reliable methods for the oxidation of alcohols into carbonyl compounds and is widely utilized in organic synthesis [18]. This method seems to be highly useful for the production of fine chemicals and drugs. Although various methods for the activation of DMSO have been developed so far, the activation with trifluoroacetic anhydride (TFAA) is frequently employed in modern organic synthesis [19]. The reaction is usually carried out at low temperatures (–50 °C or below) because it is well known that activation of DMSO at higher temperature leads to an inevitable side reaction: Pummerer rearrangement. However, requirement of such low temperatures causes severe limitations in industrial applications.

The MCPT research group found that the use of microflow systems consisting of three micromixers and three microtube reactors was quite effective for conducting Swern oxidation at higher temperatures (Fig. 15.15) [20]. It is noteworthy that the reaction can be conducted in a highly selective manner even at room temperature by reducing the residence time in R1 to 0.01 s (see Table 15.1).

The durability of the process was checked for scale-up directing toward industrial production. The reaction of cyclohexanol was conducted for 3 h (0.36 mol scale) at 20 °C. The alcohol conversion and the product selectivity did not change (Table 15.2).

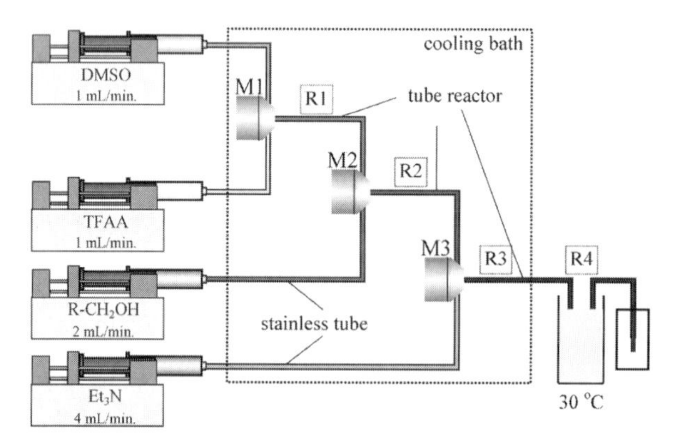

Fig. 15.15 Schematic diagram of the microscale flow system for the Swern oxidation.

Table 15.1 Oxidation of cyclohexanol to cyclohexanone.

Method	Residence time in R1 (s)	Temperature (°C)	Conversion (%)	Selectivity of cyclo-hexanone (%)	Selectivity of MTM ether (%)	Selectivity of TFA ester (%)
microflow	2.4	–20	88	88	6	5
	0.01	0	90	89	7	1
	0.01	20	81	88	5	2
macrobatch		–20	86	19	2	70
		–70	88	83	10	5

Table 15.2 Continuous Swern oxidation of alcohols using microflow system.

Time (h)	Conversion (%)	Selectivity of cyclohexanone (%)	Selectivity of MTM ether (%)	Selectivity of TFA ester (%)
0	83	92	5	4
0.5	84	92	5	4
1.0	85	89	5	4
1.5	81	89	4	4
2.0	86	91	5	3
2.5	87	91	5	3
3.0	84	92	5	4

15.2.3.2 Oxidation with H_2O_2

Hydrogen peroxide (H_2O_2) is an excellent oxidant from environmental and economic points of view. The oxidation reactions with hydrogen peroxide are usually carried out under very mild conditions: lower temperature, lower concentration of the oxidant, and nearly atmospheric pressure. From the viewpoint of productivity, however, higher temperature and higher concentration of the oxidant are preferable, but such severe conditions also accelerate unfavorable side reactions such as consecutive overoxidation, parallel side-chain oxidation, decomposition of hydrogen peroxide, and so on. Therefore, in traditional semi-batch processes, hydrogen peroxide is usually added dropwise in order to maintain high selectivity of the desired products. Both temperature and concentration have to be kept low throughout the operation, which gives rise to long reaction time.

2-Methyl-1,4-naphthoquinone (vitamin K_3) has been synthesized by the oxidation of 2-methylnaphthalene with CrO_3 in a commercial production, and the use of H_2O_2 as an oxidant oxidation would be much more beneficial. Thus, the use of a microflow system for H_2O_2 (60% aqueous solution) oxidation of 2-methylnaphthalene was examined by a research group of MCPT [21].

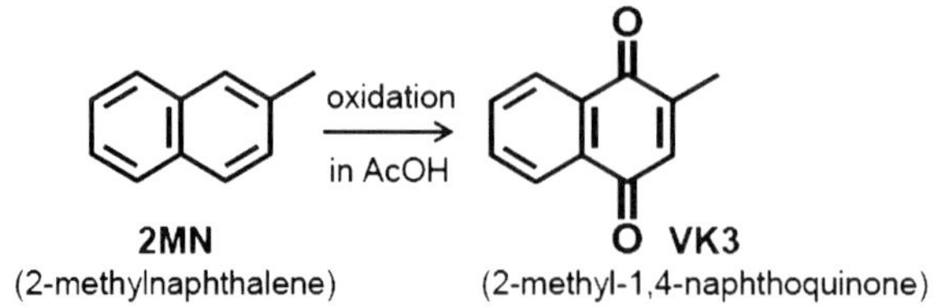

oxidation
in AcOH

2MN
(2-methylnaphthalene)

VK3
(2-methyl-1,4-naphthoquinone)

Fig. 15.16 Oxidation of 2-methylnaphthalene.

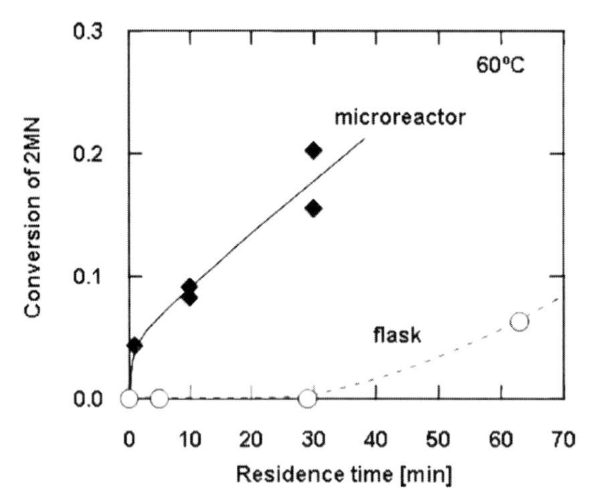

Fig. 15.17 Conversions of 2-methylnaphthalene in acetic acid with 60 wt% aqueous hydrogen peroxide solution as oxidant at 60 °C.

By the use of a microsystem, the reaction time was greatly shortened compared with the case of the semibatch process, but the selectivity of 2-methyl-1,4-naphthoquinone was low. The use of peracetic acid, which was prepared by acetic acid and H_2O_2 as an oxidant, however, gave rise to good selectivity of 2-methyl-1,4-naphthoquinone.

$$H_2O_2 + AcOH \rightarrow H_2O + AcOOH \quad (2)$$

$$2MN + 3AcOOH \rightarrow VK3 + 3AcOH \quad (3)$$

15.2.3.3 **Friedel–Crafts Alkylation Reaction**

Friedel–Crafts alkylation reactions are one of the most popular reactions in organic chemistry [22]. It is well recognized that the initial product is, in principle, more reactive than the starting aromatic compounds because of the electron-donating nature of an alkyl group that is introduced in the first reaction.

Thus, the second alkylation takes place more readily than the first alkylation. This inevitable dialkylation (or polyalkylation) problem is usually avoided by the use of a large excess amount of a starting aromatic compound (sometimes an aromatic compound is used as solvent) [23]. Therefore, selective monoalkylation using one equivalent of an aromatic compound has remained as one of the most challenging problems in organic synthesis over the years. The dramatic effect of micromixing for the selectivity of Friedel–Crafts reactions of aromatic compounds with N-acyliminium ions has been reported [24]. In this case, the second alkylation seems to be slower than the first alkylation. But the reactions of highly reactive aromatic compounds lead to the formation of significant amounts of dialkylation products, because the reaction is faster than mixing. This problem was solved by the use of micromixing.

From an industrial point of view, Friedel–Crafts reactions are significantly important. Bisphenol-F, which is a material for epoxy resin having superior heat resistance and low viscosity, is synthesized by the Friedel–Crafts reaction of phenol and formaldehyde with an acidic catalyst (Fig. 15.19). In this synthetic scheme the intermediate, hydroxybenzylalcohol (HBA), forms immediately and is then converted into bisphenol-F. At the same time, further reaction forming higher-order condensates such as tris-phenols also proceeds, by the reaction of HBA and the bisphenol formed [25, 26].

Since these byproducts significantly impair the desired low viscosity of bisphenol-F, the overreaction must be suppressed. To achieve this, the mole ratio of phenol to formaldehyde is fixed at an extremely high value, between 30 and 40, in the current process for the production of general-grade bisphenol-F (purity 90–94%). This value is more than 15 times larger than the theoretical mole ratio required. The excess feed causes problems such as increased expense for recovery of the unchanged phenol, and low productivity. There is a strong desire to improve this situation through development of an efficient and environmen-

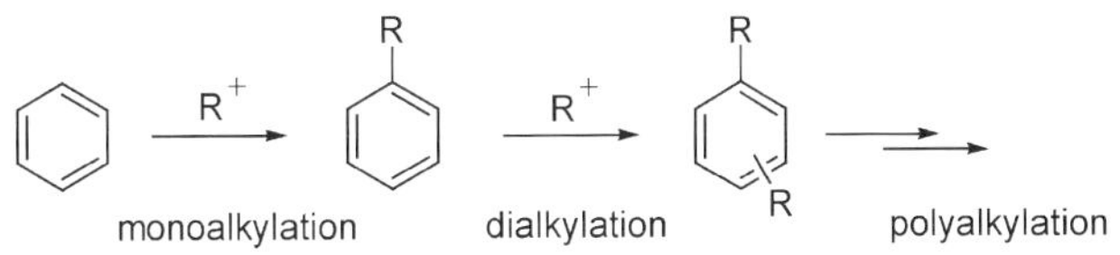

R
R^+
R
R^+
R
monoalkylation
dialkylation
polyalkylation

Fig. 15.18 Scheme of consecutive alkylation.

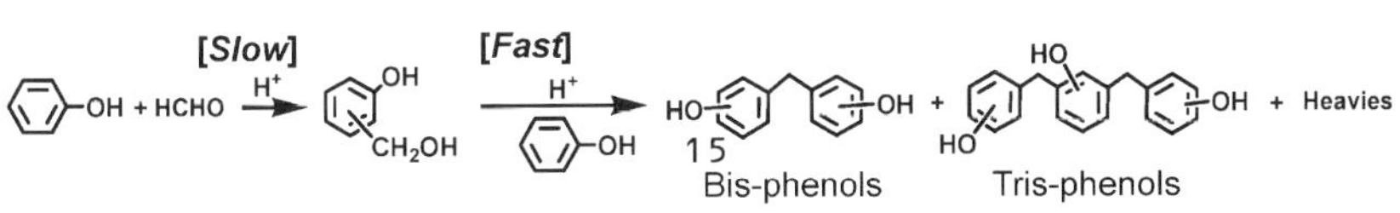

[Slow]
H^+
OH + HCHO
OH
CH_2OH
[Fast]
H^+
OH
HO
OH
15
Bis-phenols
HO
HO
OH
Tris-phenols
+ Heavies

Fig. 15.19 Reaction scheme of bisphenol-F synthesis.

tally benign production method to increase the selectivity for bis-phenols using the theoretical molar ratio of reactants.

A research group of MCPT has studied the Friedel–Crafts reaction of phenol and formaldehyde to improve the selectivity of bis-phenols by suppressing the consecutive overreaction, the selectivity for bis-phenols under various phenol/formaldehyde mole ratios was compared between a microflow system and a batch system using a flask, in order to clarify the validity of microflow systems for increasing the selectivity for bis-phenols.

Three types of micromixers, a) a T-shaped mixer with 500 μm ID, b) a multi-lamination-type micromixer with 40-μm channels (Institut für Mikrotechnik Maintz GmbH), and c) a K–M mixer were used. The K–M(Kyoto–MCPT) mixer is a new model of the so-called center-collision type for achieving an instant mixing that was developed by MCPT. The mixing concept is based on the collision of microsegments in the center of the mixer that consisted of three cylindrical components. As shown in Fig. 15.20, the two fluids are fed into the ring header on the inlet plate (1) to be uniformly distributed to the plural channels on the inner plate (2). The fluids then converge at the center of the inner plate (2). Mixing is promoted by both the diffusion and the shear rate at the center. Finally, the mixed fluid is discharged through the hole of the exit plate (3) [27, 28]. In this experiment, an inner plate having eight channels of 200 μm width and 200 μm depth was used.

The reactions were carried out using a microsystem shown in Fig. 15.21. Solution A (HBA/phenol) and solution B (p-toluenesulfonic acid (PTS)/phenol) were fed into the mixer and the products were analyzed by HPLC, the selectivity for bis-phenols is defined according to Eq. (4):

$$\text{Bis-phenols selectivity} = \text{Bis-phenols}/(\text{Bis-phenols} + \text{Tris-phenols}) \tag{4}$$

Figure 15.22 shows the effect of the phenol/HBA ratio on the selectivity for bis-phenols. The selectivity increased with an increase in the phenol/HB ratio. The selectivity using the K–M mixer was the highest at all the phenol/HBA ratios tested, and its advantage was greatest at low phenol/HBA ratios. Comparing the selectivity to the commercial process, it was found that the phenol/HBA ratio could be halved by using the K–M mixer. This clearly shows that micromixing is useful for obtaining bis-phenols in high yields with economic advantages [29].

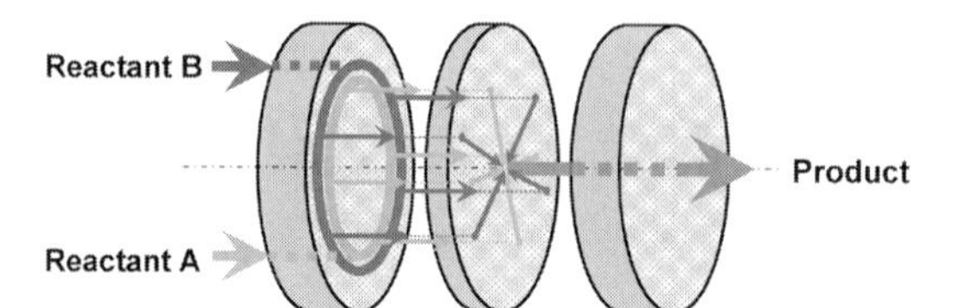

Fig. 15.20 Structure of K–M mixer.

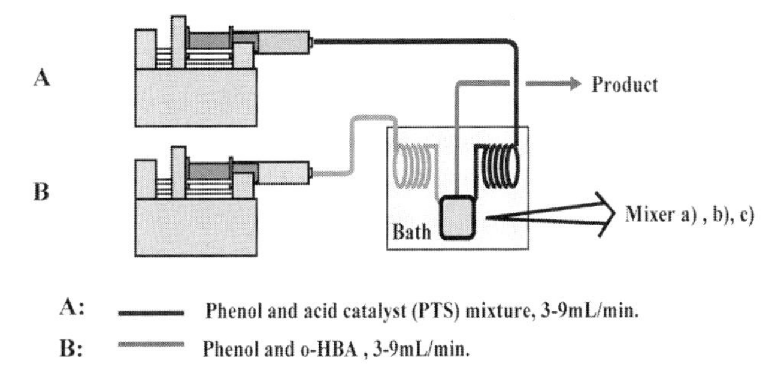

Fig. 15.21 Reaction apparatus.

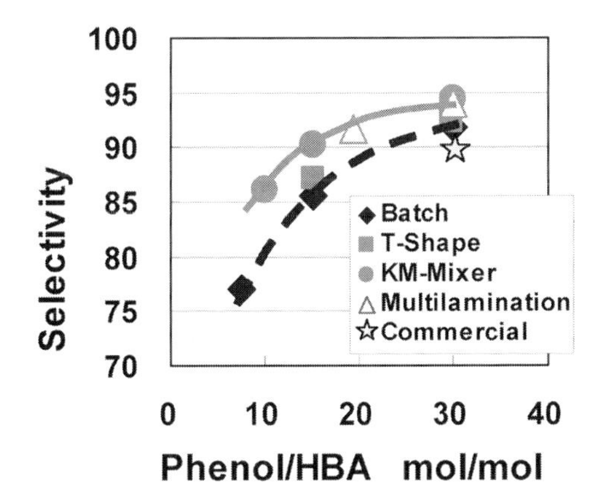

Fig. 15.22 Influence of phenol/HBA mol ratio to the bisphenols selectivities: the top value of each mixer.

15.2.4 Polymerization Process

15.2.4.1 Radical Polymerization Process

Polymerization in microflow systems is an attractive field from the viewpoints of both academia and industry. There are, however, only a few reports on the polymerization in microreactors. For example, it has been reported that the premixing using a micromixer is quite effective for radical [30] and anionic [31] polymerization of acrylates to reduce the amount of high molecular weight polymer fractions. Beers and coworkers [32] reported that the atom transfer radical polymerization (ATRP) can be conducted in a microchip reactor. Yoshida and coworkers [33] reported carbocationic polymerization using a microflow system, in which the molecular weight and molecular-weight distribution can be controlled effectively by fast mixing, temperature control, and residence-time control without decelerating the propagation by the active species/dormant species equilibrium.

Free-radical polymerization is an important process for the industrial synthesis of macromolecules since free radicals are compatible with a wide variety of functional groups that are not compatible with ionic and metal-catalyzed polymerization [34]. As free-radical polymerization is usually highly exothermic, the precise temperature control is essential for carrying out free-radical polymerization in a highly controlled manner. Therefore, polymerization in conventional macroscale bath reactors often suffers from a low level of molecular-weight distribution control because of inefficient heat removal and the lack of homogeneity of the reactor temperature. Therefore, the controllability of the reaction temperature is a major concern with free-radical polymerization from both academic and industrial viewpoints. As a matter of fact, heat-removal capacity often becomes a limiting factor in polymerizations in macroscale batch reactors.

The advantage of free-radical polymerization in microreactors is obvious, because efficient heat transfer is one of the most important features of microreactors. Generally, the heat-generation rate in polymerization decreases with decreasing reactor volume, while the heat-removal capacity increases with decreasing reactor volume. Therefore, as the reactor size decreases, the temperature control becomes more efficient, leading to better control of molecular-weight distribution.

A research group of MCPT reported the control of radical polymerization by virtue of efficient heat transfer of microsystems. The outline of the microsystem they used is shown in Fig. 15.23 [35].

For the polymerization of butyl acrylate (BA), Mw/Mn of the polymer obtained using the microflow reactor was much smaller than that obtained with a macroscale batch reactor. The result can be explained in terms of a much higher heat-removal efficiency of the microflow reactor compared with the macroscale batch reactor. For the polymerization of benzyl methacrylate (BMA) and methyl methacryate (MMA), the effect of the microreactor on Mw/Mn was

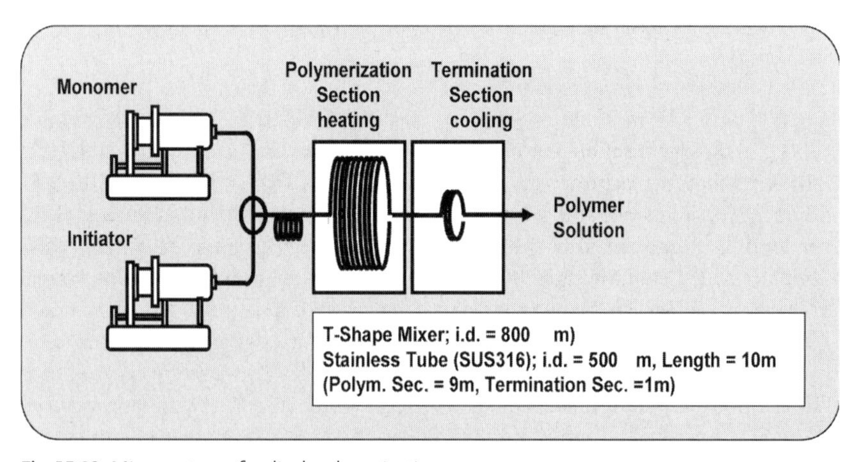

Fig. 15.23 Microsystem of radical polymerization.

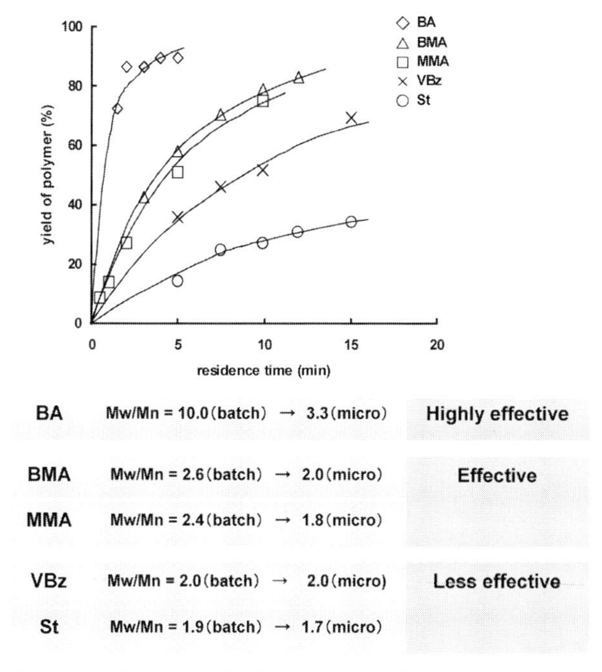

Fig. 15.24 Relative rate of polymerization in the microreactor.

smaller than the case of BA. For the polymerization of vinyl benzoate (VBz) and styrene (St), no appreciable affect of the microflow reactor on Mw/Mn was observed. These experimental results are summarized in Fig. 15.24, which indicates that the microflow reactor is quite effective for molecular-weight distribution control for highly exothermic free-radical polymerizations (BA, BMA, and MMA), but that it is not so effective for less-exothermic polymerizations (VBz and St). On the basis of these data, a pilot plant of radical polymerization has been built.

15.2.4.2 **Polycondensation Process**

Polycondensation reactions are important processes for the synthesis of highly thermally resistant polymers such as polyamide and polyimide. Because these reactions are quite exothermic, it is generally difficult to control molecular weights and molecular-weight distributions precisely in a macroscale batch reactor. There is, however, no report on the polycondensation reactions in a microreactor. Thus, the MCPT research group examined the two-step reaction, polycondensation and terminal modification reactions in a microreactor, in which the reaction temperature can be precisely controlled (Fig. 15.25). The rate of the polymerization and the molecular-weight distribution were compared with those for batch-type polymerization (see Fig. 15.26) [36].

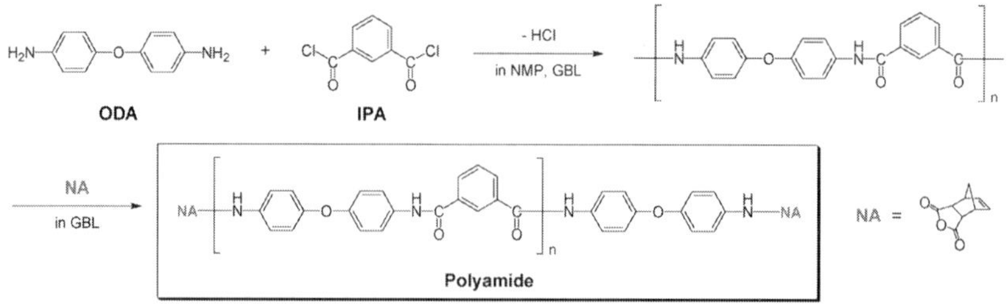

Fig. 15.25 Synthesis of terminally modified polyamide.

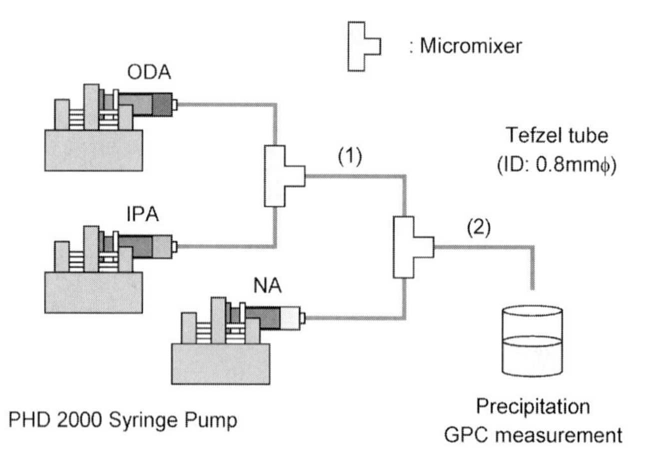

Fig. 15.26 Experimental apparatus for synthesis of terminally modified polyamide.

The reaction in the microreactor was found to be faster than that in the batch system. A higher mixing efficiency of monomers seems to be responsible for the faster reaction. It is also important to note that the molecular-weight distribution of terminally modified polymers obtained in the microreactor system is slightly narrower than that obtained in the batch reactor. Precise control of reaction temperature in the microreactor seems to be responsible for the narrower molecular-weight distribution.

15.3 Possible Future Developments

From the examples described above, we are already seeing the evidence that microreactors can enjoy industrial applications by the virtue of inherent advantages due to microstructures. One commercial plant is already working to produce fine particles in Japan. Three pilot plants have also been built and tested to examine the feasibility and durability of microchemical processes. The information obtained from these pilot plants should profit future development of this field to realize commercial plants, although the market may often be conservative. Pilot plants for some other processes described in the previous sections are now being built to take off from the level of an innovating new technology. It is also noteworthy that many other promising processes based on the advantages of microreactors have been developed by several research groups in both industry and academia [37].

The following points should be stressed from the viewpoint of possible future developments of this field.

(1) The potentiality of microreactors to effect extremely fast chemical transformations that are difficult to perform in a controlled manner using macro scale batch reactors has been well recognized in recent studies. Extremely fast mixing and precise temperature control seem to be responsible for this feature. Therefore, future applications of microchemical plants may focus on such transformations, which include fine-particle synthesis, polymerization, and extremely fast organic reactions.

(2) Short residence times together with precise temperature control in reactors may also be advantageous from the viewpoint of the control of highly reactive intermediates. This feature enables chemical transformations that are difficult to achieve in conventional macroscale batch reactors. Many examples will hopefully be developed and enjoy industrial production in future.

(3) The potentiality of microreactors in producing the desired chemical substances in more selective, safe, and energy-saving ways than conventional methods should be pointed out in terms of green sustainable chemistry. It is also noteworthy that microchemical plants may reduce the need to store and transport hazardous or reactive chemicals.

In summary, significant progress in microchemical technology has already been made to meet the demands of the chemical industry, although more progress is strongly needed for the numbering-up of microreactors and continuous operation of microchemical plants under commercial conditions [38]. Microchemical technology can now be developed more rapidly to face its more challenging quest to realize "ideal reactions" to produce desired chemical substances, without making any undesired byproducts, in a time- and cost-efficient and environmentally benign manner.

Acknowledgement

We thank Tosoh Corporation and The Research Association of Microchemical Process Technology (MCPT) for providing the information on their microchemical processes and for the permission to use their photographs and figures.

References

1 A. J. Bard, *Integrated Chemical Systems*, **1994**, Wiley, New York.

2 H. Masuhara, ed.: *Microchemistry – Operate the Reaction in Micro-space*, **1993**, Kagaku-dojin, Kyoto (in Japanese).

3 M. Freemantle, *Chem. Eng. News*, February 22, **1999**, 27–36.

4 O. Geschke, H. Klank, P. Telleman, *Microsystem Engineering of Lab-on-a-chip Devices*, **2004**, Wiley-VCH.

5 T. Kitamori ed., *Integrated Chemistry – Science and Technology developed by Microchemical Tip*, **2004**, CMC, Tokyo (in Japanese).

6 A. van den Berg, P. Bergveld, *Micro Total Analysis System*, Kluwer Academic Publishers, **1995**.

7 For example, (a) P. D. I. Fletcher, S. J. Haswell, E. Pombo-Villar, B. H. Warrington, P. Watts, S. Y. F. Wong, X. Zhang, *Tetrahedron* **2002**, *58*, 4735. (b) K. Jähnisch, V. Hessel, H. Löwe, M. Baerns, *Angew. Chem., Int. Ed. Engl.* **2004**, *43*, 406. (c) J. Yoshida, A. Nagaki, T. Iwasaki, S. Suga, *Chem. Eng. Technol.* **2005**, *28*, 259.

8 W. Ehrfeld ed., *Proc. of the 1st Int. Conf. on Microreaction Technology (IMRET1)*, **1998**, Springer, Berlin-Heidelberg.

9 (a) Kinki Chemical Society ed., *Microreactor – Current Status and Perspective*, Sumika Technical Information Center, Osaka, **1999** (in Japanese). (b) J. Yoshida ed., *Microreactors, Epoch-making Technology for Synthesis*, CMC, **2003**, Tokyo (in Japanese). (c) J. Yoshida, ed., *Recent Progress in Robotic Synthesis and Microreactor Synthesis*, **2004**, Kagakudojin, Kyoto (in Japanese).

10 URL, http://www.scej.org/en_html/index-e.htm

11 URL, http://www.shokubai.org/index_e.html

12 URL, http://www.mcpt.jp/eindex.html

13 (a) A. Kawai, S. Matsumoto, H. Kiriya, T. Oikawa, K. Hara, T. Ohkawa, T. Futami, K. Katayama, K. Nishizawa, *Tosoh Res. Tech. Rev.*, **2003**, *47*, 3–9; (b) K. Nishizawa, M. Fukuda, Kagaku-Souchi (Chemical Equipment), **2003**, *45*, 79–83 (in Japanese).

14 Ch. Wille, V. Autze, H. Kim, U. Nickel, S. Oberbeck, Th. Schwalbe, L. Unverdorben, *Proc. of the 6th Int. Conf. on Microreaction Technology (IMRET6)* **2002**, 7–17.

15 H. Nagasawa, H. Mae, *Proc. of the 7th Int. Conf. on Microreaction Technology (IMRET7)* **2003**, 127–129.

16 H. Wakami, J. Yoshida, *Org. Process. Res. Dev.* **2005**, *9*, 787.

17 S. Hikage, J. Yoshida, *Proc. of the 8th Int. Conf. on Microreaction Technology (IMRET8)* **2005**, 132f, AIChE, Atlanta.

18 For example, T. T. Tidwell, *Org. React.* **1990**, *39*, 297–572.

19 (a) K. Omura, A. K. Sharma, D. Swern, *J. Org. Chem.* **1976**, *41*, 957–962; (b) K. Omura, D. Swern, *Tetrahedron* **1978**, *34*, 1651–1660.

20 T. Kawaguchi, H. Miyata, K. Ataka, K. Mae, J. Yoshida, *Angew. Chem. Int. Ed.* **2005**, *44*, 2413.

21 K. Yube, K. Mae, *Chem. Eng. Technol.* **2005**, *28*, 331.

22 G. A. Olah, *Acc. Chem. Res.*, **1971**, *4*, 240.

23 R. Bruckner, *Advanced Organic Chemistry: Reaction Mechanisms*, Harcourt/Academic Press, San Diego, **2002**.

24 (a) S. Suga, A. Nagaki, J. Yoshida, *J. Chem. Commun.* **2003**, 355; (b) A. Nagaki, M. Togai, S. Seiji, N. Aoki, K. Mae, J. Yoshida, *J. Am. Chem. Soc.* **2005**, *127*, 11666.

25 R. W. Martin, *The Chemistry of Phenolic Resins*, **1956**, J. Wiley, New York.

26 H. C. Malhotra, *J. Appl. Polym. Sci.* **1976**, 461.

27 H. Nagasawa, N. Aoki, K. Mae, *Proc. 3rd Workshop on Microchemical Plants*, **2005**, 42, Awaji, Japan.

28 H. Nagasawa, N. Aoki, K. Mae, *J. Chem. Eng. Tech.*, **2005**, *28*, 324.

29 N. Daito, J. Yoshida, K. Mae, *Proc. of the 8th Int. Conf. on Microreaction Technology (IMRET8)* **2005**, 133 f, AIChE, Atlanta.

30 (a) T. Bayer, D. Pysall, O. Wachsen, *Proc. of the 5th Int. Conf. on Microreaction Technology (IMRET5)* **2000**, Springer, Berlin, 165–170. (b) D. Pysall, O. Wachsen, T. Bayer, S. Wulf, WO 99/54362.

31 B. Vuillemin, S. Nowe, US Patent 005886112.

32 T. Wu, Y. Mei, J.T. Cabral, C. Xu, K.L. Beers, *J. Am. Chem. Soc.* **2004**, *126*, 9880.

33 A. Nagaki, K. Kawamura, S. Suga, T. Ando, M. Sawamoto, J. Yoshida, *J. Am. Chem. Soc.* **2004**, *126*, 14702.

34 For example, (a) K. Matyjaszewsky, T.P. Davis, eds, *Handbook of Radical Polymerization*; **2002**, Wiley, New York. (b) K. Matyjaszewsky, *Controlled/Living Radical Polymerization*; **1998**, American Chemical Society, Washington, DC. (c) K. Matyjaszewski, J. Xia, *Chem. Rev.* **2001**, *101*, 2921–2990. (d) V. Coessens, T. Pintauer, K. Matyjaszewski, *Prog. Polym. Sci.* **2001**, *26*, 337–377. (e) M. Kamigaito, T. Ando, M. Sawamoto, *Chem. Rev.* **2001**, *101*, 3689–3745. (f) T.E. Patten, K. Matyjaszewski, *Acc. Chem. Res.* **1999**, *32*, 895–903. (g) C.J. Hawker, A.W. Bosman, E. Harth, *Chem. Rev.* **2001**, *101*, 3661–3688. (h) C.J. Hawker, *J. Acc. Chem. Res.* **1997**, *30*, 373–382.

35 T. Iwasaki, J. Yoshida, *Macromolecules*, **2005**, *38*, 1159.

36 T. Kuboyama, J. Yoshida, *Proc. of the 8th Int. Conf. on Microreaction Technology (IMRET8)* **2005**, 132d, AIChE, Atlanta.

37 For example, (a) S. Suga, M. Okajima, K. Fujiwara, J. Yoshida, *J. Am. Chem. Soc.* **2001**, *123*, 7941. (b) H. Hisamoto, T. Saito, M. Tokeshi, A. Hibara, T. Kitamori, *Chem. Commun.* **2001**, 2662. (c) T. Fukuyama, M. Shinmen, S. Nishitani, M. Sato, I. Ryu, *Org. Lett.* **2002**, *4*, 1691. (d) M. Ueno, H. Hisamoto, T. Kitamori, S. Kobayashi, *Chem. Commun.* **2003**, 936. (e) K. Mikami, M. Yamanaka, M.N. Islam, K. Kudo, N. Seino, M. Shinoda, *Tetrahedron Lett.* **2003**, *44*, 7545. (f) S. Suga, A.Nagaki, Y. Tsutsui, J. Yoshida, *Org. Lett.* **2003**, *5*, 945. (g) J. Kobayashi, Y. Mori, K. Okamoto, R. Akiyama, M. Ueno, T. Kitamori, S. Kobayashi, *Science* **2004**, *304*, 1305. (h) S. Liu, T. Fukuyama, M. Sato, I. Ryu, *Org. Process. Res. Dev.* **2004**, 8, 477. (i) R. Horcajada, M. Okajima, S. Suga, J. Yoshida, *Chem. Commun.* **2005**, 1303. (j) N. Yoswathananont, K. Nitta, Y. Nishiuchi, M. Sato, *Chem. Commun.* **2005**, 40.

38 (a) S. Hasebe, *Comp. Chem. Eng.*, **2004**, *29*, 57. (b) O. Tonomura, S. Tanaka, M. Noda, M. Kano, S. Hasebe, I. Hashimoto, *Chem. Eng. J.*, **2004**, *101*, 397.

16
Laboratory Applications of Microstructured Devices in Student Education

Walther Klemm and Bernd Ondruschka, Institute of Technical Chemistry and Environmental Chemistry, Friedrich Schiller University Jena, Germany
Michael Köhler and Mike Günther, Institute of Physics, Technical University Ilmenau, Germany

Abstract
In this chapter, first experiences with the use of microstructured devices for the education of future scientists and engineers at universities and technical universities are presented based on experimental lab classes that have already been or are currently being established. The different experiments are intended for both undergraduate and graduate students. Considering the learning goal to be achieved and the basic theoretical knowledge, experimental set-ups are described and experiments are discussed. Experimental results are presented in order to evaluate if the anticipated goals of the experiments have been achieved. This evaluation then serves as the basis of the future extension and intensification of experimental lab classes.

Keywords
Unit operations, unit reactions, photochemical reactions, undergraduate education, graduate education, curriculum

Advanced Micro and Nanosystems Vol. 5. Micro Process Engineering. Edited by N. Kockmann

ISBN: 3-527-31246-3

16.1
Introduction, Aims of Education

Ever since the introduction of microstructured devices for industrial applications in the 1990s, these devices have experienced a growing importance. The reason for this development being a constant growth of application fields in the chemical industry and related industries (pharmaceutical industry, food industry, biotechnology and fine chemicals). Prerequisites for the use of this new technology are:
- Thorough investigation of the microfluidics of the individual devices as well as the entire process including all measuring and control equipment
- Compound specific analysis of processes and sequences in microstructured devices
- Development and testing of microapparatuses for pumping and dosing substances
- Development and testing of microanalytical online systems (sensors).

In the scientific community, with microchemical engineering a new subdiscipline of chemical engineering has emerged. *Microchemical engineering is the application of the well-known chemical engineering principles of unit operations and unit reactions to microstructured technical devices such as mixers, heat exchangers, separators and reactors as well as combinations thereof in microsystems and microplants.* The special characteristic of the microdevices is that all diffusion and conduction-controlled mass- and heat transfer processes are intensified by orders of magnitudes in some cases. For chemical reactions this offers new production technologies for small-scale and special products as well as a safer production of dangerous chemicals due to small hold-up in the microstructured devices. In order to investigate the mass transfer and macrokinetics of chemical reactions in microstructured devices, standardized microstructured lab equipment is required.

During the last few years several different approaches for experiments using microstructured devices and chips or chip carriers have been developed. The developments were driven by three main intentions:

1. Realization of well-controlled reaction conditions by short distances for diffusion and heat transfer,

2. Microparallelization for multiparameter analytics and combinatorial chemistry, and
3. Setup of complex flow-through processes in tabletop arrangements (lab-on-a-chip concept, μTAS concept).

Indeed, the miniaturization leads to new devices, new processes and partially to new substances as well as to miniaturized substance libraries. However, one important expectation is not yet fulfilled by the recent state of microprocess engineering technology: experiments did not get easier. Among most microprocesses, completely automated ones and optimized systems are the exceptions. In most cases, the use of microstructured or other chip devices for chemical or biomolecular operations requires new strategies in experimental procedures, adaptation of protocols for the use of these microdevices and a cautious handling.

Meanwhile, the speed of introduction of microstructured devices and lab scale experiments is no longer limited by the availability of various devices. There exist many principle solutions. However, there still remains a serious lack of experimental experience and optimization. Presently we are in a situation that urgently requires a deeper and wider application of microdevices in chemistry in order to learn how processes and microdevices could be adapted and improved. For the development and application of microstructured devices in industry, qualified scientists and engineers are required. According to industry leaders, to date there is still a lack of personnel qualified in this area. It is therefore the duty of universities to train the scientists and engineers required. Most universities have just acknowledged this new challenging task.

While so far the development of microstructured equipment has been mostly done for isolated unit operations and reactions on the lab-scale (results have been published in monographs [1, 2] and in progress reports [3–10]), little has been reported on the use of these devices in education. Experiments that have been published so far include:

- Experiments on heat transfer [11–13]
- Experiments on stirring and mixing (still under development) [12]
- Experiments on heterogeneous catalysis in the gas phase [14]
- Experiments on photoinitiated liquid phase reactions for the decomposition of water pollutants [12, 15]
- Experiments on methane steam reforming [13]
- Electrochemical standard experiments in micro flow-through devices [16].

Some of the aforementioned experiments will be discussed in further detail in the following sections. The aim of these experiments is the extension of already existing experiments on selected unit operations and unit reactions in lab equipment for physical chemical experiments. Thus, students learn about the application of microstructured equipment and about the qualitative and quantitative differences compared to conventional equipment.

16.2
Concept of Education

16.2.1
Contents of Education

The industry platform microchemical engineering founded by DECHEMA in 2000 [17] proposed the introduction of microreaction technology to curricula in science (preferably chemistry) and engineering (chemical engineering). The goal is the education of experts to introduce these microstructured devices with their multiple advantages to the chemical industry. However, these future experts should also know about the limitations of this new technology. For this purpose, teaching material was developed [18], which mostly refers to the unit operations of chemical technology listed in Table 16.1.

The lack in microexperimentation culture gives the universities, and preferably technical universities, the motivation to apply miniaturized devices in educational experiments. The concept of introduction of microdevices in educational experiments will help in the following:

1. Defining of conditions for standard operations. The researcher and the teacher learn to design chemical experiments in microstructured devices and protocols as readily as possible.
2. Learning a great deal about the transferability of chemical operations in these devices, their reproducibility and the reliability of protocols with regard to different users, if the experiments are carried out multiple times under equal or similar conditions.

Table 16.1 Advantages of microstructured devices.

Unit operation	Advantages	Result
Mixing	Fast mixing	• small energy input • reduction of mixing times, thus faster and better homogenization • fast and even dispersion with small regular droplets and formation of fine disperse emulsions
Heat transfer	Improved heat transfer	• high heat-transfer coefficients • improved and faster temperature control
Reaction technology	Increased conversion	• high conversions in small reactors • isothermal reaction conditions for highly exothermic reactions • avoiding of unwanted hotspots in reactors
	Increased safety	• small hold-up • avoiding of critical process situations • gas reactions in the explosive regime possible

3. Advantages and disadvantages of special devices become clear by multiple checks during the experiments. Standard microdevices can much more efficiently be improved than in case of some single special application.
4. The periphery for experiments in microstructured devices will be economically optimized. It is expected that the types of peripheral devices and their cost will be reduced in many cases, if educational experiments are carried out multiple times.
5. *The most important aspect of miniaturized educational experiments* consists of the gain in knowledge and experience of young people on how to handle and utilize microstructured devices. At the moment, a generation of chemists, biochemists, microbiologists leave the universities without having any practical training with microstructured and other chip devices. The development of this kind of experimentation culture will only become successful if young scientists know about the advantages and the problems of their application in experimental and industrial practice.

Other aspects of miniaturization of chemical technical operations are more or less side-effects. They concern the saving of required space, material, energy and provide safety by reducing the amounts of substances and reaction volumes when carrying out experiments with hazardous chemicals. Students must learn these advantages. However, it must be noted that miniaturized experiments will not necessarily be cheaper than conventional experiments. The main purpose of miniaturized educational experiments is therefore not lowering educational costs but, moreover, teaching and learning about a completely new class of instrumentation and experimentation, which will evolve significant economical power in its future technical application.

16.2.2
Practical Education in Labscale Experiments

There are two typical groups of educational experiments with microstructured devices. *The first group represents miniaturized arrangements for complex operations* such as in technical chemistry, a special discipline in the education of chemists in Germany [19]. Preferably, flow-through processes connected with higher temperature and pressure, special solvents, application of catalysts or complex heterogeneous systems are involved in such experiments. Students learn about the chemistry of the process, the typical setups and procedures in chemical technol ogy, the construction of pilot plants and their down-scaling. In addition, they have to deal with the special materials, surface properties, residence behavior, chemical, temperature and pressure sensitivity of microstructured components, with process control, monitoring, analytics and with the specific flow, thermodynamic and kinetic conditions of the miniaturized process. This type of educational experiments sometimes has more the characteristics of a small research project than of a simple experiment. Students must understand the whole complex system. They have to become familiar with the basics of chemical education and should become experienced with chemical procedures to a certain extent. The time required for the experiments can be days and even weeks. However, most commonly the experimental time allotted amounts to 4 to 5 h per day. These types of experiments are required for all students of chemistry, chemical engineering, chemical technology and other specialized directions with a main orientation to chemical processes. Some experiments are basic experiments, offered for undergraduates, other more complex experiments are only suited for graduates.

The second group represents miniaturized experiments for the teaching of simple chemical or physicochemical facts. They are typically oriented to one single aspect of basic chemistry. The experimental setup is much less complex than in the first groups of experiments. It is easier to overview all components and conditions of an entire experiment. Students only need a comparatively short introduction to the experiment and its theoretical background. Experimental experience is not a prerequisite. The typical duration of this type of experiments is in the range of some hours, but not more than one day. The experiments can be done without a detailed knowledge of chemistry. Therefore, these experiments can be offered to undergraduate students of chemistry and related fields.

Besides the main chemical or physicochemical issue, also the simple experiments help to learn about the principles and problems of the application of miniaturized experimental equipments. Thus, students can operate with microchannels, see problems induced by bubbles or particles and become familiar with the strong influence of surface. They have to deal with the evaporation of solvents and the wettability of surfaces. The definition of reaction zones and monitoring zones, the application of sensors and actuators, the properties of laminar flow, fluidic dispersion, the effect of pressure drop induced by viscous liquids and other general problems of microfluidics are learned in this way. In addition, students become familiar with the periphery of chip reactor devices including fluid ports, tubes, fluid actors, measurement and control devices. Therefore, various aspects of microfluidics, chip technology and microreactor application are taught besides the pure chemical facts in a very practical and motivating way.

16.3
Examples of Experiments for Education in Technical Chemistry

In the following sections the first miniaturized experiments are described that have already been introduced or are currently being installed at the Friedrich Schiller University in Jena, Germany. Applications from other educational institutions are referred to as well.

16.3.1 Heat Transfer

Introduction

This lab class is an extension of the established heat transfer experiment performed in most technical chemical lab classes. Usually, the experiment is carried out with the simplest technical recuperator, the shell-and-tube heat exchanger [20, 21]. The aim of the experiments is to show students the great heat transfer power of microstructured devices with their large surface/volume ratios. Furthermore, students will realize that the same equations that describe the heat transfer in industrial-scale apparatuses can also be applied to microheat exchangers.

Fundamentals

Calculation of real heat transfer coefficients from heat balances

For the typical heat transfer process in a heat exchanger, consisting of heat transfer (fluid to wall), heat conduction (through the wall) and heat transfer (wall to fluid), partial heat currents are calculated from the heat transfer balance (see Eq. 3.68/69) and the overall heat transfer current is calculated from the mean value:

$$\dot{Q}_m = \frac{\dot{Q}_1 + \dot{Q}_2}{2} \tag{16.1}$$

The calculation of the real heat transfer coefficient is entirely based on the logarithmic mean temperature method for parallel flow and counterflow according to Eq. (3.76).

For plate micro heat exchangers the effective (active) heat transfer surface A [22] can be calculated (see comment at Eq. (3.71)):

$$A = n_f \cdot n_{ch} \cdot U_{ch} \cdot l_{ch} \tag{16.2}$$

n_f number of heat-exchanger platelets
n_{ch} number of channels on each platelet
U_{ch} wetted circumference of each channel with $U_{ch} = 2\ (a + h)$
l_{ch} effective length of each channel (without inlet and outlet region)

For cross-flow heat exchangers the logarithmic mean temperature difference is smaller than for parallel and counterflow and is defined as

$$\Delta T_{log,cross} = F \cdot \Delta T_{log,counter} \tag{16.3}$$

The correction factor F is a function of two dimensionless temperature differences R and P and is taken from the text book of Incropera and de Witt [23]. It can also be calculated from $F = (1 - F')$ with an auxiliary function $F' = f\,(R',\ P')$ [24]:

Table 16.2 Constants of Eq. (16.4).

$a_{i,k}$	$i = 1$	2	3	4
$k = 1$	0.0669	0.0	0.0395	0.0
2	–0.278	0.0	–0.22	0.0
3	1.11	0.0	0.454	0.0
4	0.136	0.0	–0.258	0.0

$$F' = \sum_{i=1}^{4}\sum_{k=1}^{4}\left[a_{i,k} \cdot (1 - \Delta T_{\log,\dim})^k \sin\left(2 \cdot i \cdot \arctan\frac{R'}{P'}\right)\right] \tag{16.4}$$

with

$$\Delta T^*_{\log} = \frac{\Delta T_{\log,\text{counter}}}{T_1 - T_2} \tag{16.5}$$

$$R' = \frac{T_1 - T_1'}{T_1 - T_2} \tag{16.6}$$

$$P' = \frac{T_2' - T_2}{T_1 - T_2} \tag{16.7}$$

Constants are summarized in Table 16.2.

Calculation of the theoretical heat transfer coefficient by dimensionless numbers

The heat transfer coefficient k is calculated according to Eq. (3.76).

For the calculation of the heat transfer coefficient k by dimensionless numbers the use of the following equations is recommended [25]:

a) Heat transfer for forced laminar flow (Re < 2300)

$$\text{Nu} = (\text{Nu}_1^3 + 0.7^3 + (\text{Nu}_2 - 0.7)^3 + \text{Nu}_3^3)^{1/3} \tag{16.8}$$

with

$$\text{Nu}_1 = 3.66 \tag{16.8 a}$$

$$\text{Nu}_2 = 1.615\left(\text{Re} \cdot \text{Pr} \cdot \frac{d_\text{h}}{l}\right)^{1/3} \tag{16.8 b}$$

$$\text{Nu}_3 = \left(\frac{2}{1 + 22 \cdot \text{Pr}}\right)^{1/6} \cdot \left(\text{Re} \cdot \text{Pr} \cdot \frac{d_\text{h}}{l}\right)^{0.5} \tag{16.8 c}$$

b) Heat transfer for forced turbulent flow in the transitional region (Re > 2300)

$$Nu = 0.0214 \cdot (Re^{0.8} - 100) \cdot Pr^{0.4} \cdot \left[1 + \left(\frac{d_h}{l}\right)^{2/3}\right] \quad (16.9)$$

for 0.5 < Pr < 1.5

$$Nu = 0.012 \cdot (Re^{0.87} - 280) \cdot Pr^{0.4} \cdot \left[1 + \left(\frac{d_h}{l}\right)^{2/3}\right] \quad (16.10)$$

for 1.5 < Pr < 500.

The hydraulic diameter d_h of the channels is defined analog to that for circular tubes (annular gaps, ellipse, rectangular channels):

$$d_h = \frac{4 \cdot A_{ch}}{U} \quad (16.11)$$

Experimental setup

The heat exchangers used are two microplate heat exchangers made of stainless steel (technical data are summarized in Table 16.3). The microheat exchanger MHE 1 is from the Institute of Microtechnology Mainz (IMM, Germany) and can be operated in parallel flow and counterflow mode. The microheat exchanger MHE 2 is from the Karlsruhe Research Center (FZK, Germany) and operates in cross-flow mode. The channels of MHE 1 exhibit no defined rectangular channel cross sections due to their manufacturing process. To simplify the calculations, however, the channels are considered to be rectangular. Channel cross sections of MHE 2 are geometrically rectangular.

The experimental setup as well as the two microheat exchangers are depicted in Fig. 16.1.

The experimental setup for the investigation of the heat transfer process is designed to operate in parallel flow and counterflow for MHE 1 and in cross flow for MHE 2. The heating medium is hot and cold water, which is adjusted to constant entry temperatures of 25 and 48 °C with a cryostat and thermostat, respectively. Three-way valves are used for switching between the different heat exchangers and operation modes. Hot water always flows in the same direction while the flow of cold water is alternated for MHE 1. The volume flow is adjusted by setting the rotational frequency of the pump. Fine tuning of the flow is possible with the throttle valves of the flow meters. Entry and exit temperatures are registered with digital thermometers.

Table 16.3 Technical characteristics of the microheat exchangers (channels of MHE 1′, calculated with real cross section).

Dimension	MEH1 (IMM)	MHE2 (FZK)	MHE1′ (IMM)
A [10^{-4} m^2]	73.44	51.51	71.68
A_{ch} [10^{-6} m^2]	24.48	17.00	23.39
n_{foil} [–]	10	25	10
$n_{k,ch}$ [–]	34	34	34
d_h [10^{-6} m]	266.7	133.3	261
l_{ch} [10^{-3} mm]	20	10	20
w_{ch} [10^{-6} m]	300	200	–
h_{ch} [10^{-6} m]	240	100	–
s_{ch} [10^{-6} m]	160	100	160

Fig. 16.1 Experimental setup with microheat exchangers (upper left-hand side: current and counter-current microheat exchanger (IMM), lower left-hand side: cross-flow microheat exchanger (FZK)).

Experimental results

For comparison of the experimental results between the shell-and-tube heat exchanger and the microheat exchanger, volume flows of the hot and cold water ($\dot{V}_1$ and $\dot{V}_2$) were set to 21, 39, 66 and 81 L/h. For the shell-and-tube heat exchanger hot water flow was 70 L/h and hot water flow 120 L/h. 120 L/h could not be realized in MHE 1 due to too high a pressure drop. The flow in the micro heat exchanger is for the used volume flows laminar (Re < 400) while it is always turbulent for the shell-and-tube heat exchanger.

Table 16.4 *k*-Values of the double-pipe heat exchangers.

Device	Glass heat exchanger		Metal heat exchanger	
Flow	Parallel flow	Counter flow	Parallel flow	Counter flow
$\dot{V}_1$ [l/h]	70	70	70	70
$\dot{V}_2$ [l/h]	120	120	120	120
ΔT_1 [K]	4.0	4.1	8.4	8.8
ΔT_2 [K]	3.2	3.4	6.7	7.0
Re_1 [–]	8237	8334	6982	6958
Re_2 [–]	2627	2736	2626	2629
k_{VDI} [W/m^2 K]	**698**	**703**	**2091**	**2069**
k_{ex} [W/m^2 K]	**781**	**814**	**1827**	**1854**

Table 16.5 *k*-values of the microheat exchangers MHE 1.

Flow	Parallel flow				Counter flow			
Test	1	2	3	4	1	2	3	4
$\dot{V}_1$ [l/h]	21	39	66	81	21	39	66	81
$\dot{V}_2$ [l/h]	21	39	66	81	21	39	66	81
ΔT_1 [K]	13.2	9.5	8.3	8.3	9.4	10.4	8.8	8.5
ΔT_2 [K]	9.1	8.9	8.0	8.0	8.4	10.6	8.7	8.2
Re_1 [–]	99	187	321	395	100	185	319	394
Re_2 [–]	80	148	249	302	80	151	251	307
k_{VDI} [W/m^2 K]	**4438**	**6219**	**(9810)**	**(11 840)**	**4552**	**4984**	**5580**	**5886**
k_{ex} [W/m^2 K]	**4810**	**6115**	**7329**	**8490**	**4704**	**5784**	**7087**	**8147**

For the parallel flow and counterflow shell-and-tube heat exchanger, heat transfer coefficients were determined experimentally and by computation and are summarized in Table 16.4.

For the cross-flow heat exchanger, heat transfer coefficients were determined for the same volume flows and temperatures as described in Table 16.5. The flow regime in this MHE 2 is laminar as well ($Re < 400$).

The *k*-values of the cross-flow heat exchanger are significantly higher for lower volume flows and laminar flow conditions than those of the parallel flow and counterflow shell-and-tube heat exchanger for turbulent flow conditions. In addition, the values obtained for the microheat exchangers show good agreement with the calculated values.

Conclusion and evaluation

- The comparison between microheat exchangers and the conventionally used shell-and-tube heat exchangers reveals that microheat exchangers show a better heat transfer despite their laminar flow regime. For MHE 1 (IMM) the heat transfer is increased by a factor of 3–12 and for MHE 2 (FZK) by a factor of 5–17. Thus, the educational goal of showing students the superiority of microheat exchangers over traditional heat exchangers (mostly in miniplants) has been achieved.
- If sufficient means of heating and cooling are available the experiment can also be extended to the turbulent flow regime. However, no new findings can be expected and the volume flow must be >400 L/h and would require much higher pumping power.
- Microheat exchangers could also be constructed from different materials, e.g. glass and plastics, since these materials are used in industry for corrosive fluids.
- The experiment was designed for graduate students but could also be part of an undergraduate curriculum. Commonly, the experiment is performed by two students and takes approximately 2 h.

16.3.2
Stirring and Mixing (Macrokinetics)

Introduction

This published experiment [26] was introduced to the technical chemical lab class and is an extension of the experiment "stirring and mixing of viscous fluids". The experiment consists of two parts. In the first part, performance and mixing characteristics of different stirrers are determined in a discontinuous stirring reactor for Newtonian and non-Newtonian fluids. Dispersing is investigated for the saponification of ethyl acetate with sodium hydroxide. During the experiment, the course of the concentration is followed for this fast heterogeneous liquid-phase reaction. The effective reaction rate and macrokinetics mass-transfer coefficient are determined using the theoretical droplet size and thus the mass-exchange surface. Stirring speed and phase ratio are varied.

The second part of the experiment uses a micromixer, which can also be a microreactor, to visually follow a dispersion process and to determine the mass-transfer coefficient depending on the stirring speed and the phase ratio. Furthermore, the power input can be determined and compared to that of the stirring reactor. The aim is to demonstrate to the students the intensification of mass transfer in the microstructured reactor compared to a conventional stirring reactor.

Fundamentals

The calculation of performance and mixing characteristics is based on the well-known stirrer Reynolds number $\mathrm{Re_R}$, the impeller diameter d_2, the Newton number Ne and the dimensionless mixing number Θ:

- Stirrer Reynolds number

$$\mathrm{Re_R} = \frac{n \cdot d_2^2 \cdot \bar{\rho}}{\bar{\eta}} \tag{16.12}$$

- Newton number

$$\mathrm{Ne} = \frac{P}{\bar{\rho} \cdot n^3 \cdot d_2^5} \tag{16.13}$$

- Mixing number

$$\Theta = n \cdot t_\mathrm{M} \tag{16.14}$$

For the determination of the mass-transfer surface the following is required:

- The stirrer Weber number

$$\mathrm{We_R} = \frac{n^2 \cdot d_2^3 \cdot \rho_\mathrm{c}}{\sigma} \tag{16.15}$$

- The average droplet diameter (Sauter diameter d_{32})

$$\frac{d_{32}}{d_2} = 0.047 \cdot \mathrm{We_R^{-0.6}}(1 + 2.5\Phi_\mathrm{d}) \tag{16.16}$$

- The phase ratio of the disperse phase

$$\Phi_\mathrm{d} = \frac{V_\mathrm{d}}{V_\mathrm{c} + V_\mathrm{d}} \tag{16.17}$$

where $\Phi_\mathrm{d} < 0.5$. Otherwise, phase inversion occurs and mass transfer is significantly impeded.

Knowing the Sauter diameter and the phase ratio Φ_d of the disperse phase, the specific interface area a of the material being stirred can be determined:

$$a = \frac{6 \cdot \Phi_\mathrm{d}}{d_{32}} \tag{16.18}$$

The entire interface area A_d of the disperse phase is

$$A_\mathrm{d} = a \cdot V_\mathrm{d} \tag{16.19}$$

From the experimentally determined concentration–time diagram the following data are determined:

- The effective reaction rate constant k_{eff} from the decrease of the concentration of sodium hydroxide c_{NaOH} in the reaction mixture.

$$c_{NaOH} = c_{0,NaOH} \cdot e^{-k_{eff} \cdot t} \tag{16.20}$$

- The reaction constant k and the mass-transfer coefficient β.

$$\frac{1}{k_{eff}} = \frac{1}{k} + \frac{1}{a \cdot \beta} \tag{16.21}$$

- The conversion X from the concentration c or the measured equivalent conductivities κ.

$$X = 1 - \frac{c_{t,NaOH}}{c_{0,NaOH}} = 1 - \frac{\kappa_t - \kappa_\infty}{\kappa_0 - \kappa_\infty} \tag{16.22}$$

Experimental setup stirring reactor

Except from the stirring vessel, the experiment was entirely built from commercially available components. It consists of the following parts:

- Temperature-controlled stirring vessel made of Pyrex with bottom outlet valve, reinforcement with 4 baffles, content max. 1500 mL, for experiment 1200 mL, inner diameter 115 mm, max. filling height 120 mm (diameter: height approx. 1:1)
- Laboratory stirrer LIGHTNIN, type LabMaster L1U10F with grip and shaft, revolution control 50 < 1800 rpm, power input max. 90 W, torque 48 N cm, display showing the revolution number, torque, temperature, stirrer type, local control or software remote control, axial impeller (propeller, 3-blade mixer), radial impeller (Rushton turbine)
- Temperature-measurement system with modules for PC connection, power input, 3 thermocouples type T, response time 0.8 s
- Conductivity meter
- PC, Windows 98 operating system or higher, HP-VEE software
- Thermostat.

Used fluids:

- Deionized water
- Glycerin 86%
- Hydroxyethyl cellulose solution 3%.

For the dispersion experiment:

- 0.1 N NaOH
- Ethyl acetate (stoichiometric ratio).

The experimental setup is depicted in Fig. 16.2.

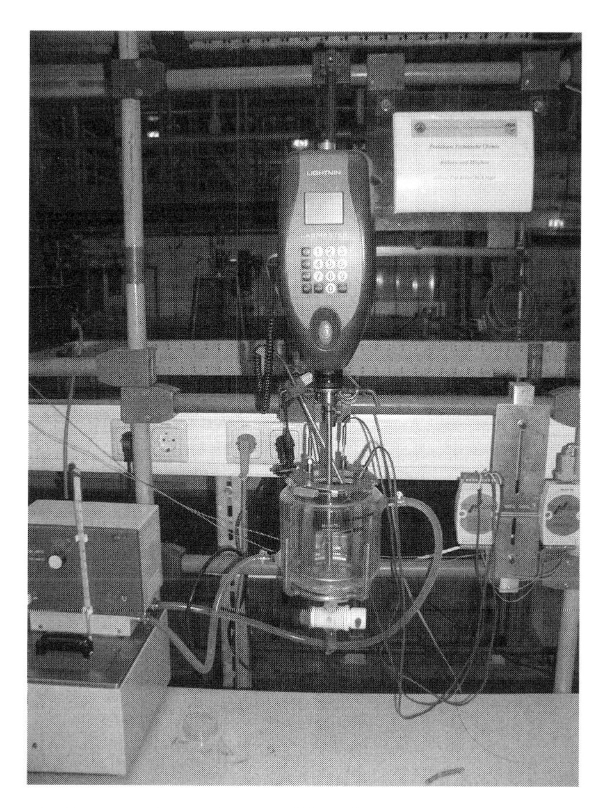

Fig. 16.2 Experimental setup of the stirring reactor.

Experimental setup microreactor

This experimental setup is currently built and will consist of the following, mostly commercially available, compounds:

- Microreactors
  T-mixer from the University of Freiburg (Germany) with channel cross sections of 1600×400, 1000×500, 600×300, 800×200 μm, mixing channel length of 16 mm
  Y-mixer SMSO 104 with 5 different channel cross sections, mixing lengths and shapes on a chip from thinXXS GmbH Mainz (Germany): 100×100×21 000, 640×640×121 000, 320×320×66 000, 320×320×50 000, 320×320×16 000 μm
- 2 product pumps ISMA-REGLO-CPF/RH00, 2.7–2700 mL/h
- 2 liquid mass flow meters X-FM-5880 with control unit from Brooks Instrument
- 3 pressure sensors, RS components
- Temperature-measurement system analog to stirring reactor
- Zeiss stereo microscope Stemi 2000C with accessories
- Digital video camera JVC GZ-MC200E
- Personal computer, operating system Windows ME, XP or higher, 64 MB RAM, 500 MB disk space, software HP VEE.

Anticipated reactions:
- Saponification analog to stirring reactor.
- Villermaux-Dushman reaction.

Future installation:
- Online conductivity measurement.

Experimental results

The experiment with the stirring reactor resulted in the expected performance and mixing characteristics as well as macrokinetics. No data is available for the experiment with the microreactor yet. The determination of macrokinetics data proofed difficult since no commercial device for online detection of physical parameters such as conductivity at the exit of the mixer/reactor is commercially available yet.

Conclusion and evaluation

- The first part of the experiment in the conventional stirring reactor, which is not applied in this form at any other institution, proved useful and gives a good basis for the second part of the experiment.
- The second part of the experiment, that has not been realized in the lab class yet, will hopefully have similar success if the online measurement of physical parameters is solved.
- The experiment is suited for undergraduate as well as graduate classes. It should be carried out in student groups since each part of the experiment takes approximately 3 to 4 h.

16.3.3
Heterogeneous Catalysis

Introduction

This experiment is an extension of the standard heterogeneous catalysis experiment in the technical chemical lab class that is carried out in a plug-flow reactor with fixed catalyst bed. The newly selected model reaction is the dimerization of isobutene to 2,5-dimethyl-1,5-hexadiene on Bi_2O_3. During the experiment, variations of the educt/oxygen ratio, residence time and different catalysts are tested. The aim of the experiment in the microreactor is to show students the effect of different volume/surface ratio and improved heat transfer on conversion and selectivity.

Fundamentals

Prerequisite for the experiment is the knowledge of the basic processes of heterogeneous catalysis that consists of the following steps:

- Mass transfer (convective transport of the educts to the catalyst grain surface, pore diffusion to the internal surfaces)
- Microkinetics (adsorption on the catalyst surface, reaction on the catalyst surface, desorption from the catalyst surface)
- Mass transfer (diffusion to the outer grain surface, convective transport through the gas boundary layer).

In addition to macrokinetics it is required to know how the effective reaction rate constant is obtained from an Arrhenius plot. From that, the activation enthalpy and indirectly the rate-determining step can be derived.

Experimental setup

To a defined nitrogen carrier gas stream containing isobutene a stoichiometric amount of air oxygen is added by means of flow sensors (type FLOMEGA®). The gas mixture is then directed either through the plug-flow reactor or the microstructured reactor type CTMR (catalytic test microreactor) from IMM Mainz

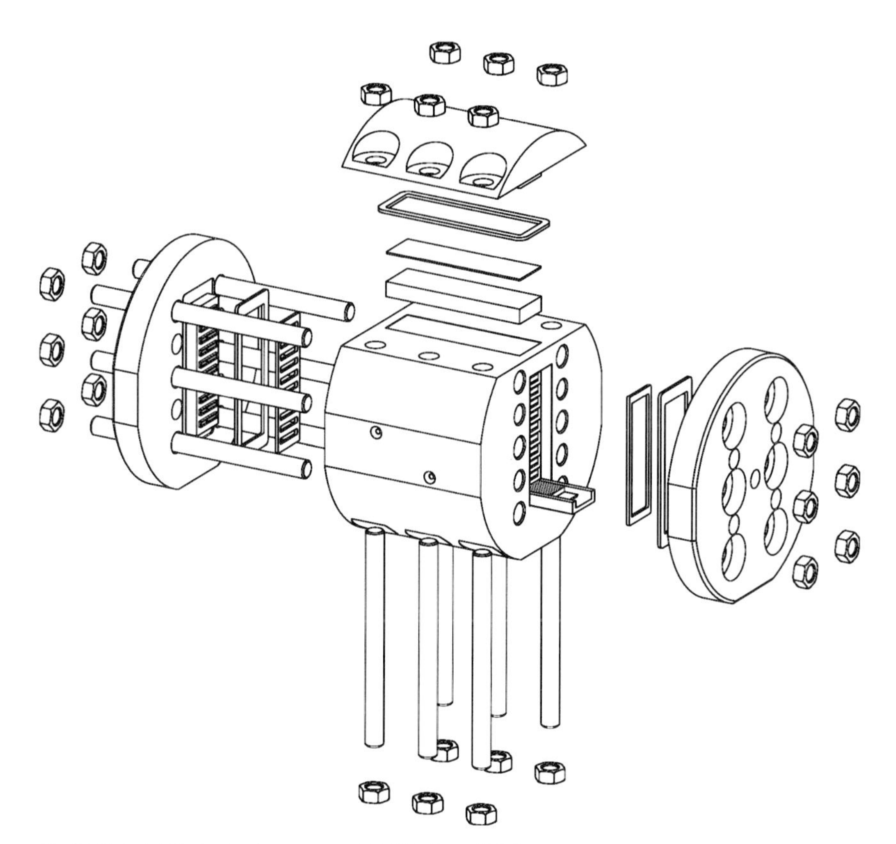

Fig. 16.3 Catalytic test microreactor (CMTR) from IMM.

(Germany), which are both equipped with the identical catalyst. An exploded drawing of the CTMR is depicted in Fig. 16.3.

Both reactors are operated under laminar flow conditions and thus changes in conversion and selectivity of the reaction are due to different volume/surface ratios and heat transfer properties or residence times of the individual reactors.

Experimental results

In the plug-flow reactor, conversions of 35% and isobutene selectivities of up to 90% to 2,5-dimethyl-1,5-hexadiene could be obtained. No publishable results for the microreactor were obtained yet.

Conclusion and evaluation

- The first part of the experiment in the plug-flow reactor was anticipated for graduate students but can be used in undergraduate courses as well. It should be carried out in groups of two students and takes approximately 4 h.
- The second part of the experiment in the microreactor is currently been used in a research class but eventually will be used for graduate classes. The experiment will be carried out in groups of two students and will take approximately 4 to 5 h.

16.3.4
Photocatalysis

Introduction

The lab class on photocatalysis in microreactors is envisioned as an addition to the already existing experiment on UV oxidation of waterborne contaminations in thin-film (falling film) reactors (producer: QVF Mainz, Germany), which students accomplish as part of their curriculum of environmental technology. The goal of the experiment is to demonstrate the advantages of photochemical microreaction systems with regard to simple operation (diodes as light source) and small light penetration depths in microchannels exhibiting high surface/volume ratios thus allowing for degradation of contaminants even in opaque solutions. Furthermore, students will be sensitized for operating photochemical reactions at low power consumption in accordance with the principles of Green Chemistry.

Fundamentals

Fundamentals to be taught in this lab class include the knowledge and application of the Beer–Lambert law and degradation kinetics. In the case of photocatalysis the Langmuir–Hinshelwood mechanism will be applied to model reactions on the solid/liquid interface. The degradation will be studied on the model substance 4-chlorophenol (cf. Eq. (16.23)).

$$ClC_6H_4OH + 6\frac{1}{2}O_2 \xrightarrow{h\nu, TiO_2} 6CO_2 + 2H_2O + HCl \quad (16.23)$$

Experimental setup

The degradation of the model substance takes place in a photocatalytic microreactor that was conceived in one of our department's research projects. An exploded drawing of the reactor is depicted in Fig. 16.4.

Results

The first results obtained for the degradation of 4-chlorophenol in the photocatalytic microreactor are published in [15].

Conclusions and evaluation

Students shall compare the results obtained for the degradation of waterborne contaminations in the UV thin-film reactor with those in the photocatalytic microreactor. The evaluation of the results should also include a consideration of energy efficiency of the different reactor types, i.e. how much photochemical en-

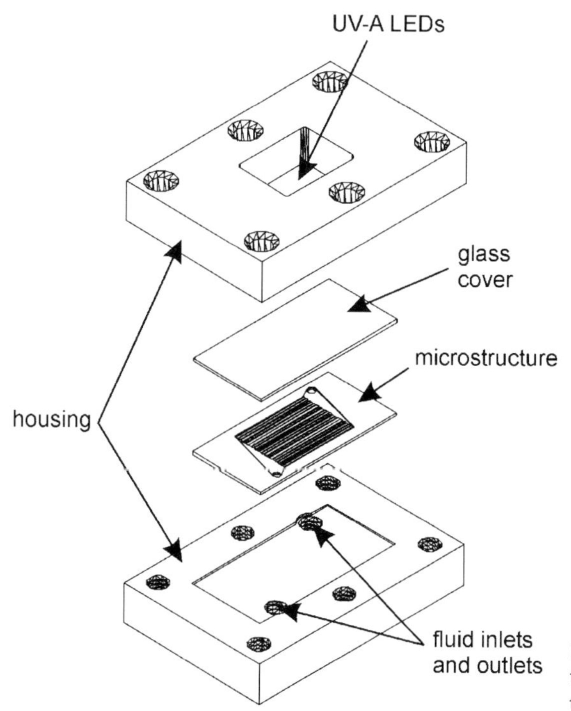

Fig. 16.4 Exploded drawing of the photocatalytic microreactor (adapted from [8]).

ergy (or electric energy) is required to degrade a given amount of model substance. The lab class will be designed for postgraduate students in environmental chemistry and chemical engineering. The time requirement for the experiment is estimated to be 4 to 5 h in groups of 2 to 3 students.

16.4
Examples of Experiments for Basic Education in Chemistry

In the following, some examples of the second group of miniaturized experiments are described. All examples are focused on one simple aspect of basic chemistry or physical chemistry. They have been designed for practical exercises in chemistry for the education of technical physicists and mechatronics engineers. All experiments have been carried out with several groups of students. Students have to pass lectures in general and inorganic chemistry and need a theoretical basic knowledge in physical and analytical chemistry. The duration of the theoretical preparation is about 2 h, the duration for carrying out of the experiments ranges between 3 and 5 h. An additional 2 h are typically needed for the preparation of the lab report.

16.4.1
Determination of Fluid Composition by Chip Transducers

Microanalysis with a capacitive stray field sensor

Introduction First, students have to calibrate the capacitive sensor with five known water/ethanol mixtures and make a calibration curve of capacitance C vs. ethanol content in %. This curve shows approximately linear behavior. All calibration measurements have to be carried out ten times for each calibration mixture. Secondly, the ethanol content of an unknown mixture is determined graphically with the calibration curve.

Fundamentals Basic for this experiment is the knowledge of permittivity in matter ε (product of the permittivity in free space ε_0 and the relative permittivity ε_r), the cause of dipole moments in molecules and which molecules show it, and how to generate an induced dipole moment in molecules. Furthermore, it should be known how the capacitance C of plate capacitors with dielectric fluid is associated with the permittivity ε. The working principle and assembly of the used capacitive stray field sensor also need to be known.

Experimental setup Required equipments and materials are:

- The capacitive stray field sensor (available from CiS Institut für Mikrosensorik gGmbH, Erfurt, Germany)
- A capacitance-measuring device (must be able to measure in the pF range), two pipettes and some beakers as well as distilled water and pure ethanol.

For analyzing the unknown mixture it is necessary to determine the capacitance C of the five known water/ethanol mixture ten times to get the mean value and the standard deviation. Students have to create a calibration straight line through linear regression with these five mean values vs. the corresponding ethanol content. The unknown ethanol content is determined by using the function $f(x) = ax + b$ of the calibration line.

Evaluation This low difficulty level experiment is suited for undergraduate students. Normally, it should take about 2–3 h to finalize this experiment. Recommended literature and information for this student experiment is found in [27–30].

Microcalorimetry

Introduction The first part of this experiment is the determination of the specific heat capacities c_p of water and ethanol as well as the molar evaporating enthalpies $\Delta_V H$ of these substances with a microchip calorimeter. The second part is the determination of the composition of water/ethanol mixtures.

Fundamentals Basics for this experiment are the knowledge of enthalpies and their definitions, knowledge of the evaporation and vapor pressure, the Clausius–Clapeyron equation and the dynamic equilibrium.

Experimental setup Required equipments and materials are:
- The microchip calorimeter with chip controller (from IPHT Jena, Germany)
- PC with analog/digital converter card, reaction tubes (Eppendorf, Germany)
- Two pipettes with 20 and 100 μL volume (Eppendorf or Abimed, Germany)
- Disposable pipette tips, distilled water and pure ethanol.

An experimental setup is shown in Fig. 16.5.

At the beginning of the practical exercise students have to prepare ten different ethanol/water mixtures. After that, they determine the specific heat capacities c_p of pure ethanol as well as distilled water. Then, the molar evaporating enthalpies $\Delta_V H$ of pure ethanol, distilled water and the evaporating enthalpies of the ethanol/water mixture are determined by measuring the heat energy. A calibration curve with graphical representation of enthalpy vs. ethanol or water content is prepared after measuring all enthalpies of evaporation. The composition of an unknown mixture is determined graphically with the calibration curve.

For determination of the specific heat capacity c_p of the distilled water and pure ethanol, it is necessary to measure the time t, ΔT and the consumed heating power required to evaporate the complete fluid drops. For analyzing the unknown mixture it is necessary to measure the heat quantity needed for complete evaporation of the different water/ethanol mixtures. Students have to create a calibration straight line through linear regression with this heat quantities vs. the corresponding ethanol content. The unknown ethanol content is obtained by using the function $f(x) = ax + b$ of the calibration line.

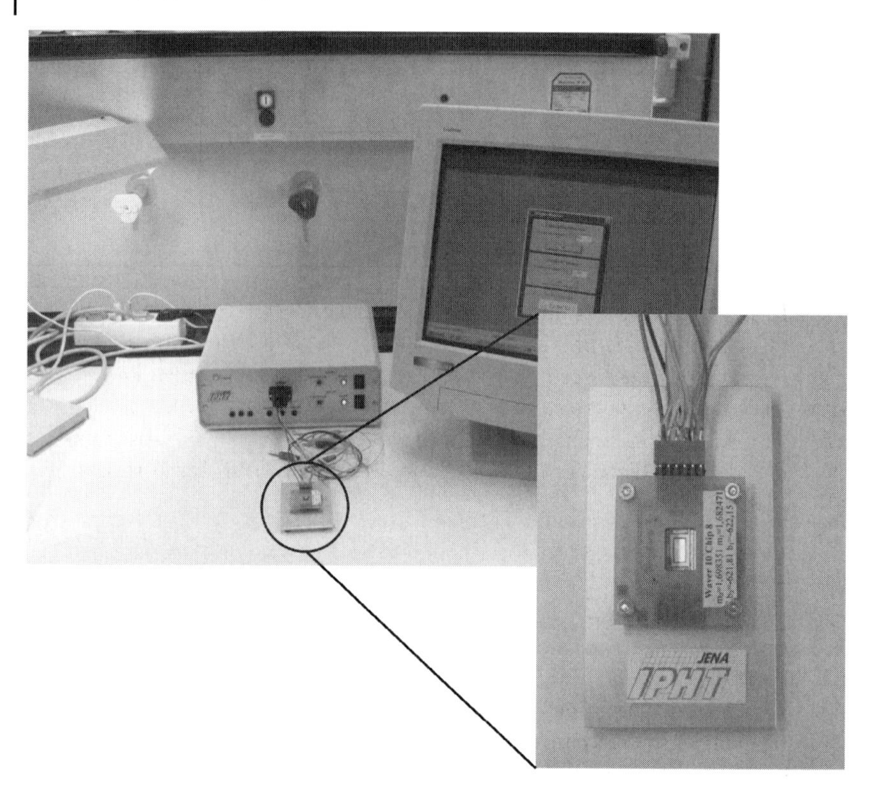

Fig. 16.5 Experimental setup microchip calorimeter.

Evaluation This low difficulty level experiment is suited for undergraduate students. Normally, it should take about 2–3 h to finalize this experiment. Recommended literature and information for this student experiment is found in [31, 32].

16.4.2
Interdiffusion in Laminar Flow

Mixing by interdiffusion and reaction after interdiffusion in a laminar flow system in microchannels

Introduction The aim of this experiment is to study the diffusion behavior of molecules under static as well as laminar flow conditions in microchannels. First, the diffusion of an organic dye molecule or a colored inorganic compound is determined under stop-flow conditions as well as under continuous-flow conditions. Second, the formation rate of a colored inorganic compound under stop-flow conditions is observed.

Fundamentals Prerequisite for this experiment is the knowledge of the Fick's first and second laws of diffusion and the concept of "random walk". Furthermore, students should be familiar with diffusivity, the Einstein–Smoluchowski equation (Eq. (16.24)) and the Beer–Lambert law (Eq. (16.25)).

$$D = \frac{x^2}{2t} \tag{16.24}$$

For analyzing the diffusion constants of the malachite green dye and the colored ferric rhodanide it is necessary to observe the motion of the borderline between the colored part of the channel and the colorless part with the microscope camera. Every ten seconds a snapshot of the channel is taken for a period of ten minutes. It is possible to get the transmittance T of the color gradient in all taken photographs with Eq. (16.26) and then to calculate the absorbance A of some selected distances in the photos with the Beer–Lambert law (Eq. (16.25)).

$$A = -\lg T \tag{16.25}$$

$$T = \frac{I}{I_0} = \frac{I_n - I_{min}}{I_{max} - I_{min}} \tag{16.26}$$

The intensity I is determined by gray scale analysis. I_{min} is the lowest intensity, I_{max} the highest intensity in each picture take by camera. I_n could be found at each selected distance in all photographs.

All distances must lie in a vertical line. The absorbance values of each photograph must be plotted against the position (distance from upper edge to bottom edge of the channel in points). Every curve of the absorbance taken at different times shows a different increase of the absorbance values due to diffusion. The diffusion constants can be calculated afterwards with the Einstein–Smoluchowski equation (16.24).

Experimental setup Required equipment and materials are:

- A one-channel foil reactor [33]
- A PC with monitor and USB interface, a microscope with internal camera (Motic DM143, VWR international, Germany)
- Two syringe pumps (SP100zi, World Precision Instruments Inc., USA)
- A color filter (LINOS BG12, LINOS Photonics GmbH & Co. KG, Germany), disposable syringes with 1 mL volume and cannulae (Braun, Germany)
- 0.2 molar ferric chloride solution, 0.2 molar ammonium rhodanide solution, pure glycerol
- Tubings and pipettes: PTFE tubing (inner diameter of 0.3 mm), silicone tubing (inner diameter of 0.3 mm), reaction tubes (Eppendorf, Germany), pipettes with 100 μL volume (Eppendorf or Abimed, Germany), a pinchcock, disposable pipette tips
- Distilled water and pure ethanol.

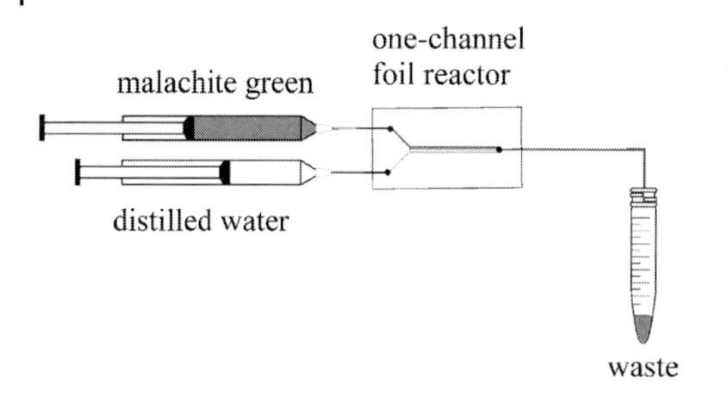

Fig. 16.6 Sketch of the experimental mixing setup.

A sketch of the experimental setup is shown in Fig. 16.6.

Before students start the experiments, they must test the reactor for leaks. In a second step, they prepare a 0.2 molar solution of ferric rhodanide solution by reaction of ferric chloride and ammonium rhodanide or prepare a malachite green solution with 1 mg of dye in 1 mL water. Following the determination of diffusion constants of ferric rhodanide or malachite green in water as well as in a solution containing 20% glycerol is carried out in the stop-flow mode. Finally, the formation of ferric rhodanide in laminar flow mode at several flow rates is measured during the reaction of ferric chloride and ammonium rhodanide in the microchannel.

Evaluation This medium difficulty level experiment is suited for undergraduate students. Normally, it should take about 4–5 h to finalize this experiment. Recommended literature and information for this student experiment is found in [27, 33–35].

16.4.3
Electrochemical Reactions in Modular Chip Reactor Arrangement

Simple oxidation: anodic formation of iodine from iodide

Introduction This experiment deals with electrochemical reactions, especially with the anodic oxidation and shows the methodology of electrochemical syntheses as well as electroanalytics.

Fundamentals Basics for this experiment include the definition of the Galvani potential and the difference between this potential and potentials of electrode reactions. Furthermore, the Faraday law and the Beer–Lambert law should be known by the students.

Experimental setup Required equipment and materials are:

- A one-channel foil reactor [16, 36]
- A PC with monitor and USB interface
- A microscope with internal camera (Motic DM143, VWR international, Germany), pumps and utilities: a double syringe pump (SP210iw, World Preci-

sion Instruments Inc., USA), a power supply (Potentiostat PS4, Meinsberg, Germany), an ampere meter for measuring current in the range of nA, disposable syringes with 1 mL volume and cannulae (Braun, Germany)
- 5 molar sodium iodide solution, freshly prepared saturated starch solution
- Tubings and pipettes: PTFE tubing (inner diameter of 0.3 mm), silicone tubing (inner diameter of 0.3 mm), reaction tubes (Eppendorf, Germany), pipettes with 100 μL volume (Eppendorf or Abimed, Germany), disposable pipette tips,
- Distilled water.

A sketch of the experimental setup is shown in Fig. 16.7.

Before starting all experiments, students have to test the reactor for leaks and prepare five sodium iodide solutions with concentrations between 1 and 5 mol/L. Thereafter, students should determine the dependency of electrochemical oxidation of iodide to iodine on different flow rates as well as different oxidation potentials. Furthermore, students have to determine the concentration of an unknown iodide solution. The electrolysis is carried out in a two-electrode arrangement. The potential range lying between +1200 and +1400 mV, measured in reference to the second electrode in the so-called quasipotentiostatic mode of operation. Students have to fill the syringes with an aqueous KI solution, to which 3–4 drops of freshly prepared starch solution should be added. A saturated starch solution is used as indicator for iodine [36, 37]. The standard flow rate of the solution is 1 mL/h. A potential of +1200 mV is used as standard potential, to suppress the H_2 evolution. The determination of the absorbance E is carried out optically. For each concentration a snapshot of the cell channel before and during the current flow is taken, at equal time intervals. With these photographs it was possible to detect the transmittance of the corresponding KI concentration. Through analysis of each photograph with graphic software the intensity according to the generated iodine amount was calculated. The snapshot before the current flow gives the intensity without the sample and the snapshot during current flow gives the intensity of the sample. After calculating the absorbance from transmittance with the Beer–Lambert law, Eq. (16.25), the absorbance in dependence on the KI concentration can be drawn. Subsequently, stu-

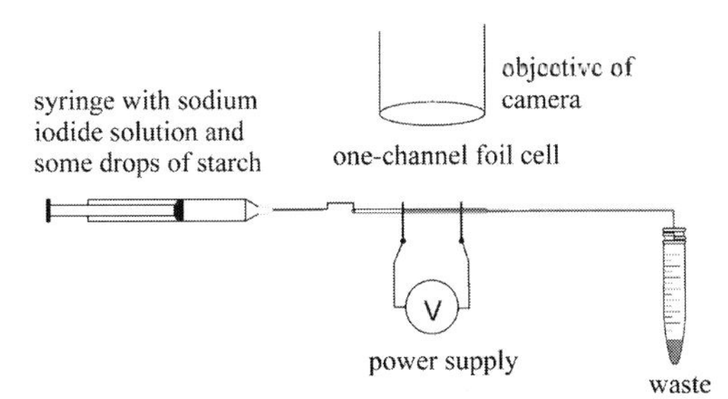

Fig. 16.7 Sketch of the experimental electrochemical reaction setup.

dents record a calibration curve of absorbance vs. concentration. This calibration curve is used to determine the concentration of an unknown KI solution.

Evaluation This medium difficulty level experiment is suited for undergraduate students. Normally, it should take about 3–4 h to finalize this experiment. Recommended literature and information for this student experiment is found in [16, 27, 33, 35–38].

Two-step process involving cathodic reduction of Fe (III) ions and complex formation

Introduction This experiment is similar to the experiment with electrochemical reactions, especially with the cathodic reduction and shows the methodology of electrochemical syntheses as well as electroanalytics.

Fundamentals Basics for this experiment include the definition of the Galvani potential and the difference between this potential and potentials of electrode reactions. Furthermore, the Faraday law and the Beer–Lambert law should be known. Additionally, students must know how metal complexes are built as well as the general structure of such complexes.

Experimental setup Required equipment and materials are:
- Two one-channel foil reactors [9, 30]
- A micromixer (IPHT Jena, Germany) [30]
- A PC with monitor and USB interface
- A microscope with internal camera (Motic DM143, VWR international, Germany), Pumps and utilities: a double syringe pump (SP210iw, World Precision Instruments Inc., USA) and a single syringe pump (SP100zi, World Precision Instruments Inc., USA), a power supply (Potentiostat PS4, Meinsberg, Germany), an ampere meter for measuring current in the range of nA, disposable syringes with 1 mL volume and cannulae (Braun, Germany)
- 0.1 molar ferric chloride solution, 0.1 molar 1,10-phenanthrolinium chloride solution
- Tubings and pipettes: PTFE tubing (inner diameter of 0.3 mm), silicone tubing (inner diameter of 0.3 mm), reaction tubes (Eppendorf, Germany), pipettes with 100 μL volume (Eppendorf or Abimed, Germany), disposable pipette tips and
- Distilled water.

Parts of the experimental setup are shown in Fig. 16.8.

At the beginning of all experiments students have to test the reactor for leaks and prepare three ferric chloride solutions with concentrations between 0.025 and 0.1 mol/L. The students should determine the dependency of electrochemical reduction of ferric ions to ferrous ions on different flow rates as well as different

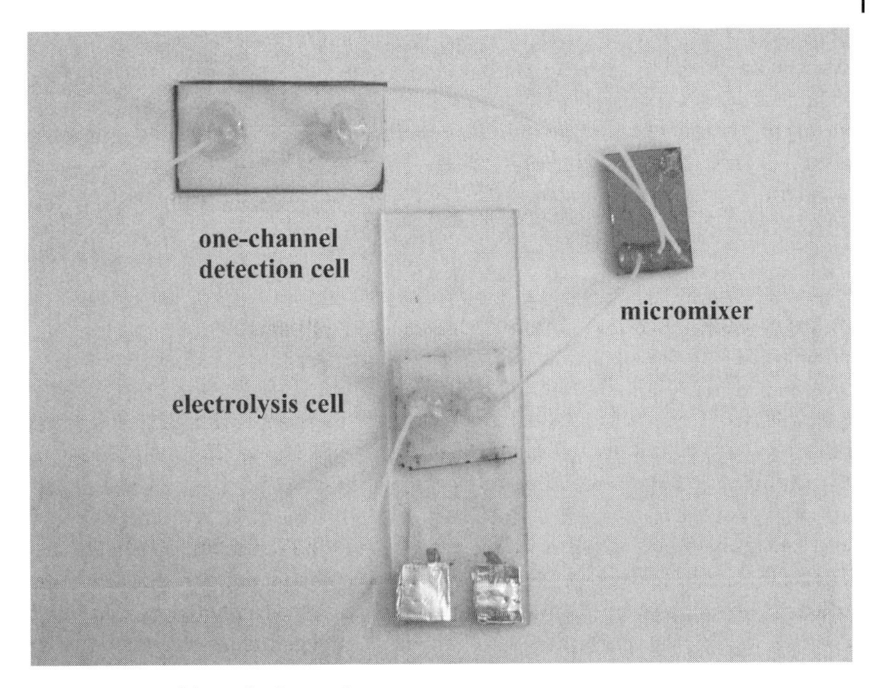

Fig. 16.8 Parts of the cathodic reaction setup.

reduction potentials. Furthermore, they have to determine the concentration of an unknown ferric chloride solution. The electrolysis is carried out in a two-electrode arrangement. The potential range lying between –800 and –1200 mV, measured in reference to the second electrode in the so-called quasipotentiostatic mode of operation. The standard flow rate of all solutions is 1 mL/h. The determination of the absorbance E is carried out optically analogous to first experiment in Section 16.4.2. For each concentration a snapshot of the one-channel cell before and during the current flow is taken, after the same time interval. With these photographs it was possible to detect the transmittance of the corresponding $FeCl_3$ concentration. Through data analysis of each photograph with a graphic software the intensity according to the generated amount of ferrous ions (visible through simultaneous addition of 1,10-phenanthrolinium chloride) was calculated. The snapshot before the current flow gives the intensity without sample and the snapshot during current flow gives the intensity of the sample. After the calculation of the absorbance from transmittance with the Beer–Lambert law (Eq. (16.25)) the absorbance in dependence on the ferrous cation concentration (depending on the $FeCl_3$ concentration) can be drawn. Subsequently, students record a calibration curve, absorbance vs. concentration. This curve is used to determine the concentration of an unknown KI solution.

Evaluation This medium difficulty level experiment is suited for undergraduate students. Normally, it should take about 3–4 h to finalize this experiment. Recommended literature and information for this student experiment is found in [9, 20, 27, 29–32].

16.4.4
Photochemical Experiments in Modular Chip Reactor Arrangement

Photoreduction of Fe(III) ions

Introduction This experiment acquaints students with photochemical reactions. It is shown that photoinduced reactions depend on the wavelength of the incident light.

Fundamentals Basics for this experiment are the knowledge on photochemical activation as well as the occurring reaction. Furthermore, it should be known what the Beer–Lambert law and what quantum yield mean. Additionally, students must know how metal complexes are built and what the general structure of such complexes is.

Experimental setup The photochemical apparatus consists of a micromixer, photochemical reactor, radiation source with which students have to prepare the light-sensitive trioxalatoferrate(III) ions through reaction of a ferric chloride or nitrate solution and oxalic acid, photolyze the photosensitive complex and detect the generated Fe(II) cations. The formation of ferrous ions is observed for several light wavelengths as well as flow rates of the trioxalatoferrate(III) ion solution.

Required equipment and materials are:

- Two micromixers (IPHT Jena, Germany) [36] in a black box
- A PC with monitor, a transmittance-measuring sensor [39]
- Pumps and utilities: a double syringe pump (SP210iw, World Precision Instruments Inc., USA) and a single syringe pump (SP100zi, World Precision Instruments Inc., USA), disposable syringes with 1 mL volume and cannulae (Braun, Germany)
- A radiation source, five edge filters with filtered wavelength from 385 nm up to 700 nm (LINOS Photonics GmbH & Co. KG, Germany)
- 0.3 molar ferric chloride or nitrate solution, 0.3 molar 1,10-phenanthrolinium chloride solution, 0.4 molar oxalic acid solution
- Tubings and pipettes: PTFE tubing (inner diameter of 0.3 mm), silicone tubing (inner diameter of 0.3 mm), reaction tubes (Eppendorf, Germany), a beaker with 50 mL volume, pipettes with 100 μL volume (Eppendorf or Abimed, Germany), disposable pipette tips and
- Distilled water.

The inside of experimental black box is shown in Fig. 16.9.

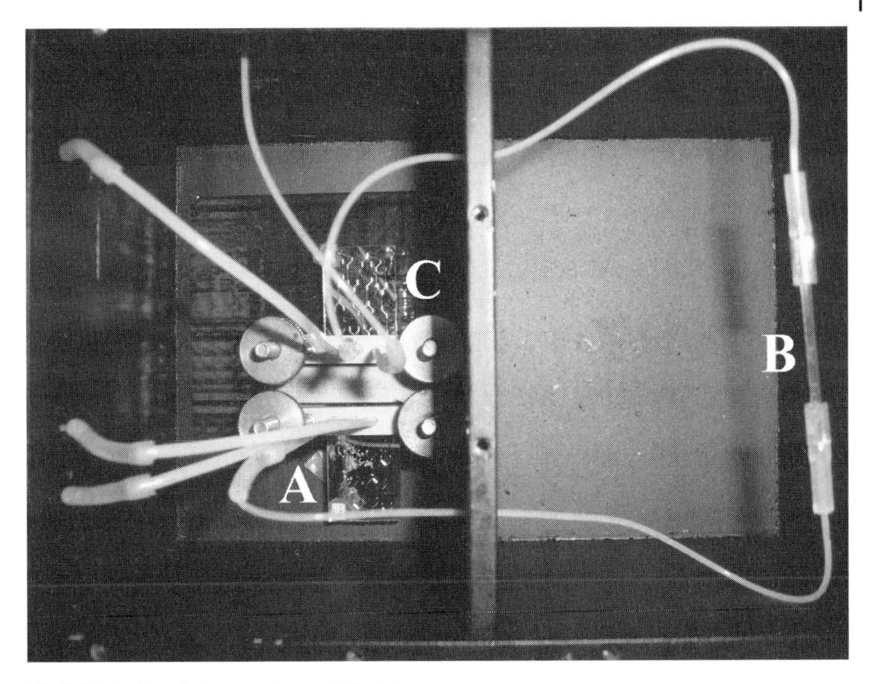

Fig. 16.9 Inside of the experimental black box.

In the first step, students have to prepare ferric salt solutions with concentrations between 0.01 and 0.001 mol/L from a 0.1 molar stock solution. Subsequently, 1-mL syringes are filled with the ferric salt solutions, the oxalic acid solution and the phenanthrolinium solution. An edge filter is inserted into a mount of the black box above the illumination capillary. Students have to start with the longest wavelength filter. After placing one syringe of the Fe(III) solutions and the syringe of oxalic acid in the double syringe pump as well as the syringe of phenanthrolinium chloride in the single pump, all pumps are started simultaneously. The flow rate of all pumps is 1.5 mL/h. Approximately one minute after starting the syringe pumps, the radiation source (white light lamp) is switched on and the data acquisition software is started. The highest measured transmittance value should be noted. This procedure has to be repeated for all edge filters ending with the filter of shortest wavelength. With this edge filter, the dependence of the photochemical conversion on the flow rate is determined. Finally, students have to create graphical diagrams of the absorbance vs. wavelength as well as absorbance vs. flow rate and discuss the results.

Evaluation This low difficult level experiment is suited for undergraduate students. Normally, it should take about 3–4 h to finalize this experiment. Recommended literature and information for this student experiment is found in [38–41].

16.4.5
Chemical Synthesis in Microflow-through Devices

Azocoupling

Introduction Chemical syntheses of organic dyes in micromixers, especially azo dyes, are the subject of this experiment. Students have to synthesize azo dyes through coupling of diazonium salts with aromatic alcohols or amines. Basic for this experiment is knowledge of the preparation and structure of diazonium salts as well as azoic dyes.

Experimental setup Required equipment and materials are:
- A micromixer (IPHT Jena, Germany) [36]
- A double syringe pump (SP210iw, World Precision Instruments Inc., USA)
- An UV-VIS spectrometer, disposable syringes with 1 mL volume and cannulae (Braun, Germany)
- Sodium naphthalenesulfonates, 2-naphthol, diazonium salts (available from Sigma-Aldrich), acetone
- Tubings and pipettes PTFE tubing (inner diameter of 0.3 mm), silicone tubing (inner diameter of 0.3 mm), reaction tubes (Eppendorf, Germany), a beaker with 50 mL volume, pipettes with 1000 μL volume (Eppendorf or Abimed, Germany), disposable pipette tips and
- Distilled water.

A sketch of the experimental setup is shown in Fig. 16.10.

Students have to prepare diazonium salt solutions immediately before the coupling with the aromatic coupling component. To obtain the diazonium salt solution it is necessary to dissolve 8×10^{-5} mol of the salt in 1 mL distilled water. All diazonium salt solutions must be freshly prepared because these salts react with water to form phenols at temperatures above 5 °C. Furthermore, a solution of 2-naphthol by dissolving 10 mg of the substance in a mixture of 900 μL water and 100 μL 1.0 molar NaOH solution is prepared. 10 mg of the coupling components sodium 1-naphthalenesulfonate as well as sodium 2-naphthalenesulfonate were dissolved in 1 mL distilled water. All solutions are

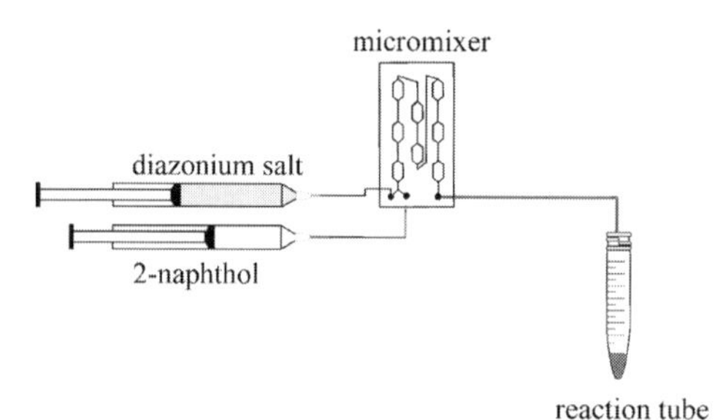

Fig. 16.10 Sketch of the synthesis experimental setup.

filled in syringes. After placing a syringe with a diazonium salt and a syringe with a coupling component in the double syringe pump, the azocoupling is achieved by starting the pump. The flow rate per syringe is 30 or 100 ml/h. UV-VIS spectra are recorded for all obtained dyes and compared with existing spectra.

Evaluation This low difficulty level experiment is suited for undergraduate students. Normally, they should take about 3 h to finalize this experiment. Recommended literature and information for this student experiment is found in [42].

16.4.6
Experiences with Miniaturized Basic Experiments

All the aforementioned experiments have been applied and tested in the chemical education of students in engineering science (Technical Physics, Mechatronics, Material Sciences). In all cases, the experiments have been carried out with a minimum of experience of students in practical chemical work. The overwhelming outcome of these exercises was that students were able to carry out the experiments completely on their own after an instruction of about 15 to 30 min. The microinstrumentation makes the experiments attractive for students. So, the motivation for engineering students dealing with chemical problems is enhanced. They see the immediate relation between physical instrumentation, electronic and mechanical devices and chemical properties and operations. So, these miniaturized experiments are an important contribution to interdisciplinary education.

All experiments have been well understood and were carried out completely correct by about 30 to 40% of students. The scattering of quantitative data is partially caused by the nature of the simple experimental arrangements and partially by the missing experimental experience of the students. Despite the observed deviations between the expected and the measured values in the case of the physicochemical and analytical experiments, in nearly all cases the students find the right qualitative relation between the different parameters in the experiments. So, the understanding of the chemical facts was very well illustrated and supported by the experiments. The experiments or measurements with analytical character reported here do not represent optimized microanalytical procedures. Therefore, they are not suited for special practical experiments in microstructured devices on quantitative analytics with high resolution or trace analytics. The experiments shown here are suited for the principle understanding of general aspects of analytical methods but can not substitute precise experiments in analytical chemistry demonstrating the whole power of microanalytical systems in a special education for analytical chemistry. For this purpose, other microexperiments have to be designed.

16.5 Future Outlook

This chapter informs on knowledge transfer of the theory and utilization of microstructured devices in student education. The universities in Ilmenau and Jena (both in Thuringia, Germany) report on their first practical experiences at training courses for basic education in chemistry and in Technical Chemistry. The resonances given by students were significantly noticeable. The progress achieved needs further development. Both academic teachers (lecturers) and partners from relevant enterprises and from sponsor offices plan a network of excellence in Germany for supporting the aim of this chapter: Laboratory applications of microstructured devices in education.

References

1 V. Hessel, St. Hardt, H. Löwe, Chemical Microprocess Engineering, Modelling and Reaction, Wiley-VCH, Weinheim, **2004**
2 V. Hessel, H. Löwe, A. Müller, E. Kolb, Chemical Microprocess Engineering, Processing and Plants, Wiley-VCH, Weinheim, **2004**
3 W. Ehrfeld (Ed.), Microreaction Technology, Proceed. 1st Int. Conf. on Microreaction Technology 1997, Springer, Berlin, **1998**.
4 Proceed. 2nd Int. Conf. on Microreaction Technology, AIChE National Spring Meeting, New Orleans, **1998**.
5 W. Ehrfeld (Ed.), Microreaction Technology: Industrial Prospects, Proceed. 3rd Int. Conf. on Microreaction Technology 1999, Frankfurt, Springer, Berlin, **2000**.
6 Proceed. 4th Int. Conf. on Microreaction Technology, AIChE National Spring Meeting, Atlanta, **2000**.
7 M. Matlosz, W. Ehrfeld, P. Baselt (Eds.), Microreaction Technology, Proceed. 5th Int. Conf. on Microreaction Technology 2001, Straßbourg, Springer, Berlin, **2002**.
8 P. Baselt, U. Eul, R. S. Wegeng (Eds.), Proceed. 6th Int. Conf. on Microreaction Technology, AIChE National Spring Meeting, New Orleans, **2002**.
9 A. Renken, M. Matlosz (Eds.), Proceed. 7th Int. Conf. on Microreaction Technology 2003, Lausanne, **2003**.
10 Proceed. 8th Int. Conf. on Microreaction Technology, AIChE National Spring Meeting, Atlanta, **2005**.
11 B. Ondruschka, P. Scholz, et al., Chem. Ing. Tech. **2002**, 74, 1577–1582.
12 R. Gorges, W. Klemm, et al., Chem. Ing. Tech. **2004**, 76, 519–522.
13 Th. Dietrich, M.A. Liauw, Chem. Ing. Tech. **2004**, 76, 517–518.
14 R. Gorges, Th. Taubert, et al., Chem. Eng. Technol. **2005**, 28, 1–4.
15 R. Gorges, S. Meyer, G. Kreisel, J. Photochem. Photobiol. A **2004**, 167, 95–97.
16 M. Günther, J.M. Köhler, Chem. Ing. Tech. **2004**, 76, 522–526.
17 A. Bazzanella, Chem. Ing. Tech. **2004**, 76, 511–513.
18 Micro Chem Tec, Foliensatz für die Aus- und Weiterbildung, DECHEMA (Ed.), Frankfurt/Main, **2005**.
19 DECHEMA (Ed.), Lehrprofil Technische Chemie, 2nd. edn, DECHEMA, Frankfurt/Main, **2002**.
20 M. Fedtke, W. Pritzkow, G. Zimmermann, Lehrbuch der Technischen Chemie, Deutscher Verlag für Grundstoffindustrie, Stuttgart, **1996**.
21 E. Patat, K. Kirchner, Praktikum der Technischen Chemie, 4th edn, de Gruyter, Berlin/New York, **1986**, 4–12.
22 J. Brandner, M. Fichtner, et al., in: Proceed. 4th Int. Conf. on Microreaction Technology (IMRET4), AIChE National Spring Meeting, Atlanta, **2000**, 244.

23 F. P. Incropera, D. P. De Witt, Fundamentals of Heat and Mass transfer, 4th edn, Wiley & Sons, New York, **1996**.

24 VDI-Wärmeatlas, Berechnungsblätter für den Wärmeübergang, 4th edn, VDI Verlag Düsseldorf, **1984**, Abschnitt Ca.

25 VDI-Wärmeatlas, Berechnungsblätter für den Wärmeübergang, 6th edn, VDI Verlag, Düsseldorf, **1991**, Abschnitt Gb4.

26 W. Reschetilowski, Technisch-Chemisches Praktikum, 1st edn, Wiley-VCH, **2002**, Weinheim, 62–70.

27 G. Wedler, Lehrbuch der Physikalischen Chemie, 3. durchges. Aufl., Wiley-VCH, Weinheim, **1987**.

28 P. W. Atkins, Physikalische Chemie, 2. korr. Nachdr. d. 1. Aufl., Übers. u. erg. von A. Höpfner, VCH, Weinheim, Basel, Cambridge, New York, **1990**.

29 For working principle and assembly of used sensors look at page: http://www.cismst.de/english/kapazitiv.html.

30 G. A. Groß, J. M. Köhler, Thermochim. Acta **2005**, in press.

31 H.-D. Försterling, H. Kuhn, Praxis der Physikalischen Chemie, 3. erg. Aufl., Wiley-VCH, Weinheim, Deerfield Beach, **1991**.

32 P. W. Atkins, J. A. Beran, Chemie einfach alles, 2. korr. Aufl., Wiley-VCH, Weinheim, New York, Brisbane, Singapore, Toronto, **1998**.

33 T. Kirner, P. Jaschinsky, J. M. Köhler, Chem. Eng. J. **2004**, 101, 163–169. [27] W. J. Moore, Physikalische Chemie, 4. durchges. und verb. Aufl., de Gruyter, Berlin, New York, **1986**.

34 A. Einstein, M. von Smoluchowski, Brownsche Bewegung: Untersuchungen über die Theorie der Brownschen Bewegung/Abhandlung über die Brownsche Bewegung und verwandte Erscheinungen, Ostwalds Klassiker der exakten Wissenschaften, 3. Auflage, Harri Deutsch, Frankfurt am Main, **1997**, Bd. 199.

35 C. H. Hamann, W. Vielstich, Elektrochemie, 3. völlig überarb. und erw. Aufl., Wiley-VCH, Weinheim, **1998**.

36 T. Kirner, J. Albert et al., Chem. Eng. J. **2004**, 101, 65–74.

37 E.-G. Jäger, K. Schöne, Elektrolytgleichgewichte und Elektrochemie; Lehrwerk Chemie: Arbeitsbuch 5, 4. durchges. Aufl., VEB Deutscher Verlag für Grundstoffindustrie, Leipzig, **1989**.

38 G. Jander, E. Blasius, Lehrbuch der analytischen und präparativen anorganischen Chemie, 12., neu bearb. Aufl., S. Hirzel, Leipzig, **1988**.

39 M. Günther, S. Schneider, et al., Chem. Eng. J., **2004**, 101, 373–378.

40 H. Labhart, Einführung in die Physikalische Chemie, Teil V, Springer, **1975**.

41 D. Wöhrle, M. W. Tausch, W.-D. Stohrer, Photochemie: Konzepte, Methoden, Experimente, Wiley-VCH, Weinheim, New York, Chichester, Brisbane, Singapore, Toronto, **1998**.

42 H. G. O. Becker, et al., Organikum: organisch-chemisches Grundpraktikum, 20., bearb. und erw. Aufl., Johann Ambrosius Barth, Heidelberg, Leipzig, **1996**.

Subject Index

Advanced Micro and Nanosystems Vol. 5. Micro Process Engineering. Edited by N. Kockmann

ISBN: 3-527-31246-3

d